MICROCLIMATE
The Biological Environment

MICROCLIMATE
The Biological Environment

Second Edition

Norman J. Rosenberg
George Holmes Professor of Agricultural Meteorology
and Director,
Center for Agricultural Meteorology and Climatology, CAMaC

Blaine L. Blad
Professor of Agricultural Meteorology, CAMaC

Shashi B. Verma
Associate Professor of Agricultural Meteorology, CAMaC
Institute of Agriculture and Natural Resources
University of Nebraska–Lincoln

A Wiley-Interscience Publication
JOHN WILEY & SONS
New York · Chichester · Brisbane · Toronto · Singapore

Copyright © 1983 by John Wiley & Sons, Inc.

All rights reserved. Published simultaneously in Canada.

Reproduction or translation of any part of this work beyond that permitted by Section 107 or 108 of the 1976 United States Copyright Act without the permission of the copyright owner is unlawful. Requests for permission or further information should be addressed to the Permissions Department, John Wiley & Sons, Inc.

Library of Congress Cataloging in Publication Data:

Rosenberg, Norman J., 1930–
  Microclimate: the biological environment.

  "A Wiley-Interscience publication."
  Includes bibliographies and index.
  1. Microclimatology. 2. Meteorology, Agricultural.
I. Blad, Blaine L. II. Verma, Shashi B. III. Title.

QH543.R6 1983   551.6'6   83-7031
ISBN 0-471-06066-6

Printed in the United States of America

10 9 8 7 6 5 4

*We dedicate this book
to our parents,
our wives,
and our children*

PREFACE

Intelligent management of our planet's environment and its resources for sustained productivity and improving quality of life is perhaps our most critical need at this time. We are faced with rapid population growth in many regions of the world where food supply is now inadequate. Far more food will be needed to sustain a projected world population of 8 billion in the year 2000. It is the purpose of this book to provide students of physical and biological science and engineering, as well as practitioners in these fields, with a basis for understanding the role of the environment in food production. If the environment is degraded it will be still more difficult to provide the food that will be needed in the years to come.

In the period following the Second World War until about the early 1970s rapid improvements were achieved in the yields of major food crops in part through applications of the results of scientific research. However, much of that success was due, as well, to the energy intensification of agriculture: the greater use of pesticides, herbicides, chemical fertilizers, and larger and more efficient machines. Of course, energy demand grew not only in agriculture but in every other industrial and civil sector during this period.

One impact of the use of fossil fuels for energy production is an increase in the concentration of carbon dioxide in the atmosphere. It is feared by many scientists that climatic changes will occur (or may have already occurred) as a result of this input of carbon dioxide into the atmosphere. Other products of fossil fuel combustion may be contributing to the formation of "acid rain," which can have deleterious effects on aquatic and terrestrial ecosystems and on crops and croplands.

The generation of power (regardless of the fuel) and other industrial processes liberate large quantities of heat. Heat and sometimes water vapor are released directly into the air or heat is discharged directly into bodies of water, altering ecological conditions in the vicinity of the power plants and factories.

Manufactured products may also threaten the stability of the environment. Chlorofluorocarbons (Freons) have been postulated to react with ozone in the upper atmosphere and reduce its concentration there. Nitrogenous fertilizers, by increasing the amount of nitrous oxides emanating from the soil, may have a similar effect. The result may be an increased flux density of ultraviolet radiation reaching the earth's surface where it can be harmful to plant and animal life.

Land overused or improperly used can be damaged, perhaps irreversibly.

Much of the semiarid land of the world is believed to be undergoing a process of "desertification." When the vegetational coverage changes, soil moisture conditions, surface temperature, and reflectivity also change. Some believe that when desertification occurs a "feedback" develops that reduces rainfall and perpetuates the desert.

Forces over which man has no control affect the condition of the atmosphere and its ability to sustain life. Recent episodes of volcanism are believed to have perturbed the weather and climate and to have altered the quantity of solar radiation received at the surface.

These are but a few of the environmental problems and issues of which we read and hear frequently. Our intent in this book is to provide the scientific basis for understanding these issues through a systematic explanation of atmospheric phenomena and their influences on plant and animal growth.

The second edition of *Microclimate* expands on the first. The scientific literature in the numerous relevant fields—agronomy, soil science, ecology, meteorology, to name a few—has been greatly enriched since the first edition appeared. Much recent material has been incorporated to update *Microclimate* as a reference work and many new subjects and concepts have been added.

Professors Blaine L. Blad and Shashi B. Verma have joined with the senior author in preparing this extensively revised and expanded second edition of *Microclimate*. We have drawn considerably on our own work and that of our students in the Center for Agricultural Meteorology and Climatology (CAMaC) of the University of Nebraska–Lincoln's Institute of Agriculture and Natural Resources.

Scientists gain knowledge and insight from the work of their colleagues as well as from their own research. The science of microclimatology developed as a synthesis of the work of many meteorologists, soil scientists, agronomists, horticulturalists, foresters, and animal scientists. Professor Rudolf Geiger's classic work *The Climate Near the Ground* (1965; first published in 1927) and Professor O. G. Sutton's *Micrometeorology* (1953) have had major impact on the development and systematization of the modern discipline.

In keeping with recommendations of the scientific societies to which we adhere we employ the *System Internationale* (SI) notation throughout the book. By now most North American students of science have been introduced to that system. One may debate the merits of SI as compared with the cgs system employed in the first edition, but debate is fruitless at this time. We yield reluctantly and, we hope, gracefully. Those who need help with the SI system can find it in a number of handbooks. One we found useful is S. H. Qasim's *SI Units in Engineering and Technology* (1977).

The *Glossary of Meteorology* (Huschke, 1959) continues to serve as the source of most of the definitions of meteorological terms used in this book. The *Smithsonian Meteorological Tables* (List, 1966) is the source for most of the physical values used.

We are indebted to many colleagues and students for specific criticisms of the first edition and suggestions for improvement. Many of these have been incorporated in this edition. We are especially indebted to Dr. Dennis D. Baldocchi for his helpful review of Chapter 8.

Mrs. Roberta Sandhorst had the primary responsibility for stenographic preparation of the manuscript and for assembling and checking the reference lists. Mrs. Nancy Brown handled the copyright search. Mrs. Betty James and Mrs. Sharon Kelly assisted with the stenography. Ms. Deborah Wood did the indexing. Mr. James Schepers prepared the figures for this edition and Mr. Bruce Sandhorst provided many of the photographs of instruments. These diligent and patient people deserve our profound appreciation.

NORMAN J. ROSENBERG
BLAINE L. BLAD
SHASHI B. VERMA

Lincoln, Nebraska
May 1983

CONTENTS

List of Symbols   xvii

Introduction   1

1   The Radiation Balance   5

    1.1   Review of Radiation Physics, 5
    1.2   Solar Energy Receipts at the Surface of the Earth: Quantitative Effects, 10
    1.3   Solar Energy Receipts at the Surface of the Earth: Qualitative Effects, 33
    1.4   Sky Radiation (Diffuse), 40
    1.5   Shortwave Reflection (Albedo), 42
    1.6   Thermal Radiation and Longwave Exchange, 49
    1.7   The Net Radiation, 51
    1.8   Relation of Net and Solar Radiation, 54
    1.9   Earth's Radiation Balance, 57
    1.10   Light Penetration into Plant Canopies and Water Bodies, 59
    1.11   Instrumentation, 71
    References, 83

2   Soil Heat Flux and Soil Temperature   94

    2.1   Introduction, 94
    2.2   Laws of Heat Conduction and Thermal Properties of Soils, 94
    2.3   Penetration of Heat into the Ground, 99
    2.4   Daily and Seasonal Patterns of Soil Temperature, 99
    2.5   Soil Temperature Profiles, 101
    2.6   Texture Influences on Soil Heat Flux and Temperature, 103
    2.7   Soil Heat Flux and Water Relations in Soils, 107
    2.8   Soil Heat and Soil Respiration, 111

xii CONTENTS

    2.9   Instrumentation, 112
           References, 115

3 Air Temperature and Sensible Heat Transfer   117

    3.1   Introduction, 117
    3.2   Adiabatic Process, Potential Temperature, 117
    3.3   The Concept of Thermal Stability, 118
    3.4   The Wet Adiabatic Lapse Rate, 119
    3.5   Temperature Profiles above Natural Surfaces, 121
    3.6   Sensible Heat Transfer in the Atmospheric Surface Layer, 123
    3.7   Resistance Approach for Estimating Sensible Heat Flux, 123
    3.8   Temperature Profiles in Plant Canopies, 126
    3.9   Daily and Annual Temperature Patterns, 126
    3.10 Influence of Elevation on Air Temperature Patterns, 129
    3.11 Instrumentation for Air Temperature Measurement, 130
           References, 132

4 Wind and Turbulent Transfer   134

    4.1   Air Flow over a Rigid Surface: Some Definitions and Concepts, 134
    4.2   Wind Speed Profile and Momentum Exchange, 135
    4.3   Internal Boundary Layer and Fetch Requirements, 139
    4.4   Atmospheric Stability, 140
    4.5   Flux Profile Relationships, 142
    4.6   Eddy Correlation Technique for Estimating Energy and Mass Fluxes, 144
    4.7   Wind Speed within Crop Canopies, 146
    4.8   Daily Wind Patterns, 147
    4.9   Seasonal Patterns of Wind Direction and Speed, 148
    4.10 Wind Speed Instrumentation, 154
           References, 164

5 Atmospheric Humidity and Dew   167

    5.1   Introduction, 167
    5.2   Physical Review, 168
    5.3   Measures of Humidity, 169

CONTENTS xiii

    5.4   The Concept of Saturation, 170
    5.5   Saturation-Based Measures of Humidity, 171
    5.6   Humidity Structure of Air, 172
    5.7   Profiles of Vapor Pressure, 174
    5.8   Dew, 175
    5.9   Instrumentation for Humidity Measurement, 178
    5.10   Instruments for Measurement of Dew, 187
    References, 187

6 Modification of the Soil Temperature and Moisture Regimes   190

    6.1   Introduction, 190
    6.2   Slope and Aspect, 191
    6.3   Mulching, 195
    6.4   Artificial Heating of the Soil, 201
    References, 206

7 Evaporation and Evapotranspiration   209

    7.1   Introduction, 209
    7.2   Importance of Evaporation and Transpiration, 212
    7.3   Soil, Plant, and Climatic Influences on Evapotranspiration, 217
    7.4   Soil–Plant–Atmosphere Continuum, 239
    7.5   Estimation of Evaporation and Evapotranspiration, 241
    7.6   Measurement of Evapotranspiration, 258
    7.7   Separation of Evaporation and Transpiration, 265
    7.8   Application of Evapotranspiration Methods to Special Situations, 267
    References, 271

8 Field Photosynthesis, Respiration, and the Carbon Balance   288

    8.1   Introduction and Definitions, 288
    8.2   Gross and Apparent Photosynthesis, 291
    8.3   Photosynthesis as a Resistance Process, 292
    8.4   Environmental Factors Affecting Photosynthesis, 293
    8.5   Environmental Influences on Respiration, 299
    8.6   Carbon Balance in the Field, 302
    8.7   Radiant Energy Conversion in Photosynthesis, 308
    8.8   Water Use Efficiency, 312
    8.9   Measurement of Photosynthesis in the Field, 318

xiv CONTENTS

    8.10 Measuring the Respiration Components, 324
References, 326

9 Windbreaks and Shelter Effects   331

9.1 Introduction, 331
9.2 Interrelations of Wind Shelter, Moisture Conservation, Plant Growth, and Yield, 334
9.3 Wind Speed and Turbulence in Shelter, 336
9.4 Microclimate in Shelter, 341
9.5 Plant Physiological Responses to Shelter, 350
9.6 Potential and Actual Water Use, 352
9.7 The Effect of Shelter on Photosynthesis, 359
9.8 The Effect of Shelter on Water Use Efficiency, 360
9.9 Some Integrative Schemes of the Spatial Differences in Shelter Effects, 361
References, 363

10 Frost and Frost Control   368

10.1 Introduction, 368
10.2 Types of Frost, 371
10.3 The Climatology of Frost Incidence, 373
10.4 Methods of Frost Protection, 375
References, 388

11 Water Use Efficiency in Crop Production: New Approaches   391

11.1 Introduction, 391
11.2 Antitranspirants, 392
11.3 Reflectants, 396
11.4 Plant Architecture, 404
11.5 Carbon Dioxide Enrichment, 409
References, 419

12 Human and Animal Biometeorology   425

12.1 Introduction, 425
12.2 Radiation Balance, 427
12.3 Energy Balance, 433
12.4 The Climate Space, 446

12.5 Effects of Climate on Humans, 449
12.6 Effects of Climate on Animals, 460
12.7 Adaptation and Acclimatization, 462
     References, 463

Author Index   **469**

Subject Index   **481**

LIST OF SYMBOLS

A	Constant or proportionality in psychrometric measurements Body area (Chapter 12)
A_h	Projected area on a horizontal surface
A_p	Projected area on a plane perpendicular to the sun
B^{-1}	Nondimensional parameter (reciprocal of sublayer Stanton number)
B_m	Basal metabolic rate
C	Cloudiness in fraction of sky cover (Chapter 1) Conductive capacity (Chapter 2) Constancy of the wind (Chapter 4) Constant in mass transport equations (Chapter 7) Heat exchange by conduction (Chapter 12)
C_a	Carbon dioxide concentration in air (Chapter 8)
C_{chl}	Carbon dioxide concentration at the chloroplast
C_p	Specific heat of air at constant pressure
C_s	Mass specific heat
C_v	Volume specific heat (Chapter 2) Specific heat of air at constant volume
CWFR	CO_2-water vapor flux ratio
D	Solar declination (Chapter 1) Characteristic dimension (Chapters 3 and 12) Percolation, deep drainage (Chapter 7)
D_0	Wet bulb depression at surface
D_z	Wet bulb depression at height z
E	Energy flux density (Chapter 1) Water vapor flux
E_1	Quantum energy content
E_c	Evaporative capacity of air
E_{cm}	Maximum evaporative capacity of air
Ei	Einstein, a mole of photons
E_0	Free water evaporation
E_{pan}	Pan evaporation
E_s	Soil evaporation

LIST OF SYMBOLS

ET	Evapotranspiration
ET_{eq}	Equilibrium evapotranspiration
ET_m	Maximum evapotranspiration
ET_p	Potential evapotranspiration
ET_r	Actual evapotranspiration
F	Cumulative leaf area index (see LAI) (Chapter 1)
	Flux of an entity
F_c	CO_2 flux above the canopy
Gr	Grashof number
Γ	Adiabatic lapse rate
H	Sensible heat flux
H_r	Respiratory sensible heat transfer
Hz	Hertz, cycles s^{-1}
HSI	Heat stress index
HTI	Humidity temperature index
H_{loc}	Local sensible heat advection
H_{reg}	Regional sensible heat advection
H_u	Humidex
I	Flux density of incident radiation
	Insulation (Chapter 12)
K	Thermal conductivity
K_1	Crop coefficient (Chapter 7)
K_c	Turbulent exchange coefficient for CO_2
	Thermal conductance (Chapter 12)
K_h	Turbulent exchange coefficient for sensible heat
K_m	Turbulent exchange coefficient for momentum
K_s/K	Ratio of exchange coefficients in shelter and open
K_w	Turbulent exchange coefficient for water vapor
L	Latent heat of vaporization (2.45 MJ kg^{-1} at 20°C)
	Obukhov length (Chapter 4)
LAI	Leaf area index
LE	Latent heat flux
LE_r	Respiratory latent heat loss
LE_s	Latent heat flux from dry skin
LW_a	Longwave radiation from atmosphere
LW_s	Longwave radiation from surroundings
M	Miscellaneous energy fluxes
	Metabolic heat production (Chapter 12)
M_a	Molecular weight of dry air

LIST OF SYMBOLS xix

M_b	Basal metabolic rate per unit area
M_w	Molecular weight of water
N_1	Number of days from the nearest equinox
Nu	Nusselt number
P	Atmospheric pressure
PAR	Photosynthetically active radiation
P_d	Partial pressure of dry air
PI	Precipitation and/or irrigation
Pr	Prandtl number
PS	Photosynthesis
aPS	Apparent photosynthesis
gPS	Gross photosynthesis
Ψ	Water potential
Ψ_a	Water potential of air
Ψ_g	Soil water potential
Ψ_l	Leaf water potential
Ψ_r	Root water potential
Q	Conductive heat flux from inner to outer surface of an organism
Q_{10}	Factor for change in respiration rate with a 10°C change in temperature
R	Universal gas constant (8.314 kJ kg^{-1} °K^{-1} mol^{-1}) Respiration rate (Chapter 2 and 8)
R_{abs}	Shortwave and longwave energy absorbed by an object
R_c	Crop respiration
R_D	Direct beam solar radiation
R_d	Dark respiration Diffuse radiation (Chapter 12)
Re	Reynolds number
R_g	Soil respiration
RH	Relative humidity
Ri	Richardson number
R_l	Photorespiration (respiration in light)
R_{lw}	Flux density of longwave radiation
R_n	Flux density of net radiation
R_{no}	Net radiation over open water
R_{ns}	Net radiation at soil surface
RO	Runoff
R_0	Respiration rate at a reference temperature

xx LIST OF SYMBOLS

R_r	Root respiration
	Reflected shortwave radiation (Chapter 12)
R_s	Solar radiation
R_{sc}	Solar constant
R_{sw}	Flux density of shortwave radiation
R_T	Flux density of effecive terrestrial radiation (Chapter 1)
	Respiration rate at temperature T
S	Soil heat flux
	Sweat rate (Chapter 12)
S_t	Heat storage
SW	Soil water content
TSI	Thermal stress index
T	Temperature
	Transmission factor (Chapter 1)
	Transpiration (Chapter 7)
T_a	Air temperature
T_b	Internal body temperature
T_{bs}	Surface body temperature
T_d	Dew-point temperature
T_e	Temperature of air expired from respiratory tract
T_{ef}	Effective temperature
T_{eq}	Equivalent blackbody temperature
T_{eqw}	Equivalent wet blackbody temperature
T_h	Humiture
THI	Temperature humidity index
T_i	Temperature of air inhaled into respiratory tract
T_l	Leaf or tissue temperature
T_0	Reference temperature for respiration
TR	Temperature range
T_s	Surface temperature
TSI	Thermal stress index
T_v	Virtual temperature
T_w	Wet bulb temperature
T_{wc}	Windchill equivalent temperature
U	Wind speed in the x direction (horizontal wind speed)
U_m	Mean day or night horizontal wind speed
U_r	Resultant wind speed
U_s/U	Ratio of wind speed in shelter and open
V	Volume
	Respiratory ventilation rate (Chapter 12)

LIST OF SYMBOLS xxi

V_w	Volume occupied by a mole of water vapor
W	Mixing ratio
	Work (Chapter 12)
WCI	Windchill index
X_d	Mole fraction of dry air in moist air
X_w	Mole fraction of water vapor in moist air
Y	Dry matter or yield
a	Atmospheric scattering or absorption coefficient
	Human or animal activity (Chapter 12)
a_m	Maximum sustainable activity
α	Absorptivity (Chapter 1)
	Thermal diffusivity (Chapter 2)
	Priestley–Taylor constant (Chapter 7)
α_h	Thermal diffusivity of air for heat
β	Solar elevation angle (Chapter 1, 12)
	Bowen ratio
	Coefficient of volumetric expansion (Chapter 12)
c	Speed of light (3×10^8 m s^{-1})
χ	Absolute humidity
d	Day of the year (January 1 = 1) (Chapter 1)
	Zero plane displacement
δ	Thickness of the internal boundary layer
δ_1	Thickness of the fully adjusted layer
e_a	Partial pressure of water vapor in air
e_0	Vapor pressure at a surface
e_s	Saturation water vapor pressure in air
e_{ss}	Saturation water vapor presssure at an evaporating surface
ϵ	Emissivity (Chapter 1)
	Ratio of M_w/M_a
ϵ_b	Emissivity of a body
exp	Base of the Naperian log system
f	Functional notation
	Cooling efficiency of sweating (Chapter 12)
g	Acceleration due to gravity
γ	Zenith distance (angle) (Chapter 1)
	C_p/C_v (Chapter 3)
	Psychrometric constant = $(PC_p)/LE$
h	Planck's constant, 6.626×10^{-34} J s (Chapter 1)
	Hour angle (Chapter 1)
	Height of crop
	Height of a wind barrier (Chapter 9)

LIST OF SYMBOLS

h_c	Convective heat transfer coefficient
i	Leaf angle with respect to the horizontal
ir	Infrared radiation
k	Boltzman constant, 1.38×10^{-23} J °K^{-1} (Chapter 1)
	von Karman's constant
k_1	Physico/physiological constant
κ	Extinction coefficient
l	Latitude
λ	Wavelength of electromagnetic radiation
m	Air mass depth (optical airmass) (Chapter 1)
	Mass of moist air
	Mass of a body (Chapter 12)
m_d	Mass of dry air
m_w	Mass of water vapor
μ	Dynamic viscosity of air
n	Number of moles
n/N	Percent possible sunshine (ratio of actual, n, to possible, N, duration of sunshine)
v	Frequency of electromagnetic radiation (Chapter 1)
	Kinematic viscosity (Chapter 3)
P	Period of temperature oscillation
π	3.1416
ϕ_m	Dimensionless stability function
q	Specific humidity
r	Reflectivity
r_a	Aerial resistance
r_b	Reflectivity of a body
r_c	Canopy resistance
	Coat resistance (Chapter 12)
r_e	Parallel equivalent resistance of r_a and r_r
r_g	Soil resistance to water flow
r_l	Leaf resistance
r_m	Mesophyll resistance
r_p	Total plant resistance to water flow
r_r	Root resistance to water flow
	Radiation resistance = $4\rho C_p \epsilon \sigma T^3$ (Chapter 12)
r_s	Stomatal resistance
	Skin resistance (Chapter 12)
r_{sb}	Stomatal resistance, bottom side of the leaf
r_{st}	Stomatal resistance, top side of leaf

LIST OF SYMBOLS

r_x	Xylem resistance to water flow
ρ	Density
ρ_a	Density of moist air
ρ_c	Density of carbon dioxide in air
ρ_{va}	Vapor density in air
ρ_{vs}	Saturation vapor density at skin surface temperature
s	Solid or aerosol content of the atmosphere (Chapter 1)
	Concentration of an entity (Chapter 4)
	Slope of the saturation vapor pressure curve at the mean wet bulb temperature of the air (Chapter 7, 12)
σ	Stephan–Boltzman constant, 5.67×10^{-8} J m^{-2} s^{-1} °K^{-4}
	Ratio of densities of water vapor and dry air (Chapter 8)
t	Time
	Transmissivity (Chapter 1)
τ	Wave number ($1/\lambda$) (Chapter 1)
	Shearing stress, momentum flux (Chapter 4)
θ	Linke's turbidity factor (Chapter 1)
	Potential temperature
	Azimuth angle (Chapter 12)
uv	Ultraviolet radiation
u_*	Friction velocity
w	Windspeed in the z direction (vertical windspeed)
	Water vapor content of the atmosphere (Chapter 1)
x	Distance, depth of the atmosphere (Chapter 1)
	Downwind distance (Chapter 4)
z	Elevation above earth's surface (Chapter 1)
	Depth (Chapter 2)
	Vertical distance
z_h	Roughness parameter for sensible heat
z_0	Roughness parameter for momentum transfer

MICROCLIMATE
The Biological Environment

INTRODUCTION

Microclimate is the climate near the ground, that is, the climate in which plants and animals live. It differs from the macroclimate, which prevails above the first few meters over the ground, primarily in the rate at which changes occur with elevation and with time. Whether the surface is bare or vegetated, the greatest diurnal range in temperature experienced at any level occurs there. Temperature changes drastically in the first few tens of millimeters from the surface into the soil or into the air. Changes in humidity with elevation are greatest near the surface. Very large quantities of energy are exchanged at the surface in the processes of evaporation and condensation. Wind speed decreases markedly as the surface is approached and its momentum is transferred to it. Thus it is the great range in environmental conditions near the surface and the rate of these changes with time and elevation that make the microclimate so different from the climate just a few meters above, where atmospheric mixing processes are much more active and the climate is both more moderate and more stable.

Solar radiation provides almost all of the energy received at the surface of the earth. Some of the radiant energy is reflected back to space. The earth also reradiates, in the thermal waveband, some of the energy received from the sun. The quantity of radiant energy remaining at the earth's surface is the net radiation R_n (the net radiant energy available at the surface when all inward and outward streams of radiation have been considered). The net radiation drives certain physical processes important to us. The energy balance may be expressed as

$$R_n + S + H + LE + PS + M = 0 \qquad (I.1)$$

where S is the flux of heat into or out of the soil, H is the flux of sensible heat between the surface and air, LE is the flux of latent heat to and from the surface through vaporization (evaporation) of water or condensation, PS is the energy fixed in plants by photosynthesis, and M is the energy involved in a number of miscellaneous processes such as respiration and heat storage in the crop canopy.

Equation I.1 is applicable on the scale of a single plant or cropped field, explaining how energy is provided to warm the soil and crop and to evaporate water. The equation is no less valid on the global scale, explaining how energy is provided to the continents and oceans where vast quantities of heat and vapor are given to or extracted from the atmosphere.

2 INTRODUCTION

This book is devoted, primarily, to an exposition of the environment in which plants and animals live. The first five chapters provide a basis for an understanding of environmental biophysics. Chapter 1 describes the radiation balance of typical surfaces. Since the soil is an important sink for energy received from the sun, the physics of heat movement by conduction and heat storage in soil is discussed next, in Chapter 2. Solar radiation warms the surface of the earth. Air coming in contact with the warmed surface gains heat from it. At night when the surface is cold the air transfers heat to it. This "sensible" heat transfer, a process of convection, is discussed in Chapter 3. A full understanding of convectional exchange of sensible heat and the exchange of matter between the surface and the atmosphere requires an understanding of wind and turbulence. The processes by which carbon dioxide is transferred in photosynthesis and respiration and water vapor in evapotranspiration are analogous to the transfer of horizontal momentum. The transfer of dust, pollen, atmospheric pollutants, and other materials can also be treated in this way. Thus insight into these processes can be gained from knowledge of wind structure in the lower layers of the atmosphere, which is described in Chapter 4. Humidity is of fundamental importance in life processes. Chapter 5 is a description and discussion of the specific physical properties of water vapor that must be understood in preparation for study of evaporation and biological responses to humidity conditions in later chapters.

Chapters 6–11 deal with prediction, manipulation, and management of the microclimate. The information is drawn mostly from agricultural microclimatology, but the concepts illustrated can be applied to other types of ecosystems as well. The modification of the soil thermal regime is discussed in Chapter 6. The water balance of earth is of great practical importance. The processes of evaporation and transpiration are major determinants of the water balance and have considerable influence on the growth of plants. Evaporation and transpiration are treated in Chapter 7.

The productivity of plants depends on the efficiency of the photosynthetic process. The availability of photosynthetically active radiation, the ambient temperature, and the rates of supply of moisture and carbon dioxide to the photosynthesizing plant are factors that control overall plant productivity. The interacting influences of microclimate, mass exchange processes, and irradiation on photosynthesis are discussed in Chapter 8. The concept of water use efficiency (the ratio of dry matter production to water consumption) is also introduced in this chapter.

Building on the physical principles explained in earlier chapters, Chapters 9–11 deal with applications of microclimatology to production problems in agriculture. Chapter 9 is a discussion of windbreaks and shelter effect. Windbreaks provide one way by which the environment can be manipulated for improved plant growth and greater water use efficiency. The physical principles learned in the early chapters and the detailed descriptions of evapotranspiration and photosynthesis are applied here and a unified theory of shelter effect is developed.

Chapter 10 deals with frost, a phenomenon of great importance in agriculture and crop production, especially in the temperate regions of the world. Types of frost are described. Many methods of protection are also explained in this chapter.

Ways to increase water use efficiency in plant production are discussed in Chapter 11. These include use of chemicals to reduce transpiration through stomatal regulation (antitranspirants) and use of reflectant materials to dispose of excess radiant energy or to redistribute it within the plant canopy. Manipulating plant architecture by means of plant breeding offers a more "natural" option for increasing water use efficiency and examples of recent attempts to do this are described. The "problem" of the increasing carbon dioxide concentration in the atmosphere is discussed at a number of points in the book. That air enriched in CO_2 has positive, beneficial effects on plant growth and water use efficiency is well known. In Chapter 11 the prospects for further artificial enrichment or "fertilization" with carbon dioxide are discussed at length.

Animals and people reside within a microclimate and are subject to many of the same physical laws that govern plant–microclimate interactions. However, because of their mobility humans and animals can often move to more favorable environments when atmospheric stresses are imposed. Man, through his reasoning power and skill, has, in many cases, created his own microclimate to reduce atmospheric stress and increase comfort. Nevertheless, animals and humans remain strongly influenced by the weather. In the past two or three decades much effort has been given to biometeorology, which is the study of the interactions between humans and other animals and their environments. A relatively brief summary of these interactions and some of the laws and principles governing the exchange of energy and mass is presented in Chapter 12. Some of the more common indices used to describe comfort levels for animals and humans are also given.

Instrumentation for measurement of solar and net radiation, soil and air temperature, soil heat flux, wind speed and direction, water vapor concentration and flux, and CO_2 concentration and flux are described in the appropriate chapters. These sections are not intended to serve as manuals but rather to illustrate principles of measurement used in microclimate research and to instruct on certain "do's and don'ts" in making observations. There are a number of fine, detailed books devoted specifically to meteorological instrumentation. Of these Fritschen and Gay (1979) and Latimer (1972) are especially useful.

The student of microclimate and its applications would also be well advised to consult references and texts that have appeared in recent years. Monteith (1973), Campbell (1977), Oke (1978), Lee (1978), and Gates (1980) provide good theoretical expositions of many of the topics covered in this

4 INTRODUCTION

book. Barfield and Gerber (1979) give a very useful compendium of information on techniques of microclimate modification.

REFERENCES

Barfield, B. J., and J. F. Gerber, eds. 1979. *Modification of the Aerial Environment of Plants*. American Society of Agricultural Engineers Monograph Number 2. 538 pp.
Campbell, G. 1977. *An Introduction to Environmental Physics*. Springer-Verlag, New York. 159 pp.
Fritschen, L. J., and L. W. Gay. 1979. *Environmental Instrumentation*. Springer-Verlag, New York. 216 pp.
Gates, D. M. 1980. *Biophysical Ecology*. Springer-Verlag, New York. 611 pp.
Geiger, R. 1965. *The Climate Near the Ground*. Transl. from the 4th German edition. Harvard Univ. Press, Cambridge. 611 pp.
Huschke, R. E., ed. 1959. *Glossary of Meteorology*. American Meteorol. Soc., Boston. 638 pp.
Latimer, J. R. 1972. *Radiation Measurement*. Intl. Field Year for the Great Lakes Technical Manual Series No. 2. Secretariat, Canadian Natl. Committee for Intl. Hydrological Decade. 53 pp.
Lee, R. 1978. *Forest Microclimatology*. Columbia Univ. Press, New York. 276 pp.
List, R. J. 1966. Smithsonian Meteorological Tables, 6th rev. ed. Smithsonian Misc. Collections Vol. 114. Smithsonian Institution, Washington, D.C. 527 pp.
Monteith, J. L. 1973. *Principles of Environmental Physics*. Contemporary Biology Series. Edward Arnold, London. 241 pp.
Oke, T. R. 1978. *Boundary Layer Climates*. Methuen, London. 372 pp.
Qasim, S. H. 1977. *SI Units in Engineering and Technology*. Pergamon Press, Oxford. 54 pp.
Sutton, O. G. 1953. *Micrometeorology*. McGraw-Hill, New York. 333 pp.

CHAPTER 1 – THE RADIATION BALANCE

1.1 REVIEW OF RADIATION PHYSICS

1.1.1 Basic Relationships

The wavelength λ of electromagnetic radiation is given by

$$\lambda = \frac{c}{\nu} \tag{1.1}$$

where λ is the shortest distance between consecutive similar points on a wave train; ν, the frequency, is the number of vibrations per second; and c, the speed of light, is a constant approximately equal to 3×10^8 m s^{-1}. The period τ (time of one vibration) is equal to $1/\nu$ and the wave number is equal to $1/\lambda$.

All waves travel with the same velocity c in empty space. When waves travel in a medium, the velocity may vary with frequency. In a glass prism, for example, visible light is split into a spectrum of different colors. Blue light travels more slowly than red light and thus suffers greater refraction or bending in passing through the prism.

The range of electromagnetic frequencies is enormous, as is shown in Table 1.1 and Fig. 1.1. Visible light occupies a very small part of the spectrum. The visible range is sharply defined: to one side we have the shorter wavelength in the ultraviolet and to the other the longer wavelength in the infrared. We do not see these types of energy, but we sense them in other ways.

1.1.2 Some Definitions

Radiant Flux Density. The radiant flux density* is defined as the amount of energy received on a unit surface in unit time. In accordance with the SI system, we use watts per square meter (W m^{-2}) as the unit of radiant flux density. The cgs unit most commonly used in meteorology prior to adaptation

---

* Meteorologists frequently use the term "intensity" for *flux density*, but the former term is not recommended since it properly denotes a measure of radiant flux per unit solid angle emanating from some source. In this book the terms *flux density* and *flux* are used interchangeably.

6 THE RADIATION BALANCE

Table 1.1 The electromagnetic spectrum

Type of radiation	Frequency range (Hz)	Wavelength range (m)
Electric waves	$0-10^4$	$\infty-3 \times 10^4$
Radio waves	10^4-10^{11}	$3 \times 10^4-3 \times 10^{-3}$
Infrared	$10^{11}-4 \times 10^{14}$	$3 \times 10^{-3}-7.6 \times 10^{-7}$
Visible	$4 \times 10^{14}-7.5 \times 10^{14}$	$7.6 \times 10^{-7}-4 \times 10^{-7}$
Ultraviolet	$7.5 \times 10^{14}-3 \times 10^{18}$	$4 \times 10^{-7}-10^{-10}$
X rays	$3 \times 10^{16}-3 \times 10^{22}$	$10^{-8}-10^{-14}$
Gamma rays	$3 \times 10^{18}-3 \times 10^{21}$	$10^{-10}-10^{-13}$

of SI is the calorie per square centimeter per minute (cal cm^{-2} min^{-1}):

$$697.93 \text{ W m}^{-2} = 1 \text{ cal cm}^{-2} \text{ min}^{-1}$$

Blackbody Radiation. A body that emits radiation at the maximum possible intensity for every wavelength is a "blackbody." Such a body will also absorb completely all radiation incident upon it. The blackbody is a physical ideal, the perfect radiator and perfect absorber.

Emissivity is defined as the ratio of the emittance of a given surface at a specified wavelength and temperature to the emittance of an ideal blackbody at the same wavelength and temperature. Thus, the emissivity ϵ of a blackbody is unity.

The **absorptivity** of a substance is defined as the ratio of the amount of radiant energy absorbed to the total amount incident upon that substance. Thus, the absorptivity α of a blackbody is also unity. For a blackbody, then, $\epsilon = \alpha = 1.0$. In contrast, for a "white body," $\epsilon = \alpha = 0$.

Natural bodies are rarely, if ever, perfect absorbers or radiators. Earth behaves, in total, as a "gray body," but in certain wavelength bands may behave effectively as a "blackbody." For example, in the 8–14 μm wavelength band dark, wet soil and vegetation have emissivities of about 0.97–0.99. The emissivity of quartz sand may be about 0.90. The emissivity of water is about 0.97, independent of its composition. However, oil spills may

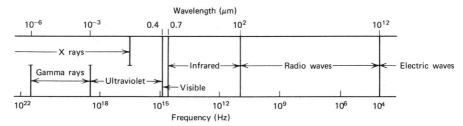

Fig. 1.1 Electromagnetic spectrum on logarithmic wavelength and frequency scales.

have an impact on water emissivity. Robinson and Davies (1972) have determined that an oil film can reduce the emissivity of water by about 3%.

A good approximation of a blackbody for experimental purposes is the interior of a box, painted flat black on the inside with only a single and extremely small aperture through which the void can be viewed.

1.1.3 Radiation Laws

Planck introduced the "particle concept," suggesting that the electromagnetic radiation consists of a stream or flow of particles or quanta. Each quantum has an energy content E_1 determined by

$$E_1 = h\nu \qquad (1.2)$$

where h is Planck's constant with value of 6.626×10^{-34} J s. The greater the frequency (shorter wavelength), the greater the energy content of the quantum. Quanta of ultraviolet light are considerably more energetic than are quanta of red light, for example.

The particle quantum concept is particularly useful when dealing with phenomena that involve the interception of radiation with consequent energy capture or conversion. Photosynthesis is an example of such a process. At the same time, such phenomena as reflection, refraction, transmission, and interference are best explained when radiation is conceived to have the properties of waves.

By (1.2) we calculate the energy content of a quantum of photosynthetically active radiation (say in the red at $\lambda = 0.7$ μm) to be about 2.84×10^{-19} J. A convenient unit for photochemical calculations is the Einstein (Ei), which is equal to one mole of photons (Avogadro's number = 6.023×10^{23} photons mole^{-1}).

Kirchoff's Law. Any "gray" object receiving radiation absorbs, reflects, or transmits a fraction of that radiation. Let α, r, and t represent absorptivity, reflectivity, and transmissivity for a specific wavelength λ. Absorptivity has already been defined. **Reflectivity** is defined as the ratio of the radiant energy reflected to the total that is incident upon the surface. **Transmissivity** is defined as the ratio of transmitted radiation to the total radiation incident upon the medium. Then

$$\alpha(\lambda) + r(\lambda) + t(\lambda) = 1 \qquad (1.3)$$

Thus the absorptivity, reflectivity, and transmissivity are each less than or equal to unity.

Kirchoff's law states that the absorptivity of a material for radiation of a specific wavelength is equal to its emissivity for the same wavelength:

$$\alpha(\lambda) = \epsilon(\lambda) \qquad (1.4)$$

8 THE RADIATION BALANCE

It is important to remember, as Lee (1978) points out, that Kirchoff's law related absorption and emission at the same wavelength. Hence, since the sun and earth emit radiation at different wavelengths the same α or ϵ cannot be used for radiation exchanged between these two bodies. In practice, t and r are measured and α and ϵ are calculated.

Stefan–Boltzmann Law (or Stefan's Law). A blackbody at absolute (Kelvin) temperature T_1, placed in an enclosure at absolute temperature T_2, gains or loses energy at a rate given by

$$\Delta E = \sigma(T_1^4 - T_2^4) \tag{1.5}$$

where σ, the Stefan–Boltzmann constant, is 5.67×10^{-8} J m^{-2} s^{-1} °K^{-4} or 5.67×10^{-8} W m^{-2} °K^{-4}. The energy flux density E of radiation from a blackbody is thus a function of the fourth power of its absolute temperature:

$$E = \sigma T^4 \tag{1.5a}$$

Because it is proportional to the fourth power of temperature, emission of radiation from earthly bodies changes considerably, even in the limited temperature range characteristic of a single day or a short season. For example, radiation flux density from a body at 303°K is 52% greater than at 273°K, whereas the increase in absolute temperature is only 30°K.

Planck's Law. Planck's law defines the energy flux density per unit wavelength emitted by a blackbody as a function of its surface temperature. The relationship is given by

$$E(\lambda) = \frac{2\pi h c^2}{\lambda^5[\exp(hc/\lambda kT) - 1]} \tag{1.6}$$

where $E(\lambda)$ is in W m^{-2} and k, the Boltzmann constant, is 1.38×10^{-23} J °K^{-1}. The wavelength distribution for blackbodies over a wide range of temperatures is shown in Fig. 1.2.

Wien's Displacement Law. The wavelength of maximum emission λ_{max} is related to absolute temperature of the radiating body by

$$\lambda_{max} \; (\mu m) = \frac{2897 \; (\mu m \; °K)}{T \; (°K)} \tag{1.7}$$

where λ_{max} is in micrometers and temperature is in °K.

From the previously stated principles, it is clear that the flux density and wavelength of maximum emission are functions of the temperature of the radiating body. If the body behaves as a perfect radiator or "blackbody," the theoretical spectral emission is calculable. Figure 1.2 shows such spectra for bodies varying in temperature from 500 to 20,000°K. Note that the relative emission of radiation increases with temperature and that λ_{max} decreases. Figure 1.3 (from Gates, 1962) shows the Planck's law distribution

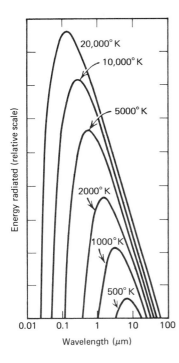

Fig. 1.2 Wavelength dependence of energy emitted by a perfect blackbody at various temperatures.

of energy flux as a function of wavelength for a body such as the sun at 6000°K as well as the actual spectrum of solar radiation at the top of the atmosphere and at the earth's surface after depletion by the atmosphere. The atmospheric constituents that absorb solar radiation are indicated in the figure. These filtering effects are discussed in greater detail in Section 1.3.

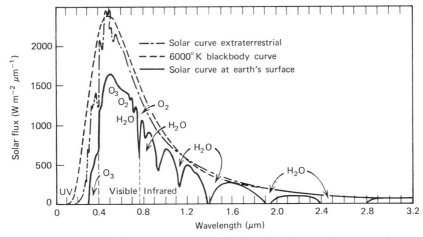

Fig. 1.3 Theoretical and actual spectra of solar radiation at the top of the atmosphere and the actual spectrum at the earth's surface (after Gates, 1962).

10 THE RADIATION BALANCE

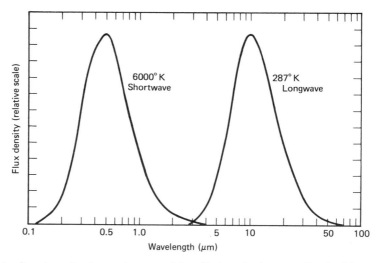

Fig. 1.4 Spectra of solar and terrestrial radiation, both normalized with respect to their peak flux density (after Reifsnyder and Lull, 1965).

The sun may be considered to behave as a blackbody whose surface temperature is ~6000°K. By Wien's law,

$$\lambda_{max, sun} = \frac{2897 \ (\mu m \ °K)}{6000°K} = 0.483 \ \mu m.$$

The earth approximates a blackbody with a mean surface temperature of about 300°K. By Wien's law,

$$\lambda_{max, earth} = \frac{2897 \ (\mu m \ °K)}{300°K} = 9.66 \ \mu m.$$

Almost all of the solar spectrum is confined to the wavelength range between about 0.15 and 4.0 μm with about 50% of the total energy in the solar spectrum delivered in the visible waveband (Fig. 1.3). Most of the earth's radiation is confined to the waveband between 3 and 80 μm. Therefore, as is shown in Fig. 1.4, the theoretical spectra for blackbodies radiating at solar and terrestrial temperatures virtually do not overlap. Solar radiation may be considered to be primarily shortwave; earth radiates primarily in the longwave.

1.2 SOLAR ENERGY RECEIPTS AT THE SURFACE OF THE EARTH: QUANTITATIVE EFFECTS

The sun supplies virtually all the energy received by the earth. This energy drives the processes of photosynthesis, heating of the soil, heating of the air, and evaporation. Solar energy provides, through the latter two processes especially, the mechanisms that drive the weather, as well.

The amount or *quantity* of solar radiation reaching a horizontal unit of the earth's surface depends on a number of factors.* These include the intensity of radiation emitted by the sun, astronomical considerations determining the position of the sun in the sky, and the general transparency of the atmosphere.

The *quality* of solar radiation reaching the earth's surface depends on some of the same factors but also, in large measure, on the composition of the atmosphere at any given time.

1.2.1 Solar Constant

The intensity of radiation emitted by the sun has been the subject of many investigations. Since the early 1900s the solar constant (flux density of the solar beam at the top of the atmosphere and at earth's mean distance from the sun) has been estimated between about 1350 and 1400 W m^{-2}. In other words, one square meter of surface normal to the source of radiation at the top of the illuminated half of the earth's atmospheric envelope will intercept radiation emitted by the sun at the rate of 1350–1400 W m^{-2}.

In recent years solar physicists and climatologists have speculated that variations in the solar constant may have been responsible, at least in part, for variations and fluctuations in the climate of planet earth. One percent of the variance in the long-term climate of the earth, as deduced by climatologists, is attributed to changes in the solar constant (see Fig. 1.9).

Variation in the intensity of solar radiation or solar activity is associated with the appearance of "spots" on the surface of the sun. These spots appear as relatively dark areas on the surface of the sun and occur in pairs. Sun spots last from a few days to several months but exhibit a general cycle with an average length of 11 years.

Sunspots have been observed for very long periods of time. Although there is a rough cyclicity to the sunspot activity, there have been periods of unusually high and unusually low activity as well. One such period of very low sunspot activity occurred during the years 1645–1715. These were the years of the so-called "Little Ice Age" which had a considerable and well-documented impact on life in Europe. Maunder, an English astronomer active in the 1890s, attributed the occurrence of the Little Ice Age to the quiessence of the sun. His theory had achieved only minimal support until recently. Eddy (1976) provides evidence drawn from archeology and other sciences that support the theory of the "Maunder minimum." Landsberg (1980) found evidence, however, that, on a hemispheric basis, the period of

---

* The total of direct beam solar radiation and diffuse sky radiation received by a unit horizontal surface is referred to as the *global radiation*.

12 THE RADIATION BALANCE

the Maunder minimum was not particularly cold and that sunspot rhythms contributed only little to the total variance in temperature.*

Measurement of the solar constant has not been an easy matter. The Smithsonian Institution's first estimates of the solar constant were based on observations made with the pyranometer of Abbott and Aldrich (1916). For a recent description of the Abbott pyranometer, still in use as a standard, see Marsh (1970).

The interposition of the atmosphere between Abbott's sensors and the sun posed certain difficulties in measurement of the solar constant. One of the first opportunities to overcome these difficulties occurred with the advent of modern rocketry. Laue and Drummond (1968) reported experiments made with the X-15 rocketship. They found the total of solar flux to be about 1360 W m^{-2} when the vehicle reached an altitude of 82 km. The ultraviolet and visible wavelengths, defined as being <607 nm, contributed about 35% of the total solar flux density.

In many subsequent manned and unmanned space missions, direct unimpeded observations have been made of the solar radiation flux density. For example, in two rocket flights separated by 30 months, measurements with an absolute pyrheliometer indicated an 0.4% increase in the solar constant (Willson et al., 1980). However, the intermittent nature of these measurements has not permitted definitive values of the solar "constant" and its variations to be determined. The Earth Radiation Balance Experiment, launched in the early 1980s, has as one of its primary purposes a constant surveillance of solar radiation. Once available, these data should have a profound impact on our understanding of atmospheric processes and climatic change.

1.2.2 Angle of Incidence (Lambert's Cosine Law)

Lambert's cosine law states that "the radiant intensity (flux per unit solid angle) emitted in any direction from a unit radiating surface varies as the cosine of the angle between the normal to the surface and the direction of radiation."

Figure 1.5 illustrates Lambert's law applied to a horizontal receiving surface. Let γ be the angle between the beam of radiation and a line normal to surface S. This angle is termed the **zenith distance.** If I_0 is the flux density of the beam received on a unit surface perpendicular to the source S_n, and I is the flux density on a unit horizontal surface S, then by Lambert's law:

$$I = I_0 \cos \gamma \qquad (1.8)$$

---

* Chinese astronomers have also challenged Eddy's findings, citing records that show 13 incidences of major sunspot activity in the Maunder period (Anonymous, 1980. U.S. astronomer challenged. *Beijing Review* No **20**:30).

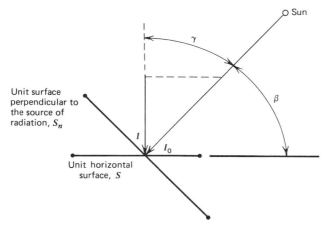

Fig. 1.5 Illustration of Lambert's law.

In terms of earth–sun geometry, the angle β (the **solar elevation**) is the angular elevation of the sun above the horizon. Zenith distance γ is the complement of the solar elevation angle β (γ = 90 − β). Thus for parallel solar radiation at a solar elevation β impinging on a horizontal surface:

$$I = I_0 \sin \beta \quad (1.9)$$

1.2.3 Astronomical Factors

Because the position of the sun in the sky is of such overriding importance in determining the flux density of solar radiation received at any location and time, some important astronomical facts and definitions are briefly reviewed below.

The sun occupies a position at one of the foci of the ellipse that is the earth's orbit. The **eccentricity** of this ellipse (ratio of the distance of one focus from the center of the ellipse to the length of one-half the major axis) is responsible for the varying earth–sun distance. Together with the phenomenon of **parallelism** (consistent inclination of the earth's axis of spin with respect to the plane of its orbit) this eccentricity determines the variability of **insolation** (solar radiation received on the earth) associated with the changes in season.

The **aphelion**, when the sun is farthest from the earth (152.1×10^6 km), occurs about July 4. The **perihelion**, when the sun is nearest the earth (147.3×10^6 km), occurs about January 3. The average earth–sun distance (149.7×10^6 km) occurs about April 4 and October 5.

The earth rotates at an essentially constant rate. This is the rational basis for recording time as

$$360°/24 \text{ h} = 15° \text{ h}^{-1} \quad (1.10)$$

14 THE RADIATION BALANCE

However, because of systematic variations in the earth's orbital path around the sun and because of the sun's changing declination (see definition below), the apparent length of day varies.

Solar noon, the moment of time when the sun crosses the meridian of observation, is a very convenient parameter in solar radiation studies. Solar and civil noon do not coincide precisely: the variance between them depends on the distance of the site from the time zone meridian and on the time of year.

The **equinox** occurs twice annually, when the sun passes directly over the equator or, more properly, when the sun's apparent annual path and the plane of the earth's equator coincide. In the Northern Hemisphere the vernal and autumnal equinoxes occur on or about March 21 and September 22, respectively.

The **solstice** occurs twice annually, when the sun's apparent path is displaced farthest north or south from the earth's equator. In the Northern Hemisphere, the summer solstice occurs on or about June 21, and the winter solstice occurs on or about December 22. Dates of the vernal and autumnal equinoxes and the summer and winter solstices are reversed in the Southern Hemisphere. Figure 1.6 illustrates the seasonal changes in earth–sun geometry.

The **solar declination** (D) is its angular distance north ($+$) or south ($-$) of the celestial equator or plane of the earth's equator. The total range, $-23.5°$ to $+23.5°$ or $47°$, is relatively large and its effects on solar radiation are profound, especially at higher latitudes. Declination angle, D, may be

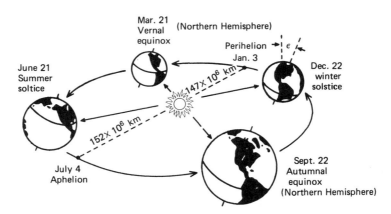

Fig. 1.6 Path of the earth around the sun. Dates of solstices and equinoxes and the associated solar declinations are illustrated. Eccentricity of the earth orbit is illustrated by reference to the earth–sun distances associated with the aphelion and perihelion. The range of the earth's tilt on its axis (now 23.5°) is also shown. (Adapted from Kerr, 1978. Copyright 1978 by the American Association for the Advancement of Science).

determined as follows:

$$D = 23.5 \cos\left[\frac{2\pi(d - 172)}{365}\right] \quad (1.11)$$

where d is the day of the year and π is in radian measure. A "rule of thumb" estimate of the declination is given by

$$D \simeq 23.5 \sin N_1 \quad (1.12)$$

where N_1 is the number of days from the nearest equinox expressed in degrees.

The change in declination of the sun proceeds constantly throughout the year but not at a uniform pace. For example, a 1° change in declination takes about 2.5 days at the time of an equinox and about 15 days at the time of a solstice.

The **solar elevation angle** β is dependent on D, as well as on the time of day (given by **hour angle** h, the angular distance from the meridian of the observer) and on **latitude** l:

$$\sin \beta = \cos(l) \cos(h) \cos(D) + \sin(l) \sin(D) \quad (1.13)$$

The maximum solar elevation angle (or solar altitude) occurs only at solar noon when the **azimuthal angle** is 180° (true south). The azimuthal angle is the angle between true north and the projection of the sun's rays onto the horizontal. A solar elevation angle of 90° occurs only at solar noon when the solar declination is in perfect correspondence with the latitude of the site, that is, when $D = 23.5°$, the solar altitude at solar noon is 90° at latitude 23.5°N (the Tropic of Cancer). This can never occur north of the Tropic of Cancer or south of the Tropic of Capricorn.

As explained above, the altitude and azimuth of the sun are functions of the latitude, time of day (hour angle), and the date or declination (season). A number of charts and nomograms are given in the *Smithsonian Meteorological Tables* (List, 1966) to facilitate the determination of solar elevation and the azimuth. Figure 1.7, taken from that publication, illustrates the patterns of apparent solar passage across the skies with changing season at the equator and at 45°N latitude. From these charts, it is possible to estimate the approximate solar altitude and azimuth for any day of the year and for any time of day. Precise values can be taken off the figures for the dates of the solstices and equinoxes.

Figure 1.8 (from Gates, 1962) illustrates the gross latitudinal effect on the annual receipts of undepleted solar radiation at different locations on earth. The total effect is the result of "cosine law" and day length factors, and it ignores atmospheric depletion. The maximum daily radiation reception is experienced at the highest latitudes, although total annual receipts are lowest there. The bimodal waves for latitudes between 0 and 23.5°N result from the twice annual passage of the sun between the Tropics of Cancer and Capricorn.

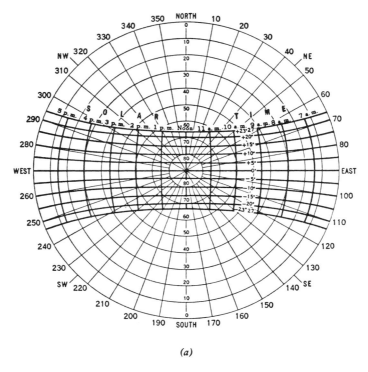

(a)

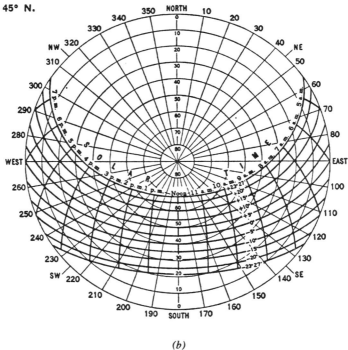

(b)

Fig. 1.7 Solar altitude and azimuth at (a) the equator and (b) at 45°N latitude (from the Smithsonian Tables, R. J. List, ed., 1966).

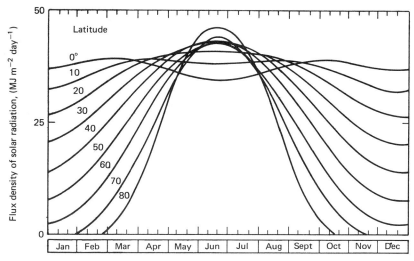

Fig. 1.8 Daily total of the undepleted solar radiation received in the Northern Hemisphere on a horizontal surface as a function of latitude and time of year. Based on a solar constant of 1354 W m^{-2} (after Gates, 1962).

The regularity of the astronomical relationships discussed permits us to define seasons and to predict changes in insolation. However, secular changes in earth–sun relationships are known to have occurred.

1. Eccentricity of the earth's orbit about the sun. Eccentricity varies between 0 and 0.06 with an average return period of 97,000 years.
2. Tilt of the earth's axis. The tilt has varied between 21.8 and 24.4° with an average period of 41,000 years.
3. Precession, the longitude within earth's orbit around the sun (i.e., the time of year) at which the perihelion occurs, varies with a return cycle of 26,000 years.

Weertman (1976) and Hays et al. (1976), among others, support the theory of Milankovitch, who proposed that the earth's ice ages occurred because of long-term variation in the solar radiation received at critical northern latitudes during certain seasons of the year. This variation is produced by changes in the eccentricity of the earth's elliptical orbit around the sun, the precession of the earth's axis, and changes in the value of the tilt of the earth's axis discussed previously.

Further evidence of the linkage between ice ages and changes in the orbital motions of earth around the sun has been reported by H. Hays of Columbia University.* Cores representing 450,000 years of deposition were extracted

---

* *New York Times,* December 5, 1976 (p. 14E). Anonymous. Earth's Orbit Influenced the Ice Ages.

18   THE RADIATION BALANCE

from the South Indian Ocean. Microorganisms in the cores, characteristic of changing climatic regimes, were examined. Cycles of about 100,000 and 42,000 years, matching orbital eccentricity and axis tilt cycles, were identified.

1.2.4   Path Length (Beer–Bouguer Law)

The atmospheric envelope of gases surrounding the earth absorbs considerable portions of the solar beam. This attenuation is a function of the constituents of the atmosphere and, because of selective absorption by these constituents, certain wavelengths are more sharply affected than others. For the moment, however, we will consider only the quantitative and not the qualitative aspects of absorption and scattering.

The Beer–Bouguer Law describes the reduction in flux density of a light beam as a function of the depth into a homogeneous absorbing medium dx:

$$\frac{I}{I_0} = \exp(-\kappa x) \tag{1.14}$$

where I_0 is the initial flux density of the beam, I is the flux density after passage through a depth x of a medium of extinction coefficient κ. The equation is easily adapted to the extinction of solar radiation in the atmosphere by substituting the solar constant R_{sc} for I_0 and the flux density of insolation at the earth's surface, R_s, for I. Then

$$R_s = R_{sc} \exp(-\kappa_a x) \tag{1.15}$$

where κ_a is the atmospheric extinction coefficient.

1.2.5   The Extinction Coefficient

The gases of air and the particles suspended in it extinguish light by two mechanisms: absorption and scattering. **Absorption** is the process by which incident radiant energy is passed to the molecular structure of a substance. Absorption depends on wavelength. For example, the longer wavelengths, near and far infrared, are absorbed effectively by water vapor and CO_2 so that solar radiation received at the earth's surface is enriched in the visible waveband as compared with the situation at the top of the atmosphere.

Scattering is the process by which molecules of a medium and small particles suspended in the medium diffuse a portion of the incident radiation in all directions. **Rayleigh's law** states that molecules intercept and scatter radiation with an efficiency proportional to $1/\lambda^4$. Thus, blue light of $\lambda \simeq 400$ nm will be scattered about 10 times more effectively than red light of $\lambda \simeq 700$ nm. Hence, the blue sky! But at a cost; the direct beam of solar radiation penetrating the atmosphere is enriched in red as a result of the scattering of the blue light.

In view of the various phenomena involved, the extinction coefficient should, properly, consider the quantities and characteristics of the major absorbing and scattering materials: gases, water droplets, and dust. Obviously, gas, dust, and so on are generic terms and can be further defined. For our immediate purposes, it is sufficient to recognize that the extinction coefficient κ_a should take a form such as that proposed by Sutton (1953):

$$\kappa_a = a_g + sa_s + wa_w \tag{1.16}$$

where a_g and a_s are the scattering coefficients for air (gaseous) molecules and for dry solid particles, respectively; a_w is the absorption coefficient for water vapor; s and w are the relative contents of dust and other solids and water vapor, respectively. These coefficients are, themselves, wavelength dependent. Values of the extinction coefficient κ_a range from about 0.01 km^{-1} in very clear air to 0.03 or 0.05 km^{-1} in turbid air.

We see, then, that two factors control the extinction of solar radiation. These are the path length through the atmosphere, which depends on the solar elevation angle and azimuth, and the extinction effects due to atmospheric gases, dusts, water vapor, and other suspended materials.

1.2.6 Effect of Clouds on Radiation Receipts

Obviously, clouds reduce the amount of solar radiation received at the earth's surface. The relationships are not easily quantified, however, because the effects differ for each type of cloudiness. High cirrus clouds have the least influence, although even the *contrails* (condensation trails) from jet aircraft are considered to have a measurable influence on solar radiation (Angell and Korshover, 1975a; Changnon, 1981). Middle clouds have some influence and low stratus clouds strongly reduce incoming radiation. Broken cumulus clouds in summer often concentrate solar radiation because of multiple reflections to the ground. Thus, values as great as the solar constant are sometimes recorded at points on the earth's surface under bright, broken cloudiness.

Haurwitz (1945) attempted to systematically adjust solar radiation receipts for both cloudiness and cloud type. He gave the amount of insolation to be expected on the average with a given cloudiness (degree of sky cover), cloud density, and solar air mass as

$$R_s = (a/m) \exp(bm) \tag{1.17}$$

where m is optical air mass (i.e., secant of the zenith distance of the sun) and a and b are constants depending on cloudiness and cloud density. Haurwitz found that the effects of cloudiness and of cloud density on the insolation are of equal importance if the sky is largely covered.

It is generally assumed that the primary mechanism by which clouds reduce solar radiation receipts at the earth's surface are reflection and absorption. Model computations indicate that absorption can approach 20% of the solar flux for the more absorbing and thicker clouds (Twomey, 1976). Maritime clouds absorb more for the same thickness than do continental clouds, due to the greater absorbing efficiency of larger moisture droplets.

Actual observations from aircraft reveal that real clouds may absorb as much as 30–40% of the solar flux incident upon them (Liou, 1976). Clouds such as thick nimbostratus and cumulonimbus may reflect 80–90% and absorb 10–20% of the solar radiation incident upon them. A thin stratus (0.1 km) reflects 45–72% and absorbs 1–8% of the solar flux. Thus, reflection, absorption, and transmission depend on the type of cloud and its thickness.

Cloudiness decreases the transmission of direct beam (unscattered) radiation to the earth's surface in proportion to cloud thickness. According to Robinson (1977), however, the scattering of radiation is increased by isolated cumulus clouds. Low clouds are effective absorbers of solar radiation and also effective emitters in the longwave. According to Allen (1971), low clouds are near perfect emitters ($\epsilon = 1.0$), but for higher clouds such as altostratus, ϵ may range from 0.30 to 1.00 with a mean of 0.79 ± 0.06. For cirrus clouds, ϵ may range from 0 to 1.00 with a mean of 0.35 ± 0.05.

The interesting possibility exists that potential changes in cloudiness caused by climate change or by the introduction of anthropogenic material into the atmosphere may compensate for other natural or anthropogenically induced alterations in the ozone concentrations of the atmosphere. Greenstone (1978) finds that sunshine duration and trends in total ozone concentration in the United States from 1963 to 1975 correlate well in winter, spring, and summer with increasing cloudiness. In fall, however, both ozone and cloudiness show a downward trend.

Empirical relations have been developed for specified locations to predict solar radiation on the basis of cloud cover. The following expression due to Black (1956) represents average monthly global conditions:

$$\frac{R_s}{R_{so}} = 0.803 - 0.340C - 0.458C^2 \tag{1.18}$$

where R_s is the radiation actually received, R_{so} is the theoretical amount of radiation reaching the earth in the absence of an atmosphere, and C is the mean monthly cloudiness in tenths of sky covered. Da Mota et al. (1977) have recently used the same approach to characterize the seasonal radiation patterns in eight Brazilian climatic regions.

Another way to relate radiation to cloudiness is by reference to measurements of **percent possible sunshine**. This is a measurable parameter defined as "the ratio of time of actual, bright (unimpeded) sunshine to the time during which sunshine is possible." Actual radiation receipts have been related to this parameter by the following type of equation:

$$\frac{R_s}{R_{so}} = a + b\frac{n}{N} \tag{1.19}$$

where n is the actual duration of the sunshine and N is the possible duration of sunshine, both in units of time, and a and b are empirical constants.

For the United States as a whole, the constants $a = 0.35$ and $b = 0.61$ were found by Fritz and MacDonald (1949). This means that, with total cloudiness, 35% of the possible radiation is received. Constants of $a = 0.24$ and $b = 0.46$ were found by Durand (1974) in France when totally cloudy days were neglected and clear day transmission of R_s is assumed to be 0.85. Katsoulis and Leontaris (1981) used an equation of the form of (1.19) above to describe sunny day solar radiation receipt in Greece. For sunless days empirical adjustments were made.

Figure 1.9 (from Rosenberg, 1964) plots actual receipts of global solar radiation on a horizontal surface as a function of percent possible sunshine at three locations in the Great Plains region of the United States. Typical curves for the seasons are shown. The low receipts of solar energy in winter and the high receipts in summer are evident. The dependence of global solar radiation receipts on the percent possible sunshine (the slope b) is greatest in summer. Both the total energy receipts and the dependence on percent possible sunshine are intermediate during the transitional seasons. Typical patterns of percent possible sunshine for the continental United States are shown in Fig. 1.10.

Angell and Korshover (1975b) have shown that percent possible sunshine tended to decrease in the contiguous United States from 1955 to 1972 which they attributed to an increase in daytime cloudiness. The downward trend was most pronounced in fall (Fig. 1.11). The long-term variations observed (change of 10% or more depending on season and over a period of record of 20 years or more) were thought significant enough to be considered in the planning of solar energy applications.

Angell and Korshover (1978) noted an abrupt reversal in the earlier trend described above. Between 1972 and 1976 annual sunshine duration increased so that by 1976 it was 2% greater than the long-term average. Autumn sunshine, which had decreased by 11% from 1953 to 1972, had increased by 7% in 1976. This reversal of trends suggests strongly that the prior decrease was due to normal climatic fluctuations rather than to anthropogenic effects such as jet aircraft contrails or increasing atmospheric turbidity.

Thompson (1976) has found that the relationship of solar radiation to cloudiness over long periods is essentially parabolic as shown in Fig. 1.12. This parabolic relationship is best expressed by an equation of the form

$$y = a + (1 - a)(1 - C)^b \qquad (1.20)$$

where y is the ratio of observed radiation to clear sky radiation, C is sky cover in hundredths, a is a constant, and b is a variable parameter less than 1.0.

The interannual variations in cloud cover based on percent possible sunshine over the continental United States have been found to be small, about

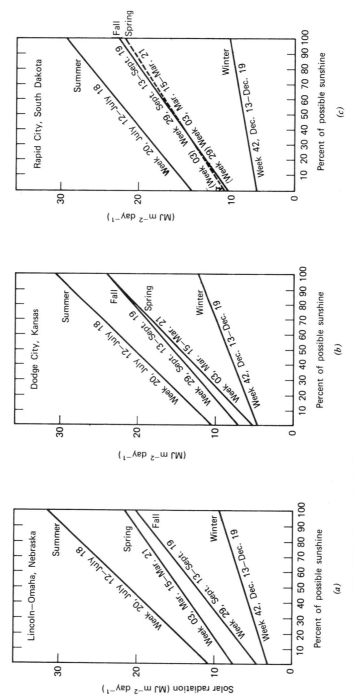

Fig. 1.9 Seasonal relationships between percent of possible sunshine and daily solar radiation at (a) Lincoln–Omaha, Nebraska, (b) Dodge City, Kansas, and (c) Rapid City, South Dakota (after Rosenberg, 1964).

SOLAR ENERGY AT THE EARTH'S SURFACE—QUANTITATIVE

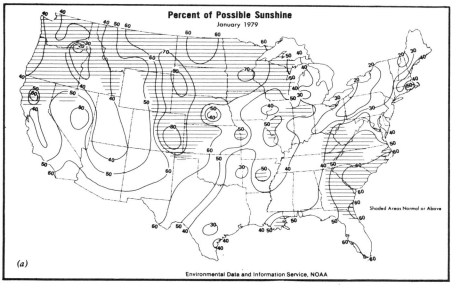

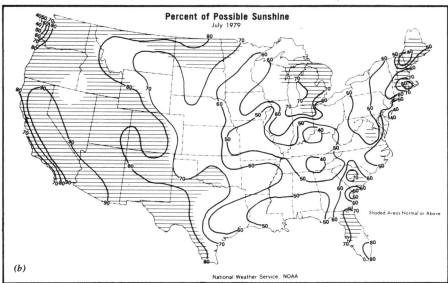

Fig. 1.10 Percent possible sunshine in the continental United States in (*a*) January 1979 and (*b*) July 1979. From *Weekly Weather and Crop Bulletin*, December 1979.

4% from year to year (Hoyt, 1978). This variation in cloudiness can account for variations in the annual receipt of solar radiation of about 1.2%.

Cloud observations are among the most difficult to make systematically. Satellite photographs are improving our knowledge of the frequency, type, and extent of cloud cover over large regions. In one recent study, satellite

24 THE RADIATION BALANCE

photographs were used on a daily basis to inventory cumulus clouds over western Kansas during May–September to assist in weather modification research and in energy balance studies of the region (Marotz and Henry, 1978). Tarpley (1979) reports results of a comparison of incident solar radiation estimates from a geosynchronous satellite with measurements from ground based pyranometers place in a network from 20 to 49°N latitude in the U.S. Great Plains. Cloud induced bias led to errors as great as 5 MJ on thickly overcast days, but in general the standard error of estimate was less than 10% of the mean daily insolation. Studies such as these will make the prediction of solar radiation receipts through remote sensing more reliable than is now possible.

1.2.7 Turbidity

Much of what has been discussed thus far concerning the quantitative depletion in solar radiation can be considered in terms of the atmosphere's turbidity.

Turbidity is defined as "any condition of the atmosphere which reduces its transparency to radiation, especially to visible radiation." Ordinarily, the term is applied to a cloud-free portion of the atmosphere. Dusts, pollens, water vapor, and all suspended materials affect the atmosphere's turbidity. The term **aerosol** is used to describe those materials of dispersed solid or liquid particles suspended in that mixture of gases we call air.

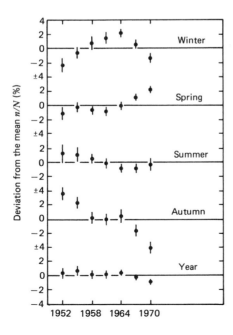

Fig. 1.11 Long-term trend in n/N within the contiguous United States, by season and mean for the year. The vertical bars extend 2 standard deviations of the mean either side of the mean, based on the six regional values (after Angell and Korshover, 1975b).

SOLAR ENERGY AT THE EARTH'S SURFACE—QUANTITATIVE

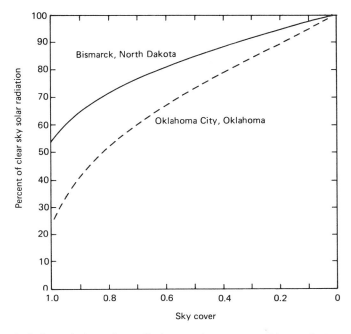

Fig. 1.12 Relation of clear sky radiation to sky cover at Bismarck, North Dakota, and Oklahoma City, Oklahoma. (Thompson, E. S. 1976. Computation of solar radiation from sky cover. *Water Resour. Res.* **12**:859–865, copyrighted by the American Geophysical Union.)

There are a number of methods for describing turbidity. The *Glossary of Meteorology* (Huschke, 1959) defines one of the more common methods: Linke's turbidity factor θ. θ is a measure of the atmospheric transmission of incident solar radiation and is given by

$$\theta = \frac{\ln I_0 - \ln I}{\ln I_0 - \ln I_m} \tag{1.21}$$

where I_0 is the flux density of the solar beam just outside the earth's atmosphere, I is the flux density at normal incidence measured at the earth's surface with the sun at a zenith distance that defines an optical air mass m, and I_m is the flux density that would be observed at the earth's surface for a pure atmosphere of depth m (under a set of standard conditions). The value of I_m is obtained from prepared tables.

Linke's turbidity factor θ can also be defined (following Sutton, 1953) in terms of (1.16) as

$$\theta = \frac{\kappa_a}{a_g} = 1 + \frac{wa_w}{a_g} + \frac{sa_s}{a_g} \tag{1.22}$$

26   THE RADIATION BALANCE

in order to more specifically express the variation of transparency with water vapor and dust content. Combination with the Beer–Bouguer law yields

$$I_m = I_0 \exp(-a_g \theta_x) \tag{1.23}$$

where $x = mx_m$ and x_m is the thickness of the atmosphere at m = 1 ($m = 1$ when the sun is at zenith and the point of observation is at sea level. For other solar elevations m is proportional to the secant of the zenith distance).

Natural Aerosols and Turbidity.   Natural sources of aerosols that control turbidity include wind scouring erosive processes, forest and range fires, slash and burn agriculture, volcanism, and plant exudation of terpenes, waxes, and other biogenic materials. Many complicated photochemical processes also act on natural and industrial gaseous compounds converting them to aerosol size. Some of these processes are described in this section. For further discussion of these processes, see Hobbs et al. (1974).

Wind Erosive Processes.   Gillette et al. (1978) have found that dust eroded from the southwestern United States ranges in size from 1 to 100 nm but is bimodally distributed with one peak between 1 and 30 and a second between 30 and 100 nm. Joseph et al. (1973) estimate, by means of normal incidence spectral solar radiation measurements, that "Khamsin" storms in the Mideast put as much as $3.2 \pm 1.6 \times 10^6$ tons of dust into the air in a single year. Extending their findings to areas of the world affected by similar conditions, they estimate a dust loading of $128 \pm 64 \times 10^6$ tons per year.

The quantities of dust carried aloft by winds are sometimes of a magnitude detectable from satellites. A dust storm originating at the Texas–New Mexico borden in spring of 1977 is shown in Fig. 1.13 (from Kessler et al., 1978). The long distance transport of dust is evident in this photograph. Direct mesurements of a decreased solar radiation penetration at Mauna Loa, Hawaii, were attributed by Shaw (1980) to dust originating thousands of kilometers away in the Gobi desert of China. The concentration of insoluble mineral aerosols in the lower atmosphere of the western Equatorial Atlantic Ocean tripled in the decade of the mid-1960s to mid-1970s (Prospero and Mees, 1977), perhaps as a result of the widespread and long-lasting Sahelian drought that diminished in severity by 1974.

Agricultural Burning.   "Slash and burn" agriculture and other land burning operations also load a considerable amount of aerosol into the atmosphere. For example, Pueschel and Langer (1973) reported that the effects of sugar cane fires were detected 100 miles from the source.

Plant Exudations.   Plants, too, emit substances of aerosol size. Fish (1972) believes that the blue haze over heavily forested areas may be due to submicrometer-sized wax particles generated by electrical processes that occur at the tips of pine needles and other wax-covered plant surfaces. Went

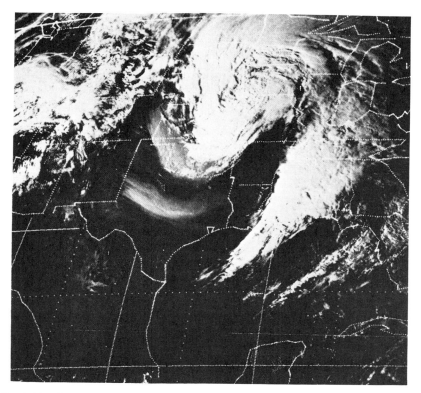

Fig. 1.13 Satellite view of dust storm, February 23, 1977 (Kessler et al., 1978).

(1955) believes that much of the blue haze observed in a number of regions of the world is due to the special optical properties of terpene materials. Schnell (1974) lists a wide range of additional plant-emitted materials that have been found dispersed in the atmosphere.

Ellis and Pueschel (1971) found annual cycles in the flux density of solar radiation at Mauna Loa observatory between 1958 and 1970 with minima occurring in summer. They attributed these to the exudation of volatile materials of plant origin that contribute to turbidity or to effects of the seasonal variations in general circulation patterns. Ellsaesser (1972) has reviewed evidence concerning the cyclicity in annual turbidity and considers that this may be due to two factors: the spread of volcanic dust and seasonal changes in biological processes that contribute to the emission of aerosols.

Volcanic Activity. Figure 1.14 (from Flowers and Viebrock, 1965) describes the effects of volcanic eruptions in Bali during March 1963 on the global solar radiation. This evidence shows convincingly that a significant

28 THE RADIATION BALANCE

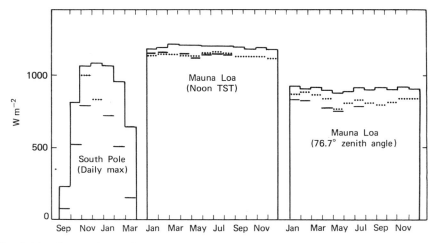

Fig. 1.14 Comparison of the 1963 and 1964 normal incidence solar radiation with the 1957–1962 monthly normals at the South Pole and at Mauna Loa, Hawaii. The outside lines represent the 1957–1962 monthly normals, the inside lines represent those for 1964, and the dots represent those for 1963 (after Flowers and Viebrock, 1965).

decrease in solar radiation receipts resulted from the volcanic eruptions and that the spread of dust was global by 1964.

The years 1956–1963 were relatively quiet with respect to volcanism. The Bali (Mt. Agung) eruption of 1963 injected dusts at 8°S latitude into the stratosphere to a height of 22–23 km. Although most of the dust remained in the Southern Hemisphere during the first year, its poleward spread could be calculated as about 0.4 m s^{-1} once it reached 30° of latitude. Dyer and Hicks (1968) collected data that indicated maxima in turbidity resulting from the Mt. Agung explosion moving toward the poles with a pulse rate of approximately 1 yr. Near the equator the maxima occurred in summer but successively later, so that maxima occurred in winter in the mid latitudes.

The Mt. St. Helens eruption in 1980 had a smaller and shorter effect on solar radiation receipt, but one that was measurable nonetheless. Analyses of materials injected into the stratosphere and the troposphere by that eruption indicate that the reflection of visible light back to space would have been increased (Ogren et al., 1981). Direct measurements of the attenuation of solar radiation made in Richland, Washington, and Billings, Montana, as the ash clouds from Mt. St. Helens passed over these sites, were reported by Howard (1981). At Billings an immediate reduction from about 1000 to 200 W m^{-2} was observed to occur when the cloud arrived at 1400 hours. The effect lasted for about $2\frac{1}{2}$ h.

It is important to realize that most of the materials that have been described above are concentrated in the troposphere. Volcanic eruptions are probably the chief source of very large quantities of dust and other materials that spread worldwide and particularly of particles that are injected into the stratosphere.

Implications of Increased Atmospheric Turbidity. Much has been written in recent years concerning the possible impacts of increased turbidity (due either to man's activities or to natural causes) on climatic events. Bryson and Goodman (1980) have correlated radiocarbon dates of volcanic activity with evidence of climatic changes. Dust, smoke, and other aerosols in the atmosphere have also been considered by some (e.g., Bryson, 1973) to effect a global cooling that could initiate a drought. The cold winters of 1977 and 1978 in the eastern United States were attributed to a thickening of atmospheric dust. Measurements made at the Mauna Loa observatory in Hawaii do not confirm this supposition.* There have been no long term changes in the amount of small particles suspended in the air at this clean air location which is unimpacted by industries, dust storms, etc.

Data on global turbidity (Heidel, 1972) obtained at the Mauna Loa, Hawaii, observatory show that turbidity decreased from the peak that occurred in 1963–1965 after a high degree of volcanic activity. By January 1970, turbidity had decreased to the pre-1963 level. At Tucson, Arizona, turbidity had also declined after the 1963–1965 period. However, after January 1970, the decline halted, possibly because of the local effects of air pollution from nearby smelters.

Mendonca et al. (1978) found that 20 years of atmospheric transmission data from Mauna Loa show secular decreases at irregular intervals. There is a regular annual variation as well. Transient decreases in transmission are strongly correlated with explosive volcanic erruptions. Recovery usually takes as much as 8 yr. In 1977 atmospheric transmission of direct incidence solar irradiation at Mauna Loa returned to values of the 1958–1962 period prior to the Mt. Agung eruption of 1963. Annual cycles show minimum transmission during April to June and maxima during November to February. There is some evidence of a biennial variation as well.

The possibility that an increase in turbidity could lead to a global cooling has been mentioned above. It may also be possible that warming may occur. For example, Idso (1974) noted a long-term warming in the weather records of Phoenix, Arizona, after 1948. He attributed the warming to a buildup in pollution in the lower layers of the atmosphere which causes an increased interception of longwave radiation. Idso and Brazel (1977) provide evidence that dust-loading events that occurred on otherwise cloudless days at Phoenix, Arizona, altered the receipts of solar and net radiation at the surface, but more importantly, altered the ratio of diffuse to direct incidence radiation. It is this ratio that Idso and Brazel consider to be the "forcing function" for atmospheric temperature change. Initial increments of dust should cause

---

* NOAA Public Affairs Office, Rockville, MD, News Release 77-50, Feb. 27, 1977. NOAA climate monitors detect no long-term increases in atmospheric dust. Mimeo, 3 pp.

a warming effect. Man's inputs of dust are much smaller than the amounts needed to cause a cooling, they hold.

Herman et al. (1978) challenge these conclusions contending that Idso and Brazel are extrapolating from the case of clear skies over land to the entire globe, which is primarily ocean, is about 50% cloud covered at any time and has a haze of dusts, terpenes, salts, and other materials.

Baldwin et al. (1976) considered the impact on climate of aerosols injected into the stratosphere. They conclude, on the basis of heat balance calculations, that aerosols generated by volcanic explosions (sulfurous gases and silicate particles) have led to climatic changes by creating dust veils that reduce the penetration of solar radiation. Their model suggests that aerosols injected by supersonic transports and space shuttles (sulfur-containing gases in the former case; aluminum oxide particles in the latter) are unlikely to have a significant climatic impact in the next few decades.

Geographic Distribution of Turbidity. The United States National Weather Service established a network to observe turbidity in about 1961. Results of some of the first systematic observations were reported by Flowers et al. (1969). The turbidity index used in this work is one due to Volz and is based only on the extinction of 0.5 μm radiation. Turbidity is seen in Fig. 1.15 to be a function of season and air mass dominance. For example, when continental polar (cP) air dominates in winter, turbidity is low. When cP air dominates in summer, turbidity is higher than in winter because of more dust, pollen, and vapor, but still the air is considerably less turbid than when the humid maritime tropical (mT) air dominates in summer.

Figure 1.16 shows very clearly the major seasonal differences in turbidity at all the observational sites in the United States. Turbidity in the Southeast is as high, despite low industrialization, as it is in the industrial North Central areas or in the smog bowl of Los Angeles. This is due to the dominance of humid mT air masses. In the central United States region of the Great Plains and the Prairie States, turbidity is relatively low, even if the observation site is only 80–150 km away from major centers of air pollution. The most severe turbidity conditions exist in the industrialized zones of the mid-Atlantic States.

Air mass influence on turbidity is well recognized. Monteny and Gosse (1978), for example, have distinguished differences in turbidity due to the origin of air masses (maritime or continental) over the Ivory Coast in western Africa. Similarly Peterson et al. (1981) find that turbidity in the continental United States is directly proportional to humidity and dew point. Southern origin air masses are the most turbid.

Cities provide much of the aerosol that increases turbidity on a local scale. Komp and Auer (1978) measured horizontal visibility and aerosol content of air upwind, over, and downwind of St. Louis, Missouri. Visibility was reduced by as much as 50% during the work week but only by 20% during weekends, indicating that industrial activity and in-city traffic were primarily

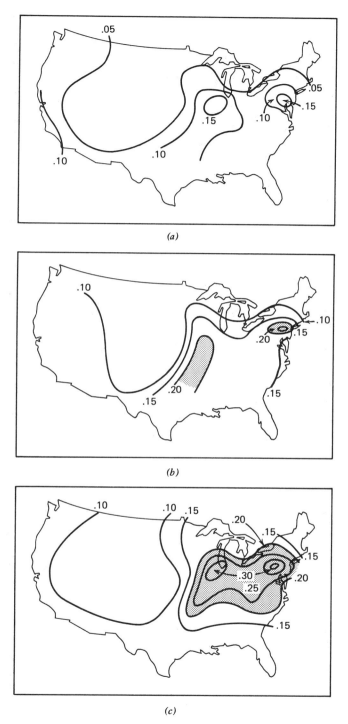

Fig. 1.15 Mean turbidity for various air masses: (*a*) cP air in winter; (*b*) cP air in summer; and (*c*) mT air in summer (after Flowers et al., 1969).

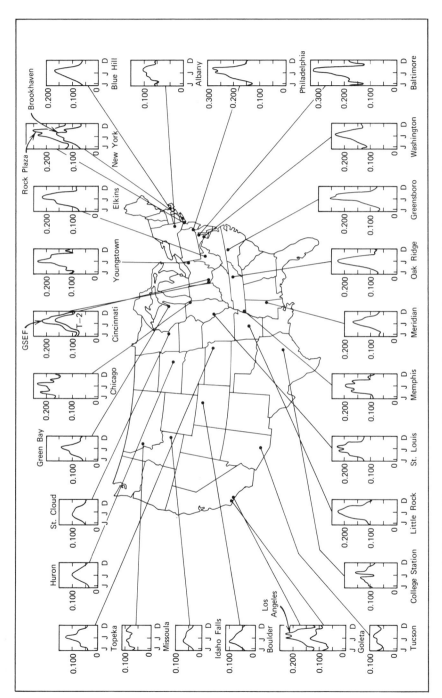

Fig. 1.16 Monthly average turbidity at U.S. National Weather Service network stations (after Flowers et al., 1969).

responsible. The receipts of all wave solar radiation are also reduced by urban air pollution. Peterson and Stoffel (1980) note that in St. Louis solar irradiation is reduced by about 3% annually (4.5% in winter and 2% in summer) as compared with adjacent rural areas.

1.3 SOLAR ENERGY RECEIPTS AT THE SURFACE OF THE EARTH: QUALITATIVE EFFECTS

1.3.1 Selective Absorption by Constituents of the Atmosphere

Earth's atmosphere is unique in its composition and hence in its qualitative and quantitative filtering effects on the solar radiation that passes through it. The major gaseous components of the dry atmosphere on earth are given in Table 1.2. Observations made possible by the Space Age have shown us just how different our atmosphere is from that of our sister planets in the solar system. The atmosphere of Mars, for example, is composed predominantly of carbon dioxide (Kliore et al., 1969).

From the point of view of absorption of solar radiation, carbon dioxide and ozone are the atmospheric constituents of major importance. Water

Table 1.2 Constituents of clean, dry air

Constituent	Volume (%)
Nitrogen	78.08
Oxygen	20.95
Argon	0.93
Carbon dioxide[b]	0.032
Neon	1.8×10^{-3}
Helium	5.24×10^{-4}
Krypton	1×10^{-4}
Hydrogen	5×10^{-5}
Xenon	8.0×10^{-6}
Ozone[a]	Variable; about 1.0×10^{-6}
Radon[a]	6.0×10^{-18}
Plus	
Water vapor	0–3%+
Dusts	?
Pollens	?
Industrial emissions	?

SOURCE: R. J. List, ed., Smithsonian Meteorological Tables, 1966, p. 389.
[a] Variable
[b] The mean global concentration reached about 0.034 in 1980.

34 THE RADIATION BALANCE

vapor, which constitutes, depending on the weather situation, from nearly 0% to more than 3% of the atmosphere by volume, is also extremely important. Dust, pollens, and industrial emissions (often complex chemical products) also interact with the solar and terrestrial radiation, but less is known of their effects.

Figure 1.17 (after Gates, 1965) shows the spectral distribution of solar radiation received at the top of the atmosphere. The inner line shows the spectral distribution received at the earth's surface through a distance twice the depth of the normal atmosphere (two air masses; a solar elevation of 30°). We already know that a quantitative reduction in irradiance results from atmospheric absorption and scattering. Changes in the quality of the solar radiation are also clearly evident in this figure.

Much of the ultraviolet (uv) radiation is screened out by the atmosphere. This is due largely to absorption by ozone and oxygen. Qualitative changes in the visible range are few except at the end of the red. In the far red and infrared (ir), there are major depletions in radiation intensity. These are caused by water vapor and carbon dioxide absorption.

Figure 1.18 gives very detailed information on the absorption spectra of the major atmospheric constituents. Included are methane (CH_4) and nitrous oxide (N_2O) which deserve special mention. Ozone is seen to be an almost perfect absorber of uv to 0.3 μm. Ozone and oxygen have other absorption

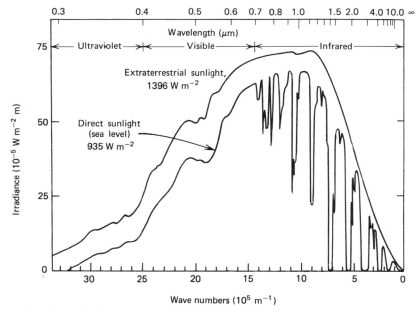

Fig. 1.17 Spectral distribution of solar radiation, extraterrestrial and at sea level, for a clear day. Each curve represents energy incident on a horizontal surface (adapted from Gates, 1965).

SOLAR ENERGY AT THE EARTH'S SURFACE—QUALITATIVE

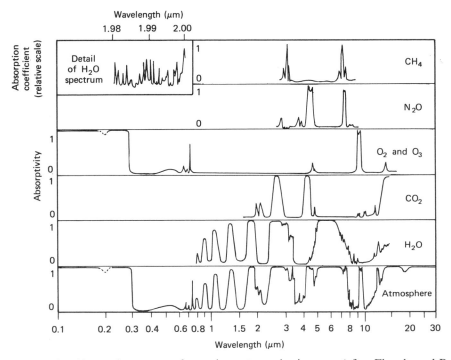

Fig. 1.18 Absorption spectra for various atmospheric gases (after Fleagle and Businger, 1963).

peaks with a major one due to ozone at 9.6 μm. Carbon dioxide has important absorption peaks at about 2.3 and 4 μm and a major peak at about 15 μm. Water vapor is a selective absorber in the far red and ir. A major absorption peak occurs between 5 and 8 μm and a "window" occurs between 8 and 14 μm, which is especially important in the energy balance of earth. Total absorptivity is shown at the bottom of the chart and is a composite of the effects of all the atmospheric components involved. Changes now occurring in the composition of the atmosphere and their implication on the quality of the radiation penetrating the atmosphere deserve further comment.

Oxygen. There have been suggestions that industrial processes and other processes leading to environmental pollution might result in a decrease in the concentration of oxygen in the atmosphere. Observations made in clean air, mainly over the oceans between 50°N and 60°S yield an almost constant value of 20.946%. Machta and Hughes (1970) indicate that changes since 1910 in oxygen concentration are either zero or smaller than the difference detectable.

Ozone. Ozone absorbs much of the uv light from the solar spectrum. Since uv is germicidal, its removal is essential for the existence of life on earth. A quotation from Basuk (1975) explains the problem most succinctly. "It has been assumed, and no alternative view has been presented, that the evolution of life on the exposed surface of the planet began with formation of the ozone screen. The removal of this screen would mean that conditions of uv illumination at the surface would become similar to those of pre-Cambrian days, and would presumably eliminate all surface life. In other words, the biosphere has adapted to all available uv illumination."

Cutchis (1974) has calculated potential increases in the erythemal dose (sunburning uv radiation) that would occur at specific latitudes for specific decreases in the ozone concentration. A 50% depletion might result in a tripling of the erythemal radiation at certain latitudes.

The ozonosphere is a layer within the stratosphere about 32 km above the earth's surface which is rich in ozone. The concentration in that layer ranges from about 5 to 10 ppm. The total amount of ozone in the stratosphere is not very great. If compressed at STP (standard temperature and pressure) at the earth's surface, the ozone would occupy a layer only 2–5 mm in thickness (Basuk, 1975).

Ozone varies naturally with latitude and with season. According to Cutchis (1974) ozone increases from fall to spring by about 20% over the Continental United States. General increases are smaller near the equator and larger near the poles. The highest latitudes receive much less uv at the surface because of the greater path length for solar radiation penetration through the atmosphere.

Recent concerns about the ozonosphere center on a number of phenomena and materials that can destroy or catalyze the destruction of ozone. In fact, total global ozone increased by about 7.5% per decade from 1957 to 1970 (London and Kelley, 1974). However, recent satellite data (7 yr of Nimbus-4 and 2 yr of Nimbus-7) reported in 1981 indicate a year to year decrease of about 0.5% per year at an altitude of 40 km. This decrease is believed to be due to anthropogenic causes (Norman, 1981).

The materials of greatest concern as potential reactants with ozone are nitrous oxides, chlorofluorocarbons, and other anthropogenic gases.

Nitrous Oxides. There is evidence (Whitten et al., 1975) that nuclear explosions, by injecting nitric oxide into the ozonosphere caused significant, but temporary, reductions in ozone concentration. Solar proton events, according to Crutzen et al. (1975), cause natural production of NO so that temporal fluctuations of ozone concentration are to be expected.

Concern has also been expressed that a fleet of supersonic transports (SSTs) flying in the stratosphere might have a similar effect since nitrogenous materials are emitted in the combustion products of jet engines. It is also feared that water vapor injected into the stratosphere by jet engines might react with the ozone as well as increase the capture of ir radiation, leading to an alteration of the earth's energy balance. Johnston (1971) calculated that a fleet of SSTs injecting NO and NO_2 into the atmosphere could reduce the ozone concentration of the atmosphere by a factor of 2. These calculations played an important role in the U.S. decision of the early 1970s to defer the construction of a fleet of SSTs.

Most recently nitrogenous fertilizers have been implicated in the possible destruction of the ozonosphere (CAST, 1977). Nitrous oxide is released from soils to the atmosphere during the nitrification of ammonium-producing fertilizers under aerobic conditions and by denitrification of fertilizer nitrogen under anaerobic conditions (Bremner and Blackmer, 1978). A possible mechanism by which nitrous oxide may react to destroy ozone is proposed by Basuk (1975).

Ellsaesser (1977), on the other hand, considers that agriculture may actually have reduced the threat to the ozone layer. Land drainage, alterations in soil pH, aeration, destruction of bacteria-culturing humus, and other actions would have resulted in a reduction of anaerobic denitrification processes despite opposing effects such as increased use of N_2O in irrigation, rice paddy cultivation, etc. In support of these ideas, Ellsaesser noted a general tendency toward an increase in total global ozone since systematic observation began in the 1920s.

It is also possible that soil acts as a sink for NO_2 as well as other atmospheric pollutants such as ethylene and SO_2. Microbial and chemical reactions in soil can immobilize these compounds (Abeles et al., 1971; Freney et al., 1978).

Actual measurements of N_2O emissions are very scarce. One recent micrometeorological study of emission from an irrigated cornfield in Colorado reports that 2.6 kg of nitrogen per hectare (about 1.3% of the total applied) was emitted by the soil as N_2O during the growing season (Hutchinson and Mosier, 1979).

Chlorofluorocarbons. In the mid-1970s it became apparent that chlorofluorocarbon compounds (commonly called Freons) might also be involved in destruction of the ozone layer. Freons are chemically inert synthetic gases used in air conditioning and as propellants for spray cans. After release, Freons eventually diffuse into the stratosphere where photochemical breakdown products can react with ozone. Some of the postulated reactions are given by Basuk (1975).

A number of analyses of Freon production trends, rates of diffusion to the stratosphere, and other factors (e.g., Howard and Hanchett, 1975; Cicerone et al., 1974)) led to recommendations by the U.S. National Academy of Sciences that Freon production be halted. This has indeed been ordered by appropriate agencies of the U.S. government.

Not all agree that the predicted effects of Freons on the ozone layer are realistic, however. For example, Parry (1977) argues that long-term trends in ozone concentration are upward, that one-dimensional models of Freon diffusion to the stratosphere are inapplicable, that there is inadequate knowledge of the other constituents of the ozonosphere which might be involved, with or without Freon, in ozone destructive reactions, and that there is inadequate knowledge of the physics of ozonosphere and radiation processes. However, the most recently documented decrease in ozone concentration has been attributed, on the basis of atmospheric models, to the activity of chlorofluoromethanes (Norman, 1981).

Anthropogenic Gases. Other man-made (anthropogenic) materials have been implicated as possibly impacting on the concentration of ozone in the stratosphere. At 1973 release rates, methochloroform was predicted by McConnell and Schiff (1978) to have an affect about 20% as great as that of the Freons. Carbon tetrachloride was also present in the atmosphere in 1975 in a concentration of about 76 ppt according to Singh et al. (1976). The major sink for these and other man-made pollutants has been shown by Singh et al. (1979) to be the stratosphere rather than the oceans or land. Such gases as methane, ammonia and nitrous oxide absorb thermal radiation in the 7–14 μm range and, hence, as Wang et al. (1976) suggest, extensive use of fertilizers and combustion processes may contribute to a global warming.

Water Vapor. The atmosphere has little effect on the visible spectrum, but a considerable amount of the near ir radiation of the solar spectrum to 4.0 μm is filtered out by water vapor and carbon dioxide. Since the earth radiates in the ir with the wavelength of maximum emission at about 10 μm, water vapor has a major influence in retaining terrestrial radiation and in reducing its escape to space. The bulk of terrestrial radiation is in the range of the atmospheric window of water vapor. However, the quantity of water vapor in air does have a major effect on the earth's radiation balance because of absorption in other wavebands in the ir, including bands at the shoulders of the atmospheric window. It has also been noted recently that water vapor in the form of ground fog attenuates uv strongly (Baker-Blocker, 1980).

Carbon Dioxide. Figure 1.18 shows that carbon dioxide has major absorption peaks in the near ir at about 2.5 μm and in the ir at about 4 and 15 μm. Carbon dioxide thus augments water vapor in retaining ir radiation within the earth–atmosphere system. The increase in global carbon dioxide concentration that is now occurring is worrisome to atmospheric scientists, particularly because of a potential warming of the lower layers of the atmosphere which might follow additional absorption of ir radiation by carbon dioxide.

Since the first edition of this book was issued, the accumulation of carbon dioxide in the atmosphere has continued and has, in fact, accelerated somewhat. At this writing, the rate of increase appears to be about 1 ppm year^{-1}. Models of future increase have been developed and indicate that a doubling of the pre-Industrial Revolution concentration (about 280–290 ppm) is likely by the year 2025. The impact of these concentration changes on global climate have been modeled in a number of ways. Perhaps remarkably, almost all models predict a global warming of the lower atmosphere of an unprecedented kind and several models predict large regional changes in climate. Some implications of the CO_2 increase in the atmosphere are discussed in Chapters 8 and 11.

Methane. Methane, another combustion product of fossil fuels, is also an effective absorber in the ir range. Most recent information suggests that methane is a natural constituent of the atmosphere, resulting primarily from vegetative decomposition and that its concentration is probably not increasing significantly as a result of man's activities.

Carbon Monoxide. Another atmospheric constituent that has recently received considerable attention is carbon monoxide. The primary reason for this attention is the suspicion that combustion processes are leading to a major increase in atmospheric concentration of this gas.

CO actively absorbs solar radiation and may also reduce OH concentrations in the atmosphere affecting the natural $CO-OH-CH_4$ cycle.

In samples taken near the ground, within the United States, CO concentrations range from near zero to >40 ppm. Wang and Shaw (1970) show that the concentration varies with traffic activity with maxima occurring near 0700 and 1700 hours. Sze (1977) has calculated that if current emission rates continue to the year 2025, CO and CH_4 may increase by 50 and 25%, respectively. However, current evidence indicates that CO produced by natural processes far exceeds combustion rates (Weinstock and Niki, 1972). Large amounts are produced by methane oxidation in the atmosphere. Swinnerton et al. (1970) have shown that the surface waters of the South Atlantic are supersaturated with respect to the partial pressure of CO in the atmosphere and, hence, the net transfer must be from sea to air. The ocean is likely the major source of atmospheric CO according to Galbally (1972). Inman et al. (1971) report, on the other hand, that acidic soils high in organic matter content rapidly deplete CO from the air.

The Atmospheric Envelope. Aside from the specific absorption effects described above, the existence of an atmosphere leads to other qualitative changes in the solar radiation reaching the earth's surface. Because of scattering and absorption the percentage of the total solar energy in the visible band decreases with increasing path length through the atmosphere, which causes an enrichment in the near ir.

The ratio of blue to red light in the visible wavelengths also decreases as path length increases because of increased scattering and also because the blue light is refracted more completely away from the horizon. This accounts for the "redness" of the sky at sunrise and sunset when the sun's rays must pass through the greatest thickness of atmosphere. There is a marked enrichment in blue light, however, immediately after sunset or before sunrise because of the predominance of diffuse rather than direct radiation when the sun is below the horizon. This effect persists for only a few minutes (Johnson et al., 1967).

1.3.2 The "Greenhouse Effect"

The term **greenhouse effect** is frequently used in discussions of the possible impacts of various pollutants on the earth's radiative balance. Water vapor, carbon dioxide, some of the chlorofluoromethanes and other materials of organic origin are essentially transparent to shortwave radiation, but are effective absorbers in the thermal ir region of the electromagnetic spectrum. Hence, any increase in the concentration of these substances in the atmosphere is expected to increase global temperature through a trapping of the terrestrial emitted longwave radiation. The analogy has been made with the behavior of a greenhouse or more properly a "glasshouse" since glass is essentially permeable to radiation in the visible waveband but impermeable to ir radiation.

This analogy is defective, however. Lee (1973) has pointed out that greenhouse warming is due primarily to the suppression of convection rather than to the trapping of ir radiation. In fact, a glass surface trapping longwave radiation from within the greenhouse will itself be heated considerably and reradiate intensely in the ir waveband.

Nonetheless, the term "greenhouse effect" continues to be used. Ramanathan (1975) speaks of a "greenhouse effect" due to chlorofluoromethane; Mercer (1978) speaks of the threats to the Antarctic ice sheets from a CO_2 "greenhouse effect."

1.4 SKY RADIATION (DIFFUSE)

Thus far, in discussing depletion of solar radiation by the atmosphere, we have concentrated on the direct beam from the sun. An object shaded from the direct beam would be illuminated by the scattered radiation or "sky" radiation and would not be in the dark. **Sky radiation** is defined as that portion of radiation that reaches the earth's surface after having been scattered from the direct beam by molecules or suspensoids in the atmosphere.

Particularly in high latitudes diffused solar radiation is very important. Even in the mid-latitudes, diffuse sky radiation may contribute 30–40% of the total solar radiation. The diffuse contribution is greatest during the winter months and when the solar angle is low (path length is great). Cloudiness also greatly increases the ratio of diffuse to direct radiation. Before sunrise and after sunset all of the solar radiation received is sky (diffuse) radiation. The biological effects of the diffuse radiation may be considerably more significant than is the energy content of that radiation. For example, diffuse radiation penetrates into plant communities more effectively than does the direct beam.

SKY RADIATION (DIFFUSE)   41

In many applications, sky radiation is assumed to be isotropically (directionally uniform) distributed in the sky hemisphere. Direct observations with radiation sensors indicate that true isotropy is rare. McArthur and Hay (1978) used fisheye photography of the sky with the disk of the sun occulted. Their films demonstrate that, especially in the clear sky case, the sky radiation is not isotropically distributed (Fig. 1.19). Peterson and Dirmhirn (1981) have found, however, that the ratio of diffuse to direct (normal incidence) irradiation varies very little throughout the course of the day in rural settings. In urban and polluted areas there may be a considerable diurnal trend.

Stanhill (1966) measured diffuse sky radiation at a site in Israel. The ratio of diffuse to total radiation is given on a mean monthly basis in Fig. 1.20.

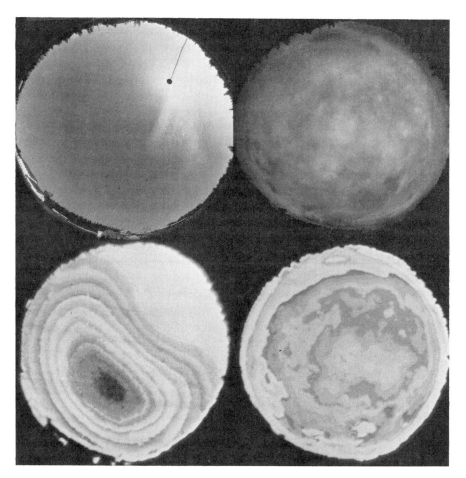

Fig. 1.19 All-sky camera photographs (top) and their corresponding electronic density slices giving 32 levels of equal light intensity. Left, clear sky condition; right, overcast sky (from McArthur and Hay, 1978).

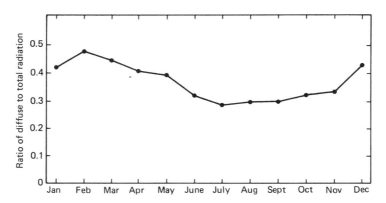

Fig. 1.20 Mean monthly ratio of diffuse to total radiation at Gilat, Israel. Reprinted with permission from *Solar Energy*, **10**, G. Stanhill, Diffuse sky and cloud radiation in Israel, copyright 1966, Pergamon Press, Ltd.

Values range from 0.3 to 0.5 with a higher fraction in the winter season. Citing data from a number of other locations, Stanhill concludes that the proportion of diffuse to total radiation is remarkably constant and, hence, that the solid constituents of the atmosphere, which do vary considerably from location to location, are of limited importance as radiation scattering agents.

1.5 SHORTWAVE REFLECTION (ALBEDO)

A portion of the total solar radiation that reaches the atmosphere and the earth is reflected back to space. The degree of reflection is a function of wavelength within this range. The term **albedo** is used to describe the reflection of the solar beam (0.3–4.0 μm), regardless of wavelength, or sometimes to describe reflection of the visible portion (0.4–0.7 μm) only. Properly the term "shortwave albedo" is used for the total solar spectrum and "visible albedo" for the visible waveband only but, as will be seen in the following pages, the literature is not consistent in the use of these terms.

Interest in the subject of albedo has grown in recent years. Satellites permit systematic observation of changes occurring in the reflection of solar radiation over large regions of the world. Kukla and Kukla (1974) noted an increase in Northern Hemisphere surface albedo in the early 1970s. Increases in the reflection of the entire planet due to aerosol loading of the atmosphere are suggested by Atwater (1970) who considers that this effect might, in itself, lead to a global cooling. However, the increased reflection could be balanced by increased absorption by the aerosols of radiation in the thermal waveband.

An intentional alteration of albedo has been proposed through the application of asphalt coatings to large areas of soil adjacent to seas or lakes in arid regions. By so doing albedo is decreased and the absorption of solar radiation is increased with a consequent surface heating. This is expected to increase thermal updrafts, thus promoting condensation and producing rainfall (Black and Tarmy, 1963). Moomen and Barney (1981) propose applying asphalt to dunes in Libya after rainfall and planting stabilizing vegetation through the coating. If this is done on a significant scale, a test of Black and Tarmy's hypothesis may be possible.

Charney (1975) and Charney et al. (1975) propose that the albedo of the Sahara and Sahelian zones of Africa may have increased because of a decrease in plant cover caused by overuse of the land. Based on global circulation models, they propose that this increased albedo can cause a decrease in rainfall, essentially by initiating subsidence and a pushing southward of the intertropical convergence zone. Thus, a change in albedo might explain the initiation or perpetuation of a drought. Whether this mechanism is real we do not yet know, but the signficance of albedo is highlighted by the "Charney hypothesis."

Otterman (1974) has noted a change in albedo coincident with the Sinai–Negev border between Egypt and Israel. The Negev soils have a lower albedo due to controlled grazing, which has permitted regrowth of vegetation in a previously overgrazed region.

It has also been suggested by Ayers and Glaeser (1974) that oil spills in the Arctic Ocean could lead to a change of albedo and consequent changes in regional energy balance. New applications of shortwave and visible albedo observations have been suggested: land use mapping (Oguntoyinbo, 1974), soil mapping (Cipra et al., 1971), and soil moisture availability (Idso et al., 1975).

It is often assumed that surfaces reflect diffusely, behaving as irregular surfaces. When the surfaces involved are relatively smooth (the scale of irregularities on the reflecting surface is small compared to the wavelength), specular or simple reflection results (the surface behaves more like a mirror than a diffuser). An example of specular reflection would be the bright spot seen on roads on sunny days. When the solar elevation is small, most vegetated surfaces manifest specular reflection. Brown et al. (1970) showed that specular reflection of the solar radiation by an alfalfa crop is especially obvious at low sun angles.

Table 1.3 gives typical values of shortwave reflectivity, assembled from several sources. Clouds are very effective reflectors, so that little solar radiation reaches the earth's surface in cloudy weather. For example, Hanson and Viebrock (1964) in flights at 7.62 km found albedo of the area between Atlantic City, New Jersey, and Erie, Pennsylvania, to be 0.158 with no clouds and 0.538 with complete altocumulus undercast.

Snow is a very effective reflector, particularly when new. Dirmhirn and Eaton (1975) found that spring snow cover in Utah exhibited specular reflectance of solar radiation. Thus, reflectance data measured at low solar elevation angles, especially, need be treated with caution.

Table 1.3 Shortwave reflectivity of natural surfaces

Surface	Shortwave reflectivity (r)
Fresh snow	0.80–0.95
Old snow	0.42–0.70
Lake ice, clear	–0.10
Lake ice with snow	–0.46
Sea surface, calm	0.07–0.08
Sea surface, windy	0.12–0.14
Niger river water, clear	–0.06
Niger river water, dirty	–0.12
Dry sandy soil	0.25–0.45
Bare dark soil	0.16–0.17
Dry clay soil	0.20–0.35
Peat soil	0.05–0.15
Savannah	–0.22
Most field crops	0.18–0.30
Deciduous forest	0.15–0.20
Coniferous forest	0.10–0.15
Mangrove swamp	–0.12
Vineyard	0.18–0.19

Sources: Many including Sutton, 1953; Geiger, 1961; Reifsnyder and Lull, 1965; Bolsenga, 1969; Kalma and Badham, 1972; Willis, 1971; Federer, 1971; Davies and Buttimor, 1969; Oguntoyinbo, 1974; Mayer, 1981.

Snow covered forest canopies were found by Leonard and Eschner (1968) to have lower shortwave albedos than normal snow. Twenty percent was the highest value measured. Apparently, greater trapping of solar radiation occurs in these irregular surfaces.

Ice absorbs effectively in the near ir portion of the solar spectrum. Bolsenga (1969) found the reflectivity in the waveband 0.3–3.0 μm to range from 10% for clear ice to 46% for "snow-ice" on the Great Lakes.

The reflectivity of water is low. Willis (1971) gives values of 7–8% for the sea under calm conditions. For very windy conditions, reflectivity may increase to 12–14%. Thus, water surfaces and the sea, poor reflectors, serve as a good sink for solar energy. Waviness and the solar elevation angle have a major influence on reflection by bodies of water.

Most rock, sand, and soil surfaces reflect 10–30% of the incident solar radiation although the reflectivity of desert sands may be considerably higher. Idso et al. (1975) have shown that surface reflectivity of a test soil is an almost perfectly linear function of water content in the top 0.02 m. Cipra et al. (1971) show higher reflectance for dry than for wet soils in Indiana and higher reflectance for crusted soils of any kind or degree of wetness.

The shortwave albedo of vegetation does not range very widely. Most crops reflect about 20–30% of the incident solar radiation. Kalma and Badham (1972) found the seasonal mean shortwave reflectance for Townsville stylo to be about 19% and for mixed grass about 22%. Oguntoyinbo (1974) measured 22% reflection of shortwave radiation over savannah. Davies and Buttimor (1969) report reflectance of about 25% for a wide range of crops grown in southern Ontario.

Forests have lower shortwave reflectivity than do crop lands. DeWalle and McGuire (1973) report 16–18% in a deciduous forest during the growing season. Despite a color change in the fall, reflectance did not change. In winter, reflectance dropped to 12–14%. With snow on the ground, reflectivity rose to 25–30%. DeWalle and McGuire found that the forest acts as a diffuse reflector. Oguntoyinbo (1974) found reflectivity of a mangrove swamp in Nigeria to be about 12%.

Thus, it appears that most rock, sand, soil, and vegetation reflect from 10–30% of the incident solar energy. The reflective effects of these various materials were observed on a continental scale by Kung et al. (1964). Shown in Fig. 1.21 are data on mean shortwave albedo during different seasons, as observed in flights from northern Mexico to above the Arctic Circle in North America. Shortwave reflectance increased with latitude during winter. This was true whether the winter was of average, maximum, or minimum snow cover. Reflection was maximal at about 80% near 70°N latitude. In midlatitudes, the range of reflectivities varied from 20 to 50%, depending on the severity of the winter. Shortwave reflectance varied markedly with latitude during spring and fall. During the summer season reflectivity was surprisingly invariant at about 18% from Mexico to the Arctic. This was probably due to the fact that soil, crop and native vegetation all have nearly the same reflectivity for shortwave radiation.

A plot of shortwave albedo, typical of many that will be found in the literature, is given in Fig. 1.22 (from Kalma and Badham, 1972). The figure shows that shortwave albedo varies with the season, with the nature of the ground cover, and with the time of day. The proportion of incident radiation reflected at dawn and dusk is greater than that reflected during midday. Thus, the daily course of shortwave albedo is parabolic.

The major cause of the parabolic relationship is probably the fact that most natural surfaces reflect specularly rather than diffusely, particularly at low sun angles. It is likely that the upper canopy layer has a higher brightness when observed obliquely than when observed directly from above. From above, many gaps of darker appearance and some soil will be seen, so that the brightness will be less.

It must also be recognized that experimental errors may cause or, at least, exaggerate the parabolic nature of the diurnal course of shortwave albedo. Brown et al. (1970) measured reflection over a sparse alfalfa stand at Mead, Nebraska, with inverted pyranometers. One of these was shielded to minimize internal reflections, the other was unshielded. At noon, the unshielded

46 THE RADIATION BALANCE

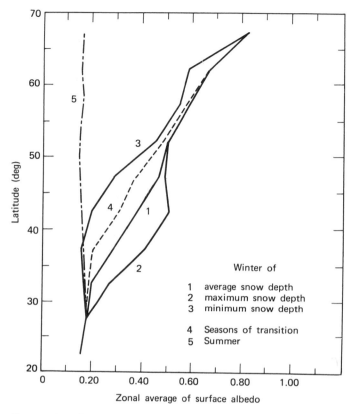

Fig. 1.21 Seasonal variation of meridional profile of continental surface albedo over North America (after Kung et al., 1964).

instrument measured a 4% lower reflectance; near dawn the difference was about 16%. The differences in shortwave albedo, particularly at low solar altitudes, illustrates the magnitude of experimental error possible in the measurement of reflected radiation. Stanhill et al. (1971) also noted serious errors in albedo measurement attributable to sensor, time of day, and proportion of diffuse to direct beam (specular) reflection. At high solar elevation angles errors of 10–20% were found; at lower angles errors were 20–30%. Coulson and Reynolds (1971) measured albedo in five wavebands of the visible spectrum for a number of materials including vegetation. A maximum reflection occurred at 10–20° sun elevation angle which may be due to direct and diffuse components being reflected differently.

The reflection of solar radiation by vegetation varies with wavelength. This is shown in Fig. 1.23 for the case of a fully developed soybean canopy grown at Mead, Nebraska. The spectral distribution of incoming solar radiation at 1358 hours (solar time) is shown as the greater of the two curves. The wavelength of maximum spectral intensity is at about 0.55 μm. Total

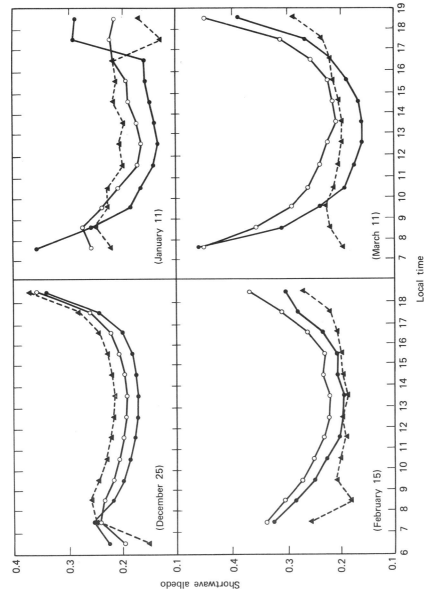

Fig. 1.22 Diurnal changes in albedo of Townsville stylo (solid circles), grasses (open circles), and bare soil (solid triangles) on 4 days of comparable daily ratios of diffuse to total radiation. Katherine, Northern Territory, Australia, 1969–1970 (after Kalma and Badham, 1972).

48 THE RADIATION BALANCE

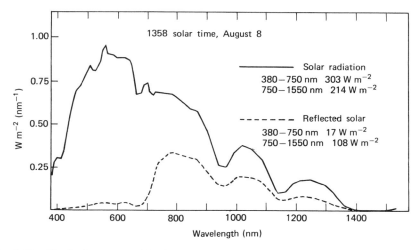

Fig. 1.23 Typical spectra of incident solar radiation and reflection from a soybean crop on a day in early August. Mead, Nebraska.

flux density of solar radiation in the visible range (0.38–0.75 μm) was 303 W m^{-2}, and in the near ir range (0.75–1.55 μm), it was 214 W m^{-2}.

The pattern of reflected radiation from the soybean canopy at about the same time is shown as the lower curve in Fig. 1.23. Maximum reflection occurred at 0.80 μm in the near ir. Far more radiation was reflected in the near ir (108 W m^{-2}) than in the visible range. Note that in the visible waveband the greatest reflection occurs in the green portion (around 0.55 μm) of the spectrum. Reflection of blue and red is reduced because these are the wavelengths most strongly absorbed by the crop. Thus, about 5% of the visible and about 50% of the near ir radiation incident on the soybean canopy was reflected.

Carlson et al. (1971) demonstrated another important influence on albedo, namely, the moisture status of the leaves. In the spectral region 0.80–2.6 μm, the leaf structure and water content of corn, sorghum, and soybeans interact with radiant energy to alter reflectivity patterns. Sinclair et al. (1973) demonstrated still another complicating factor: reflectance from the ventral side of soybean, apple, and other leaves was found greater than from the dorsal side.

Soil fertility also influences reflectivity according to Stanhill et al. (1972). A wide range of fertilizer treatments was imposed on a wheat crop. Visible and near ir reflectance was altered: in the first case by a color change; in the latter case by the influence of fertilizer on the abundance of vegetative material. The effects of fertilizer on reflectance have been used as the basis of measurements of turf quality (e.g., Birth and McVey, 1968).

Blad and Baker (1972a) found that degree of ground cover under soybeans is also a factor determining shortwave reflectivity. Albedo was 25% and increased 1.3% for each 10% increase in ground cover. Since the leaf cover fluctuated with time of day and leaf turgor, the reflectance should have been affected by changing cover and by anisotropy.

1.6 THERMAL RADIATION AND LONGWAVE EXCHANGE

To this point, we have discussed radiation fluxes from the sky earthward: direct beam solar and diffuse solar (sky) radiation. Little of the direct and diffuse solar radiation is of a wavelength greater than 4 μm. It is evident, however, that the atmospheric envelope of gases and aerosols that absorbs some portion of the solar beam will itself radiate energy. The atmosphere also absorbs longwave radiation emitted by the earth's surface and shortwave radiation reflected by the surface.

All objects at temperatures greater than absolute zero emit radiation at intensities proportional to the fourth power of their absolute temperature (1.5a). Only outer space is at or near 0°K. The atmosphere ranges from extreme cold to terrestrial temperatures of up to about 323°K (50°C). Therefore, the atmosphere radiates to earth (and also to space) predominantly in the long (ir) wavelengths.

About 90% of the infrared energy radiated to space by earth is absorbed by the atmosphere, particularly by water vapor, clouds, and carbon dioxide. Much of this longwave radiation is back- or counterradiated to the earth. Often the longwave radiation is incorrectly termed nocturnal radiation. Although longwave emission is essentially the only radiative flux at night, it also prevails, of course, during the day. In fact, the flux density of longwave radiation will be greater during the daytime because of the higher terrestrial temperatures then.

All layers of the atmosphere participate in the absorption and emission of radiation but the processes are quantitatively more important in the lower layers where the effective absorbers of longwave radiation (water vapor, carbon dioxide, nitrous oxides, methane, etc.) are most concentrated.

The combined radiational flux from the sky (shortwave direct and diffuse plus the longwave radiation) is termed the **total hemispherical radiation.** The hemispherical flux of longwave radiation (R_{lw}) was found by Swinbank (1963) to be predictable from the air temperature measured in standard shelters or "screens" about 1.5 to 2 m aboveground:

$$R_{lw} = 5.31 \times 10^{-13} T^6 \qquad (1.24)$$

where R_{lw} is in W m^{-2} and T is in °K. This expression was developed from observations made at night since measurements are more simply made when solar radiation is absent.

Even though this equation lacks a humidity term and humidity strongly absorbs in the thermal waveband, Paltridge (1970) found it very accurate at night. However, significant deviations of measured and predicted values were found under daytime conditions. The mean hemispheric flux of longwave radiation at Aspendale, Australia, was found approximately constant throughout the day and not greatly influenced by diurnal variation in screen temperatures. Swinbank's calculations must be reduced by about -30 W m^{-2} at the summer noon situation. The range at that time is about 300–330 W m^{-2}. The use of screen temperature which does not always truly represent surface or atmospheric temperatures is the cause of systematic errors during both day and night. Idso (1972) found that Paltridge's exceptions to Swinbank's formula applied in Phoenix, Arizona, as well. The intensity of terrestrial radiation depends on the earth's surface temperature and its emissivity, which is near 1.0 for most of the wavelength range of earth radiation emission.

It is interesting to speculate on the effects of man on the longwave radiation through his alterations of the earth's surface. The transition from forests and prairies to agricultural lands has probably affected the emissivity only little since most soils and vegetation behave as nearly blackbody emitters. However, the effects of changes in the moisture balance due to changing land use could be somewhat more important. Bare soil will become hot when the surface dries. A similar area under crop or other vegetation that extracts water from the entire root zone will be considerably cooler. Emission in the thermal waveband will increase in the bare area by a factor related to the fourth power of absolute surface temperature [Stefan's law (1.5a)]. Even the thermal radiation of the oceans may be affected by man's activity. The emissivity of water is near 1.0 but oil slicks on water can reduce emissivity by about 3% (Robinson and Davies, 1972).

The hemispherical flux of longwave irradiation is also influenced by cloud cover. For all weather data at Aspendale, Australia, Paltridge (1970) found an increase of 6 W m^{-2} per tenth of sky cover. About 70% of the flux under complete cover is attributable to the temperature of the underlying atmosphere and 30% is due to radiation from the cloud base.

The difference between the outgoing ir terrestrial radiation of the earth's surface and the downcoming counterradiation from the atmosphere is termed the **effective terrestrial radiation**. The form of Brunt's (1934) equation for the effective terrestrial radiation (R_T) is instructive:

$$R_T = \sigma T^4 (a - b\sqrt{e})(1 - cC) \qquad (1.25)$$

where R_T is in W m^{-2}; e, the water vapor pressure, is in kPa; a, b, and c are constants (the constant c depends on cloud type); and C is the cloud cover in tenths. The equation demonstrates that the greater the water vapor content and the greater the cloud cover, the less will be the escape of longwave terrestrial radiation. Under clear skies some 35–40% of the total hemispherical radiation is longwave. With increasing cloud cover that proportion increases since the shortwave radiation is more effectively attenuated (Pochop et al., 1968).

For a detailed record of long-term measurements of the various components of atmospheric radiation (shortwave direct beam and diffuse, longwave hemispherical, total hemispherical, etc.) consult the annual reports of the University of Bergen, Norway (e.g., Schieldrup-Paulsen, 1980).

1.7 THE NET RADIATION

The reflected portion of the total direct and diffuse solar radiation (R_{sw}) is determined by the reflectivity r of the underlying surface, that is, $(R_{sw\uparrow}) = r(R_{sw\downarrow})$. Thus, the shortwave radiation balance can be written as

$$R_{swbal} = (R_{sw\downarrow}) - (R_{sw\uparrow}) = (1 - r)R_{sw\downarrow} \qquad (1.26)$$

The arrows indicate the direction of the radiation streams. The effective longwave balance is

$$R_{lwbal} = (R_{lw\downarrow} - R_{lw\uparrow}) \qquad (1.27)$$

The net of all the fluxes of radiant energy R_n can, then, be written as

$$R_n = R_{swbal} + R_{lwbal} = (1 - r)(R_{sw\downarrow}) + R_{lwbal} \qquad (1.28)$$

The net radiation is the difference between total upward and downward radiation fluxes and is a measure of the energy available at the ground surface. The importance of this parameter is that it is the fundamental quantity of energy available at the earth's surface to drive the processes of evaporation, air and soil heating, as well as other, smaller energy-consuming processes such as photosynthesis.

The diurnal course of daily net and solar radiation on cloudy and clear days at Mead, Nebraska, is shown in Fig. 1.24. The net radiation was meas-

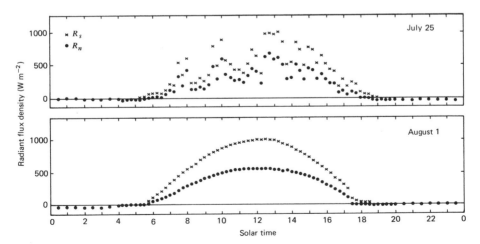

Fig. 1.24 Solar (R_s) and net (R_n) radiation over soybeans at Mead, Nebraska, on cloudy and clear days in midsummer.

ured over a fully developed soybean crop. The waves are similar in shape. Solar radiation can only be positive in sign, however. Net radiation is positive by day and negative by night. Some time must elapse after sunrise before the net radiation becomes positive, and the net radiation becomes negative some time before sunset as the flux density of longwave radiation away from the surface exceeds that of the declining solar radiation.

Net radiation and solar radiation in latitudes outside the tropics describe annual waves similar to the daily waves in Fig. 1.24. The amplitude of the waves is determined largely by latitude. Daily mean net radiation is positive during summer, reaching its peak value at about the time of the summer solstice. In winter the daily mean net radiation is negative since daytime solar energy receipts are low and the nights are long. Blad and Baker (1971) found, in a 3-year study at St. Paul, Minnesota, that net radition over a soil surface was positive in sign from April through October and negative from November through February. The average annual value of R_n was 22.5×10^5 W m^{-2}. This is about the same as the annual average found for Copenhagen, Denmark, by Jensen and Aslyng (1967).

Net radiation has been measured over many types of crop surfaces. Decker (1959) measured R_n over canopies of short grass turf, alfalfa, and corn and found significant differences over each surface. Grass had the lowest R_n because of either a higher temperature, a higher reflectivity, or both. Denmead et al. (1962) and Yao and Shaw (1964) measured R_n over and within corn canopies. Impens and Lemeur (1969a) measured R_n over oats, beans, sunflowers, and corn. There have been numerous other measurements reported [e.g., Thorpe (1978) for an apple orchard; Federer (1968) for a forest stand; Davies and Buttimor (1969) for a large number of field crops in southern Canada; Glover (1972) for sugar cane in South Africa; Kalma (1972) for a typical pasture in Australia; Campbell et al. (1981) for a sweet corn crop under drought conditions].

Denmead et al. (1962) found a 75% depletion of net radiation within a corn crop canopy. Twenty-five percent of the net radiation reached the ground with corn in 1 m rows. Seventy-three percent of all captured energy was expended in the top half of the canopy as is shown in Fig. 1.25.

Denmead et al. (1962) estimated that photosynthesis in corn might be increased 15–20% if closer row spacings were used to eliminate ground illumination and increase energy capture by the crop. Yao and Shaw (1964) tested this hypothesis by manipulating row spacing and population in a corn field. Higher population rates reduced the ratio of R_n at the ground to R_n above the crop. The same effect was achieved by narrowing the row spacing with population held constant. The water use efficiency (production of dry matter per unit of water consumed in evapotranspiration) was improved by the use of narrow rows in the same population. The yield of corn in 0.53 m rows was better than in 0.81 m and 1.06 m spacings. The orientation of the rows E–W or N–S had no effect.

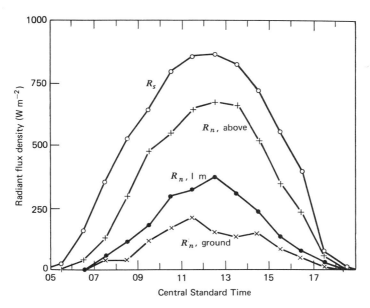

Fig. 1.25 Average hourly solar radiation above, and net radiation at various heights within a corn canopy on a typical summer day at Ames, Iowa (after Denmead et al., 1962).

Within-canopy measurements of net radiation were made in a soybean crop throughout the growing season at Mead, Nebraska. To minimize the need for a large number of sensors, special oblong net radiometers were constructed (see Instrumentation, Section 1.11). Results shown in Fig. 1.26 are typical for a sunlit and a cloudy day. On clear days, the capture of radiant energy occurs primarily in the upper third of the soybean canopy.

Impens and Lemeur (1969b) described the extinction of net radiation within the canopies of four crops: oats, beans, sunflower, and corn using an empirical expression:

$$(R_n)_z/(R_n)_h = \exp[-(\kappa_{c1}F + \kappa_{c2}F^2)] \tag{1.29}$$

where $(R_n)_z/(R_n)_h$ is the fraction of net radiation at the top (h) of the canopy transmitted to a given level (z), κ_{c1} and κ_{c2} are empirically derived constants and F is the downward cumulative leaf area index (LAI). Figure 1.27 from Impens and Lemeur is a fit of data from various sources for corn using (1.29).

We have seen that the greatest depletion of R_n usually occurs in the upper parts of the canopy. It is possible to manipulate the capture of radiation by changes in plant spacing and population and by changes in the architecture of the plant canopy itself. The objective should be to maximize the absorption of useful energy within the canopy, especially at the lower canopy levels, which are not light saturated, and to minimize the amount that reaches the

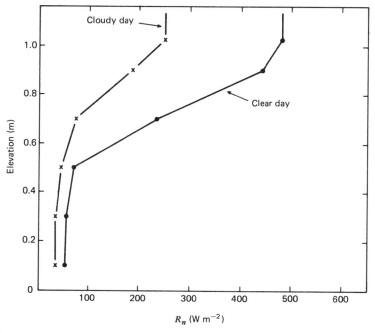

Fig. 1.26 Net radiation at solar noon above and within the canopy of soybeans on clear and cloudy days in late summer. Mead, Nebraska.

ground surface. However, caution is necessary. Aubertin and Peters (1961) have shown, for example, that when corn plants are grown very closely together under conditions of moisture shortage, the efficient capture of radiant energy may lead to greatly increased transpiration and, hence, to severe wilting. Drought influence is also demonstrated by Campbell et al. (1981) who found that R_n at 0.25 m in a sweet corn canopy increased from 32 to 46% and then to 51% of that above the canopy as water became short.

Shading provides another way to manipulate net radiation. Martsolf and Decker (1970) accomplished a reduction in net radiation and consequent reduction in evapotranspiration by a soybean crop under a louvered shade system. The application of reflectant materials to leaves also affects R_n. This subject is discussed in detail in Chapter 11.

1.8 RELATION OF NET AND SOLAR RADIATION

In many biological and engineering applications, it is the net radiation rather than the solar radiation about which information is needed. However, whereas solar radiation is measured regularly at an increasing number of locations throughout the world, net radiation is measured infrequently, usually only by research scientists in short-term studies. Many attempts have

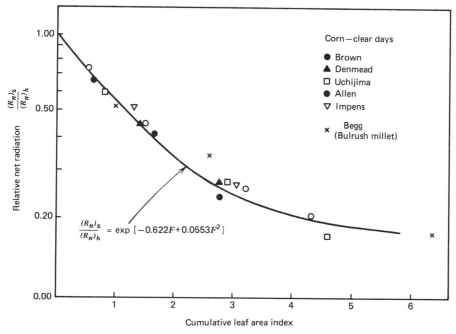

Fig. 1.27 Daytime net radiation transmission vs. downward cumulative leaf area index (see Impens and Lemeur, 1969b, for sources of the data).

been made to relate net radiation to measured solar radiation by using equations of the form

$$R_n = aR_s - b \tag{1.30}$$

A summary of some of these equations is given in Table 1.4. Gay (1971), however, considers that such simple regression models relating R_n and R_s are inadequate if they do not include a correction factor for longwave exchange as a function of shortwave exchange.

Linacre (1968) developed another type of empirical equation for estimating R_n using total global radiation, shortwave albedo, percent possible sunshine (n/N), and temperature (°C). Linacre's equation is

$$R_n = (1 - r)R_s - 1.11(0.2 + 0.8\,n/N)(100 - T) \tag{1.31}$$

where R_n and R_s are in W m^{-2}.

The ratio R_n/R_s will be generally high over oceans and lakes because the surface is cool (low $R_{lw\uparrow}$) and the shortwave reflectivity is low. Over desert regions the ratio will be low since shortwave reflectivity of light-colored materials is high, as are the surface temperatures (high $R_{lw\uparrow}$). In tropical regions, R_n/R_s will be high since reflectivity of vegetation is low and surface temperatures are moderate. Stanhill et al. (1966) found the ratio (R_n/R_s) to vary from 0.58 for open water to 0.25 for a desert and its vegetation.

Table 1.4 Net radiation as a linear function of total global solar radiation[a]

Location	Reference	Surface and comments	Equation	Units	r
Ames, Iowa	Shaw (1956)	Clipped grass, cloudy day	$R_n = 0.75R_s - 0.88$	MJ m^{-2} day^{-1}	0.97
Ames, Iowa	Shaw (1956)	Clipped grass, clear day	$R_n = 0.87R_s - 0.35$	MJ m^{-2} day^{-1}	0.98
Mead, Nebraska	Rosenberg (1969)	Alfalfa, 16 days (24-h totals)	$R_n = 0.69R_s - 0.34$	MJ m^{-2} day^{-1}	0.98
Simcoe, Ontario	Davies and Buttimor (1969)	All crops with $r = 0.23$ assumed	$R_n = 0.72R_s - 0.36$	MJ m^{-2} day^{-1}	
Tel Amara, Lebanon	Sarraf and Aboukhaled (1970)		$R_n = 0.73R_s - 0.38$	MJ m^{-2} day^{-1}	0.96
Ivory coast	Monteny et al. (1981)		$R_n = 0.67R_s - 0.47$	MJ m^{-2} day^{-1}	0.98
New Hampshire	Federer (1968)	Hardwood forests	$R_n = 0.83R_s - 91$	W m^{-2}	
Israel	Stanhill et al. (1966)	Various surfaces	$R_n = 0.85R_s - 98$	W m^{-2}	
Arizona	Fritschen (1967)	Barley	$R_n = 0.81R_s - 118$	W m^{-2}	

[a] Updated from Rosenberg and Powers (1970).

1.9 EARTH'S RADIATION BALANCE

To this point, we have discussed the separate radiation fluxes that make up the radiation balance: total solar radiation (direct beam and diffuse), shortwave reflection, upward terrestrial, and atmospheric longwave radiation. We have shown how various factors deplete, augment, or otherwise influence these components. A look at the integrated result of these interactions on regional climate and on global radiation balance is appropriate.

Satellites have made possible observation of the various radiation components from such great altitudes so that local effects no longer dominate the data. Satellite derived estimates of effective longwave (terrestrial) radiation, shortwave albedo, and net absorption of solar radiation on a planetary scale were reported by the National Center for Atmospheric Research (NCAR, 1970). Results are given in Table 1.5. Figure 1.28 (from Vonder Haar and Suomi, 1969) shows how these radiation terms vary as a function of latitude. Reflectivity is maximal at the poles and diminishes to about 25% near the equator. The mean annual longwave energy loss to space is lowest at the poles and greatest at about 20°S latitude. A secondary minimum occurs in the zone 10°S–10°N latitude.

Global net radiation has also been mapped from satellite-sensed data. An Earth Radiation Budget Experiment (ERBE) was carried out with the Nimbus-6 satellite. In Fig. 1.29 the global net radiation for July 1975 is displayed. The map shows a large surplus of radiant energy in the Northern Hemisphere where it was summer and a large deficit in the Southern Hemisphere where it was winter, as well as in the polar regions of the Northern Hemisphere where outgoing radiation exceeds incoming radiation.

Satellite observations and numerical estimates of the major radiation components shown in Table 1.5 demonstrate that the measured and calculated

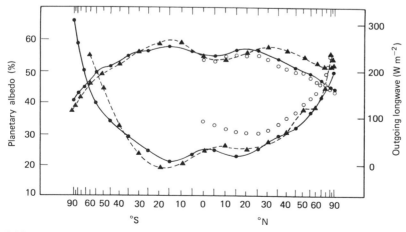

Fig. 1.28 Mean meridional profiles of outgoing infrared radiation (upper portion) and planetary albedo. (Data from various sources; see Vonder Haar and Suomi, 1969.)

Table 1.5 Calculated and satellite-measured global and hemispherical net absorption of solar radiation, emission of terrestrial radiation and mean albedo

	Net absorption of solar radiation (W m^{-2})		Emission of terrestrial radiation (W m^{-2})		Mean albedo	
	Calculated values	Satellite values	Calculated values	Satellite values	Calculated values	Satellite values
Northern hemisphere	242	251	240	230	0.306	0.29
Southern hemisphere	236	244	232	230	0.324	0.23
Global	239	244	236	230	0.315	0.29

Source: National Center for Atmospheric Research Quarterly No. 27 (1970).

LIGHT PENETRATION INTO PLANT CANOPIES AND WATER BODIES

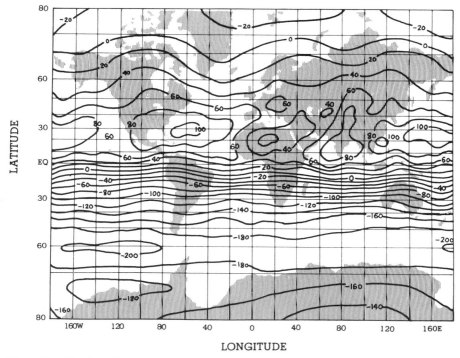

Fig. 1.29 Earth radiation budget experiment. Global net radiation in W m^{-2} from Nimbus-6 data, July 1975. A large surplus of radiant energy in the Northern Hemisphere (summer) and a large deficit in the Southern Hemisphere (winter) are seen. A deficit is also seen at the Northern Hemisphere polar region where outgoing radiation exceeds incoming.

data are in close agreement and that the values of incoming and outgoing radiant energy streams are very close. The radiation exchanges are probably not in exact balance over short periods, even over a year at a time. The earth and the atmosphere, however, have storage capacity for energy and slight imbalance between incoming and outgoing radiation need not lead to major or rapid climatic change.

1.10 LIGHT PENETRATION INTO PLANT CANOPIES AND WATER BODIES

1.10.1 General

Radiation impinging on a body will be absorbed, reflected or transmitted so that

$$\alpha(\lambda) + r(\lambda) + t(\lambda) = 1 \tag{1.3}$$

60 THE RADIATION BALANCE

Soil, for example, transmits little of the solar spectrum (penetration of light may occur to a depth in the order of millimeters). Reflection and absorption are important in soils.

1.10.2 Penetration in Water

Still water is characterized by slight reflection, strong absorption, and significant transmission. The penetration of visible light through water bodies follows well the Beer–Bougher law relationship (1.14).

Since water strongly absorbs ir radiation, that portion of the solar spectrum will be sharply depleted on penetration through water bodies. Furthermore, turbidity due to suspensoids (silt, algae, and other pigmented materials) will strongly deplete visible light. The prediction of visible light attenuation in pure water is relatively simple. Maguire (1975a) presents data showing the absorption of photosynthetically active radiation (PAR) by distilled water to a depth of 100 m. Ultraviolet and near ir radiation are rapidly depleted. Blue penetrates well so that at a depth of 100 m only blue, a little green, and a little violet remain. In natural water bodies such as lakes and oceans, actual absorption and transmission are much more complex because of great variability in the nature and content of suspended materials and thus simple Beer–Bougher law relationships may not be easily applied.

1.10.3 Penetration in Snow and Ice

It has already been shown that snow is highly reflective as is white or crystaline ice. When ice is clear, however, its albedo is relatively low and its absorption and transmission is greater than in the case of snow and white ice.

Since snow and opaque ice transmit little visible light, phytoplankton in the waters below such surfaces undergo reduced photosynthesis. Of course, ice on frozen lakes also reduces the diffusion of oxygen into the water and further retards photosynthesis. Maguire (1975b) made measurements of light 0.4–0.7 μm) transmission through water, snow, and ice. Powder snow and cloudy ice were most effective in absorbing PAR. Extinction coefficients for absorption of PAR are given by Maguire (1975b). The absorption follows predictions of the Beer–Bougher law.

1.10.4 Light Penetration into Plant Canopies

The penetration of net radiation into plant canopies was introduced and described above as a function of depth into the canopy expressed in terms of cumulative LAI. The penetration of light (quantitative and qualitative) is of considerable ecological significance.

In many crops, as they are now managed, light penetration may be inadequate or may be the factor actually limiting productivity. Honda and Fisher (1978) find that the branching angles in trees that, theoretically, result in maximum effective leaf area are close to those actually occurring in nature. Nevertheless, Halverson and Smith (1974) show that the light regime in forests can be controlled by selective cuttings where size, shape, and orientation of openings can be designed to encourage reproduction and to control soil surface temperature. In another example, Pendleton et al. (1966) found that reflective plastic coverings between rows of corn plants increased yield by 7–12% whereas black coverings had no effect. Special reflectors to direct light into corn canopies from the side were found in another study (Pendleton et al., 1967) to produce plants with more ears, more tillers, and shorter stalks and to boost theoretical production levels to as much as 23.66 t ha^{-1} under conditions conducive to high production in Illinois.

Sakamoto and Shaw (1967) have also speculated that a limit to soybean productivity follows from the fact that only the periphery of the plant canopy is fully illuminated while much of the interior (especially in bushy cultivars) is "light starved." Wahua and Miller (1978) have found that shading affects the nitrogen relations and ultimately reduces yield of soybeans. Height of the plants varied from about 1.0 to 1.8 m depending on degree of shading imposed. For substantial nitrogen fixation, the crop, they found, should be grown (without shade) in a configuration that allows 80% of the ambient illumination level at a height of 0.5 m in the crop.

From these few examples, it should be clear that there are good reasons to understand how light penetrates and is distributed within plant canopies.

The problem of light in plant canopies is complicated by the geometrical complexity of plant architecture as well as by the great degree of natural variation in leaves of different species, age, and history. For example, aquatic leaves transmit 4–8% of incident light, whereas deciduous tree leaves and grasses may transmit 5–10%. In spring most plants transmit more light than they do later as the leaves thicken probably because of the greater chlorophyll content of the thicker leaves.

If all leaves were horizontally displayed, it would be relatively easy to predict light penetration. However, leaves are complex with many forms and shapes and with variable orientation with respect to azimuth and with respect to inclination. Monsi and Saeki (1953) calculated that only 44% as much light would be intercepted by leaves oriented vertically as by leaves oriented horizontally. Other complicating factors are that many plants display an azimuthal preference. Some plants are also heliotropic (track the sun as it moves across the sky) and many plants change leaf orientation in response to time of day and to internal water potential.

Nonetheless, the penetration of light into plant canopies can be described or approximated in mathematical terms. One of the first attempts at such an approximation is due to Monsi and Saeki, who adapted the Beer–Bougher law (1.14) as follows:

$$I/I_0 = \exp(-\kappa_c F) \tag{1.32}$$

62   THE RADIATION BALANCE

where κ_c is an extinction coefficient for plant leaves and F is the downward cumulative leaf area index.

The Monsi–Saeki model assumes that the plant canopy is a homogeneous medium and that all incident light is absorbed by the leaf. It further assumes that the sky is isotropic (i.e., that all radiation is diffuse) and that κ_c is constant. Errors inherent in this model are caused by the heteorogeneous nature of plant communities with variable leaf inclinations, and by the fact that (1) light is reflected and scattered as well as absorbed, (2) the sky is not isotropic (even under heavy cloudiness true "isotropy" does not prevail (see Fig. 1.20), (3) the light quality changes, and (4) sunflecks occur. It is also likely that κ_c varies with geometry and with leaf condition (age and turgidity) as well. In order to use the model, it is necessary to know the LAI distribution as a function of height and the values of κ_c. κ_c may range from 0.3 to 0.5 in upright leaf communities and from 0.7 to 1.0 in communities with horizontal leaves.

There have been a considerable number of important additions and improvements in the modeling of light distributions in plant canopies since Monsi and Saeki's seminal paper. A detailed recounting of these is beyond the scope of this book; however, a few notable improvements and findings are given below.

Extinction Coefficients. In order to improve estimations of the extinction coefficient, Chartier (1966) proposed it be calculated as a function of light transmission for specific leaves and wavelengths:

$$\kappa_c = (1 - T) \cos i \quad \text{for } i < \beta \tag{1.33}$$

where T is a transmission factor for the leaf and wavelength of light considered and i is the angle of the leaf with respect to the horizontal; β is the solar elevation angle. Lemeur (1973) has shown that extinction coefficients in sunflower and soybean vary predictably with the solar elevation angle.

Sunflecks and Intermittency. The effect of plant motion and the generation of sunflecks within canopies has been studied by Norman and Tanner (1969) and Miller and Norman (1971). Statistical distributions were developed to describe these phenomena. Desjardins et al. (1973) observed fluctuations in photosynthetically active radiation (PAR) within corn of different cultivars by means of traversing light sensors. They found that fluctuations were affected by windiness and that the most rapid fluctuations were on the order of 10 Hz. Sinclair and Lemon (1974) found that, at certain levels within a canopy, the total PAR irradiance level can be nearly equal that above the crop because of the occurrence of sunflecks. Reifsnyder and Furnival (1969) speculated that shade tolerance in the understory of forests may be due to the speed with which the photosynthetic apparatus of these plants responds to sudden increases in illumination. Their observations indicated a linear increase in the maximum radiant flux density with increasing duration of observed sunflecks.

Leaf Orientation. Lang (1973) studied leaf angles and orientation in cotton. The plant was clearly heliotropic, the leaves facing east in the morning and west in the afternoon. Fukai and Loomis (1976) also observed heliotropism in cotton. "Sun tracking" increased photosynthesis in June, but later in the season had a negative effect. They concluded that a more even leaf distribution in cotton would result in greater photosynthesis. Shell and Lang (1976) measured the heliotropic behavior of sunflowers and found that the plants lag the sun by about 48 min throughout the day. Because of heliotropism, sunflowers receive about 40% more solar radiation than would a plant with a fixed leaf distribution. Lang and Shell (1976) also found that sunflower leaves develop different angular distributions depending on whether they are sunlit or shaded. Sunlit leaves are more concentrated in location. Ehleringer and Forseth (1980) studied solar-tracking behavior in a number of desert species that move diurnally, orienting their leaves either perpendicular or parallel to the sun's rays. In desert annuals solar tracking become more frequent as the length of the season decreases. Leaves perpendicular to the sun's rays photosynthesize rapidly throughout the day; those that are parallel to the sun's rays have a reduced leaf temperature and reduced transpirational loss.

The difficulty of studying leaf orientation in controlled environments is illustrated by the work of Blad and Baker (1972b). They found two field grown varieties of soybeans to be essentially planophile (leaves horizontally displayed) with most of the leaf area near the top of the plant. In a growth chamber, because of weak illumination, the leaves were more cylindrically distributed. They also found no preferred azimuthal distribution of the leaves in the climate of Minnesota. Lemeur (1973), on the other hand, found a distinct "bulge" to the north and east in azimuthal density of soybeans grown in north–south rows at Mead, Nebraska. The data are shown in Fig. 1.30. Similar distributions for sunflower, Jerusalem artichoke, and corn grown in Belgium are also shown. Further evidence that leaves are not randomly distributed azimuthally was demonstrated by Lugg et al. (1981) who grew sorghum in rows oriented E–W, N–S, NE–SW, and NW–SE. They found a strong bimodal azimuthal orientation either east-west or perpendicular to the rows. Additionally, more leaves were found to the south of the east–west line than to the north of it.

Leaf Angles. Leaf angle distributions with respect to the sun also differ by species. Figure 1.31 (from Lemeur, 1973) shows cumulative frequency distributions for the upper, lower and total canopy of the four crops given in Fig. 1.30. Sunflower was found to be essentially planophile (horizontally displayed) with leaves of 0–30° having maximum probability. Jerusalem artichoke has an almost uniform distribution for leaf angles between 0 and 55°.

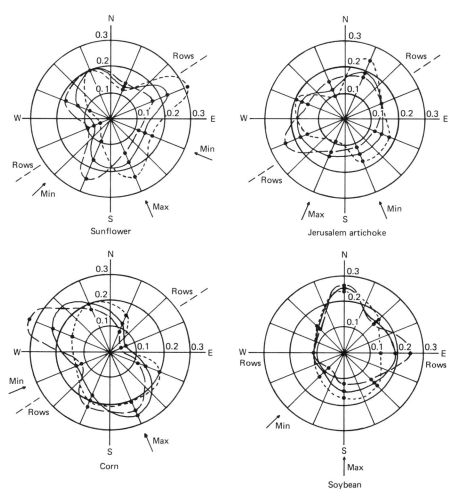

Fig. 1.30 Azimuthal density functions for sunflower, Jerusalem artichoke, corn, and soybean. Max and Min indicate directions of maximum and minimum azimuthal density (after Lemeur, 1973).

Corn is extremely plagiophile (obliquely displayed) especially in the lower leaves. The soybean cultivar used in this study (Amsoy) has typically erectophile leaves (vertically displayed) with little difference between upper and lower parts of the canopy.

Models of Light Penetration. Because of the complexity of the problem of predicting light penetration into vegetation many efforts have been made since Monsi and Saeki (1953) to model the process. Lemeur and Blad (1975) classified light penetration models according to whether they are based on geometric or statistical considerations. In the geometric approach, a stand of plants is conceived as being of regularly arranged shapes with characteristic dimensions. The relative position of each shape with respect to its neighbors is specified. In the statistical approach, the arrangement of plant elements is more random. The position of an element is unspecified but the probability of its appearance in a limited space is indicated. No individual plant exists in statistical models; rather, the canopy is composed of a display of leaves and stems.

In the geometric approach, a tree may be simulated by a cone, for example, and a corn plant by upright cylinders. The stand in which the extinction of light occurs is composed of a regular arrangement of these shapes. Geometric models generally are used to predict light interception on daily, seasonal, or yearly bases for a specified location. Statistical models are

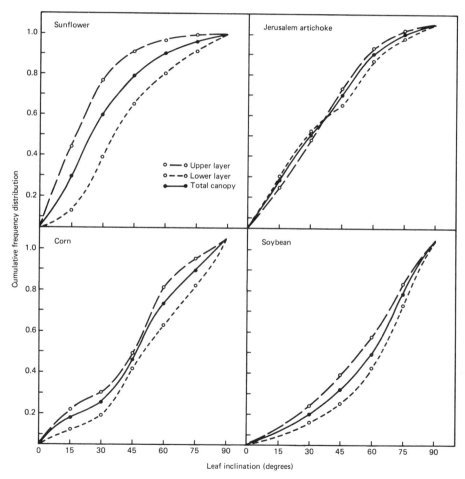

Fig. 1.31 Cumulative leaf angle distribution functions for sunflower, Jerusalem artichoke, corn, and soybean (after Lemeur, 1973).

66 THE RADIATION BALANCE

useful in providing information on light penetration to different levels within a canopy for different times of day. The geometric models, then, are best suited for climatological purposes; the statistical models apply more directly in the study of the microclimate of real plant communities.

Another contribution to modeling penetration of solar radiation into crop canopies was made by Allen (1974) who considered the effects of atmospheric transmissivity, slope, row direction, row spacing, crop height, leaf area density, distribution, and mean leaf angle. Szwarcbaum and Shaviv (1976) used a "Monte Carlo" approach, essentially, tracing the history of a parcel of light through the leaves of a canopy. They consider the effects of reflection by leaves, distribution and orientation of leaves, geometry of the canopy, and ground reflection.

Lemeur and Rosenberg (1979) attempted to model the shortwave energy flow within and above vegetative canopies, taking into consideration the interactions between the external radiation field and the morphological and architectural characteristics of the leaf biomass. Their model, SHORTWAVE, starts from simulated or actual data of the atmospheric shortwave radiation components and of their spectral energy distribution. Amount of leaf biomass, its angular distribution and the spectral properties of the leaves and underlying soil are also taken into account. Some quantitative and qualitative predictions of this model are given in connection with forests in Section 1.10.6.

The modeling of radiation distribution and radiation balance is essential in the development of models for simulating plant growth. A comprehensive review of the principles involved in modeling plant growth including the modeling of the radiation regime is that of Norman (1979).

1.10.5 Light Quality in Plant Canopies

Figure 1.32 (from Gates, 1965) is a classic description of the optical properties of a single leaf exposed to light in the wavelengths characteristic of the solar spectrum. In the visible wavelengths approximately 85–90% of the light is absorbed, 5–10% is reflected, and 5% is transmitted. Near ir reflection and transmission are much greater than in the visible range and absorption is low from about 0.70 to 1.40 μm. From about 1.50 to 2.60 μm, transmission and reflection decrease and absorption is considerable.

It is evident from Fig. 1.32 that light within plant canopies will be enriched with respect to the near ir and the green waveband and depleted with respect to the other visible wavebands.

Evidence for effects in the field was given by Allen and Brown (1965) in the data shown in Fig. 1.33. Above the 2.50 m tall corn crop the flux density of the visible, total shortwave (solar spectrum), and near ir radiation is taken as 100%. At successive depths within the canopy (described in terms

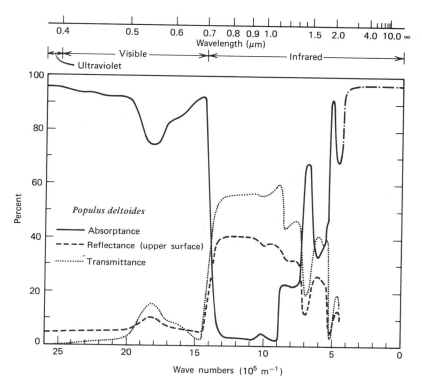

Fig. 1.32 Spectral reflectance, transmittance and absorptance of leaves of *Populus deltoides* (after Gates, 1965).

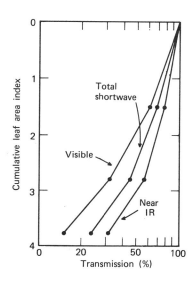

Fig. 1.33 Percent transmission of visible (left line), total shortwave (center line), and near ir (right line) radiation into the canopy of a corn crop at noon. Ellis Hollow, New York, early September (after Allen and Brown, 1965).

of cumulative LAI) the visible light is most strongly depleted and the near ir is least depleted. The resultant depletion of the total solar spectrum is, of course, intermediate.

Chartier's (1966) model (cited above) provides for this changing light quality (and quantity) as a function of sunleaf geometry as well as of leaf transmittance characteristics.

1.10.6 Radiation Penetration into Forests

The forests of the world pose a special and, perhaps, even more complex study of quantitative and qualitative radiation exchange. Forests are frequently of mixed species that differ in age, height, color, leaf shape, density, and orientation. The mixture of coniferous and deciduous species adds further complications. The changing penetration of solar radiation with season is well represented by photographs from Hutchison and Matt (1977). These are hemispherical canopy photographs (fisheye lens) of a deciduous *Liriodendron* forest in Tennessee in winter (Fig. 1.34*a*) and summer (Fig. 1.34*b*). One expects that the forest and its radiation distribution will be far different during these two seasons. But the differences are subtle.

Figure 1.35 is a plot of the changing radiation regime throughout the year measured with an array of sensors at a number of elevations within and over the forest. The amplitude of the annual solar radiation wave is considerably damped within the canopy and the largest share of its radiation is absorbed and/or reflected by the upper canopy. The forest floor receives very little radiation during the summer after the forest is fully leafed out and during winter because of the low solar elevation then.

In a summary of their annual data Hutchison and Matt show that a higher proportion of the total incident radiation penetrates the forest during the spring leafless period and the least in the summer and fall when the forest is fully leafed.

The forest canopy alters the distribution of direct beam and diffuse radiation as well. Gay et al. (1971) found, for example, that the leaves of a loblolly pine forest altered the proportion of diffuse radiation from 15% of the total global radiation above the canopy to 46% at the forest floor. Anderson (1970) noted that the proportion of diffuse to direct beam solar radiation in a forest increases with decreasing solar elevation. Hutchison et al. (1980) found that most of the diffuse radiation that penetrates a *Liriodendron* (tulip poplar) forest does so through openings that are within 10° of the solar disk. Hence, the directional distribution of diffuse radiation in the forest is strongly controlled by solar position on clear to partly cloudy days.

According to Reifsnyder et al. (1971) the extinction coefficient for Beer–Bougher law calculation of light penetration in forests is different in kind for pine and hardwood canopies. In the former case, an exponential extinction is appropriate; in the latter case, a constant ratio law is applicable.

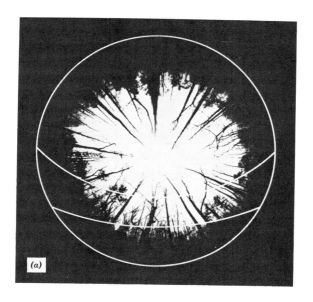

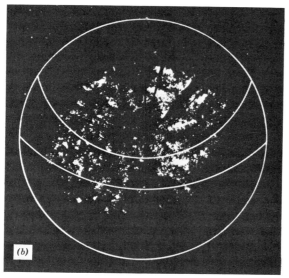

Fig. 1.34 (a) Hemispherical canopy photograph of the winter leafless forest taken from the forest floor. Outermost arc represents solar path on the winter solstice. Innermost arc represents solar path on the equinoxes. (b) Hemispherical canopy photograph of the summer fully leafed forest taken from the forest floor. Outermost arc represents solar path on the equinoxes. Innermost arc represents solar path on the summer soltice. From Hutchison and Matt (1977).

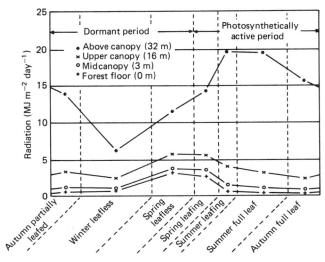

Fig. 1.35 Approximation of the daily radiation regime throughout the year above and within a *Liriodendron* forest. From Hutchison and Matt (1977).

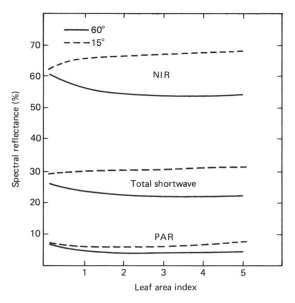

Fig. 1.36 The simulated spectral reflectance of near ir (NIR), total shortwave, and PAR radiation by poplar as a function of leaf area index at two solar elevation angles. The underlying soil is assumed to be covered with a green, dense vegetation (Lemeur and Rosenberg, 1979).

Based on the model SHORTWAVE, Lemeur and Rosenberg (1979) predict the reflectance of total shortwave, near ir, and photosynthetically active radiation from a poplar forest as a function of leaf area index and solar elevation angle. In this calculation, plotted in Fig. 1.36, the underlying soil is assumed to be covered with green, dense vegetation. The low sun angle increases reflection in the NIR waveband considerably more than it does in the PAR waveband with a consequent increase in total shortwave reflection. Leaf area index has a greater effect on NIR than on PAR but the effect is not great after an LAI of 2 is achieved.

1.11 INSTRUMENTATION

1.11.1 Total Solar Radiation

The instruments most commonly used to measure the fluxes of the direct and diffuse solar radiation are called pyranometers. The Eppley pyranometer is shown in Fig. 1.37. The sensing element is a thermopile with black and white segments. Since black absorbs and white reflects radiation these seg-

Fig. 1.37 The Eppley pyranometer used widely in the United States, courtesy of the Eppley Laboratory Inc., Newport, Rhode Island.

ments develop different temperatures. The greater the flux density of radiation, the greater is the temperature difference that develops between the white and black areas. Temperature difference is sensed by a differential thermopile whose output is nearly linear with the flux density of the solar radiation.

The glass cover (dome) of a pyranometer must be of a material that permits transmission of the major portion of the solar spectrum. Standard glass domes are permeable to radiation from about 0.28 to 2.8 μm. Special quartz glass is permeable to the wavelength range 0.20–4.5 μm.

Important instrument characteristics to consider in choosing a pyranometer include the following:

1. Sensitivity should be sufficient to detect biologically or physically significant differences in flux density of radiation.
2. Temperature dependence should be known and correctable by electronic or by computational means. It is desirable that the same kinds of sensors be used over widely varying climates. Thus, temperature dependence must be considered. Circuits to compensate for changing temperatures can be incorporated into pyranometers.
3. The response time, or time constant, the time required for the instrument to make a 63.2% adjustment to new environmental condition* should be appropriate to the needed measurement. If sensors are too slow, the measurements may be incorrect; if they are too fast, the record may contain far more detail than is needed and may be costly to the user.
4. An instrument that senses light or radiation must respond appropriately, according to the cosine of the angle of incidence of the impinging radiation stream (the cosine response). Distortions due to the optical properties of the dome or due to specular reflectance (bright spots) of the receiving surface must be minimized as far as possible. If the flux density of a normal beam (90°) is unity, the flux density of the same beam impinging at 0° should be sensed as zero. At all other angles the flux density sensed should be in proportion to the cosine of the angle between the ray and a normal to the surface. Deviations from cosine response may lead to serious measurement errors.

Pyranometers were formerly called pyrheliometers. This is now incorrect usage. Pyrheliometers are instruments for measuring the direct beam (normal incidence) solar radiation. Such sensors track the sun and are always oriented at right angles to it.

There are many other types or styles of electrical pyranometers. Nonelectrical pyranometers have also been widely used. Some work on the same

---

* The time constant is determined as $(1 - 1/e) = 0.632$ where e is the base of the Naperian logarithm system with numerical value 2.7182.

principle as the Eppley pyranometer, that is, development of a differential temperature between white and black-coated areas. The temperature differences are registered by nonelectrical means. Other pyranometers are based on somewhat different principles, such as, for example, the Bellani pyranometer in which absorbed solar radiation is converted to heat that vaporizes alcohol. The quantity of alcohol condensed at the base of the instrument is proportional to the solar energy received.

1.11.2 Sky (Diffuse) Radiation

Pyranometers, when normally exposed, measure the total incoming solar radiation, which is comprised of the direct solar beam and the diffuse, scattered sky radiation. To measure the latter component alone, the pyranometer must be shaded from the direct beam and exposed only to the sky radiation. This can be done by means of a shadow band or "occulting ring," such as is shown in Fig. 1.38 (Horowitz, 1969). These rings are designed to cast a narrow shadow over the sensing disk of the pyranometer. The ring must be freely adjustable to accommodate changing solar elevation and changing azimuth of sunrise and sunset as the year advances, or it must be sufficiently wide to accommodate the range of solar elevation angles and the azimuths of sunrise and sunset throughout the year. The output of the shaded pyr-

Fig. 1.38 Pyranometer and occulting ring according to a design by Horowitz (1969). Photo courtesy of Prof. Albert Weiss, University of Nebraska Panhandle Station, Scottsbluff.

74 THE RADIATION BALANCE

anometer must also be adjusted upward to account for the scattered radiation from that portion of the sky that the shadow band blocks from view. Single, small occulting disks can also be programmed to move with the sun in order to continually shade only the sensing surface of a pyranometer. In this case, the above-mentioned correction should be minimal.

With total solar radiation and scattered radiation measured, the direct beam fraction may be calculated as

$$R_s(\text{direct beam}) = R_s(\text{total global}) - R_s(\text{diffuse}) \tag{1.33}$$

1.11.3 Reflectivity (Albedo)

The reflection of shortwave solar radiation is also measured with pyranometers. The sensors are inverted so as to see the reflecting surfaces, but they are kept high enough off the ground so that the area of the shadow they cast is relatively insignificant. Figure 1.39 shows an Eppley pyranometer

Fig. 1.39 Albedometer, a pyranometer shielded and inverted to measure reflection of solar radiation from a crop canopy.

mounted above a crop surface. The sensor is placed in a shade to limit the field of view to only the surface of interest; otherwise the instrument will "see" to the horizons. The shield also minimizes the incidence of light above the surface of the inverted sensor. This penetration and internal reflection has been shown by Brown et al. (1970) to cause significant errors in measurement of shortwave reflection.

Any type of pyranometer can be adapted to measure reflectivity. If some mechanical reason precludes actually inverting the sensor, a reflecting surface may be placed above the pyranometer. Kuhn and Suomi (1958), for example, placed an Eppley pyranometer in the prime focus of a parabolic reflector for use under an aircraft.

1.11.4 Sunshine Duration

Only sparse networks of pyranometers for direct measurement of solar radiation are maintained throughout the world. It is, however, possible to estimate solar radiation from data on sunshine duration or cloudiness. One type of "sunshine switch," used extensively by the U.S. National Weather Service, is shown in Fig. 1.40. The instrument reported by Foster and Foskett (1953) transmits a signal when light intensity is sufficiently strong to activate an electrical circuit.

Another interesting nonelectrical sunshine duration meter is the Campbell–Stokes sunshine recorder. The instrument consists of a glass globe that, when sunlight is sufficiently strong, acts as would a magnifying glass, focusing the beam onto a special recording paper. A trace is burned on this paper as the sun moves through the sky. The trace indicates the duration of bright sunlight. The depth of the burn is also a rough indicator of the intensity of the solar radiation. The Campbell–Stokes meter was used ex-

Fig. 1.40 Photoelectric sunshine recorder used by the U.S. National Weather Service.

76 THE RADIATION BALANCE

tensively by British meteorologists and is still used in locations lacking electrical service.

1.11.5 Net Radiometers

The net radiometer ideally absorbs radiation of all wavelengths directed downward toward the earth's surface and upward away from it. The absorbing surfaces should, therefore, behave as blackbody absorbers for all wavelengths.

The sensing element of the net radiometer is a differential thermopile separated by an insulating material so that each blackened absorbing surface (top and bottom) develops a temperature proportional to the flux density of radiation impinging on it. The difference in temperature is translated into a difference in voltage output of the thermopile.

Most net radiometers in use today are shielded with plastic domes that are transparent to both the visible and ir wavelengths. Such a net radiometer is shown in Fig. 1.41. Inflated plastic domes of polyethylene or polystyrene cover the sensor, protecting it from rainfall. The transparency of the plastic materials is not identical for longwave and shortwave radiation. Idso (1970) has found, however, that net radiometers with polyethylene coverings can be used without special correction for the slight differences in optical transmission of longwave and shortwave radiation. A few net radiometers that are not covered with domes but are instead ventilated by a stream of air on both sides of the sensing plate are still in use.

To gain a true understanding of the distribution of net radiation within plant canopies is very difficult. Impens et al. (1970) have shown that the

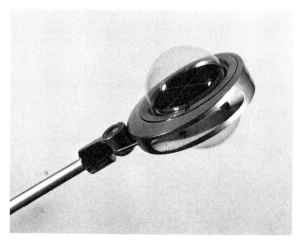

Fig. 1.41 Net radiometer, Swissteco-type S-1. Courtesy of Science Associates, Princeton, New Jersey.

normal lateral variation of net radiation with depth and with time in canopies of corn, beans, and sunflowers requires that very large numbers of sensors must be used to obtain valid averages. Observations are usually biased toward low values of R_n.

One way to minimize the number of sensors necessary for true averages within plant canopies is by mounting the net radiometer on a moving track that traverses and integrates. Another way is by the use of large radiometers, which can be inserted into the canopy and which integrate the net radiation over large areas. An array of such "oblong" radiometers placed at varying levels within a crop canopy is shown in Fig. 1.42.* A close view of one radiometer is also shown.

As with pyranometers, the performance of net radiometers must be judged on the basis of their sensitivity, temperature dependence, response time, and cosine response. The calibration of net radiometers is usually less stable than that of pyranometers. This is due primarily to the aging of the blackened sensing surfaces and of the plastic domes, as well. Frequent checks against a standard instrument and field or laboratory calibrations are desirable. Fritschen (1962) described a laboratory procedure and Sellers (1965) described a field procedure for calibrating net radiometers. Anderson (1971) gives instructions for calibrating oblong type radiometers. These should be arranged in the field with their long axis E-W to avoid errors arising from sensor geometry.

1.11.6 Total Hemispherical Radiometers

Net radiometers of the shielded or unshielded type can be adapted to measure the total radiation (shortwave and longwave) on one side or the other of the sensor. To measure the total hemispherical radiation from the sky, for example, the bottom surface of a shielded net radiometer is covered with a metallic dome blackened on the inside. A thermocouple measures the internal surface temperature of the metal dome and the longwave radiation impinging on the lower surface of the sensor can be calculated from (1.5a). The total (all wave) hemispherical radiation is then:

$$R_{lw\downarrow} + R_{sw\downarrow} = \text{measured net radiation} + \text{calculated longwave radiation} \quad (1.34)$$

If solar radiation measurements are made simultaneously with a pyranometer, the longwave radiation $R_{lw\downarrow}$ from the sky can then be calculated by subtracting solar from total hemispherical radiation:

$$R_{lw\downarrow} = (R_{lw\downarrow} + R_{sw\downarrow}) - R_{sw\downarrow} \quad (1.35)$$

* Measurement of the total solar radiation or net shortwave radiation within plant canopies can be accomplished similarly by using long strip radiometers encased in glass, as has been proposed by Luxmoore et al. (1971).

78 THE RADIATION BALANCE

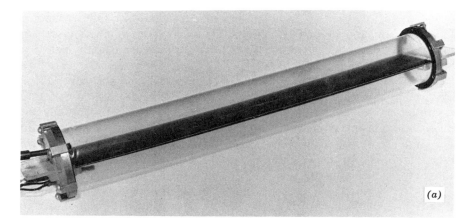

Fig. 1.42 Oblong radiometer (*a*) close up and (*b*) installed in a crop canopy.

1.11.7 Light Quality

In many biological applications, the researcher is interested in the flux density of radiation of a specific wavelength or a defined waveband: uv, PAR, near ir, for example, or blue, green, red. Many specific sensors have been designed for these purposes. Only a few are mentioned here.

Pyranometers can be used in conjunction with filters to measure the intensity of light in discrete wavelength bands. Platt and Griffiths (1964, p. 71) give details of available filters for this purpose.

The intensity of light in the visible wavelength band is often measured with photographic-type light meters. These are composed of sensitive coatings that upon illumination cause a current to flow. The selenium cell is most commonly used. Spectral response of the selenium cell is almost identical to that of the human eye with a peak sensitivity between 0.50 and 0.60 μm. Selenium cells change calibration with time, do not respond linearly, and are not sensitive to low light intensities. The major objection to the use of such sensors is, however, the spectral response, which differs so markedly from the action spectrum of photosynthesis, which peaks at about 0.45 and 0.67 μm.

There is a relationship between the number of molecules photochemically changed in photosynthesis and the number of photons absorbed, regardless of the energy of the photon, as long as it is within the requisite waveband of 0.40–0.70 μm. Norman et al. (1969) developed a sensor tailored to this waveband. A commercially produced sensor for this purpose is shown in Fig. 1.43a. This sensor responds closely to the ideal for all photons in the selected waveband. Figure 1.43b shows the actual response of the sensor.*

Instruments that can measure spectral radiation continuously or in very narrow waveband increments through the visible and near ir regions have become increasingly popular. Such instruments are known as spectroradiometers. One type of spectroradiometer that measures radiation in 6 nm bandwidths in the 0.30–1.10 μm wavelength range is shown in Fig. 1.44. This instrument has a fiberoptic extension head, which can be placed in cramped spaces or within plant canopies to measure light quality and quantity. The LI-1800 spectroradiometer also can be equipped with an integrating sphere to measure diffuse reflectance or transmittance of individual plant leaves.

Interest in remote sensing has led to the development of multiband or multispectral radiometers, which measure radiation in narrow wavebands in certain important wavelength regions in the visible, near ir, middle ir, and thermal ir regions. Some of these sensors, such as the one shown in Fig. 1.45, can be handheld, placed on booms for measurements at heights several meters above the surface, or mounted in helicopters or airplanes. Other instruments including multispectral scanners are carried aboard satellites and aircraft. These instruments measure the radiation in spectral bandwidths ranging from about 0.05 to 0.50 μm in the visible through the middle ir region and 2 to 6 μm in the thermal ir portion of the spectrum.

1.11.8 Surface Temperature

Contact thermometry is difficult especially with such irregular surfaces as those of soil or plant canopies. The Stefan–Boltzman relation (1.5a) provides

---

* A line quantum sensor (Licor model LI-1915B) is available from Lambda Instruments, Lincoln, Nebraska.

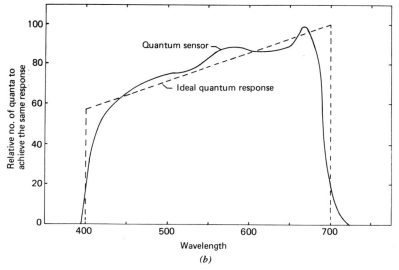

Fig. 1.43 (a) Cosine-corrected photosynthesis response (Quantum) sensor and (b) Spectral response of the Quantum sensor and that of an ideal quantum sensor. Courtesy LI-COR Instruments, Lincoln, Nebraska.

Fig. 1.44 LI-1800 Portable Spectroradiometer Research System. Courtesy of LI-COR Instruments Inc., Lincoln, Nebraska.

Fig. 1.45 Multiband radiometer Model 12-1000. Courtesy of Barnes Engineering Co., Instruments Division, Stamford, Connecticut.

82 THE RADIATION BALANCE

the basis for remote measurement of the temperature of any surface. Emissivities from most crops and soil surfaces range from 0.97 to 0.99. Corrections are needed to account for the non blackbody emission of the surface and for the reflection of longwave radiation from the sky.

Most radiation thermometers (often called infrared thermometers) are constructed to sense radiation in the 8–14 μm waveband. With a device known as a thermistor bolometer detector, the sensor compares the flux density of radiation from the target and from an internally controlled cavity. A handheld radiation thermometer is shown in Fig. 1.46. The same principle of measurement is used in multispectral thermal scanners carried on in aircraft and satellites.

1.11.9 Leaf Architecture

The measurement of leaf architecture is required in many studies of radiation penetration into plant canopies. These measurements, because of their difficulty, are often neglected and simplifying assumptions are substituted for facts. The instrument shown in Fig. 1.47 can be used to measure leaf inclination and azimuth. One of the more ingenious techniques for measuring canopy structure was developed by Norman et al. (1979). This technique utilizes the measurement of sunfleck areas at different levels within a canopy with a quantum sensor mounted on a model train, which traverses the plant canopy. Using mathematical inversion techniques data obtained with this system can be converted to give an average leaf angle inclination value and to provide reliable estimates of the cumulative LAI.

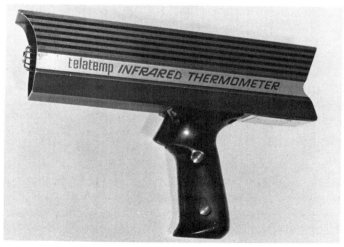

Fig. 1.46 An infrared thermometer used in agricultural research and for field crop management. Courtesy Telatemp Corp., Fullerton, California.

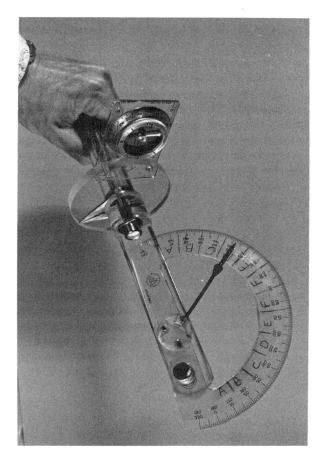

Fig. 1.47 A device to measure leaf inclination and azimuth. Designed by Prof. Raoul Lemeur, Chair of Plant Ecology, University of Ghent, Belgium.

REFERENCES

Abeles, F. B., L. E. Craker, L. E. Forrence, and G. R. Leather. 1971. Fate of air pollutants: Removal of ethylene, sulphur dioxide and nitrogen dioxide by soil. *Science* **173**:914–916.

Abbot, C. G., and L. B. Aldrich. 1916. The pyranometer—An instrument for measuring sky radiation. *Smithsonian Misc. Collections* **66**(7), 9 pp.

Allen, J. R. 1971. Measurements of cloud emissivity in the 8–13 μ waveband. *J. Appl. Meteorol.* **10**:260–265.

Allen, L. H., Jr., and K. W. Brown. 1965. Shortwave radiation in a corn crop. *Agron. J.* **57**:575–580.

Allen, L. H., Jr. 1974. Model of light penetration into a wide row crop. *Agron. J.* **66**:41–47.

Anderson, M. C. 1970. Interpreting the fraction of solar radiation available in forest. *Agric. Meteorol.* **7**:19–28.

Anderson, M. C. 1971. Radiation and crop structure. *Plant Photosynthetic Production: Manual of Methods* (Z. Sestak, J. Catsky, and P. G. Jarvis, eds.), W. Junk N.V. Publishers, The Hague, pp. 431–432.

Angell, J. K., and J. Korshover. 1975a. Decrease in sunshine noted. *Bull. Am. Meteorol. Soc.* **56**:556–557.

Angell, J. K., and J. Korshover. 1975b. Variation in sunshine duration over the contiguous United States between 1950 and 1972. *J. Appl. Meteorol.* **14**:1174–1181.

Angell, J. K., and J. Korshover. 1978. A recent increase in sunshine duration within the contiguous United States. *J. Appl. Meteorol.* **17**:819–824.

Anonymous. 1980. U.S. astronomer challenged. *Beijing Rev.* **20**:30.

Atwater, M. A. 1970. Planetary albedo changes due to aerosols. *Science* **170**:64–66.

Aubertin, G. M., and D. B. Peters. 1961. Net radiation determinations in a cornfield. *Agron. J.* **53**:269–272.

Ayers, R. C., and J. L. Glaeser. 1974. Oil spills in the Arctic Ocean: Extent of spreading and possibility of large-scale thermal effects. *Science* **186**:843–844.

Baker-Blocker, A. 1980. The effects of sunshine, cloudiness and haze on received ultraviolet radiation in New York. *J. Appl. Meteorol.* **19**:889–893.

Baldwin, B., J. B. Pollack, A. Summers, O. B. Toon, C. Sagan, and W. Van Camp. 1976. Stratospheric aerosols and climatic change. *Nature (London)* **263**:551–555.

Basuk, J. 1975. Freons and ozone in the stratosphere. *Bull. Am. Meteorol. Soc.* **56**:589–592.

Birth, G. S., and G. R. McVey. 1968. Measuring the color and growing turf with a reflectance spectrophotometer. *Agron. J.* **60**:640–643.

Black, J. N. 1956. The distribution of solar radiation over the earth's surface. *Arch. Meteorol. Geophys. Bioklimatol. B* **7**:165–189.

Black, J. R., and B. L. Tarmy. 1963. The use of asphalt coatings to increase rainfall. *J. Appl. Meteorol.* **2**:557–564.

Blad, B. L., and D. G. Baker. 1971. A three year study of net radiation at St. Paul, Minnesota. *J. Appl. Meteorol.* **10**:820–824.

Blad, B. L., and D. G. Baker. 1972a. Reflected radiation from a soybean crop. *Agron. J.* **64**:277–280.

Blad, B. L., and D. G. Baker. 1972b. Orientation and distribution of leaves within soybean canopies. *Agron. J.* **64**:26–29.

Bolsenga, S. J. 1969. Total albedo of Great Lakes ice. *Water Resour. Res.* **5**:1132–1133.

Bremner, J. M., and A. M. Blackmer. 1978. Nitrous oxide: Emission from soils during nitrification of fertilizer nitrogen. *Science* **199**:295–296.

Brown, K. W., N. J. Rosenberg, and P. C. Doraiswamy. 1970. Shading inverted pyranometers and measurements of radiation reflected from an alfalfa crop. *Water Resour. Res.* **6**:1782–1786.

Brunt, D. 1934. Physical and Dynamical Meteorology. Cambridge Univ. Press, London. 441 pp.

Bryson, R. A. 1973. Drought in Sahelia—Who or what is to blame? *Ecologist* **3**:366–371.

Bryson, R. A., and B. M. Goodman. 1980. Volcanic activity and climatic changes. *Science* **207**:1041–1044.
Campbell, R. B., D. C. Reicosky, and C. W. Doty. 1981. Net radiation within a canopy of sweet corn during drought. *Agric. Meteorol.* **23**:143–150.
Carlson, R. E., D. N. Yarger, and R. H. Shaw. 1971. Factors affecting the spectral properties of leaves with special emphasis on leaf water status. *Agron. J.* **63**:486–489.
CAST (Council for Agricultural Science and Technology), P. F. Pratt, Chairman. 1977. Effect of increased nitrogen fixation on stratopheric ozone. *Climatic Change* **1**:109–135.
Changnon, S. A. Jr. 1981. Midwestern cloud, sunshine and temperature trends since 1901: Possible evidence of jet contrail effects. *J. Appl. Meteorol.* **20**:496–508.
Charney, J. G. 1975. Dynamics of deserts and drought in the Sahel. *Q. J. Roy. Meteorol. Soc.* **101**:193–202.
Charney, J. G., P. H. Stone, and W. J. Quirk. 1975. Drought in the Sahara: A biogeophysical feedback mechanism. *Science* **187**:434–435.
Chartier, P. 1966. Etude du microclimat lumineux dans la vegetation. *Ann. Agron.* **17**:571–602.
Cicerone, R. J., R. S. Stolarski, and S. Walters. 1974. Stratospheric ozone destruction by man-made chlorofluoromethanes. *Science* **185**:1165–1166.
Cipra, J. E., M. F. Baumgardner, E. R. Stoner, and R. B. MacDonald. 1971. Measuring radiance characteristics of soil with a field spectroradiometer. *Soil Sci. Soc. Am. Proc.* **35**:1014–1017.
Coulson, K. L., and D. W. Reynolds. 1971. The spectral reflectance of natural surfaces. *J. Appl. Meteorol.* **10**:1285–1295.
Crutzen, P. J., I. S. A. Isaksen, and G. C. Reid. 1975. Solar proton events: Stratospheric sources of nitric oxide. *Science* **189**:457–458.
Cutchis, P. 1974. Stratospheric ozone depletion and solar ultraviolet radiation on earth. *Science* **184**:13–20.
Davies, J. A., and P. H. Buttimor. 1969. Reflection coefficients, heating coefficients and net radiation at Simcoe, Southern Ontario. *Agric. Meteorol.* **6**:373–386.
da Mota, F. S., M. I. C. Beirsdorf, and M. J. C. Acosta. 1977. Estimates of solar radiation in Brazil. *Agric. Meteorol.* **18**:241–254.
Decker, W. L. 1959. Variations in the net exchange of radiation from vegetation of different heights. *J. Geophys. Res.* **64**:1617–1619.
Denmead, O. T., L. J. Fritschen, and R. H. Shaw. 1962. Spatial distribution of net radiation in a cornfield. *Agron. J.* **54**:505–510.
Desjardins, R. L., T. R. Sinclair, and E. R. Lemon. 1973. Light fluctuations in corn. *Agron. J.* **65**:904–908.
DeWalle, D. R., and S. G. McGuire. 1973. Albedo variations of an oak forest in Pennsylvania. *Agric. Meteorol.* **11**:107–113.
Dirmhirn, I., and F. D. Eaton. 1975. Some characteristics of the albedo of snow. *J. Appl. Meteorol.* **14**:375–379.
Durand, R. 1974. Estimation du rayonnement global a partir de la duree d'insolation. *Ann. Agron.* **25**:779–795.
Dyer, A. J., and B. B. Hicks. 1968. Global spread of volcanic dust from the Bali eruption of 1963. *Q. J. Roy. Meteorol. Soc.* **94**:545–554.
Eddy, J. A. 1976. The Maunder minimum. *Science* **192**:1189–1202.

Ehleringer, J., and I. Forseth. 1980. Solar tracking by plants. *Science* **210**:1094–1098.

Ellis, H. T., and R. F. Pueschel. 1971. Turbidity of the atmosphere: Source of its background variation with season. *Science* **172**:845.

Ellsaesser, H. W. 1972. Turbidity of the atmosphere: Source of its background variation with the season. *Science* **176**:814–815.

Ellsaesser, H. W. 1977. Has man increased stratospheric ozone? *Nature (London)* **270**:592–593.

Federer, C. A. 1968. Spatial variation of net radiation, albedo and surface temperature of forests. *J. Appl. Meteorol.* **7**:789–795.

Federer, C. A. 1971. Solar radiation absorption by leafless hardwood forests. *Agric. Meteorol.* **9**:3–20.

Fish, B. R. 1972. Electrical generation of natural aerosols from vegetation. *Science* **175**:1239–1240.

Fleagle, R. G., and J. A. Businger. 1963. *An Introduction to Atmospheric Physics.* Academic Press, New York. 346 pp.

Flowers, E. C., and H. J. Viebrock. 1965. Solar radiation: An anomalous decrease of direct solar radiation. *Science* **148**:493–494.

Flowers, E. C., R. A. McCormick, and K. R. Kurfis. 1969. Atmospheric turbidity over the United States, 1961–1966. *J. Appl. Meteorol.* **8**:955–962.

Foster, N. B., and L. W. Foskett. 1953. A photoelectric sunshine recorder. *Bull. Am. Meteorol. Soc.* **34**:212.

Freney, J. R., O. T. Denmead, and J. R. Simpson. 1978. Soil as a source or sink for atmospheric nitrous oxide. *Nature (London)* **273**:530–532.

Fritschen, L. J. 1962. Construction and evaluation of a miniature net radiometer. *J. Appl. Meteorol.* **2**:165–172.

Fritschen, L. J. 1967. Net and solar-radiation relations over irrigated field crops. *Agric. Meteorol.* **4**:55–62.

Fritz, S., and T. H. MacDonald. 1949. Average solar radiation in the United States. *Heating and Ventilation* **46**:61–64.

Fukai, S., and R. S. Loomis. 1976. Leaf display and light environments in row-planted cotton communities. *Agric. Meteorol.* **17**:353–379.

Galbally, I. E. 1972. Production of carbon monoxide in rain water. *J. Geophys. Res.* **77**:7129–7132.

Gates, D. M. 1962. *Energy Exchange in the Biosphere.* Harper & Row, New York. 51 pp.

Gates, D. M. 1965. Radiant energy, its receipt and disposal. *Agricultural Meteorology* (P. E. Waggoner, ed.), Chap. 1 (*Meteorol. Monogr.* **6**:1–26) Am. Meteorol. Soc., Boston.

Gay, L. W. 1971. The regression of net radiation upon solar radiation. *Arch. Meteorol. Geophys. Bioklimatol.* B **19**:1–14.

Gay, L. W., L. R. Knoerr, and M. O. Braaten. 1971. Solar radiation variability on the floor of a pine plantation. *Agric. Meteorol.* **8**:39–50.

Geiger, R. 1961. The climate near the ground. Transl. from the fourth German edition. 1965. Harvard Univ. Press, Cambridge, Massachusetts. 611 pp.

Gillette, D. A., R. N. Clayton, T. K. Mayeda, M. L. Jackson, and K. Sridhar. 1978. Tropospheric aerosols from some major dust storms of the Southwestern United States. *J. Appl. Meteorol.* **17**:832–845.

Glover, J. 1972. Net radiation over tall and short grasses. *Agric. Meteorol.* **10**:455–459.

Greenstone, R. 1978. The possibility that changes in cloudiness will compensate for changes in ozone and lead to natural protection against ultraviolet radiation. *J. Appl. Meteorol.* **17**:107–109.

Halverson, H. G., and J. L. Smith. 1974. Controlling solar light and heat in a forest by managing shadow sources. USDA Forest Service Res. Paper PSW-102. 14 pp.

Hanson, K. J., and H. J. Viebrock. 1964. Albedo measurements over the N.E. United States. *Mon. Weather Rev.* **92**:223–234.

Haurwitz, B. 1945. Insolation in relation to cloudiness and cloud density. *J. Meteorol.* **2**:154–166.

Hays, J. D., J. Imbrie, and N. J. Shackleton. 1976. Variations in the earth's orbit: Pacemaker of the ice ages. *Science* **194**:1121–1132.

Heidel, K. 1972. Turbidity trends at Tucson, Arizona. *Science* **177**:882–883.

Herman, B. M., S. A. Twomey, and D. O. Staley. 1978. Atmospheric dust: Climatological consequences. *Science* **201**:378.

Hobbs, P. V., H. Harrison, and E. Robinson. 1974. Atmospheric effects of pollutants. *Science* **183**:909–915.

Honda, H., and J. B. Fisher. 1978. Tree branch angle: Maximizing effective leaf area. *Science* **199**:888–890.

Horowitz, J. L. 1969. An easily constructed shadow-band for separating direct and diffuse solar radiation. *Solar Energy* **12**:543–545.

Howard, P. H., and A. Hanchett. 1975. Chlorofluorocarbon sources of environmental contamination. *Science* **189**:217–219.

Howard, R. G. 1981. Attenuation of terrestrial solar radiation by the eruption of Mt. St. Helens. *Bull. Am. Meteorol. Soc.* **62**:241–242.

Hoyt, D. V. 1978. Interannual cloud-cover variations in the contiguous United States. *J. Appl. Meteorol.* **17**:354–357.

Huschke, R. E., ed. 1959. *Glossary of Meteorology.* Am. Meteorol. Soc., Boston. 638 pp.

Hutchinson, G. L., and A. R. Mosier. 1979. Nitrous oxide emissions from an irrigated cornfield. *Science* **205**:1125–1127.

Hutchison, B. A., and D. R. Matt. 1977. The annual cycle of solar radiation in a deciduous forest. *Agric. Meteorol.* **18**:255–265.

Hutchison, B. A., D. R. Matt, and R. T. McMillen. 1980. Effects of sky brightness distribution upon penetration of diffuse radiation through canopy gaps in a deciduous forest. *Agric. Meteorol.* **22**:137–147.

Idso, S. B. 1970. The relative sensitivities of polyethylene shielded net radiometers for short and long wave radiation. *Rev. Sci. Instrum.* **41**:939–943.

Idso, S. B. 1972. Systematic deviations of clear sky atmospheric thermal radiation from predictions of empirical formulae. *Q. J. Roy. Meteorol. Soc.* **98**:399–401.

Idso, S. B. 1974. Thermal blanketing: A case for aerosol-induced climatic alteration. *Science* **186**:50–51.

Idso, S. B., R. D. Jackson, R. J. Reginato, B. A. Kimball, and F. S. Nakayama. 1975. The dependence of bare soil albedo on soil water content. *J. Appl. Meteorol.* **14**:109–113.

Idso, S. B., and A. J. Brazel. 1977. Planetary radiation balance as a function of atmospheric dust: Climatological consequences. *Science* **198**:731–733.

Impens, I., and R. Lemeur. 1969a. The radiation balance of several field crops. *Arch. Meterol. Geophys. Bioklimatol. B* **17**:261–268.

Impens, I., and R. Lemeur. 1969b. Extinction of net radiation in different crop canopies. *Arch. Meterol. Geophys. Bioklimatol. B* **17**:403–412.

Impens, I., R. Lemeur, and R. Moermans. 1970. Spatial and temporal variation of net radiation in crop canopies. *Agric. Meteorol.* **7**:335–337.

Inman, R. E., R. B. Ingersoll, and E. A. Levy. 1971. Soil: A natural sink for carbon monoxide. *Science* **172**:1229–1231.

Jensen, E., Sr., and H. C. Aslyng. 1967. Net radiation and net long-wave radiation at Copenhagen 1962–1964. *Arch. Meteorol. Geophys. Bioklimatol. B* **15**:127–140.

Johnson, T. B., F. B. Salisbury, and G. I. Connor. 1967. Ratio of blue to red light: A brief increase following sunset. *Science* **155**:1663–1665.

Johnston, H. 1971. Reduction of stratospheric ozone by nitrogen oxide catalysts from supersonic transport exhaust. *Science* **173**:517–522.

Joseph, J. H., A. Manes, and D. Ashbel. 1973. Desert aerosols transported by Khamsinic depressions and their climatic effects. *J. Appl. Meteorol.* **12**:792–797.

Kalma, J. D. 1972. The radiation balance of a tropical pasture. II. Net all-wave radiation. *Agric. Meteorol.* **10**:261–275.

Kalma, J. D., and R. Badham. 1972. The radiation balance of a tropical pasture. I. The reflection of short wave radiation. *Agric. Meteorol.* **10**:251–259.

Katsoulis, B. D., and S. N. Leontaris. 1981. The distribution over Greece of global solar radiation on a horizontal surface. *Agric. Meteorol.* **23**:217–229.

Kerr, R. A. 1978. Climate control: How large a role for orbital variations? *Science* **201**:144–146.

Kessler, E., D. Y. Alexander, and J. F. Rarick. 1978. Duststorms from the U.S. High Plains in late winter 1977—Search for cause and implications. *Proc. Okla. Acad. Sci.* **58**:116–128.

Kliore, A., G. Fjeldbo, B. L. Seidel, and S. I. Rasool. 1969. Mariners 6 and 7: Radio occultation measurements of the atmosphere of Mars. *Science* **166**:1393–1397.

Komp, M. J., and A. H. Auer, Jr. 1978. Visibility reduction and accompanying aerosol evolution downwind of St. Louis. *J. Appl. Meteorol.* **17**:1357–1367.

Kuhn, P. M., and V. E. Suomi. 1958. Airborne observations of albedo with a beam reflector. *J. Meteorol.* **15**:172–174.

Kukla, G. J., and H. J. Kukla. 1974. Increased surface albedo in the Northern Hemisphere. *Science* **183**:709–714.

Kung, E. C., R. A. Bryson, and D. H. Lenschow. 1964. Study of a continental surface albedo on the basis of flight measurements and structure of the earth's surface cover over North America. *Mon. Weather Rev.* **92**:543–564.

Landsberg, H. E. 1980. Variable solar emissions, the "Maunder minimum" and climatic temperature fluctuations. *Arch. Metereol. Geophys. Bioklimatol. B* **28**:181–191.

Lang, A. R. G., and G. S. G. Shell. 1976. Sunlit areas and angular distributions of sunflower leaves for plants in single and multiple rows. *Agric. Meteorol.* **16**:5–15.

Lang, A. R. G. 1973. Leaf orientation of a cotton plant. *Agric. Meteorol.* **11**:37–51.

Laue, E. G., and A. J. Drummond. 1968. Solar constant: First direct measurements. *Science* **161**:888–891.

Lee, R. 1973. The "greenhouse" effect. *J. Appl. Meteorol.* **12**:556–557.

Lee, R. 1978. *Forest Microclimatology*. Columbia Univ. Press, New York. 276 pp.

Lemeur, R. 1973. A method for simulating the direct solar radiation regime in sunflower, Jerusalem artichoke, corn and soybean canopies using actual stand structure data. *Agric. Meteorol.* **12**:229–247.

Lemeur, R., and N. J. Rosenberg. 1979. Simulating the quality and quantity of short wave radiation within and above canopies. *Comparison of Forest Water and Energy Exchange Models* (S. Halldin, ed.), International Society for Ecological Modelling, Copenhagen, pp. 77–100.

Lemeur, R., and B. L. Blad. 1975. A critical review of light models for estimating the shortwave radiation regime of plant canopies. *Plant Modification for More Efficient Water Use* (J. F. Stone, ed.) Elsevier, Amsterdam, pp. 255–286.

Leonard, R. E., and A. R. Eschner. 1968. Albedo of intercepted snow. *Water Resour. Res.* **4**:931–935.

Linacre, E. T. 1968. Estimating net radiation flux. *Agric. Meteorol.* **5**:49–63.

Liou, K-N. 1976. On the absorption, reflection and transmission of solar radiation in cloudy atmospheres. *J. Atmos. Sci.* **33**:798–805.

List, R. J., ed. 1966. *Smithsonian Meteorological Tables*, 6th rev. ed. Smithsonian Misc. Collections, Vol. 114. Smithsonian Institution, Washington, D.C. 527 pp.

London, J., and J. Kelley. 1974. Global trends in total atmospheric ozone. *Science* **184**:987–989.

Lugg, D. G., V. E. Youngman, and G. Hinze. 1981. Leaf azimuthal orientation of sorghum in four row directions. *Agron. J.* **73**:497–500.

Luxmoore, R. J., R. J. Millington, and A. R. Aston. 1971. Modified tube solarimeters for additive and net measurements of visible, infrared and solar radiation. *Agron. J.* **63**:329–331.

McArthur, L. B., and J. E. Hay. 1978. On the anisotropy of diffuse solar radiation. *Bull. Am. Meteorol. Soc.* **59**:1442–1443.

McConnell, J. C., and H. I. Schiff. 1978. Methyl chloroform: Impact on stratospheric ozone. *Science* **199**:174–176.

Machta, L., and E. Hughes. 1970. Atmospheric oxygen in 1967 to 1970. *Science* **168**:1582–1584.

Maguire, R. J. 1975a. Effects of ice and snow cover on transmission of light in lakes. Scientific Ser. 54. Inland Waters Directorate, Environment Canada. 24 pp.

Maguire, R. J. 1975b. Light transmission through snow and ice. Tech. Bull. 91, Inland Waters Directorate, Environment Canada. 7 pp.

Marotz, G. A., and J. A. Henry. 1978. Satellite-derived cumulus cloud statistics for western Kansas. *J. Appl. Meteorol.* **17**:1725–1736.

Marsh, V. 1970. The Abbot pyranometer as a reference standard for calibration of pyranometers. *J. Appl. Meteorol.* **9**:136–142.

Martsolf, J. D., and W. L. Decker. 1970. Microclimate modification by manipulation of net radiation. *Agric. Meteorol.* **7**:197–216.

Mayer, H. 1981. Strahlungs Messungen über einem Ackerboden, einem Maisfeld und einem Weingarten. *Agric. Meteorol.* **23**:317–330.

Mendonca, B. G., K. J. Hanson, and J. J. DeLuisi. 1978. Volcanically related secular trends in atmospheric transmission at Mauna Loa Observatory, Hawaii. *Science* **202**:513–515.

Mercer, J. H. 1978. West Antarctic ice sheet and CO_2 greenhouse effect: A threat of disaster. *Nature (London)* **271**:321–325.

Miller, E. E., and J. M. Norman. 1971. A sunfleck theory for plant canopies. I. Lengths of sunlit segments along a transect. *Agron. J.* **63**:735–738.

Monsi, M., and T. Saeki. 1953. Über der lichtfaktor in den Pflanzengesell-schaften und seine Bedeutung fur die Stoffproduktion. *Japan. J. Bot.* **14**:22–52.

Monteny, B., and G. Gosse. 1978. Trouble atmospherique et rayonnement solaire en basse Cote d'Ivoire (atmospheric turbidity and solar radiation in Ivory Coast). *Agric. Meteorol.* **19**:121–136.

Monteny, B., J. Humbert, J. P. L'homme, and J. M. Kalms. 1981. Le rayonnement net et l'estimation de l'evapotranspiration en Cote d'Ivoire. *Agric. Meteorol.* **23**:45–59.

Moomen, S. E., and C. W. Barney. 1981. A modern technique to halt desertification in the Libyan Jamahiriya. *Agric. Meteorol.* **23**:131–136.

National Center for Atmospheric Research Quarterly No. 27, 1970.

Norman, J. M., and C. B. Tanner. 1969. Transient light measurements in plant canopies. *Agron. J.* **61**:847–849.

Norman, J. M., C. B. Tanner, and G. W. Thurtell. 1969. Photosynthetic light sensor for measurements in plant canopies. *Agron. J.* **61**:840–843.

Norman, J. M. 1979. Modeling the complete canopy. (B. J. Barfield and J. F. Gerber, eds.). *Modification of the Aerial Environment of Crops*. pp. 249–277. Amer. Soc. Agric. Engin. St. Joseph, Mich.

Norman, J. M., S. G. Perry, A. B. Fraser, and W. Mach. 1979. Remote sensing of canopy structure. Extended Abstracts 14th Conference on Agricultural and Forest Meteorology. Am. Meteorol. Soc., pp. 184–185.

Norman, C. 1981. Satellite data indicate ozone depletion. *Science* **213**:1088–1089.

Ogren, J. A., R. J. Charlson, L. F. Radke, and S. K. Domonkos. 1981. Absorption of visible radiation by aerosols in the volcanic plume of Mt. St. Helens. *Science* **211**:834–836.

Oguntoyinbo, J. S. 1974. Land use and reflection coefficient (albedo) map for southern parts of Nigeria. *Agric. Meteorol.* **13**:227–237.

Otterman, J. 1974. Baring high-albedo soils by overgrazing: A hypothesized desertification mechanism. *Science* **186**:531–533.

Paltridge, G. W. 1970. Daytime long-wave radiation from the sky. *Q. J. Roy. Meteorol. Soc.* **96**:645–653.

Parry, H. D. 1977. Ozone depletion by chlorofluoromethanes? Yet another look. *J. Appl. Meteorol.* **16**:1137–1148.

Pendleton, J. W., D. B. Peters, and J. W. Peek. 1966. Role of reflected light in the corn ecosystem. *Agron. J.* **58**:73–74.

Pendleton, J. W., D. B. Egli, and D. B. Peters. 1967. Response of *Zea mays* L. to a "light rich" field environment. *Agron. J.* **59**:395–397.

Peterson, J. T., and T. L. Stoffel. 1980. Analysis of urban-rural solar radiation data from St. Louis, Missouri. *J. Appl. Meteorol.* **19**:275–283.

Peterson, J. T., E. C. Flowers, G. J. Berri, C. L. Reynolds, and J. H. Rudisill. 1981. Annual average turbidity 0.147 (0.336 aerosol optical thickness) among highest non-urban turbidity in U.S. *J. Appl. Meteorol.* **20**:229–241.

Peterson, W. A., and I. Dirmhirn. 1981. The ratio of diffuse to direct solar irradiance (perpendicular to the sun's rays) with clear skies—A conserved quantity throughout the day. *J. Appl. Meteorol.* **20**:826–828.

Platt, R. B., and J. F. Griffiths. 1964. *Environmental Measurement and Interpretation*. Reinhold, New York. 235 pp.

Pochop, L. O., M. D. Shanklin, and D. A. Horner. 1968. Sky cover influence on total hemispheric radiation during the daylight hours. *J. Appl. Meteorol.* **7**:484–489.

Prospero, J. M., and R. T. Nees. 1977. Dust concentration in the atmosphere of the Equatorial North Atlantic: Possible relationship to the Sahelian drought. *Science* **196**:1196–1198.

Pueschel, R. F., and G. Langer. 1973. Sugar cane fires as a source of ice nuclei in Hawaii. *J. Appl. Meteorol.* **12**:549–551.

Ramanathan, V. 1975. Greenhouse effect due to chlorofluorocarbons: Climatic implications. *Science* **190**:50–51.

Reifsnyder, W. E., and H. W. Lull. 1965. Radiant energy in relation to forests. Tech. Bull. No. 1344. USDA, Forest Service, Washington, D.C. 111 pp.

Reifsnyder, W. E., and G. M. Furnival. 1969. Power-spectrum analysis of the energy contained in sunflecks. Proc. 3rd Forest Microclimate Symp., Kananaskis Forest Expt. Sta., Seebe, Alberta, Canada, Sept. 23–26, 1969, pp. 117–118.

Reifsnyder, W. E., G. M. Furnival, and J. L. Horowitz. 1971. Spatial and temporal distribution of solar radiation beneath forest canopies. *Agric. Meteorol.* **9**:21–37.

Robinson, P. J., and J. A. Davies. 1972. Laboratory determinations of water surface emissivity. *J. Appl. Meteorol.* **11**:1391–1393.

Robinson, P. J. 1977. Measurements of downward scattered solar radiation from isolated cumulus clouds. *J. Appl. Meteorol.* **16**:620–625.

Rosenberg, N. J. 1964. Solar energy and sunshine in Nebraska. *Neb. Agr. Exp. Station Res.* Bull No. 213, 29 pps.

Rosenberg, N. J. 1969. Seasonal patterns in evapotranspiration by alfalfa in the central Great Plains. *Agron. J.* **61**:879–886.

Rosenberg, N. J., and W. L. Powers. 1970. Potential for evapotranspiration and its manipulation in the Plains region. Evapotranspiration in the Great Plains, Great Plains Agricultural Council Publication No. 50, pp. 275–300.

Sakamoto, C. M., and R. H. Shaw. 1967. Apparent photosynthesis in field soybean canopies. *Agron. J.* **59**:73–75.

Sarraf, S., and A. Aboukhaled. 1970. Rayonnement solaire, rayonnement net et eclairement au Liban. Institut de Recherches Agronomiques, Tal-Amara, Liban. 7 pp.

Schieldrup-Paulsen. 1980. Radiation observations in Bergen, Norway. U. of Bergen, Geophys. Inst. 55 pp.

Schnell, R. C. 1974. Biogenic and inorganic sources for ice nuclei in the drought-stricken areas of the Sahel—1974. Interim report to Directors, Rockefeller Foundation, NCAR, Boulder, Colorado.

Sellers, W. D. 1965. *Physical Climatology*. Univ. Chicago Press. 272 pp.

Shaw, R. H. 1956. A comparison of solar radiation and net radiation. *Bull. Am. Meteorol. Soc.* **37**:205–206.

Shaw, G. E. 1980. Transport of Asian desert aerosol to the Hawaiian Islands. *J. Appl. Meteorol.* **19**:1254–1259.

Shell, G. S. G., and A. R. G. Lang. 1976. Movements of sunflower leaves over a 24-hr period. *Agric. Meteorol.* **16**:161–170.

Sinclair, T. R., M. M. Schreiber, and R. M. Hoffer. 1973. Diffuse reflectance hypothesis for the pathway of solar radiation through leaves. *Agron. J.* **65**:276–283.

Sinclair, T. R., and E. R. Lemon. 1974. Penetration of photosynthetically active radiation in corn canopies. *Agron. J.* **66**:201–205.

Singh, H. B., D. P. Fowler, and T. O. Peyton. 1976. Atmospheric carbon tetrachloride: Another man-made pollutant. *Science* **192**:1231–1234.

Singh, H. B., L. J. Salas, H. Shigeishi, and E. Scribner. 1979. Atmospheric halocarbons, hydrocarbons and sulfur hexafluoride: Global distributions, sources and sinks. *Science* **203**:899–903.

Stanhill, G. 1966. Diffuse sky and cloud radiation in Israel. *Solar Energy* **10**:96–101.

Stanhill, G., G. J. Hofstede, and J. D. Kalma. 1966. Radiation balance of natural and agricultural vegetation. *Q. J. Roy. Meteorol. Soc.* **92**:128–140.

Stanhill, G., M. Fuchs, and J. Oguntoyinbo. 1971. The accuracy of field measurements of solar reflectivity. *Arch. Metereol. Geophys. Bioklimatol. B* **19**:113–132.

Stanhill, G., V. Kafkafi, M. Fuchs, and Y. Kagan. 1972. The effect of fertilizer application on solar reflectance from a wheat crop. *Israel J. Agric. Res.* **22**:109–118.

Sutton, O. G. 1953. *Micrometeorology*. McGraw-Hill, New York. 333 pp.

Swinbank, W. C. 1963. Long-wave radiation from clear skies. *Q. J. Roy. Meteorol. Soc.* **89**:339–348.

Swinnerton, J. W., V. J. Linnenbom, and R. A. Lamontagne. 1970. The ocean: A natural source of carbon monoxide. *Science* **167**:984–986.

Sze, N. D. 1977. Anthropogenic CO emissions: Implications for the atmospheric $CO-OH-CH_4$ cycle. *Science* **195**:673–675.

Szwarcbaum, I., and G. Shaviv. 1976. A Monte-Carlo model for the radiation field in plant canopies. *Agric. Meteorol.* **17**:333–352.

Tarpley, J. D. 1979. Estimating incident solar radiation at the surface from geostationary satellite data. *J. Appl. Meteorol.* **18**:1172–1181.

Thompson, E. S. 1976. Computation of solar radiation from sky cover. *Water Resour. Res.* **12**:859–865.

Thorpe, M. R. 1978. Net radiation and transpiration of apple trees in rows. *Agric. Meteorol.* **19**:41–57.

Twomey, S. 1976. Computations of the absorption of solar radiation by clouds. *J. Atmos. Sci.* **33**:1087–1091.

Vonder Haar, T., and V. E. Suomi. 1969. Satellite observations of the earth's radiation budget. *Science* **163**:667–669.

Wahua, T. A. T., and D. A. Miller. 1978. Effects of shading on the N_2-fixation, yield and plant composition of fieldgrown soybeans. *Agron. J.* **70**:387–392.

Wang, V., and J. H. Shaw. 1970. Variations in the abundance of atmospheric carbon monoxide near the ground. *J. Appl. Meteorol.* **9**:180–182.

Wang, W. C., Y. L. Yung, A. A. Lacis, T. Mo, and J. E. Hansen. 1976. Greenhouse effects due to man-made perturbations of trace gases. *Science* **194**:685–690.

Weertman, J. 1976. Milankovitch solar radiation variations and ice age ice sheet sizes. *Nature (London)* **261**:17–20.

Weinstock, B., and H. Niki. 1972. Carbon monoxide balance in nature. *Science* **176**:290–293.

Went, F. W. 1955. Air pollution. *Sci. Am.* **192**:62.

Whitten, R. C., W. J. Borucki, and R. P. Turco. 1975. Possible ozone depletions following nuclear explosions. *Nature (London)* **257**:38–39.

Willis, J. 1971. Some high values for the albedo of the sea. *J. Appl. Meteorol.* **10**:1296–1302.

Willson, R. C., C. H. Duncan, and J. Geist. 1980. Direct measurement of solar luminosity variation. *Science* **207**:177–179.

Yao, A. Y. M., and R. H. Shaw. 1964. Effect of plant population and planting pattern of corn on water use and yield. *Agron. J.* **56**:147–152.

CHAPTER 2 – SOIL HEAT FLUX AND SOIL TEMPERATURE

2.1 INTRODUCTION

The soil mantle of the earth is indispensable for the maintenance of plant life, affording mechanical support and supplying nutrients and water. Soil constitutes a major storage location for heat, acting as a sink for energy during the day and a source to the surface at night. In annual terms, the soil stores energy during the warm season and releases it to air during the cold portions of the year.

The flux of heat into and out of the soil is due to the process of conduction in which lively molecular motion is transmitted to adjacent, more sluggish molecules. This transmission of energy occurs while the bodies involved remain at rest. Heat is conducted through the soil, which need not change its physical shape or condition for this to occur.

2.2 LAWS OF HEAT CONDUCTION AND THERMAL PROPERTIES OF SOILS

Some definitions follow:

The **heat capacity** of a substance is the ratio of heat absorbed (or released) to the corresponding temperature rise (or fall). Heat capacity has units of J °K^{-1}.

The **mass specific heat** (C_s) of a substance is the amount of heat required to raise the temperature of 1 kg of that substance by 1°C. C_s has units of J kg^{-1} °K^{-1}. C_s is the heat capacity of a substance per unit mass.

The **volume specific heat** (C_v) of a substance is the amount of heat required to raise the temperature of a unit volume of the material by 1°K. C_v has units of J l^{-1} °K^{-1}. C_v is the heat capacity of a substance per unit volume.

The mass and volume specific heats are related by:

$$\rho C_s = C_v \qquad (2.1)$$

where ρ is the **density** or volume weight of the substance, with units of kg l^{-1}.

LAWS OF HEAT CONDUCTION AND THERMAL PROPERTIES OF SOILS

Soil materials vary greatly in their specific heats. C_s of a coarse sandy loam, for example, may be about 830 J kg^{-1} °K^{-1}, whereas that of humus may be as high as 2000 J kg^{-1} °K^{-1}. The mass and volume specific heats of water are 4.188 × 10^3 J kg^{-1} °K^{-1} and 4.188 × 10^3 J l^{-1} °K^{-1}. The density of soil materials may range from 0.8 kg l^{-1} for a peat to 1.8 kg l^{-1} for a compacted sand. Depending then on the constituents of the soil, its physical condition (density) and its water content, the specific heat and heat capacity may range widely. As an example, dry kaolin has a specific heat of about 840 J °K^{-1} l^{-1}. When 50 and 100% saturated with water, the specific heats are 2090 and 3350 J °K^{-1} l^{-1}, respectively.

The rate of heat flux into a soil is determined by the temperature gradient and the **thermal conductivity** (K). The latter is defined as the quantity of heat flowing in unit time through a unit cross section of soil in response to a specified temperature gradient. K has units of W m^{-1} °K^{-1}.

The **soil heat flux** S, or heat flow into or out of the soil, is given by

$$S = K \, dT/dz \tag{2.2}$$

where dT/dz is the temperature gradient within the soil. Here we adopt a convention that will hold in all following chapters. All energy fluxes to the surface will be considered positive and all away from the surface will be negative.

Thermal conductivity depends on porosity, moisture content, and organic matter content of the soil. At similar moisture contents conductivity decreases from fine sand to silt loam to clay soil because of the increasing porosity in this sequence of textures. This is very well illustrated by the work of Al-Nakshabandi and Kohnke (1965), shown in Fig. 2.1a. They found, however, that when the soil moisture is plotted as a function of the tension with which the moisture is held in the soil (soil moisture potential) rather than as a function of the percentage of moisture, the thermal conductivities for the three soil textures are nearly identical (Fig. 2.1b). This occurs because, at similar moisture potentials for the soils represented, the moisture film thicknesses are nearly identical.

The thermal conductivity of soils determines the rate of heat transfer. The temperature change with time experienced by the body as a result of the heat transfer will vary with its heat capacity. Therefore, a parameter that considers heat capacity can be useful. The **thermal diffusivity** α is the quotient of the thermal conductivity and the volume specific heat:

$$\alpha = \frac{K}{C_v} = \frac{K}{\rho C_s} \tag{2.3}$$

and has units of m^2 s^{-1}.

Table 2.1, based primarily on data from Sutton (1953), shows the range of density, specific heat, heat capacity, thermal conductivity, and diffusivity

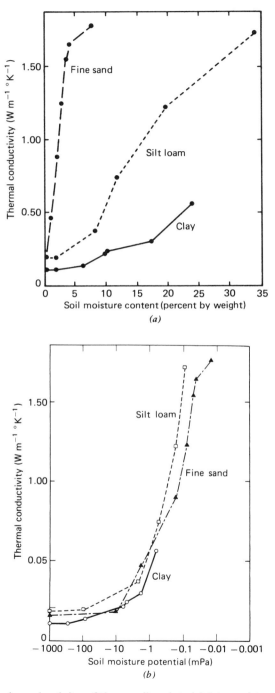

Fig. 2.1 Thermal conductivity of three soils related (*a*) to moisture content and (*b*) to moisture potential (after Al-Nakshabandi and Kohnke, 1965).

Table 2.1 Density and thermal properties of some soil and reference materials[a]

	Density, ρ (kg l^{-1})	Specific heat		Thermal conductivity, K (W m^{-1} °K^{-1})	Thermal diffusivity, α (m^2 s^{-1})
		Mass, C_s (J °K^{-1} kg^{-1})	Volume, C_v (J °K^{-1} l^{-1})		
Water @ 20°C	1.00	4.188×10^3	4.188×10^3	5.862×10^{-1}	1.4×10^{-7}
Dead air	0.013	1.005×10^3	1.307×10^1	1.6×10^{-6}	1.6×10^{-5}
Silver metal	10.50	0.251×10^3	2.636×10^3	4.523×10^2	1.72×10^{-4}
Clay	1.80	3.350×10^3	6.030×10^3	1.206	2.0×10^{-7}
Light soil with roots	0.30	1.256×10^3	3.768×10^2	1.131×10^{-1}	3.0×10^{-7}
Wet sandy soil	1.60	1.675×10^3	2.680×10^3	2.68	1.0×10^{-6}

[a] Based on data from Sutton (1953) and others.

for some natural soil materials and for the reference materials silver metal, dead air, and water.

Thermal diffusivity of a soil is a parabolic function of moisture content. A small amount of water reduces the insulating effect of the pore space filled with air, but further increases in water content markedly increase the heat capacity. This is so because the heat capacity of water, which is high, is substituted for that of air, which is almost negligible.

Soil organic matter lowers thermal conductivity because of its influence in increasing porosity. Compaction increases the thermal conductivity by decreasing the volume of the insulating pore space.

Moench and Evans (1970) evaluated thermal parameters in moist soils. For a sandy loam, they found that the thermal conductivity increased rapidly with water content at low percentages of saturation and proportionately to the volumetric heat capacity above 30% of saturation. Diffusivity, on the other hand, was nearly constant from about 30 to 100% of saturation.

From the foregoing, one may readily see that soil tillage and soil water management may profoundly affect the flow of heat into and out of soil, and the soil temperature regime, as well. One may also see that prediction of the net influence of any soil tillage or irrigation practice can be very difficult.

Determination of the soil physical properties that affect its thermal behavior is difficult and will not be described in detail here. The basic methods for determining soil physical properties were reviewed by van Wijk (1964, 1965). More recently, Hadas (1973) developed methods to determine K, C_v, and α of topsoil. Thermal conductivity was determined by packing soil into a box and maintaining steady heat flow. Thermal diffusivity and C_v were calculated from measurements of soil temperature changes at various depths in a soil-filled Lucite tube when the tube was brought into contact with a heat source.

For the purposes of most microclimate studies in the field, soil heat flux S can be measured directly with instruments designed specifically for that purpose (see Section 2.9.1). For more refined evaluations of S in soil physics research, more complex laboratory and computational procedures are needed. Hanks and Jacobs (1971), for example, describe a calorimetric procedure for determining S.

Kimball and Jackson (1975) have a "null-alignment" method that can be used in the field for computing S from measurements of soil temperature and moisture in the upper 0.2 m of the soil. Kimball et al. (1976) tested the null alignment method and others against a "temperature gradient" method involving computed values of thermal conductivity for particular reference depths. The latter was found to be adequate if the reference depth is 0.20 m below the surface but specific calibrations are needed for each soil if the reference depth is closer to the surface. The temperature gradient method is computationally much simpler than the null-alignment method and, also, avoids problems of thermal and moisture barriers that may occur when flux plates are used in the soil.

DAILY AND SEASONAL PATTERNS OF SOIL TEMPERATURE 99

2.3 PENETRATION OF HEAT INTO THE GROUND

The thermal diffusivity of soil is small, considerably less than that of air at rest. For example, α in soil ranges from about 2×10^{-7} to 1.0×10^{-6}, while α for dead air is about $1.6 \times 10^{-5}\,m^2\,s^{-1}$. Thus soil temperature will change slowly and will be observed as waves during the course of a day. The amplitude of the temperature wave at the ground surface will be great, but it will diminish with depth below the surface. The range of temperature at any depth in the soil (if it is assumed that soil properties including porosity, water content, and organic matter content are uniform with depth) is given in theory by

$$TR_z = TR_s \exp\left[-z\left(\frac{\pi}{\alpha p}\right)^{1/2}\right] \quad (2.4)$$

where TR_z and TR_s are the temperature ranges at depth z and at the surface, respectively, α is the thermal diffusivity, and p is the period of oscillation in seconds.

2.4 DAILY AND SEASONAL PATTERNS OF SOIL TEMPERATURE

The pattern of decreasing amplitude of the soil temperature wave is well illustrated in Fig. 2.2. The amplitude of soil temperature in summer and winter decreases with increasing depth into both bare and sod covered soil at St. Paul, Minnesota. At the 0.40 m depth, the wave is virtually damped out, especially in winter, and at 0.80 m no diurnal wave occurs. Soil temperature decreases with depth during the daytime in summer. Thus, temperature gradients direct heat into the soil. At night, however, the temperature is highest between 0.20 and 0.40 m; from that level heat is directed both upward and downward.

Figure 2.2 also illustrates the fact that in winter the 0.80 m depth is warmest and the diurnal wave is still only barely evident at 0.40 m. At Copenhagen, Denmark, the daily temperature wave under bare cropped soil is also damped at about 0.50 m according to Kristensen (1959). Heat is transferred upward from those levels throughout the day and night. Because of the low flux density of solar radiation in winter, the daily amplitude of surface temperatures is very small. However, the amplitude is still greater at the surface than at any level below. Only during the hours near noon does heat penetrate the soil from the surface. The damping influence of sod cover on the daily temperature wave in both winter and summer is also seen in Fig. 2.2.

There are other ways to describe the cyclical behavior in soil temperature. Parton and Logan (1981), for example, have used a truncated sine wave to predict daytime temperatures and an exponential function is used for nighttime temperatures. Their models were developed from observations made at a shortgrass prairie site in eastern Colorado. Soil temperatures were measured at the surface and at a depth of 0.10 m.

100 SOIL HEAT FLUX AND SOIL TEMPERATURE

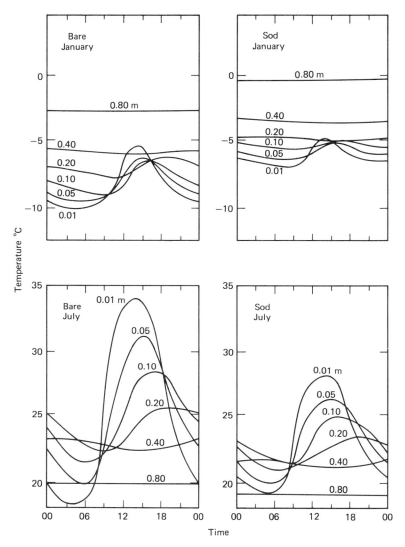

Fig. 2.2 Average hourly soil temperature under bare and sod-covered soil at St. Paul, Minnesota in January (top) and July (bottom). Soil depth is shown in m (after Baker, 1965).

By reference to (2.4), it can be seen that the amplitude (range) of the temperature wave must decrease exponentially with depth. It is also seen that the greater the diffusivity or the greater the period, the slower the depth function decrease in amplitude of the soil temperature wave. As a "rule of thumb," it may be assumed that the annual wave is extinguished at a depth that is approximately $(365)^{1/2}$ or about 19 times the depth of the diurnal wave. In Fig. 2.2 the daily wave is seen to be extinguished at about 0.40 m. It may be assumed that in a uniform material the annual wave would penetrate to about 7.6 m.

Figure 2.2 also illustrates the fact that the achievement of maximum and minimum temperatures lags as a function of depth. The time lag between levels can be determined by the relation

$$t_2 - t_1 = \frac{z_2 - z_1}{2} \left(\frac{p}{\alpha \pi}\right)^{1/2} \quad (2.5)$$

where t_2 and t_1 are the times at which the maxima (or minima) are detected at depths z_2 and z_1.

Since, as Fig. 2.2 illustrates, the thermal lag is so great that soil at great depths may reach its highest temperature by midwinter and its lowest temperature in summer—a good energy-saving reason to build human and animal habitations as far below ground level as is consistent with comfort and psychology.

2.5 SOIL TEMPERATURE PROFILES

The foregoing illustrates that heat is continually moving into or out of the soil and that thermal energy is being continually redistributed in the soil. Heat will not flow under **isothermal** conditions (no temperature gradient).

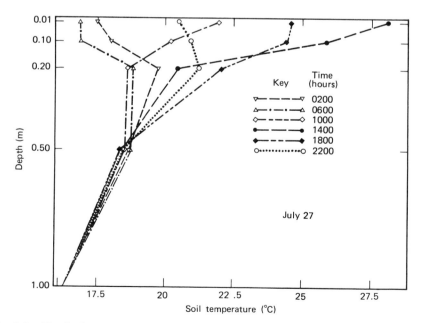

Fig. 2.3 Vertical temperature profiles in soil during the course of a typical summer day at Argonne, Illinois (after Carson and Moses, 1963).

102 SOIL HEAT FLUX AND SOIL TEMPERATURE

The pattern of soil temperature profiles changes rapidly during a normal day, as is illustrated in Fig. 2.3 for measurements made at Argonne National Laboratory by Carson and Moses (1963).

The soil surface is the coldest level in the early morning and the warmest in the early afternoon. At midday, heat is directed downward through the upper 1 m of soil. The profiles show that heat exit from the middle of the layer begins after sunset but some heat flow continues downward throughout the night. During most of the day the temperature profiles indicate a downward heat flux. Thus in summer there is a net daily gain or storage of heat in the soil.

Figure 2.4 (from Kimball and Jackson, 1975) describes soil temperature profiles in the upper 0.20 m in greater detail. The profiles are somewhat symmetrical and illustrate the diurnal cycle of warming and cooling. The points indicated by arrows indicate zero temperature gradient and, hence, zero soil heat flux, when the flow from below and the flow from above are equal. Determination of the location of this "null" point is essential in Kimball and Jackson's (1975) method of soil heat flux calculations.

The diurnal range in surface temperature depends on a number of physical parameters including reflectivity and soil thermal conductivity. Saltzman and Pollack (1977) modeled soil surface temperature range and found that the reflectivity r and "conductive capacity" were major determining factors. Conductive capacity C is defined as

$$C = \rho C_s \alpha^{1/2} \tag{2.6}$$

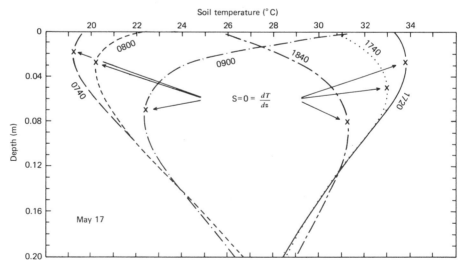

Fig. 2.4 Soil temperature profiles obtained at 0740, 0800, 1720, 1740, and 1840 hours in Avondale loam. All the profiles exhibit a null point where the temperature gradient is zero and hence where it can be assumed the soil heat flux is also zero (after Kimball and Jackson, 1975).

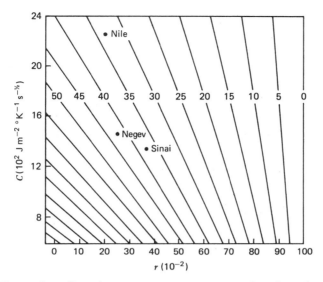

Fig. 2.5 The surface diurnal temperature range D as a function of conductive capacity C and surface reflectivity r, holding all other parameters constant, for summer conditions in the Sinai–Negev region. Units are °K. Plausible values for the Sinai, Negev, and Nile Valley are shown (from Saltzman and Pollack, 1977).

where ρ is the density of the soil at the surface, C_s is the specific heat, and α is the thermal diffusivity. A plot for predicting diurnal temperature range is given as Fig. 2.5. Plausible values for locations in Sinai, the Negev Desert of Israel, and the Nile Valley are shown in this figure.

For purposes of long-term planning, soil temperature patterns can be modeled as a function of ambient weather conditions (e.g., radiation, air temperature). However, as should be clear from the discussion above, soil moisture also plays a major role. For this reason, Ouellet (1973) in Canada prepared a model for estimating monthly soil temperatures under shortgrass. Variables include rainfall, snowfall, potential evapotranspiration, and soil temperature of the previous month.

2.6 TEXTURE INFLUENCES ON SOIL HEAT FLUX AND TEMPERATURE

Soil texture has a distinct influence on soil heat flux and, hence, on soil temperature. Figure 2.6a,b show the diurnal temperature wave at the surface and at a depth of 50 mm in sand, loam, peat, and clay in Hokkaido, Japan (Yakuwa, 1946). The amplitude of the daily temperature wave decreases in the order sand > loam > peat > clay. This order can be attributed to the greater heat capacity of the finer textured materials and their normally greater thermal conductivity.

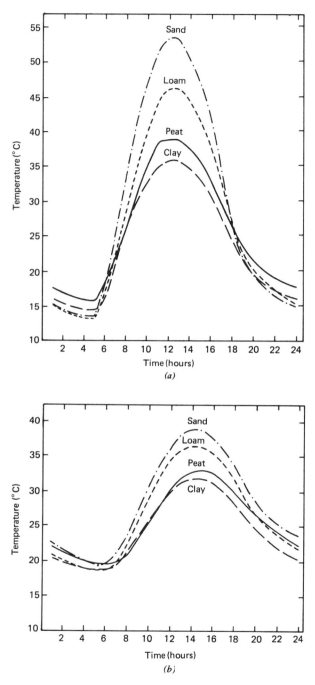

Fig. 2.6 Daily course of temperature (*a*) at the surface and (*b*) at a depth of 50 mm on clear summer days at Sapporo, Japan (after Yakuwa, 1946).

Table 2.2 Soil temperature conditions for vegetable seed germination[a]

	Minimum temp. (°C)	Optimum temp. (°C)	Maximum temp. (°C)
Asparagus	0.0	23.8	35.0
Bean, lima	15.5	29.4	29.4
Beet	4.4	29.4	35.0
Cauliflower	4.4	26.6	37.7
Cucumber	15.5	35.0	40.6
Lettuce	1.6	23.9	29.4
Onion	1.6	23.9	35.0
Pea	4.4	23.9	29.4
Radish	4.4	29.4	35.0
Tomato	10.0	29.4	35.0
Watermelon	15.5	35.0	40.6

[a] After Lorenz and Maynard (1980).

It is instructive to speculate on the patterns of warming in a sandy (coarse textured) soil and a silt loam (fine textured) in spring. At the end of winter both soils would likely be saturated. Their thermal conductivity may be about equal at saturation. Because the fine sand holds less water and usually drains better, it will dry more rapidly than will the silt. Within a few days, the thermal conductivity of the sand will decrease sharply as the remaining pore space is filled with air, a poor conductor. Its overall thermal capacity will decrease since water has the highest thermal capacity of any substance in soil. Furthermore, the evaporative cooling at the surface will cease when water becomes unavailable. For these reasons, a sandy soil may warm much more rapidly in spring. It will, likely, cool more rapidly in fall than will silty or clay soils under the same weather conditions because of the lower moisture content and thermal capacity.

The change in soil temperature is probably of greatest importance in the spring season in temperate and cool climates. It is then that the sowing of crops depends on a suitable seedbed temperature being reached and maintained. Seeds planted in cold soil may not germinate and may rot or become unviable, instead. Seeds planted in soil that is too warm may also not germinate. A list of germination temperatures for the seeds of common vegetable species adapted from Lorenz and Maynard (1980) is given in Table 2.2 Days required for seedling emergence at various soil temperatures, also after Lorenz and Maynard, is given in Table 2.3.

In the United States, soil temperature is observed regularly in a small number of locations, generally at agricultural experiment stations. The advancing temperatures are followed closely in spring by many agriculturalists. Figure 2.7 is typical of maps published in late winter and spring by the U.S. Department of Agriculture in cooperation with the National Oceanic and Atmospheric Administration.

Table 2.3 Days required for seedling emergence at various soil temperatures from seed planted 125 mm deep[a]

	Soil temperature (°C)							
	0	5	10	15	20	25	30	35	40
Asparagus	NG[b]	NG	53	24	15	10	12	20	28
Bean, lima	—[c]	—	NG	31	18	7	7	NG	—
Beet	—	42	17	10	6	5	5	5	—
Cauliflower	—	—	20	10	6	5	5	—	—
Cucumber	NG	NG	NG	13	6	4	3	3	—
Lettuce	49	15	7	4	3	2	3	NG	NG
Onion	136	31	13	7	5	4	4	13	NG
Pea	—	36	14	9	8	6	6	—	—
Radish	NG	29	11	6	4	4	3	—	—
Tomato	NG	NG	43	14	8	6	6	9	NG
Watermelon	—	NG	—	—	12	5	4	3	—

[a] After Lorenz and Maynard (1980).
[b] No germination.
[c] Not tested.

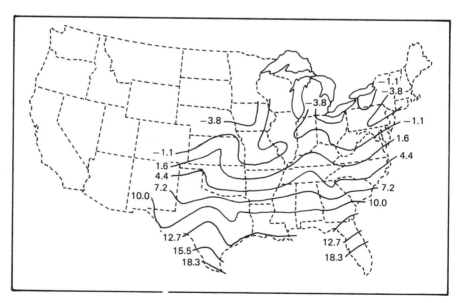

Fig. 2.7 Average soil temperature °C at the 10 mm depth in bare fields, February 12–18, 1979. *Weekly Weather and Crop Bulletin,* NOAA and U.S. Dept. of Agriculture, February 21, 1979.

2.7 SOIL HEAT FLUX AND WATER RELATIONS IN SOILS

The moisture in soil moves in response to water potential gradients. Moisture also moves as a result of temperature gradients. The water in soil is in thermal equilibrium with the soil matrix at the same depth. We know that liquid or vapor flows from areas where its kinetic energy is high to areas where it is low, that is, from warm to cold. Therefore, the temperature profile may influence the patterns of moisture distribution in soil.

The possible magnitude of such flows is given by Cary (1966) for various soils and commercial materials (Table 2.4). Vapor flow in soils in response to thermal gradients of $0.01°C$ m^{-1} are seen to vary from about 0.4 to 2.0 mm day^{-1}, the same flow rates that would require pressure heads of 0.05–2.50 m of water. Since vapor flow out of wet soils (evaporation) may range from near zero to perhaps 12 mm day depending on climatic conditions, one can see that temperature-induced vapor flow is not of small consequence.

Jackson et al. (1973) studied time–depth patterns of soil water flux in the 90 mm zone of a bare field soil 3, 7, 16, and 37 days after irrigation (Fig. 2.8). The variation in soil water flux at the 10 and 90 mm depths is also shown in this figure. Downward flux of water was observed below 10–30 mm during several hours between sunrise and early afternoon on all days. At first, only a single period of downward flux was observed, but as the soil dried as many as four periods occurred during a day. Of course, as time progressed the surface dried and more and more of the water moving through the soil originated at the deeper level.

Jackson et al. (1973) explain the bidirectional drying by noting that soil water flux in the vapor phase is proportional to vapor pressure gradients in the soil. Vapor pressure gradients are controlled, in turn, by temperature and the water content. During the night, both temperature and water content cause water to flow toward the surface, but after sunrise the temperature gradient reverses causing water to move downward. Temperature gradients in the upper layers are directed downward during many hours of the day and upward during the late afternoon and night, so that direction of thermally induced water flow will be reversed a number of times during 24 h. On an annual basis, however, the transport of water by temperature-induced flow must be considerable. Raudkivi and Van U'u (1976) found, by warming the ends of soil columns, that redistribution of water is negligible when soils are relatively dry but relatively important with soils wet to about 20% by volume.

Gardner and Hanks (1966) have shown that the measurement of soil heat flux at various depths can give a good indication of the location of the zone in which evaporation is taking place. This is so since the amount of heat required to raise or lower the temperature of soil and its moisture is very much less per unit volume than that required for evaporation. In Fig. 2.9 the movement of the evaporation zone into soil becomes evident with the diurnal fluctuation of heat applied to the surface. Gardner and Hanks interpret their results in this way: where heat flux is the same at 10 as at 20

Table 2.4 Comparisons of moisture flow due to thermal gradients and due to head gradients in various porous materials[a]

Type of porous material	Moisture content or suction (MPa)	Thermal flow [(mm H$_2$O day^{-1})/ °C (m × 10^{-2})$^{-1}$]	m H$_2$O head equivalent to a gradient of 1°C	Mean temperature (°C)	Type of solution flowing	Phase of flow
Sintered glass, 1-μm mean pore diameter	Saturated	0	0	25	Distilled water	Liquid
DuPont 600 cellophane	Saturated	0.1[b]	—	40	Double distilled	Liquid
Cellulose acetate, 27 nm mean pore diameter	Saturated	—	7 × 10^{-2}	20	Distilled water	Liquid
Millipore filter, 10 nm mean pore diameter	Saturated	—	1 × 10^{-3}	20	Distilled water	Liquid
Wyoming bentonite paste 18%	Saturated	0.1[c]	4 × 10^{-2}	35	0.1 N NaCl	Liquid
Yolo loam soil, ρβ = 1.4[d]	0.006	2.0	5.5 × 10^{-2}	35	0.01 N CaSO$_4$	Liquid + vapor
Houston black clay	0.066	0.7	—	38	Distilled water	Liquid + vapor
Houston black clay	0.140	0.8	—	37	Distilled water	Liquid + vapor

Material						Liquid	
Millville silt loam, $\rho\beta$ = 1.5	Saturated	—	—	7×10^{-4}	25	Water	Vapor + liquid
Millville silt loam	2.2	—		1.4×10^{1}	25	Water	Vapor
Millville silt loam	12.6	—		8.3×10^{2}	25	Water	20% vapor 80% liquid
Columbia loam, $\rho\beta$ = 1.2	0.07	1.8		3×10^{-2}	19	Distilled water	33% vapor 67% liquid
Columbia loam, $\rho\beta$ = 1.2	0.24	0.9		1.3×10^{-1}	8	Distilled water	55% vapor 45% liquid
Columbia loam, $\rho\beta$ = 1.2	0.24	2.0		1.4×10^{-1}	33	Distilled water	Vapor + liquid
Columbia loam, $\rho\beta$ = 1.2	0.45	1.6		2.5 m	25	Distilled water	
Stable air	100% porosity	0.4		8×10^{2}	27	Distilled water	Vapor

[a] All thermal flow is from warm to cool (after Cary, 1966). See the original paper for sources of the data.
[b] Cross-sectional area estimated as .04 m² and membrane thickness estimated as 1 mm.
[c] Results changed with aging.
[d] $\rho\beta$ = dry bulk density.

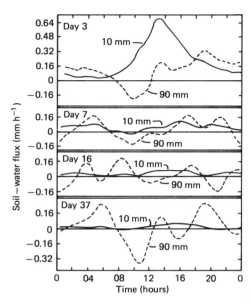

Fig. 2.8 Soil water flux at 10 and 90 mm depths for days 3, 7, 16, and 37 after irrigation (after Jackson et al., 1973). Reprinted from *Soil Science Society of America Proceedings,* Volume **37,** page 508, 1973, by permission of the Soil Science Society of America.

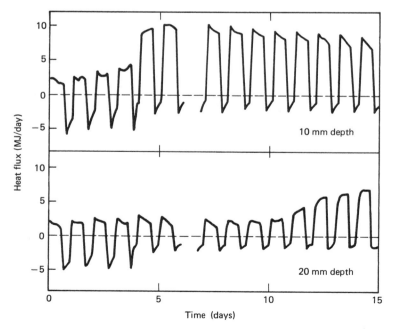

Fig. 2.9 Heat flux at several depths in soil under a diurnal variation in heat supply (after Gardner and Hanks, 1966).

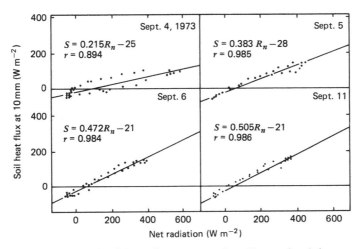

Fig. 2.10 Regressions of soil heat flux measured at 10 mm depth in a smooth bare field of Williams loam at Sidney, Montana, in September, vs. net radiation measured at 0.25 m above the field, following rains supplying a total of 91 mm of water on 1, 2, and 3 September (after Idso et al., 1975). Copyright © 1975 by D. Reidel Publishing Company.

mm, most evaporation is occurring above 10 mm. When the 10-mm curve recorded its maximum heat flux, moisture loss took place below 10 mm. When the 20-mm curve had not yet started to rise, moisture loss was taking place between 10 and 20 mm.

Idso et al. (1975) studied the relations between soil heat flux and radiation balance. Water was applied in 100-mm increments to loam soils at Phoenix, Arizona, and Sidney, Montana. The regression of soil heat flux on net radiation changed significantly as the soil dried, with the difference between them linearly related to the water content of the uppermost 0.02–0.40 m of soil. From these observations and further observations with air-dried soil, Idso et al. (1975) were able to evaluate what the net radiation − soil heat flux difference would be from wet surfaces at any time of year, knowing net radiation, air temperature, and soil water content. The relation of soil heat flux and net radiation is given in Fig. 2.10 for Sidney, Montana.

2.8 SOIL HEAT AND SOIL RESPIRATION

The sources of carbon dioxide used by plants in photosynthesis will be discussed in detail in Chapter 8. Second only to the atmosphere itself as a source of carbon dioxide is the soil in which living roots and organisms and decaying organic matter consume oxygen and respire. The rate of carbon dioxide supply to the air above is a matter of importance for sustained productivity. Respiration is a temperature-dependent function that can be de-

112 SOIL HEAT FLUX AND SOIL TEMPERATURE

scribed by an equation of the following form:

$$R_T = R_0 Q_{10}^{(T-T_0)/10} \tag{2.7}$$

where R_T and R_0 are the respiration rates at temperature T and reference temperature T_0. Q_{10} is a factor for the change of respiration rate per 10°C change in temperature.

Soil respiration R_g under a barley crop was measured by Monteith et al. (1964) in England. The Q_{10} of respiration was about 3. Mogensen (1971) measured soil respiration in an alfalfa field at Mead, Nebraska. He found the Q_{10} of respiration to be 2.84. Da Costa (1981) working with soybeans found Q_{10} values for the full crop to range typically from 1.60 to 2.36.

The concentration of CO_2 and other gaseous constituents of the soil is also a function of soil temperature. Yamaguchi et al. (1967), for example, show that CO_2 concentration decreases with decreasing temperature whereas N_2 concentration follows the opposite trend. Soil moisture content affects respiration rates as well as gaseous diffusion and, hence, interacts with temperature in determining gaseous composition of the soil. The subject of soil respiration, as it is affected by soil temperature and other factors, is discussed in greater detail in Chapter 8.

2.9 INSTRUMENTATION

2.9.1 Soil Heat Flux

The flux of heat into and out of soils can be measured directly by means of "thermal flux plates," or it can be calculated from

$$S = K \, dT/dz \tag{2.2}$$

if the thermal conductivity and temperature gradient are known.

The measurement of K is complex since it is a constantly changing function of soil aeration, moisture content, and compaction. The direct measurement of soil temperature is not difficult. Details of appropriate sensors for temperature measurement are given in Chapter 3 in connection with air temperature instrumentation.

Heat flux plates measure the soil heat flux directly. These sensors are identical in principle with the sensing element of net radiometers. A differential thermopile is connected between the top and bottom portions of the sensing plate, and the temperature drop is measured across a known thermal impedance. So that the meter itself will not impede heat flow, it should have an emissivity of the same order of magnitude as that of the material in which it is placed. Soil emissivity is nearly that of a blackbody, so that materials such as anodized aluminum are appropriate. The heat flux plate should also have a low heat capacity so as to reach steady state quickly and to respond rapidly.

INSTRUMENTATION 113

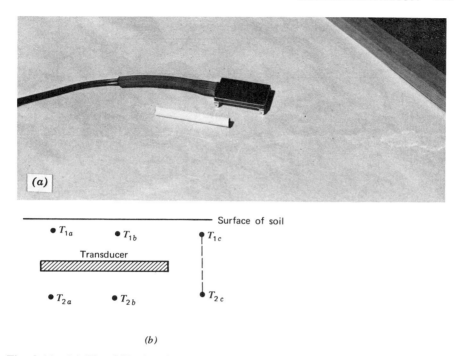

Fig. 2.11 (*a*) The CSIRO soil heat flux plate, courtesy of Middleton Instruments, Melbourne, Australia. (*b*) placement of temperature sensors and heat flux plate in soil for testing thermal compatability of the plate.

Soil heat flux plates may be either disk or wafer shaped. Designers try to keep the size of the plates small so as to minimize the distortion in the soil temperature field and to minimize the vapor trapping that these impermeable objects must cause when emplaced in soil. A typical model of a flux plate is shown in Fig. 2.11*a*. Figure 2.11*b* illustrates a good test of the thermal compatibility of a soil flux plate. If the soil temperatures within level 1 and level 2 and if the gradients of temperature in transects *a, b, c* are in reasonably good agreement, it may be assumed that the flux plate is causing no major restriction to the flow and distribution of heat.

Detailed discussions of the design and calibration of flux plates are given by Fuchs and Tanner (1968) and by Mogensen (1970). A new apparatus for calibrating soil heat flux plates is described by Biscoe et al. (1977).

2.9.2 Soil Temperature

The soil temperature may be measured directly with any appropriate type of thermometer. Thermocouples and thermistor probes are most frequently used for placement in the soil. These require electronic meters for continuous

114 SOIL HEAT FLUX AND SOIL TEMPERATURE

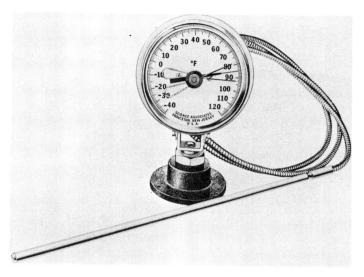

Fig. 2.12 Soil maximum–minimum thermometer, courtesy of Science Associates, Princeton, New Jersey.

or intermittent readout and are no different in principle from thermocouples or thermistors used in many other scientific and commercial applications. However, in soil these sensors should be encased in materials that are good electrical insulators (e.g., Teflon, epoxy resin, etc.) to avoid the development of "ground loops" that introduce spurious electrical signals into the measurement circuits. Thermocouples and thermistors are discussed in more detail in Chapter 3.

One useful, nonelectrical device is the Palmer soil maximum and minimum thermometer, shown in Fig. 2.12. A mercury activated temperature-sensitive bulb with a direct-drive Bourdon spring registers the current temperature on a dial. The active pointer pushes markers to indicate the highest and lowest temperatures that have occurred since the previous reading. Markers are reset after each reading (usually once a day). These instruments are quite rugged and reliable, with the dials usually covered to protect them against moisture seepage. The probes should be inserted horizontally in the soil. Mercury in glass thermometers, if suitably protected against radiation and suitably installed to avoid moisture seepage, can also be used to measure soil temperature.

A way of determining depth of freezing without the use of thermometers was proposed by Caprio (1977). A soil freeze depth tube consisting of a length of 20-mm PVC pipe is sealed at the bottom. Within the PVC tube is an acrylic tube filled with coarse sand. The sand is saturated with water to which the dye fluorescein has been added. Upon freezing the dye changes color from green to brown. The depth and thickness of the frozen layer can be read with a scale glued to the side of the tube when the tube is pulled out for examination.

REFERENCES

Al-Nakshabandi, G., and H. Kohnke. 1965. Thermal conductivity and diffusivity of soils as related to moisture tension and other physical properties. *Agric. Meteorol.* **2**:271–279.

Baker, D. G. 1965. Factors affecting soil temperature. *Minn. Farm Home Sci.* **22**:11–13.

Biscoe, P. V., R. A. Saffell, and P. D. Smith. 1977. An apparatus for calibrating soil heat flux plates. *Agric. Meteorol.* **18**:49–54.

Caprio, J. 1977. Soil freeze depth tubes. Abstracts, National Conference on Agricultural and Forest Meteorology. Am. Meteorol. Soc., Tucson, Arizona, October, 1977.

Carson, J. E., and H. Moses. 1963. The annual and diurnal heat-exchange cycles in upper layers of soil. *J. Appl. Meteorol.* **2**:397–406.

Cary, J. W. 1966. Soil moisture transport due to thermal gradients: Practical aspects. *Soil Sci. Soc. Am. Proc.* **30**:428–433.

Da Costa, J. M. N. 1981. Dark respiration of soybeans. Chapter VI in Progress Report to National Science Foundation on Grant ATM-7901017. Center for Agricultural Meteorology and Climatology, University of Nebraska-Lincoln.

Fuchs, M., and C. B. Tanner. 1968. Calibration and field test of soil heat flux plates. *Soil Sci. Soc. Am. Proc.* **32**:326–328.

Gardner, H. R., and R. J. Hanks. 1966. Evaluation of the evaporation zone in soil by measurement of heat flux. *Soil Sci. Soc. Am. Proc.* **30**:425–428.

Hadas, A. 1973. Evaluation of the block method for determining the thermal properties of the top soil. *Agric. Meteorol.* **11**:269–276.

Hanks, R. J., and H. S. Jacobs. 1971. Comparison of the calorimetric and flux meter measurements of soil heat flow. *Soil Sci. Soc. Am. Proc.* **35**:671–674.

Idso, S. B., J. K. Aase, and R. D. Jackson. 1975. Net radiation-soil heat flux relations as influenced by soil water content variations. *Boundary-Layer Meteorol.* **9**:113–122.

Jackson, R. D., B. A. Kimball, R. J. Reginato, and F. S. Nakayama. 1973. Diurnal soil-water evaporation: Time-depth-flux patterns. *Soil Sci. Soc. Am. Proc.* **37**:505–509.

Kimball, B. A., and R. D. Jackson. 1975. Soil heat flux determination: A null-alignment method. *Agric. Meteorol.* **15**:1–9.

Kimball, B. A., R. D. Jackson, F. S. Nakayama, S. B. Idso, and R. J. Reginato. 1976. Soil-heat flux determination: Temperature gradient method with computed thermal conductivities. *Soil Sci. Soc. Am. Proc.* **40**:25–28.

Kristensen, K. J. 1959. Temperature and heat balance of soil. *OIKOS (Copenhagen)* **10**:103–120.

Lorenz, O. A., and D. N. Maynard. 1980. Knott's Handbook for Vegetable Growers, 2nd edition. Wiley-Interscience, New York.

Moench, A. F., and D. D. Evans. 1970. Thermal conductivity and diffusivity of soil using a cylindrical heat source. *Soil Sci. Soc. Am. Proc.* **34**:377–381.

Mogensen, V. O. 1970. The calibration factor of heat flux meters in relation to the thermal conductivity of the surrounding medium. *Agric. Meteorol.* **7**:401–410.

Mogensen, V. O. 1971. An improved method for measuring soil and root respiration. Progress Report to NOAA-Environmental Data Service Grant No. E-293-68-(G), Univ. of Nebraska, Dept. of Horticulture.

Monteith, J. L., G. Szeicz, and K. Yabuki. 1964. Crop photosynthesis and the flux of carbon dioxide below the canopy. *J. Appl. Ecol.* **1**:321–337.

Ouellet, C. E. 1973. Macroclimatic model for estimating monthly soil temperatures under short grass cover in Canada. *Can. J. Soil Sci.* **53**:263–274.

Parton, W. J., and J. A. Logan. 1981. A model for diurnal variation in soil and air temperature. *Agric. Meteorol.* **23**:205–216.

Raudkivi, A. J., and N. Van U'u. 1976. Soil moisture movement by temperature gradient. *J. Geotechnical Eng. Div. (ASCE)* **102**:1225–1244.

Saltzman, B., and J. A. Pollack. 1977. Sensitivity of the diurnal surface temperature range to changes in physical parameters. *J. Appl. Meteorol.* **16**:614–619.

Sutton, O. G. 1953. *Micrometeorology.* McGraw-Hill, New York. 333 pp.

van Wijk, W. R. 1964. Two new methods for the determination of the thermal properties of soil near the surface. *Physica* **30**:387–388.

van Wijk, W. R. 1965. Soil microclimate, its creation, observation and modification. Chap. 3 in P. E. Waggoner, ed. *Meteorol. Monogr.* **6**:59–73. *Am. Metorol. Soc.* Boston.

Yamaguchi, M., W. J. Flocker, and F. D. Howard. 1967. Soil atmosphere as influenced by temperature and moisture. *Soil Sci. Soc. Am. Proc.* **31**:164–167.

Yakuwa, R. 1946. Über die Bodentemperaturen in den verschiedenen Bodenarten in Hokkaido (On soil temperature in the different soil types of Hokkaido). *Geophys. Mag. (Tokyo)* **14**:1–12.

CHAPTER 3–AIR TEMPERATURE AND SENSIBLE HEAT TRANSFER

3.1 INTRODUCTION

In Chapter 2 we discussed one of the mechanisms by which energy at the earth's surface is consumed, namely, by the flux of heat into the soil, a process of conduction.

Very large quantities of energy are also transferred between the surface and the air by the process of convection. **Convection** is defined in the *Glossary of Meteorology* (Huschke, 1959) as "mass motions of fluid (air, in this case) resulting in transport and mixing of the properties of that fluid." We speak of the flow of heat between surface and air as "sensible" heat flux because it is that transfer that determines air temperature, a property of air that we sense.

There are two types of convection usually distinguished, both of which are important in micrometeorology. They are defined as follows:

Forced convection is motion induced by mechanical forces, flow caused by friction of the fluid, or motion caused by applied external forces such as the wind.

Free convection is the motion caused only by density differences within the fluid. Still air in contact with a warm surface, for example, will be heated and will rise because of its increased bouyancy.

3.2 ADIABATIC PROCESS, POTENTIAL TEMPERATURE

A process (change of state) that takes place without any addition or withdrawal of heat is called an *adiabatic process*. If the pressure of a volume of air is changed adiabatically (from P_1 to P_2), the resulting change in temperature (from T_1 to T_2) can be given by

$$\frac{T_2}{T_1} = \left(\frac{P_2}{P_1}\right)^{(\gamma-1)/\gamma} \tag{3.1}$$

118 AIR TEMPERATURE AND SENSIBLE HEAT TRANSFER

where T_1 and T_2 are absolute temperatures (°K); $\gamma = C_p/C_v$, where C_p and C_v are the specific heats of air at constant pressure and constant volume, respectively. Equation 3.1 can be derived by using the first law of thermodynamics and the ideal gas law (see Sutton, 1953, for details).

For example, air at a pressure of 100 kPa and a temperature of 27°C (300°K) will drop in temperature to 281°K (8°C) if lifted adiabatically to a level where the pressure is 80 kPa. Therefore, if a parcel of air is moved up adiabatically (receiving no heat from or losing no heat to the surrounding air), it will cool in accordance with (3.1). This decrease in temperature with height is called the *adiabatic lapse rate*. The adiabatic lapse rate (Γ) for dry air is a very important quantity in meteorology. Its value is about 1°C/100 m. The lapse rate in saturated air (the wet lapse rate) is not constant and is smaller than the dry adiabatic lapse rate for reasons given in detail in Section 3.4.

Next we introduce another concept useful in meteorology, the *potential temperature*. This is the temperature a parcel of air would attain if brought adiabatically from its actual pressure (P) to a standard pressure (generally taken as 100 kPa). Potential temperature θ is related to the actual air temperature T by

$$\theta = T\left(\frac{100}{P}\right)^{(\gamma-1)/\gamma} \simeq T\left(\frac{100}{P}\right)^{0.288} \qquad (3.2)$$

3.3 THE CONCEPT OF THERMAL STABILITY

The definitions of adiabatic process and adiabatic lapse rate in the previous section lead us to the concept of thermal stability in the atmosphere. A parcel of air moved adiabatically from one level to another will always be at the same temperature as the environment and will therefore have the same density as the surrounding air. Consider the situation, shown in Fig. 3.1a, of the actual temperature (solid lines) decreasing at a rate greater than the adiabatic lapse rate (the dotted line indicates temperature change at the dry adiabatic rate). Meteorologists call this kind of variation of temperature with height a *lapse* temperature profile. In this case, a parcel of air moved up adiabatically by an infinitesimal amount (from point A to point B) will be warmer and lighter than the surrounding air (B') and will continue to rise. Such a condition of the atmosphere is termed *unstable*.

Next let us consider a case when the actual temperature increases (Fig. 3.1b or when it decreases at a rate less than the adiabatic lapse rate. This kind of temperature profile is termed an *inversion*. In such a case a parcel of air moved up adiabatically by an infinitesimal amount will be colder and denser than the surroundings. The parcel will have a tendency to return to its original position. This condition of the atmosphere is called *stable*.

To summarize then, the lapse and inversion profiles of actual temperature

THE WET ADIABATIC LAPSE RATE 119

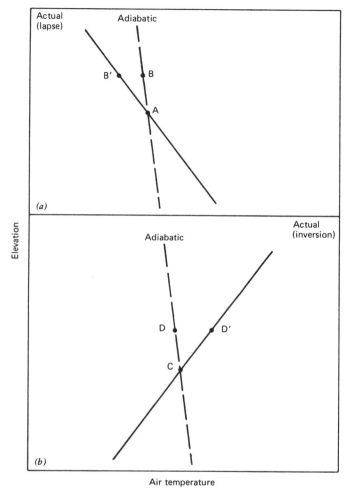

Fig. 3.1 A schematic representation of actual vs. adiabatic temperature profiles illustrating atmospheric thermal stability conditions.

shown in Fig. 3.1 correspond, respectively, to unstable and stable conditions of the atmosphere. Following similar reasoning, an adiabatic temperature profile (shown by dotted lines in Fig. 3.1) corresponds to a *neutral* stability condition of the atmosphere.

3.4 THE WET ADIABATIC LAPSE RATE

When air rises, it cools and becomes saturated with respect to water vapor. Then condensation occurs, leading to cloud formation and to precipitation.

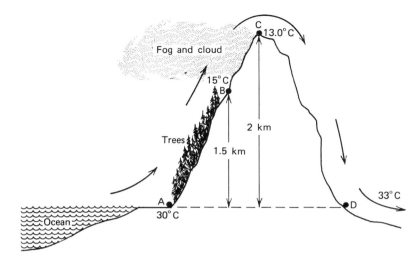

Fig. 3.2 Influence of a mountain on air temperature on the windward and leeward sides.

When air is saturated, the saturated lapse rate differs considerably from the dry adiabatic rate, except when temperatures are very low.

The condensation of water vapor liberates considerable heat (about 2.44 MJ kg^{-1}) into the atmosphere. This release of energy retards the adiabatic cooling process. The moist (wet) adiabatic lapse rate is also much more dependent on temperature and pressure of the surroundings than is the dry adiabatic lapse rate. The following are some approximate values: at 30°C the moist adiabatic lapse rate is 0.36°C/100 m at 100 kPa; at 0°C the rate is 0.69°C/100 m at 100 kPa and 0.54°C/100 m at 50 kPa.

The difference in dry and wet lapse rates accounts for some very interesting orographic phenomena. Consider, for example, the effect of a mountain on air temperature on the windward and the lee sides as shown in Fig. 3.2. At point A, moist air at 30°C is blowing off a water body. As the air rises because of the topography, it cools at the dry adiabatic lapse rate of about 1°C/100 m. At point B, the air is chilled to a temperature at which the water vapor condenses. The air is now saturated and it has been cooled to 15°C. As air continues to rise, it cools at the wet adiabatic lapse rate of, say, 0.4°C/100 m, so that at point C the temperature is 13°C. On its descent to D, the air warms and is no longer saturated. The dry lapse rate applies, and in this example, the descending air would achieve a temperature of 33°C. If saturation is reached, air passing over a mountain will be dried and warmed in the process. The Chinook (snow-eating) winds that blow out of the Rocky Mountains and the Foehn winds of alpine Europe are extreme manifestations of this process.

3.5 TEMPERATURE PROFILES ABOVE NATURAL SURFACES

During the night, the temperature of ground and crop surfaces falls rapidly because of radiational cooling so that the surface is the coldest location in the profile. Air in contact with the surface loses energy to the surface, becoming chilled and heavy in the process. Therefore, a temperature inversion develops and the lower atmosphere is thermally stable. The temperature measurements shown in Fig. 3.3 were made above land planted with grass. The profiles shown are typical of conditions in late July. The existence of the nocturnal temperature inversion is seen in this figure (earliest and latest profiles). Inversion layers may be local and reach to only a few meters during the night, or they may be very large scale and dominate the weather of vast regions for considerable periods of time.

Lapse conditions (temperature decreasing with increasing elevation) and a thermally unstable atmosphere are normal during the daytime hours. This is seen clearly in Fig. 3.3 during the morning hours.

Temperature inversions can also develop during daytime when rapid evaporative cooling makes the surface of the soil or the crop canopy cold relative to the air passing over the field (see Fig. 3.4). During most of the daytime hours after 1000 hours in Fig. 3.4, temperature profiles were inverted. This occurred because warm, dry air was passing over a rapidly transpiring, relatively cool alfalfa crop supplying energy for transpiration. This effect of consumption rather than generation of sensible heat is generally referred to as evidence of *sensible heat advection* and is discussed in greater detail in Chapter 7.

It is important to realize that the atmosphere is commonly made up of a number of layers of different thermal stability. During the daytime the lower unstable layer is usually capped by an inversion layer. The inversion cap

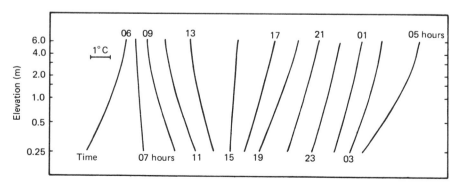

Fig. 3.3 Hourly air temperature profiles above grass (0.12 m tall), late July at Davis, California (adapted from Brooks et al., 1963).

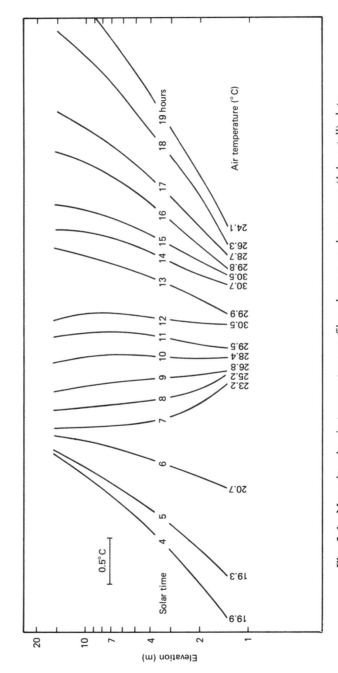

Fig. 3.4 Mean hourly air temperature profiles above a soybean crop (1.1 m tall), late spring to midsummer at Mead, Nebraska.

impedes the vertical motion and the heat and water vapor transport is limited to the lower layer (Oke, 1978).

3.6 SENSIBLE HEAT TRANSFER IN THE ATMOSPHERIC SURFACE LAYER

Objects that absorb radiation or to which energy is supplied by other means so that they become warmer than their surroundings may dispose of some of that energy by convection. Normally during the daytime heat will be transferred from the warm ground or crop surface to the cooler air above. At night, when the air is warm and the surface is cool, the converse situation prevails and heat will be transferred to the surface.

Air flow over a rigid surface is discussed in Chapter 4. Except for a very thin layer of air immediately above the surface (called the laminar sublayer in which transfer occurs by conduction or molecular diffusion), the transfer of sensible heat is largely controlled by turbulence. The vertical flux of sensible heat H can be estimated by

$$H = \rho_a C_p K_h \frac{\partial \theta}{\partial z} \quad (3.3)$$

where ρ_a is the air density, C_p is the specific heat of air at constant pressure, K_h is the turbulent exchange coefficient, and $\partial \theta/\partial z$ is the vertical gradient of potential temperature. In the first 2–3 m above the ground, $\partial \theta/\partial z$ can be approximated by $\partial T/\partial z$, the vertical gradient of actual temperature. It is apparent from (3.3) that the sign of H or the direction of sensible heat transfer is determined by the sign of $\partial \theta/\partial z$. During lapse conditions, $\partial \theta/\partial z$ is negative and, therefore, H is negative.* Conversely, during inversion conditions $\partial \theta/\partial z$ and H are positive.

Equation 3.3 and other turbulent flux equations, as well as the techniques for estimating the turbulent exchange coefficients, are discussed in detail in Chapter 4.

3.7 RESISTANCE APPROACH FOR ESTIMATING SENSIBLE HEAT FLUX

3.7.1 Concept

Monteith (1963) showed that it is possible to simplify the computation of sensible heat flux from natural surfaces by means of a "resistance" approach. We may conceive of the flux of sensible heat as a process analogous to the flow of electrical current. In an analogy to Ohm's law, we can write

---

* The sign convention employed in this work is such that all fluxes toward the crop or ground surface are positive and all away from the surface are negative.

124 AIR TEMPERATURE AND SENSIBLE HEAT TRANSFER

$$\text{heat flux} = \frac{\text{temperature gradient (driving force)}}{\text{aerial resistance to the flow of sensible heat}}$$

$$H = \frac{\rho_a C_p (T_a - T_s)}{r_a} \tag{3.4}$$

where T_s and T_a are the surface and air temperatures, respectively, and r_a is the aerial (aerodynamic or boundary layer) resistance to the flow of sensible heat.

The effect of surface to air temperature difference $(T_a - T_s)$ on sensible heat flux is evident from (3.4). Sensible heat flux will increase with decreasing aerial resistance r_a.

3.7.2 Application to Small Objects and Plant Leaves

Several empirical relationships, based on correlations using dimensionless groups (such as the Nusselt number, Reynolds number, Prandtl number, and Grashof number) derived from wind tunnel and other engineering experiments, are available in the literature. Monteith (1973) and Campbell (1977) present excellent discussions on these relationships. Other major contributions to the theory of heat dissipation by plant leaves have been made by Raschke (1960), Linacre (1964), Gates (1968), Parkhurst et al. (1968), and Pearman et al. (1972). In forced convection, the aerial (aerodynamic) resistance to heat transfer can be written as

$$r_a = \frac{D}{\alpha_h \text{Nu}} \tag{3.5}$$

where D is a characteristic dimension (usually the mean width for a leaf), α_h is the thermal diffusivity of air, and Nu is the Nusselt number given by

$$\text{Nu} = \frac{HD}{\rho_a C_p \alpha_h (T_a - T_s)} \tag{3.6}$$

Nu can be interpreted as a ratio of heat actually transferred to that that would be transferred under the circumstances of pure conduction (Huschke, 1959). For a flat plate in laminar flow, the heat transfer (from one side of the plate) can be given by

$$\text{Nu} = 0.66 \, \text{Re}^{1/2} \, \text{Pr}^{1/3} \tag{3.7}$$

where Re is the Reynolds number and Pr is the Prandtl number. Re represents the ratio of inertial forces (producing changes in velocity) to viscous forces (tending to oppose changes in velocity) in a fluid (Monteith, 1973). Re = UD/ν, where U is the wind speed and ν is the kinematic viscosity. Pr is the ratio of kinematic viscosity to the thermal diffusivity (Pr = ν/α_h). Combining these equations Campbell (1977) gives the following relationship for r_a in the case of forced convection (for air at 20°C and 100 kPa)

$$r_a = 307(D/U)^{1/2} \tag{3.8}$$

where r_a, D, and U are in s m^{-1}, m, and m s^{-1} respectively.

Similarly, in the case of free convection (for air at 20°C and 100 kPa) Campbell (1977) derives

$$r_a = 840 \left(\frac{D}{T_s - T_a}\right)^{1/4} \tag{3.9}$$

where r_a and D are in s m^{-1} and m, respectively. T_a and T_s are in °C.

It is important to note that the relationships given above were developed from wind tunnel tests generally made under laminar flow conditions. In turbulent flow in the open atmosphere, the aerodynamic resistance is smaller. For typical conditions of outdoor turbulence, the resistance is around 60–70% that predicted by (3.8) and (3.9).

As can be seen from (3.8) and (3.9), the larger the object, the greater will be r_a and consequently the heat transfer per unit area of the surface will be less effective [(3.4)]. This is primarily because the most intense turbulent mixing occurs at the leading edge. From this observation, we might conclude that small-leaved plants such as grasses (if it is assumed that all other factors are equal) are better heat exchangers than are broad-leaved plants. Also the greater the wind speed, the more effective is the heat transfer in forced convection [see (3.4) and (3.8)].

3.7.3 Application to Crop Canopies

The aerial resistance for application of (3.4) to crop canopies can be derived from measurements of wind speed profiles. Using the wind profile theory discussed in Chapter 4 in conjunction with (3.3) and (3.4) the aerial resistance that occurs between a height z above the ground and the apparent source or sink for heat can be derived as (see Monteith, 1973, for details)

$$r_a = \frac{\ln[(z-d)/z_h]}{ku_*} = \frac{\ln[(z-d)/z_0]}{ku_*} + \frac{\ln(z_0/z_h)}{ku_*} \tag{3.10}$$

where k is the von Karman constant, z_h is a parameter for heat transfer which is analogous to the roughness parameter z_0 for momentum transfer, d is the zero plane displacement (assumed to be the same for heat as for momentum transfer), and u_* is the friction velocity. u_* and z_0 are defined and discussed in Chapter 4. The apparent source or sink for heat is located at a lower level in the canopy than the apparent sink for momentum, that is, $(d + z_h) < (d + z_0)$. A detailed discussion of the relative sizes of z_0, z_h, and the relative magnitudes of resistances to momentum and sensible heat transfer can be found in Chamberlain (1966), Thom (1972), Monteith (1973), and Verma and Barfield (1979).

The ratio z_0/z_h is generally related to a nondimensional parameter B^{-1} (originally given by Owen and Thomson, 1963)

$$\frac{z_0}{z_h} = \exp(kB^{-1}) \tag{3.11}$$

B^{-1} can be estimated from the results reported by Thom (1972) and by Garratt and Hicks (1973). With the knowledge of wind speed profile r_a can be estimated by using (3.10) and (3.11). In case of $z_0 \approx z_h$, using the wind profile theory (Chapter 4), (3.10) simplifies to

$$r_a = \frac{\ln[(z - d)/z_0]}{ku_*} = \frac{[\ln(z - d)/z_0]^2}{k^2 U(z)} \tag{3.12}$$

3.8 TEMPERATURE PROFILES IN PLANT CANOPIES

Temperature profiles measured within crop canopies are quite different from those above the crop. Some examples of temperature profiles measured over the course of a day in a well-watered soybean field are shown in Fig. 3.5. Note that the slope of the profile indicates the direction of sensible heat flow at that level. Often during daytime there is a temperature maximum near the mid to upper portions of the canopy. The temperature maximum occurs near the level of maximum leaf area where most of the solar radiation is absorbed. Above this level the profile is generally lapse, as it is typically above the canopy during daytime. Below this level, there is a temperature inversion because the canopy is warmer than the soil surface underneath.

At night, temperature profiles (Fig. 3.5) in the lower levels of the canopy are close to isothermal since the canopy traps outgoing longwave radiation. Temperature profiles are inverted in the upper canopy since longwave radiation is being transmitted to the nocturnal sky.

The profiles described above help us illustrate the phenomenon of within-canopy heat transfer in very general terms. The "real" situations are complicated by several factors. As the day progresses, for example, the stomatal resistances vary and the sources and sinks (locations and strengths) of sensible and latent heat undergo considerable change.

3.9 DAILY AND ANNUAL TEMPERATURE PATTERNS

The daily pattern of air temperature describes a sine curve, with the minimum normally occurring in the early morning hours near sunrise and the maximum occurring sometime after the peak of solar and net radiation has been reached. The air temperature wave is not necessarily so regular on

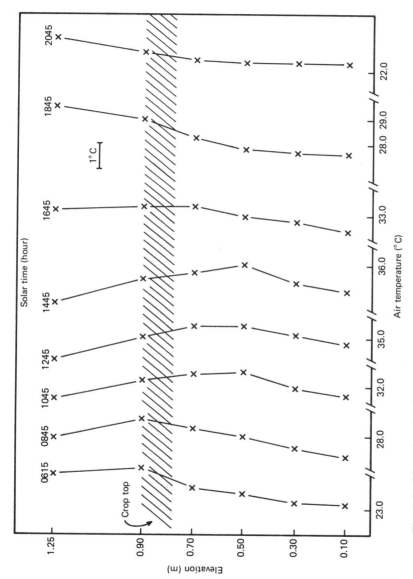

Fig. 3.5 Mean hourly air temperature profiles within and above a well-watered soybean crop, early August at Mead, Nebraska.

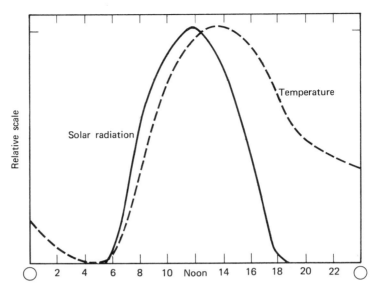

Fig. 3.6 Normalized solar radiation and air temperature above bare soil, early September at Scottsbluff, Nebraska (after Neild et al., 1967).

individual days, especially in climates that feature frequent frontal passages, irregular cloudiness, or strong advection of sensible heat generated in other regions. However, as averages of hourly air temperature are calculated over increasingly long periods, the regularity of the waves becomes more and more evident, so much so that one can with confidence fit smooth, harmonic curves to the data for prediction purposes.

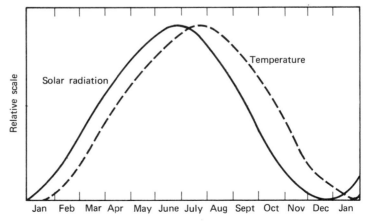

Fig. 3.7 Normalized and smoothed monthly mean solar radiation and monthly mean temperature at Lincoln, Nebraska (after Neild et al., 1967).

The lag of temperature behind radiation is illustrated in Fig. 3.6 (from Neild et al., 1967). These data for a single, totally clear day at Scottsbluff, Nebraska, are normalized from peak values of solar radiation and air temperature of 860 W m^{-2} and 30.5°C, respectively. The peak solar radiation occurred at noon, but the peak temperature did not occur until 1400 hours. Although the sun set at 1823 hours, the lowest air temperatures did not occur until shortly before sunrise, as is predicted by radiation theory.

The afternoon temperature lag is a result, primarily, of the balance between net incoming and outgoing radiation. From sunrise on, a considerable amount of radiant energy is required to heat the soil and crop, which are coldest at that time. Until these surfaces become warm relative to the air above, no net sensible heat flux to air occurs. More and more energy goes into warming the air as the surfaces become hotter. Even though the balance between net incoming and outgoing radiation is the primary determinant, other factors such as convection, conduction, advection, and evapotranspiration can sometimes be important in determining the afternoon temperature lag (Sellers, 1980).

The annual wave of air temperature follows a similar pattern with respect to the wave of solar radiation. As is shown in Fig. 3.7 (from Neild et al., 1967), the peak solar radiation occurs in June at Lincoln, Nebraska. The peak of the temperature lags that of the radiation by about 1 month at this location. Similarly, the minimum annual air temperature occurs 1 month after the minimum of solar radiation at this location.

The reasons for the annual lag are similar to those for the daily lag. During spring and early summer, a large portion of the incident solar energy flows into the soil, which has reached its lowest temperature by the end of winter. As that portion of the energy flux decreases and the soil becomes warm relative to the environment, more energy is converted to sensible heat.

3.10 INFLUENCE OF ELEVATION ON AIR TEMPERATURE PATTERNS

Figure 3.8 illustrates the normal course of temperature near the ground. Data are from a 16-m mast at Mead, Nebraska. The amplitude of the temperature wave is greatest near the ground and decreases with increasing elevation.

As can be seen from the foregoing and by reference to information in Chapter 2 on soil heat flux, the daily and annual patterns of soil and air temperature are intimately related. In fact, they constitute a sort of mirror image of one another. The highest and lowest temperatures occur nearest the surface. Heat flux (soil or sensible heat) is normally away from the surface during the day and toward the surface at night. The daily and yearly maxima and minima in temperature lag radiation increasingly with increasing distance from the surface. The daily and yearly temperature waves also become more and more damped with distance from the surface.

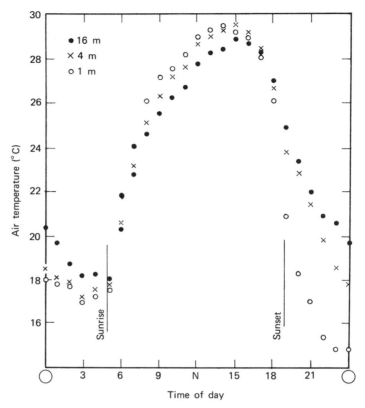

Fig. 3.8 Daily pattern of air temperature at three elevations over 0.5 m tall soybeans, early July at Mead, Nebraska.

3.11 INSTRUMENTATION FOR AIR TEMPERATURE MEASUREMENT

The measurement of air temperature is relatively simple although a number of important precautions are necessary if accuracy is to be achieved. Virtually any sensor normally used for temperature measurement can be properly adapted for air temperature measurement in micrometeorology. Such sensors include mercury or alcohol in glass thermometers, thermocouples, thermistors, platinum resistance thermometers, and temperature-sensitive diodes.

Liquid-in-glass thermometers are well known to every student. The expansion of the liquid with increasing temperature is calibrated, and a precision of about 0.5°C is achievable with most reasonably priced instruments. Thermocouples are junctions of dissimilar metals, which generate an electromotive force (emf) proportional to the temperature of their surroundings. Copper–constantan is the most convenient metal pair for measurement of air temperature in micrometeorology. In certain types of differential thermocouples constantan–manganin or constantan–evanohm are used. The latter is especially useful in contact with plant leaves since its thermal conductivity is much lower than copper. Thermocouples must be referenced against an ice bath or an electronic reference in order for the generated emf to be properly related to temperature. A wiring diagram for a single thermocouple is shown in Fig. 3.9. Usually temperatures can be directly measured with thermocouples with an accuracy of between 0.1 and 0.25°C. It is possible, however, to measure temperature differences with considerably greater accuracy when thermocouples are wired differentially. A diagram for a differential thermocouple is also shown in Fig. 3.9.

Thermocouples make stable temperature sensors. However, they produce a small signal, which requires expensive meters for amplification and readout. The signal can be increased by wiring a large number of thermocouple junctions in series. If the junctions are placed in different locations and sample different conditions, the resulting signal gives an integration of the average temperature condition. Thermocouple junctions can also be wired in parallel, in which case current is increased and the voltage measured gives the average temperature for all points sampled.

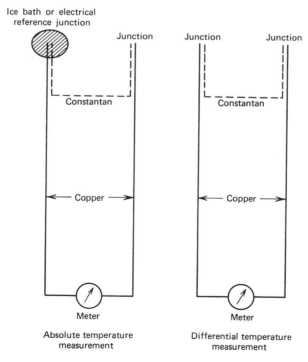

Fig. 3.9 Wiring for absolute and differential temperature measurement with thermocouples.

Thermistors are crystals whose resistance to current flow varies with ambient temperature. Because of the normally large resistances involved, inexpensive meters are adequate to measure the response due to changes in ambient temperature. With new technological developments, thermistors are becoming more popular in meteorological applications.

Whatever the sensor used, it is important for the field scientist to realize that the sensor measures only its own temperature. For that reason great care must be taken to see that the sensor is not exposed to direct solar or reflected radiation and that it does not remain in stagnant air. Similarly, temperature sensors should not be exposed to the sky at night. National Weather Service mercury-in-glass thermometers are placed in standard louvered shelters constructed of wood. There they are shaded, and air movement through the louvers ensures some degree of ventilation. Thermocouples and thermistors in the field should be placed in a radiation shield of some type. Normally, metal sheets above and below the sensor are used. These are made highly reflectant on the out-facing sides and are usually blackened on the in-facing sides to reduce internal reflection. Temperature sensors should be aspirated (forced ventilation) as well, whenever possible, so as to assure that they are measuring the temperature of the ambient air.

REFERENCES

Brooks, F. A., W. O. Pruitt, and D. R. Nielsen. 1963. Investigation of energy and mass transfers near the ground including the influences of the soil-plant-atmosphere system. Final Rpt. to U.S. Army Electronic Proving Ground, Task 3A99-27-005-08. 285 pp.

Campbell, G. S. 1977. *An Introduction to Environmental Biophysics*. Springer-Verlag, New York. 159 pp.

Chamberlain, A. C. 1966. Transport of gases to and from grass and grasslike surfaces. *Proc. R. Soc. London* **290**:236–265.

Garratt, J. R., and B. B. Hicks. 1973. Momentum, heat and water vapor transfer to and from natural and artificial surfaces. *Q. J. Roy. Meteorol. Soc.* **99**:680–687.

Gates, D. M. 1968. Energy exchange in the biosphere. *Nat. Resour. Res.* **5**:33–43. UNESCO, Paris.

Huschke, R. E., ed. 1959. *Glossary of Meteorology*. American Meteorol. Soc., Boston. 638 pp.

Linacre, E. T. 1964. Determination of the heat transfer coefficient of a leaf. *Plant Physiol.* **39**:687–690.

Monteith, J. L. 1973. *Principles of Environmental Physics*. Edward Arnold, London. 241 pp.

Neild, R. E., R. Myers, and N. J. Rosenberg. 1967. Temperature patterns and some relations to agriculture in Nebraska. *Nebr. Agric. Exp. Stn. Misc. Bull.* **16**.

Oke, T. R. 1978. *Boundary Layer Climates*. Methuen, London. 372 pp.

Owen, P. R., and W. R. Thomson. 1963. Heat transfer across rough surfaces. *J. Fluid Mech.* **15**:321–334.

Parkhurst, D. F., P. R. Duncan, D. M. Gates, and F. Kreith. 1968. Wind tunnel modelling of convection of heat between air and broad leaves of plants. *Agric. Meteorol.* **5**:33–47.

Pearman, G. I., H. L. Weaver, and C. B. Tanner. 1972. Boundary layer heat transfer coefficients under field conditions. *Agric. Meteorol.* **10**:83–92.

Raschke, K. 1960. Heat transfer between the plant and the environment. *Annu. Rev. Plant Physiol.* **11**:111–126.

Sellers, W. D. 1980. A comment on the cause of the diurnal and annual temperature cycles. *Bull. Am. Meteorol. Soc.* **11**:1410–1411.

Sutton, O. G. 1953. *Micrometeorology.* McGraw-Hill, New York. 333 pp.

Thom, A. S. 1972. Momentum, mass and heat exchange of vegetation. *Q. J. Roy. Meteor. Soc.* **98**:124–134.

Verma, S. B., and B. J. Barfield. 1979. Aerial and crop resistances affecting energy transport. *Modification of the Aerial Environment of Crops* (B. J. Barfield and J. F. Gerber, eds.), Am. Soc. Agric. Engr., St. Joseph, Michigan pp. 230–249.

CHAPTER 4 – WIND AND TURBULENT TRANSFER

4.1 AIR FLOW OVER A RIGID SURFACE: SOME DEFINITIONS AND CONCEPTS

In considering the flow of air over the earth's surface, we can envision a very thin layer of air immediately above the surface where the transfer processes are controlled primarily by molecular diffusion. This layer is called the **laminar sublayer**. The laminar sublayer may be only a few millimeters thick and may, sometimes, be even thinner, especially under windy conditions. Above the laminar sublayer is the **turbulent surface layer** (or simply the **surface layer**) which extends up to about 50–100 m and is dominated by strong mixing or eddying motion. The wind structure in this layer is primarily determined by the nature of the underlying surface and the vertical gradient of air temperature. The effect of the earth's rotation, the Coriolis force, is small and may be neglected as the frictional effects of the surface dominate. The **planetary boundary layer**, which envelops the surface layer and extends to about 1 km above the surface, is a zone of transition from the disturbed flow near the surface to the frictionless or smooth flow of the free atmosphere (Sutton, 1953).

In the surface layer, the layer of greatest interest in micrometeorology, the air motion is highly irregular and is characterized by fluctuations, vortices, or eddies. Under lapse conditions, for example, small and relatively rapid and irregular fluctuations in air motion appear to be superimposed on larger and much slower irregular fluctuations. The small fluctuations associated with high frequencies are primarily due to **mechanical turbulence** generated by the frictional effects of the surface. The larger fluctuations associated with low frequencies are the result of **thermal turbulence** arising from the effects of buoyancy.

It is important to note that, even though the air flow in the surface layer is not regular, the overall behavior of average properties such as wind speed, air temperature, humidity, and carbon dioxide concentration can be examined systematically on a statistical basis. For this purpose averaging periods of about 30 min to 1 h are considered adequate.

4.2 WIND SPEED PROFILE AND MOMENTUM EXCHANGE

Knowledge of the shape of the wind profile (variation of wind speed with height) is necessary for, at least, two reasons. From the profile description, it is possible to estimate the effectiveness of vertical exchange processes. With the knowledge of wind speed at a fixed or reference level, it is also possible to estimate wind speed at other levels for various applications.

When a fluid flows along a rigid surface, fluid particles close to the surface are slowed down because of the drag exerted on the flow by the underlying surface. The drag or flow retardation forces cause marked variation in mean horizontal wind speed with height (Fig. 4.1). Momentum is defined as the product of mass and velocity. Employing an analogy to heat transfer, which is directed from a warmer to a cooler location, an examination of Fig. 4.1. leads to the conclusion that horizontal momentum (mass × horizontal wind speed) is transferred from higher to lower elevations or from airflow to the underlying surface.

The total drag force per unit of surface area is called the **shearing stress** τ. The dimensions of shearing stress are the same as those of momentum per unit area per unit time (i.e., momentum flux). Thom (1975) suggests that the wind drag on a surface is simply a manifestation of the continuous downward transport of horizontal momentum from the airflow to the surface. The magnitude of the momentum flux is indicative of the effectiveness of turbulence in exchanging water vapor, sensible heat, carbon dioxide, and other entities between the surface and the atmosphere.

Typical shapes of mean wind profiles are given in Figs. 4.1 and 4.2. Under conditions of neutral atmospheric stability, the mean wind speed profile over an open, level and relatively smooth site can be described as a logarithmic function of elevation (Fig. 4.1):

$$U(z) = \frac{u_*}{k} \ln \frac{z}{z_0} \tag{4.1}$$

where $U(z)$ is the mean wind speed at height z, k is von Karman's constant (with a value of about 0.4), u_* is the friction velocity, and z_0 is the roughness parameter.

A surface is said to be "rough" if it is corrugated or covered with protuberances, usually referred to as **roughness elements**. For mean wind speed profiles over rough surfaces,* such as crop canopies, a zero plane displacement d is introduced (see Fig. 4.2) and (4.1) transforms into

$$U(z) = \frac{u_*}{k} \ln \frac{z - d}{z_0} \tag{4.2}$$

---

* Some recent observations (e.g., Garratt, 1980) over very rough surfaces, such as forests, indicate that the logarithmic law may not be completely applicable, particularly very close to the top of the roughness elements.

136 WIND AND TURBULENT TRANSFER

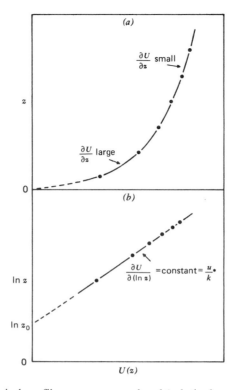

Fig. 4.1 Typical wind profile over an open, level (relatively smooth) site: (*a*) plotted linearly against height z; (*b*) plotted against the logarithm of z. From A. S. Thom, 1975, Momentum, mass and heat exchange of plant communities, In *Vegetation and the Atmosphere,* Vol. 1, *Principles,* J. L. Monteith (ed.). Copyright by Academic Press, Inc. (London), Ltd.

Equations 4.1 or 4.2 are often referred to as the logarithmic law of wind speed profiles.

The **friction velocity** u_* is given by

$$u_* = \left(\frac{\tau}{\rho_a}\right)^{1/2} \quad (4.3)$$

where τ is the shearing stress and ρ_a is the air density. u_* represents a characteristic velocity of the flow and relates to the effectiveness of turbulent exchange over the surface.

The **roughness length** or **roughness parameter** z_0 is a measure of the aerodynamic roughness of the surface over which the wind speed profile is measured. z_0 is determined by extrapolating measured $U(z)$ and $\ln z$ to the point where $U = 0$ (see Fig. 4.1). In the case of crops and other rough surfaces $\ln z$ is replaced by $\ln(z - d)$ as shown in Fig. 4.2.

WIND SPEED PROFILE AND MOMENTUM EXCHANGE

The roughness parameter for crops is about an order of magnitude smaller than the crop height h. Szeicz et al. (1969) summarized a number of studies and empirically related z_0 to crop height h by

$$\log_{10} z_0 = 0.997 \log_{10} h - 0.883 \qquad (4.4)$$

where z_0 and h are in meters.

The **zero plane displacement** d can be considered to indicate the mean level at which momentum is absorbed by the individual elements of the plant community, that is, the level of action of the bulk aerodynamic drag on the community (Thom, 1971, 1975). In general, d/h ranges between 0.5 and 0.8. Figure 4.2b shows a graphical procedure used to determine d and z_0 from measured wind speed profiles under nearly neutral conditions. By trial and error, one finds a value of d such that a plot of U (on the linear scale) versus

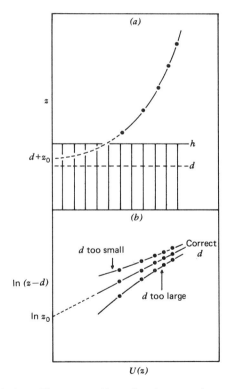

Fig. 4.2 Typical wind profile over uniform level vegetation of height h: (a) plotted linearly against z; (b) plotted against the logarithm of distance above the zero plane displacement level. From A. S. Thom, 1975, Momentum, mass and heat exchange of plant communities, In *Vegetation and the Atmosphere*, Vol. 1, *Principles*, J. L. Monteith (ed.). Copyright by Academic Press, Inc. (London), Ltd.

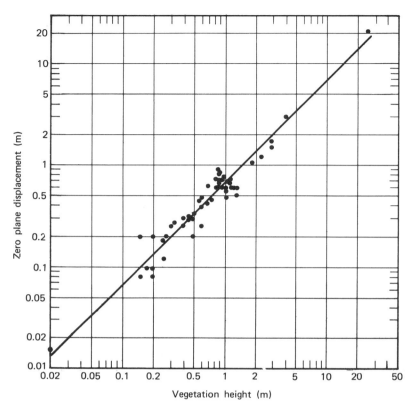

Fig. 4.3 Relationship between zero plane displacement d and vegetation height h for different forms of vegetation and bare ridged soil (after Stanhill, 1969).

$z - d$ (on the log scale) becomes a straight line. The intercept on the $z - d$ axis gives z_0, and the slope of this straight line is u_*/k.

Stanhill (1969) fitted an expression giving zero plane displacement d as a function of crop height h for a wide range of crops (Fig. 4.3):

$$\log_{10} d = 0.979 \log_{10} h - 0.154 \qquad (4.5)$$

where h and d are in meters.

The ratios d/h and z_0/h depend on the spacing of roughness elements and on the ratio of the accumulated area of each element to a unit area of the underlying surface (Monteith, 1973). The problem of accurately estimating z_0 and d is increased by the fact that crops, tall or short, adjust to the mechanical force of the wind. Sometimes bending occurs, as in small grains. Some crops become "streamlined" by the force of the wind (see a review by Makkink and van Heemst, 1970, for further details).

From the above discussion it is clear that, with knowledge of z_0 and d, the complete wind profile above the canopy can be constructed from the value of U at a fixed or reference level

$$\frac{U_2}{U_1} = \frac{\ln(z_2 - d) - \ln z_0}{\ln(z_1 - d) - \ln z_0} \qquad (4.6)$$

where U_1 and U_2 are the mean wind speeds at elevation z_1 and z_2, respectively (here z_1 can be considered the reference level). It is important to note that validity of the logarithmic wind profile equations (4.1 or 4.2) is subject to two major assumptions: (1) the existence of neutral atmospheric stability and (2) the availability of adequate "fetch" or upwind distance of uniform crop cover or roughness. Stability conditions and fetch requirements are discussed in the following sections. Also note that to obtain adequate mean wind speed values for use in (4.1), (4.2), and (4.6) averaging should be done over a period of at least 15 min and preferably over 30–60 min.

4.3 INTERNAL BOUNDARY LAYER AND FETCH REQUIREMENTS

Each field or surface feature of varying roughness or height affects the airstream passing over it. In moving downwind after a change in surface roughness is encountered at the "leading edge", air begins to adjust to the new surface boundary conditions (see Fig. 4.4). The layer of air downwind of the leading edge, affected by the new underlying surface, is called the **internal boundary layer**.* Thickness δ of the internal boundary layer increases with fetch or distance downwind of the leading edge. Wind tunnel experiments and other micrometeorological studies suggest that only the lowest 10% of the internal boundary layer is "fully adjusted", that is, in complete equilib-

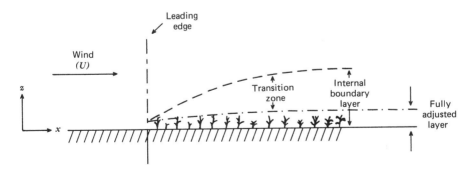

Fig. 4.4 The development of an internal boundary layer as air flows from a smooth to a rougher vegetation surface (adapted from Oke, 1978).

---

* In the discussion presented here, we have concentrated on the leading edge effect on wind flow resulting from a change in surface roughness. Similar concepts can be applied to leading edge effects resulting from changes in surface temperature or humidity.

rium with the new boundary conditions. The thickness of this fully adjusted layer δ_1, measured above the zero plane displacement, can be approximated according to Munro and Oke (1975) by

$$\delta_1(x) = 0.1 x^{4/5} z_0^{1/5} \qquad (4.7)$$

where x is the distance downwind from the leading edge and z_0 is the roughness parameter of the new underlying surface.

The air flow is fully equilibrated with the new surface to the depth $\delta_1(x)$. Therefore, a logarithmic wind profile characteristic of the underlying surface should develop to that depth. The flux of momentum should be independent of height in the fully adjusted layer. One may derive, using (4.7), the height to fetch ratio that determines the limit below which sensors must be placed to properly represent conditions in the fully adjusted layer. Appropriate values of surface roughness z_0 must be applied, however. Equation 4.7 yields a height to fetch ratio of about 1:50 for agricultural crops. For safety sake, a height to fetch ratio of 1:100 is usually considered adequate for studies made over agricultural crop surfaces. This means that in a field with uniform crop cover and upwind fetch of 200 m a layer about 2 m deep will develop in which the logarithmic wind profile relationships prevail. In this layer flux is almost independent of height.

4.4 ATMOSPHERIC STABILITY

The concept of atmospheric thermal stability was inroduced in Chapter 3. Turbulence is enhanced by buoyancy forces under unstable conditions and is suppressed under stable conditions. The effects of thermal stability on the shape of the wind speed profile and on the turbulent exchange rate (discussed in the next section) are generally expressed in terms of two nondimensional parameters: the **Richardson number** (Ri) and the **Monin–Obukhov parameter** z/L. The Richardson number (often called the gradient Richardson number) is given by

$$\mathrm{Ri} = \frac{g(\partial\theta/\partial z)}{T(\partial U/\partial z)^2} \qquad (4.8)$$

where g is the acceleration due to gravity, $\partial\theta/\partial z$ and $\partial U/\partial z$ are the vertical gradients of mean potential temperature and mean horizontal wind speed, and T is the mean absolute temperature (degrees Kelvin). Ri can be interpreted as a parameter describing the relative importance of buoyancy and mechanical forces, that is, the relative importance of free versus forced convection. The sign of Ri is determined by the gradient of potential temperature which is negative in lapse (or unstable), and positive under inversion (or stable) conditions. Ri approaching zero implies near neutral conditions. For application in the first 1–2 m above ground, the Richardson number may be calculated with $\partial T/\partial z$ substituted for $\partial\theta/\partial z$ in (4.8).

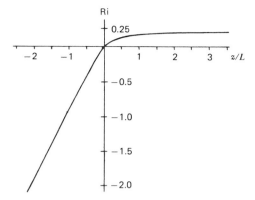

Fig. 4.5 Relationship between Richardson number Ri and the Monin–Obukhov stability parameter z/L (adapted from Businger et al., 1971).

The Monin–Obukhov stability parameter (z/L) given by

$$z/L = \frac{kzgH}{\rho_a C_p T u_*^3} \tag{4.9}$$

is derived as the ratio of buoyant production to mechanical production of turbulence, where H is the sensible heat flux* and C_p is the specific heat of air at constant pressure (for further details, see Lumley and Panofsky, 1964). Measurements of sensible heat flux H and friction velocity u_* are needed to evaluate this parameter. Theoretically, z/L is considered a more precise indicator of thermal stability than Ri. Ri, however, requires measurement of wind speed and air temperature gradients only, and is easier to evaluate in practice. A typical relationship between Ri and z/L observed in the surface layer is given in Fig. 4.5 (adapted from Businger et al., 1971).

Moist air is less dense than dry air. To account for the effect of moisture on buoyancy a virtual temperature, T_v (°K) is defined by

$$T_v = T(1 + 0.6q) \tag{4.10}$$

where T is the absolute temperature (°K) and q is the specific humidity. The use of virtual temperature (T_v) in lieu of the actual temperature makes it possible to apply the ideal gas law for moist air in the same way as for dry air. Therefore, when buoyancy forces are involved, gradients of T_v rather that T should be considered. For further details the reader is referred to Lumley and Panofsky (1964), Webb (1965), Plate (1971), and Busch (1973).

---

* Note that the sign convention employed in this book is such that all fluxes toward the ground surface are positive and all away from the surface are negative.

4.5 FLUX PROFILE RELATIONSHIPS

In aerodynamic theory, the transfer of momentum, sensible heat, water vapor, or any other entity may be estimated by analogous equations

$$\text{momentum flux:} \quad \tau = \rho_a K_m \frac{\partial U}{\partial z} \quad (4.11)$$

$$\text{sensible heat flux:} \quad H = \rho_a C_p K_h \frac{\partial \theta}{\partial z} \quad (4.12)$$

$$\text{water vapor flux:} \quad E = \frac{(M_w/M_a)}{P} \rho_a K_w \frac{\partial e_a}{\partial z} \quad (4.13)$$

where K_m, K_h, and K_w are the turbulent exchange coefficients for momentum, sensible heat, and water vapor, respectively; ρ_a is the air density; C_p is the specific heat at constant pressure; P is the atmospheric pressure; $\partial U/\partial z$, $\partial \theta/\partial z$, and $\partial e_a/\partial z$ are the vertical gradients of mean wind speed, potential temperature, and vapor pressure, respectively; M_w and M_a are the molecular weights of water vapor and air.

If these exchange coefficients are assumed to be identical, then a knowledge of any one in conjunction with the appropriate gradient measurements permits the estimation of any and all of the fluxes. This assumption of identity of exchange coefficients is called *Reynolds analogy*.

Observations suggest that the assumption of identity in exchange coefficients is valid only when the atmosphere is in a condition of nearly neutral stability. Such conditions normally prevail for only limited periods of the day. The effects of atmospheric stability on the relationship between exchange coefficients under nonneutral conditions has been determined experimentally in several micrometeorological investigations. Some recent findings are summarized below

$$\begin{aligned}
\text{unstable conditions:} \quad & \frac{K_h}{K_m} \approx \frac{K_w}{K_m} = (1 - 16 \text{ Ri})^{0.25} \\
& \qquad\qquad\qquad\quad \text{(Dyer and Hicks, 1970)} \\
& \frac{K_w}{K_m} = 1.13 \ (1 - 60 \text{ Ri})^{0.074} \\
& \qquad\qquad\qquad\quad \text{(Pruitt et al., 1973)} \\
\text{stable conditions:} \quad & \frac{K_h}{K_m} \approx \frac{K_w}{K_m} \approx 1 \\
& \qquad\qquad\qquad\quad \text{(Webb, 1970)} \\
& \frac{K_w}{K_m} = 1.13 \ (1 + 95 \text{ Ri})^{-0.11} \\
& \qquad\qquad\qquad\quad \text{(Pruitt et al., 1973)}
\end{aligned} \quad (4.14)$$

As discussed in Section 4.2, the logarithmic wind profile equations are valid only in neutral conditions. When nonneutral conditions exist, the shape of the wind profile deviates significantly from the logarithmic ideal (Fig. 4.6). The change in shape of the wind profile due to thermal stability effects

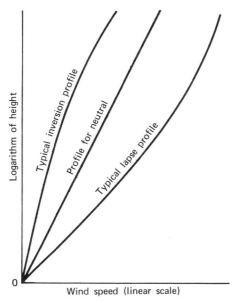

Fig. 4.6 Shape of the wind profile as a function of atmospheric stability (after Sutton, 1953).

is accounted for by introducing a dimensionless stability function ϕ_m (Monin and Obukhov, 1954) given by

$$\frac{\partial U}{\partial z} = \frac{u_*}{kz} \phi_m \qquad (4.15)$$

where ϕ_m is generally expressed as a function of Richardson number Ri or z/L (the Monin–Obukhov stability parameter). [Use $(z - d)$ instead of z in (4.15) for rough surfaces.] Note that for neutral conditions ϕ_m is equal to 1 and (4.15) converts into the logarithmic wind profile equation (4.1). There have been many efforts to precisely define the functional relationships between ϕ_m and Ri (or z/L) in stable and unstable conditions (see Holzman, 1943; Panofsky, 1963; Lumley and Panofsky, 1964; Dyer and Hicks, 1970; Webb, 1970; Oke, 1970; Swinbank, 1964, 1968; Businger et al., 1971; Pruitt et al., 1973). Some recent results are given below

unstable conditions: $\phi_m = (1 - 16 \text{ Ri})^{-0.25}$
(Dyer and Hicks, 1970)

$\phi_m = (1 - 16 \text{ Ri})^{-1/3}$
(Pruitt et al., 1973)

stable conditions: $\phi_m = (1 - 5.2 \text{ Ri})^{-1}$
(Webb, 1970)

$\phi_m = (1 + 16 \text{ Ri})^{1/3}$
(Pruitt et al., 1973)

(4.16)

For further detailed evaluation of these relationships, the reader is referred to Dyer (1974), Wieringa (1980), and Francey and Garratt (1981).

Employing (4.11) and (4.15), the equations for sensible heat (4.12) and water vapor flux (4.13) can be rewritten as

$$\text{sensible heat flux} = H = \rho_a C_p k^2 z^2 \left(\frac{\partial U}{\partial z}\right)\left(\frac{\partial \theta}{\partial z}\right)\left(\frac{K_h}{K_m}\phi_m^{-2}\right) \quad (4.17)$$

$$\text{water vapor flux} = E = \frac{(M_w/M_a)}{P}\rho_a k^2 z^2 \left(\frac{\partial U}{\partial z}\right)\left(\frac{\partial e_a}{\partial z}\right)\left(\frac{K_w}{K_m}\phi_m^{-2}\right) \quad (4.18)$$

Therefore, with knowledge of the appropriate gradients ($\partial U/\partial z$, $\partial \theta/\partial z$, $\partial e_a/\partial z$), in conjunction with the stability corrections [(4.14) and (4.16)], the fluxes of sensible heat, water vapor, or any other entity may be estimated from the flux gradient relationships described above.*

4.6 EDDY CORRELATION TECHNIQUE FOR ESTIMATING ENERGY AND MASS FLUXES

In fully turbulent flow the mean vertical flux F of an entity s per unit mass of the fluid is given by

$$F = \overline{\rho_a w s} \quad (4.19)$$

where ρ_a is the air density, w is the vertical velocity, and the overbar denotes the average value during a time period of suitable length.

In the surface layer, as discussed in Section 4.1, all atmospheric entities exhibit short-period fluctuations about their mean value (see Fig. 4.7 for an example). Therefore, the instantaneous values of w, s, and ρ_a in (4.19) can be expressed by

$$w = \overline{w} + w', \quad s = \overline{s} + s', \quad \text{and} \quad \rho_a = \overline{\rho}_a + \rho'_a \quad (4.20)$$

where the prime symbol denotes an instantaneous departure from the mean. These expressions can be substituted into (4.19) and, if we neglect fluctuations in density, the mean vertical flux F reduces† to

$$F = \overline{\rho_a}\,\overline{w}\,\overline{s} + \overline{\rho_a}\,\overline{w's'} \quad (4.21)$$

or by writing ρ_a for $\overline{\rho}_a$

$$F = \rho_a \overline{w}\,\overline{s} + \rho_a \overline{w's'} \quad (4.22)$$

---

* Note that for proper applications of the stability correction models described in (4.14) and (4.16) appropriate values of k (as suggested in the respective reports) should be used: for the Dyer and Hicks (1970) and Webb (1970) models, use $k = 0.4$ but for the Pruitt et al. (1973) model, use $k = 0.42$. For rough surfaces, use $(z - d)^2$ instead of z^2 in (4.17) and (4.18).

† Note that the following rules (the Reynolds rules of averaging) apply: $x = \overline{x} + x'$; $y = \overline{y} + y'$; $\overline{x'} = \overline{y'} = 0$; $\overline{x + y} = \overline{x} + \overline{y}$; $\overline{x'y'} \neq 0$.

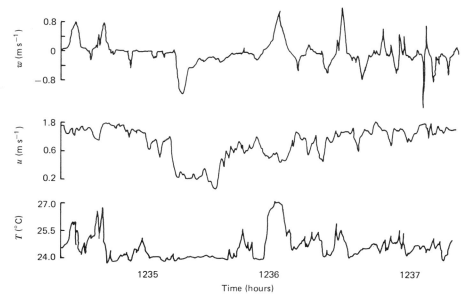

Fig. 4.7 Simultaneous recordings of signals from fast response sensors of vertical wind speed w, horizontal wind speed u and air temperature T above a corn crop. Reprinted from *Physiological Aspects of Crop Yield*, Chapter 6 by Edgar Lemon, page 132, 1969, by permission of the American Society of Agronomy and Crop Science Society of America.

The first term on the right-hand side of (4.22) represents flux due to the mean vertical flow or mass transfer. The second term represents flux due to eddying motion or eddy flux. The mass transfer term may arise from a convergence or divergence of air due to sloping surface. For a sufficiently long period of time over horizontally uniform terrain the total quantity of ascending air is approximately equal to the quantity descending and the mean value of the vertical velocity will be negligible. Therefore, (4.22) reduces to

$$F \approx \rho_a \overline{w's'} \tag{4.23}$$

Based on the above equation, the fluxes of momentum, sensible heat, and water vapor can be expressed as

momentum flux: $\quad \tau = -\rho_a \overline{u'w'} \tag{4.24}$

sensible heat flux: $\quad H = -\rho_a C_p \overline{w'T'} \tag{4.25}$

water vapor flux: $\quad E = -\dfrac{\epsilon}{P} \rho_a \overline{w'e'_a} \tag{4.26}$

where u', T', and e'_a are instantaneous departures from the mean horizontal velocity, air temperature, and vapor pressure; ϵ is the ratio of molecular weights of water vapor and air $= M_w/M_a$ and P is the atmospheric pressure.

Some recent analyses (e.g., Webb et al., 1980; Webb, 1982) indicate that atmospheric turbulent fluxes of minor constituents such as water vapor and CO_2 may be influenced by density variations due to simultaneous transfers of sensible heat and water vapor (the density effect may cause a very small but non zero mean vertical air speed). For further details on these effects the reader is referred to the original publications.

The eddy correlation technique, when properly applied, provides reasonably accurate estimates of fluxes more directly than do the other micrometeorological methods. Over rough surfaces, for example, the turbulent exchange coefficients are large and hence vertical gradients of temperature and humidity are exceedingly small. Correspondingly, vertical velocity fluctuations are quite large. Therefore, as Kanemasu et al. (1979) point out the eddy correlation technique should be more accurate than those methods that rely on measurement of vertical gradients. Instrumentation for measurement of turbulent fluctuations has been under development for a number of years. These instruments include hot wire anemometers, sonic anemometers, drag anemometers and propeller anemometers for the measurement of velocity fluctuations, fine wire thermocouples (dry and wet bulb) or sensitive thermistors to sense temperature fluctuations (see Chapter 3), and hygrometers (infrared and Lyman-alpha) to sense humidity fluctuations (see Chapter 5).

In the application of the eddy correlation technique, two points require particular attention. First, instruments must be so sensitive as to be able to detect simultaneous changes (fluctuations) in quantities such as velocity, temperature, and humidity caused by the rapid passage of different eddies. The frequency of these fluctuations increases as the surface is approached. The less sensitive the sensors, the higher above the surface they must stand and thus the greater the likelihood that they will be outside the fully adjusted layer. Second, a capability to quickly record and process the large quantities of data produced by these sensors is required. With the continuing improvements in instrumentation, data acquisition, and computing systems, the use of the eddy correlation technique for estimating energy and mass fluxes in the atmospheric surface layer is becoming more and more practical.

4.7 WIND SPEED WITHIN CROP CANOPIES

In previous sections, we have discussed the difficulties involved in accurately describing mean wind profile relationships above plant canopies. Wind speed relationships within canopies are still more difficult to establish. A good example of the complexity of the wind structure in canopies is found in Fig. 4.8 due to Allen (1968) showing the shape of a typical wind profile in a Japanese larch tree plantation in upper New York State. Campbell (1977) considers that the within-canopy flow regime is divided into three layers

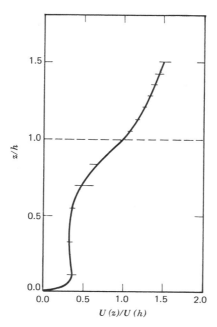

Fig. 4.8 Average normalized wind speed above and within a Japanese larch plantation, mid-November, near Ithaca, New York (after Allen, 1968).

which are evident in Fig. 4.8:

1. The top layer ($d < z < h$) is the layer that exerts most of the drag on the wind above the crop. The wind in this layer decreases exponentially with distance downward from the canopy top and has the same direction as the mean wind above the canopy.
2. The second layer (about $0.1\,h < z < d$) is comprised mainly of the stem space of the crop. There the wind may be quite unrelated, in both speed and direction, to the wind above the canopy.
3. The wind profile in the third layer ($z < 0.1\,h$) is similar in shape to the above canopy logarithmic profile. The profile in this layer is influenced, however, by the soil surface roughness instead of by the crop roughness.

4.8 DAILY WIND PATTERNS

Wind speed increases with elevation above the ground surface. Under fairly constant weather conditions a diurnal wave of wind speed can be detected at levels of measurements near the ground. A typical pattern is given in Fig. 4.9 for wind at a number of levels aboveground. Nighttime situations are normally calm, and peak wind speeds usually occur near noon.

148 WIND AND TURBULENT TRANSFER

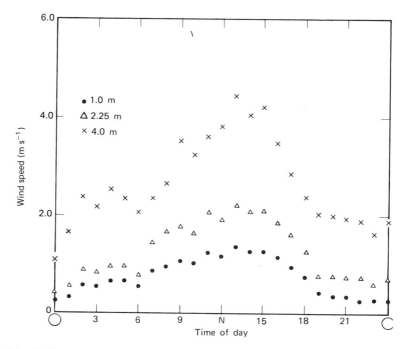

Fig. 4.9 Daily course of wind speed over 50 cm tall soybeans, early July, at Mead, Nebraska.

This normal condition is especially apparent in climatological summaries where the diurnal wind speed patterns over many days are summarized and the means are calculated. In regions of strong frontal activity or where convective storms are common, however, the actual daily patterns of wind speed are much more variable.

4.9 SEASONAL PATTERNS OF WIND DIRECTION AND SPEED

It is beyond the scope of this work to discuss the many large-scale and local seasonal patterns of winds that prevail in differing climatic regions of the world. It is important, however, to recognize that there do exist such recurring patterns and that these can be identified through climatological surveys and can be predicted with some degree of certainty by the methods of statistical climatology. Our own work in characterizing the wind regime in the central Great Plains may serve as a useful example.

The distribution of wind speed and direction at a given location is often presented in the form of a "wind rose" which is a polar plot of the frequency of wind flow as a function of wind direction (Slade, 1968). Some typical wind roses for Grand Island, Nebraska, are shown in Fig. 4.10. The data represent

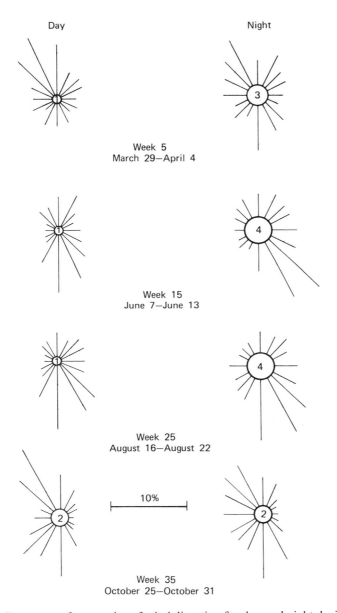

Fig. 4.10 Percentage frequencies of wind direction for day and night during 4 weeks of the growing season at Grand Island, Nebraska. Percent of calm hours indicated in center circle (after Rosenberg, 1965).

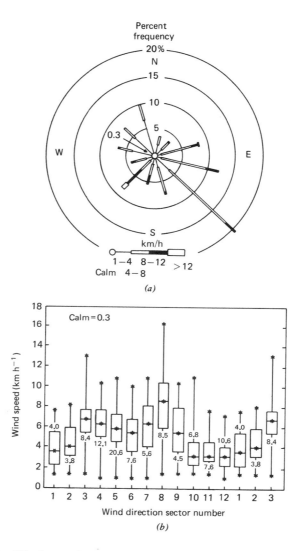

Fig. 4.11 (a) Wind rose for hourly averaged wind data, Artesia Junction, New Mexico, August–December 1971. (b) Wind box plot for the same data used in (a). The upper and lower boundaries of each box indicate the upper and lower quartiles of the distribution; the median value is shown by the asterisk and line within the box; and the extreme upper and lower data values are shown by asterisks above and below the box. Sector centers are 15°, 45°, 75°,.... The arbitrary discontinuity at 360° is avoided by duplicate plotting of sectors 1–3. Numbers on the figure refer to the frequency (in percent)) of wind flow within a given sector (adapted from Graedel, 1977).

means of the hourly wind speed recorded over a 14-year period (1948–1961) and were summarized by Rosenberg (1965) on the basis of climatic weeks. Week 1 is March 1–7.

Northwesterly flow dominates the central Great Plains in the early spring, with south to southeasterly flow dominating through late spring and summer. Although in late October the predominant flow is from the northwest, south and southeasterly winds still comprise an important portion of the total.

Graedel (1977) introduced a new method of graphically displaying the distributions of wind speed and direction. This method, called the "wind box plot," displays the characteristics of the distributions clearly and more completely than does the traditional wind rose. Selected characteristics of the distributions, such as the highest value, upper quartile, median, lower quartile and lowest value, are included in a wind box plot. An example taken from Graedel (1977) is shown in Fig. 4.11. This technique is readily adapted to computer statistics and graphics packages.

Figure 4.12 is a representation for the growing season (weeks 1–39) at Grand Island, Nebraska, of the mean day and night horizontal wind speeds (top) and the resultant horizontal wind speeds (bottom) computed for the same data used in Fig. 4.10. The mean day or night horizontal wind speed U_m is simply an average of hourly horizontal wind speeds. The resultant wind speed U_r is calculated by dividing the distance between the original

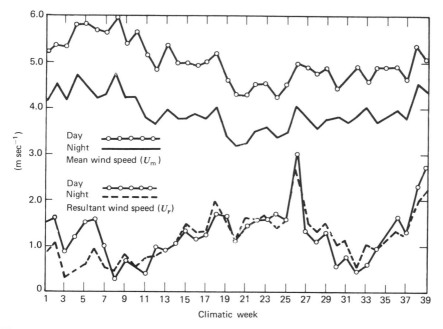

Fig. 4.12 Weekly mean and resultant wind speed at Grand Island, Nebraska (after Rosenberg, 1965).

Table 4.1 Mean and resultant wind speeds and wind directions during the day and night at Grand Island, Nebraska[a]

	Day				Night			
Week	Mean wind speed U_m (m s^{-1})	Resultant wind speed U_r (m s^{-1})	Wind direction (8 pt.)	Constancy C (%)	Mean wind speed U_m (m s^{-1})	Resultant wind speed U_r (m s^{-1})	Wind direction (8 pt.)	Constancy C (%)
01	5.4	1.6	NW	29	4.3	0.8	N	20
02	5.6	1.7	N	30	4.7	1.1	N	24
03	5.5	0.9	N	16	4.3	0.3	NE	7
04	6.0	1.3	NW	21	4.9	0.5	NW	10
05	6.1	1.6	N	26	4.6	0.6	N	13
06	5.9	1.7	N	28	4.4	1.0	N	24
07	5.9	1.0	NE	18	4.5	0.6	W	13
08	6.3	0.3	NE	5	5.0	0.4	SE	9
09	5.6	0.8	SE	14	4.4	0.8	SE	19
10	5.9	0.6	NE	11	4.4	0.6	E	13
11	5.4	0.4	SE	8	3.9	0.8	SE	19
12	5.0	1.0	SE	21	3.8	0.8	SE	21
13	5.6	0.9	SE	16	4.2	0.9	SE	23
14	5.2	1.1	SE	21	3.9	1.1	SE	29
15	5.2	1.4	SE	28	3.9	1.5	SE	39
16	5.1	1.2	S	24	4.0	1.3	SE	34

17	5.2	1.3	S	25	3.9	1.3	SE	34
18	5.4	1.8	S	34	4.2	2.1	SE	50
19	4.8	1.7	S	36	3.5	1.6	SE	46
20	4.5	1.2	SE	26	3.3	1.2	SE	36
21	4.5	1.5	SE	33	3.4	1.7	S	50
22	4.7	1.7	S	35	3.7	1.6	S	46
23	4.7	1.7	SE	36	3.8	1.7	SE	47
24	4.4	1.8	SE	41	3.5	1.4	SE	41
25	4.7	1.7	SE	36	3.6	1.6	SE	46
26	5.2	3.2	S	61	4.3	2.9	S	66
27	5.1	1.4	S	28	3.9	1.7	S	42
28	5.0	1.2	S	23	3.7	1.4	S	37
29	5.1	1.4	S	28	3.9	1.6	S	41
30	4.6	0.6	SW	12	4.0	1.1	S	26
31	4.9	0.8	S	17	3.8	1.2	S	31
32	5.1	0.5	SW	9	4.0	0.6	S	15
33	4.7	0.6	SW	13	4.2	1.1	SW	30
34	5.1	1.0	SW	20	3.8	1.0	S	26
35	5.1	1.3	W	25	4.0	1.2	W	29
36	5.1	1.7	NW	36	4.2	1.4	W	34
37	4.8	1.4	W	29	3.9	1.3	W	33
38	5.6	2.4	NW	43	4.7	2.1	W	43
39	5.2	2.9	NW	56	4.6	2.3	W	51

[a] After Rosenberg (1965).

and final positions of a hypothetical particle of air by the time elapsed in its travel considering the changes in direction that occur.

The constancy C of the wind relates the resultant and mean (day or night) wind speeds by

$$C = \frac{U_r}{U_m} \times 100 \qquad (4.27)$$

A constancy of $C = 0$ means a perfectly circular distribution of wind vectors, that is, $U_r = 0$. A constancy of $C = 100$ indicates wind blowing from one direction only, that is, $U_r = U_m$. The techniques of calculating resultant wind speeds and the directions are given by Brooks and Corruthers (1953).

Figure 4.12 and Table 4.1 show that at Grand Island, Nebraska, peak average wind speeds during the growing season occur in the spring with a gradual decrease to midsummer. Whereas wind speeds are lowest in midsummer, they are most constant during that season. Nocturnal average winds are always lower than daytime wind, but the resultant wind speed at night may exceed that during the day. A very sharp increase in constancy occurs during week 26 (August 23–29). After week 32, the constancy of westerly and northwesterly winds increases gradually.

Figures 4.10, 4.11, and 4.12 and Table 4.1 show some alternative ways of representing wind data. The application of the data should determine the methods used. It should be apparent, however, that calculations of mean wind speed and prevailing wind direction when used alone may be of limited value except for very general purposes of climatic description and may, in fact, obscure important relationships.

4.10 WIND SPEED INSTRUMENTATION

A complete discussion of wind sensors is beyond the scope of this work. Those who plan to use wind instruments should refer to authoritative works on meteorological instrumentation to determine the specific type of sensor that best meets the needs and the budget available. Only some general comments on sensor design principles are offered below. The major types of anemometers shown in Fig. 4.13 may be classified as pressure, mechanical, thermoelectric, and acoustical.

The Pitot tube is an example of a pressure anemometer. The difference in pressure developed between two tubes as wind blows into one and transverse to the other is directly related to the wind speed. Pitot tube systems are quite accurate and are usually used as standards for calibrating other types of wind speed sensors. The same principle is used in the anemoclinometer, which measures downwind, crosswind, and vertical components of wind velocity simultaneously. The anemoclinometer (Fig. 4.13a) developed by Thurtell et al. (1970) is a small metal sphere with electronic pressure sensors placed in holes facing directly into the wind and at 90° angles to the

Fig. 4.13 (*a*) Spherical sensing head of the anemoclinometer showing pressure ports (after Thurtell et al., 1970).

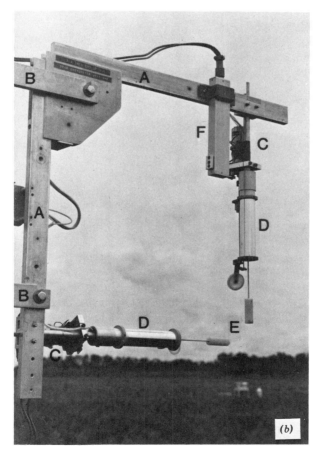

Fig. 4.13 (*b*) Drag anemometer (A, frame; B, mounting brackets; C, shroud motors; D, strain gage assembly housing and shrouds; E, wind sensing elements) and fine wire thermocouple electronics (F) (adapted from Redford et al., 1981).

Fig. 4.13 (c) Three-cup anemometer (courtesy of Science Associates, Princeton, New Jersey).

main axis. Pressure differences between the various sets of holes are used to calculate the three-dimensional wind speed components.

Rapid fluctuations of wind speed in three dimensions can also be measured with the drag anemometer. The sensor (Fig. 4.13b), is described in Norman et al. (1976). Recently Redford et al. (1981) have thoroughly tested the instrument for turbulent flux and spectra measurements over vegetated surfaces. The drag force of the wind on an aerodynamic shape is proportional to the square of the wind speed. This force can be measured by the deflection of a strain gauge attached to an object held perpendicular to the wind. Since the wind is not constant in direction, three mutually perpendicular wind sensing elements with strain gauges attached can be used to resolve the instantaneous wind speed and direction.

Cup anemometers are mechanical devices. One type is shown in Fig. 4.13c. The rate of rotation of a cup anemometer is a quadratic function of the mean horizontal wind speed and of certain geometric design parameters of the cups. Cup anemometers are intended for use in the horizontal position only. The instrument should respond to oblique winds only as the cosine of the attack angle. Cup anemometers start relatively slowly because of their inertia. Cup anemometers tend to "overspeed" primarily due to their non linear response to fluctuating winds. The instrument responds faster to an increase in wind than to a decrease of the same magnitude. Therefore, in gusty winds the mean wind speed measured by a cup anemometer is overestimated, generally by about 10–15%. For further details on the dynamic response of the cup anemometer in fluctuating winds the reader is referred to publications such as Busch and Kristensen (1976) and Wyngaard (1981). Therefore, the smaller the mass of the cups and arms, the better the sensitivity and accuracy. Mechanical contacts or electrical systems (light choppers) are used to record the rotation of the cups. The less the friction and inertia of the internal counting mechanism, the more accurate will be the sensor.

Propeller anemometers have been used to measure air flow. The helicoid-

Fig. 4.13 (*d*) UVW propeller anemometer measures directly the three orthogonal wind components (courtesy of R. M. Young Co., Ann Arbor, Michigan).

158 WIND AND TURBULENT TRANSFER

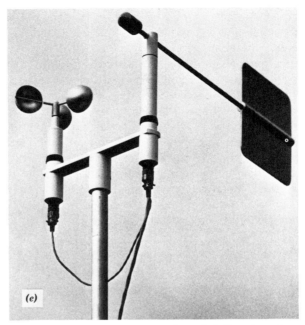

Fig. 4.13 (e) The Gill microvane for measuring wind direction, shown with a three-cup anemometer (courtesy of R. M. Young Co., Ann Arbor, Michigan).

shaped polystyrene propeller drives a miniature dc tachometer generator. A propeller responds only to the component of the wind which is parallel to its axis of rotation and thus may be positioned to measure either along wind, across wind, or vertical component of wind velocity. The UVW propeller anemometer system shown in Fig. 4.13d consists of three helicoidal propeller sensors mounted on a common mast at right angles to each other and is used to measure the three orthogonal wind components.

Wind vanes available from instrument manufacturers to measure wind direction come in various shapes and sizes. Designs differ primarily in the shape of the fin. One example is shown in Fig. 4.13e. A wind vane coupled with a propeller as shown in Fig. 4.13f, for example, can be used to measure horizontal wind direction and speed.

Bivanes are employed to obtain simultaneous measurements of azimuth angle and elevation angle of the wind. One type of bivane is shown in Fig. 4.13g. A bivane coupled with a propeller, as shown in Fig. 4.13h, for example, has been used to measure wind speed plus simultaneous azimuth and elevation angles of the wind.

Thermoelectric anemometry is represented by hot-wire-type sensors. Examples of hot-wire and hot-film anemometers are shown in Fig. 4.13i. Anemometers of this kind are particularly useful in applications where wind speed is very light, as in plant canopies. There are many industrial applications as well where the stream of air or gas is slow and mechanical sensors might interfere with the flow. The rate of convective cooling by a heated body is a function of the rate of fluid flow past that body. One type of hot-wire anemometer is maintained at a constant temperature, while the current flow needed to maintain that temperature is measured. Another type involves constant current flow while temperature fluctuations indicate the fluid flow. Sets of hot wires can be oriented so as to measure the three directional components of wind movement.

Heated thermistor anemometers (Bergen, 1971) have been used for air flow measurements in plant canopies. A heated thermistor anemometer consists of two matched thermistors (see Fig. 4.13j). One of the thermistors (A) is exposed to the air movement while the other one (B) is shielded. In still air, the resistances of A and B are equal and there is no voltage drop across the terminals d and f (bottom figure). With some air movement the thermistor

Fig. 4.13 (f) Propeller vane for measuring horizontal wind direction and wind speed (courtesy of R. M. Young Co., Ann Arbor, Michigan).

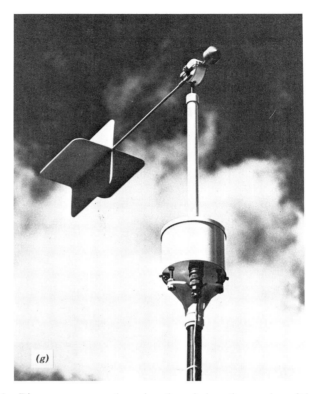

Fig. 4.13 (g) Bivane measures the azimuth and elevation angles of the wind (courtesy of R. M. Young Co., Ann Arbor, Michigan).

Fig. 4.13 (*h*) The Gill anemometer bivane measures wind speed plus simultaneous azimuth and elevation angles of the wind (courtesy of R. M. Young Co., Ann Arbor, Michigan).

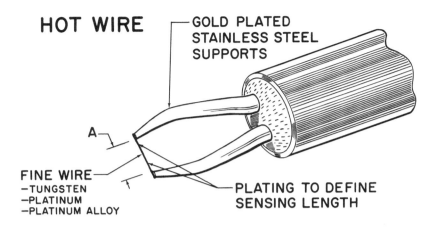
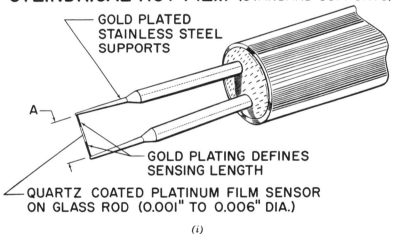

(i)

Fig. 4.13 (i) Schematic of hot-wire (top) and hot-film (bottom) anemometers (courtesy of Thermo Systems, Inc., St. Paul, Minnesota).

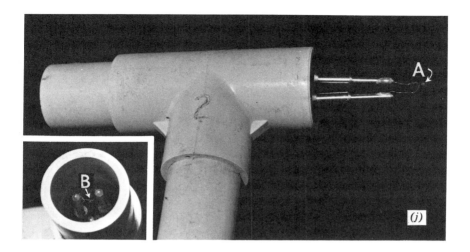

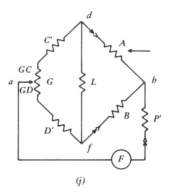

Fig. 4.13 (*j*) Heated thermistor anemometer: (top) a photograph; (bottom) a schematic bridge circuit diagram. A and B are two matched bead thermistors, C' and D' are two fixed resistors, F is a dc voltage source across the bridge, G is a rheostat subdivided into resistances GC and GD, P' represents the equivalent resistance for the connecting cables, and L represents the equivalent resistance of the output meter (adapted from Bergen, 1971; for further explanation see Bergen's paper).

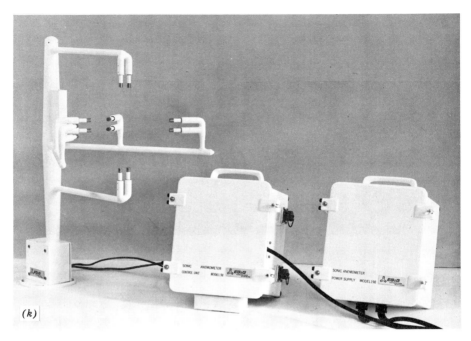

Fig. 4.13 (*k*) Triaxial sonic anemometer (courtesy of EG&G, Cambridge Systems, Inc., Newton, Massachusetts).

A cools and its resistance increases. This change in resistance gives rise to a voltage across points d and f.

Sonic anemometers are among the newest developments in anemometry. Sound travels faster from an emitter to a receiver in the direction of the wind and, conversely, travels more slowly against the wind. The speed of sound varies with such factors as air temperature, water vapor pressure, and atmospheric pressure. With three sets of emitters and receivers oriented in the x, y, and z directions, it is possible to determine three orthogonal components of wind velocity simultaneously. A three-dimensional sonic anemometer is shown in Fig. 4.13k. Campbell and Unsworth (1979) have reported on a one-dimensional (vertical) sonic anemometer. This instrument is relatively inexpensive, and because of the simplicity of its design, appears quite promising.

REFERENCES

Allen, L. H., Jr. 1968. Turbulence and windspeed spectra within a Japanese larch plantation. *J. Appl. Meteorol.* **7**:73–78.

Bergen, J. D. 1971. An inexpensive heated thermistor anemometer. *Agric. Meteorol.* **8**:395–405.

REFERENCES

Brooks, C. E. P., and N. Corruthers. 1953. Handbook of Statistical Methods in Meteorology. M.O. 538. Air Ministry, Meteorological Office, H. M. Stationary Office, London.

Busch, N. E., 1973. On the mechanisms of atmospheric turbulence. Workshop on Micrometeorology (D. A. Haugen, ed.). Am. Meteorol. Soc., Boston.

Busch, N. E., and L. Kristensen. 1976. Cup anemometer overspeeding. *J. Appl. Meteorol.* **15**:1328–1332.

Businger, J. A., J. C. Wyngaard, Y. Izumi, and E. F. Bradley. 1971. Flux-profile relationships in the atmospheric surface layer. *J. Atmos. Sci.* **28**:181–189.

Campbell, G. S. 1977. *An Introduction to Environmental Biophysics*. Springer-Verlag, New York. 159 pp.

Campbell, G. S., and M. H. Unsworth. 1979. An inexpensive sonic anemometer for eddy correlation. *J. Appl. Meteorol.* **18**:1072–1077.

Dyer, A. J., and B. B. Hicks. 1970. Flux-gradient relationships in the constant flux layer. *Q. J. Roy. Meteorol. Soc.* **96**:715–721.

Dyer, A. J. 1974. A review of flux-profile relationships. *Boundary-Layer Meteorol.* **7**:363–372.

Francey, R. J., and J. R. Garratt. 1981. Interpretation of flux-profile observations at ITCE (1976). *J. Appl. Meteorol.* **20**:603–618.

Garratt, J. R. 1980. Surface influence upon vertical profiles in the atmospheric near-surface layer. *Q. J. Roy. Meteorol. Soc.* **106**:803–819.

Graedel, T. E. 1977. The wind box plot: An improved wind rose. *J. Appl. Meteorol.* **16**:448–450.

Holzman, B. 1943. The influence of stability on evaporation. *Ann. N.Y. Acad. Sci.* **44**:13–18.

Kanemasu, E. T., M. L. Wesely, B. B. Hicks, and J. L. Heilman. 1979. Techniques for calculating energy and mass fluxes. *Modification of the Aerial Environment of Crops* (B. J. Barfield and J. F. Gerber, eds.). Am. Soc. Agric. Engineers, St. Joseph, Michigan.

Lemon, E. 1969. Gaseous exchange in crop stands. *Physiological Aspects of Crop Yield* (J. D. Eastin, F. A. Haskins, C. Y. Sullivan, and C. H. M. van Bavel, eds.). Am. Soc. of Agronomy, Madison, Wisconsin.

Lumley, J. L., and H. A. Panofsky. 1964. *The Structure of Atmospheric Turbulence*. Wiley, New York. 239 pp.

Makkink, G. F., and H. D. J. van Heemst. 1970. *Potential Evaporation from a Canopy*. Institute for Biological and Chemical Research on Crops, Wageningen, Meded. 417. In Dutch (English summary).

Monin, A. S., and A. M. Obukhov. 1954. The basic laws of turbulent mixing in the surface layer of the atmosphere. *Akad. Nauk USSR Acad. Sci.* **24**:163–187.

Monteith, J. L. 1973. *Principles of Environmental Physics*. Edward Arnold, London. 241 pp.

Munro, D. S., and T. R. Oke. 1975. Aerodynamic boundary-layer adjustment over a crop in neutral stability. *Boundary-Layer Meteorol.* **9**:53–61.

Norman, J. M., S. G. Perry, and H. A. Panofsky. 1976. Measurement of theory of horizontal coherence at a two-meter height. Preprints, Third Symposium on Atmospheric Turbulence, Diffusion and Air Quality, American Meteorological Society, October 19–22, 1976, Raleigh, North Carolina, pp. 26–31.

Oke, T. R. 1970. Turbulent transport near the ground in stable conditions. *J. Appl. Meteorol.* **9**:778–786.

Oke, T. R. 1978. *Boundary Layer Climates*. Methuen and Co. Ltd., London. 372 pp.

Panofsky, H. A. 1963. Determination of stress from wind and temperature measurements. *Q. J. Roy. Meteorol. Soc.* **89**:85–94.

Plate, E. J. 1971. Aerodynamic Characteristics of Atmospheric Boundary Layer. U.S. Atomic Energy Commission. NTIS TID-25465, 190 pp.

Pruitt, W. O., D. L. Morgan, and F. J. Lourence. 1973. Momentum and mass transfer in the surface boundary layer. *Q. J. Roy. Meteorol. Soc.* **99**:370–386.

Redford, Jr., T. G., S. B. Verma, and N. J. Rosenberg. 1981. Drag anemometer measurements of turbulence over a vegetated surface. *J. Appl. Meteorol.* **20**:1222–1230.

Rosenberg, N. J. 1965. Climate of the central Platte Valley of Nebraska. *Nebr. Agric Exp. Stn. Misc. Bull.* **11**, part B.

Slade, D. H., ed. 1968. Meteorology and Atomic Energy. U.S. Atomic Energy Commission, Office of Information Services. 445 pp.

Stanhill, G. 1969. A simple instrument for field measurement of turbulent diffusion flux. *J. Appl. Meteorol.* **8**:509–513.

Sutton, O. G. 1953. *Micrometeorology*. McGraw-Hill, New York. 333 pp.

Swinbank, W. C. 1964. The exponential wind profile. *Q. J. Roy. Meteorol. Soc.* **90**:119–135.

Swinbank, W. C. 1968. A comparison between predictions of dimensional analysis for the constant-flux layer and observations in unstable conditions. *Q. J. Roy. Meteorol. Soc.* **94**:460–467.

Szeicz, G., G. Endrodi, and S. Tajchman. 1969. Aerodynamic and surface factors in evaporation. *Water Resour. Res.* **5**:380–394.

Thom, A. S. 1971. Momentum absorption by vegetation. *Q. J. Roy. Meteorol. Soc.* **97**:414–418.

Thom, A. S. 1975. Momentum, mass and heat exchange of plant communities. *Vegetation and the Atmosphere* (J.. L. Monteith, ed.), Academic Press, pp. 57–109.

Thurtell, G. W., C. B. Tanner, and M. L. Wesely. 1970. Three-dimensional pressure sphere anemometer system. *J. Appl. Meteorol.* **9**:379–385.

Webb, E. K. 1965. *Aerial Microclimate*. Meteorological Monographs. Vol. 6, No. 28, Chapter 2, pp. 27–58.

Webb, E. K. 1970. Profile relationships: The log-linear range, and extension to strong stability. *Q. J. Roy. Meteorol. Soc.* **96**:67–90.

Webb, E. K., G. I. Pearman, and R. Leuning. 1980. Correction of flux measurements for density effects due to heat and water vapour transfer. *Q. J. Roy. Meteorol. Soc.* **106**:85–100.

Webb, E. K. 1982. On the correction of flux measurements for effects of heat and water vapour transfer. *Boundary-Layer Meteorol.* **23**:251–254.

Wieringa, J. 1980. A revaluation of the Kansas mast influence on measurements of stress and cup anemometer overspeeding. *Boundary-Layer Meteorol.* **18**:411–430.

Wyngaard, J. C. 1981. Cup, propeller, vane and sonic anemometers in turbulence research. *Annu. Rev. Fluid Mech.* **13**:399–423.

CHAPTER 5 – ATMOSPHERIC HUMIDITY AND DEW

5.1 INTRODUCTION

To the meteorologist, water vapor is the single most important constituent of the atmosphere. This is so since at terrestrial temperatures water passes easily from vapor to the liquid and solid phases with a large release or absorption of heat. For example, the evaporation of 1 kg of water at 20°C requires about 2.45 MJ (the latent heat of vaporization). The vapor is carried aloft by turbulent eddies to a level at which condensation occurs, with its concomitant release of the 2.45 MJ kg^{-1}. When 1 kg of water freezes, about 0.34 MJ is released (heat of fusion) and the same quantity is, of course, required to melt snow or ice. **Sublimation,** the direct transition of water from the solid to vapor phase, and vice versa, without passing through an intermediate liquid phase, involves the consumption or release of about 2.84 MJ kg^{-1}. Thus, the energy-consuming and -releasing processes in water phase changes provide the mechanism for the transportation of large quantities of heat to and from the surface of the earth.

Humidity in the atmosphere is of great biological importance as well. Hoffman (1973) has shown that atmospheric humidity influences the internal water potential of plants and the rate at which plants transpire water into the atmosphere. Humidity conditions and especially dew affect the growth and development of many phytopathogens, especially the fungal organisms. For example, Blad et al. (1978) have shown that incidence of dry bean infection by the organism *Sclerotinia sclerotiorum* (white mold) is affected by density of crop canopy and frequency of irrigation, both of which strongly influence the humidity of air within the plant canopy.

Human and animal adaptation are also influenced by atmospheric humidity conditions. Hadley (1979), for example, has shown that the desert beetle *Cryptoglossa verrucosa* (Le Conte) exhibits colors ranging from light blue to jet black when subjected to extremes of low and high humidity, respectively. These color changes are due to the excretion of a network of wax filaments at low humidities.

Evaporative cooling by sweating (the major mechanism by which animals maintain normal body temperature under conditions of high temperature) is controlled, in part, by the atmospheric humidity. Whereas sweating can occur in an atmosphere saturated with water vapor, evaporative cooling is reduced to virtual insignificance. Indices of comfort that consider both temperature and humidity have been developed for use in weather advisory reports. More detail on the influence of humidity on human and animal comfort is given in Chapter 12.

Other important facets of atmospheric humidity involve the formation of dew. Dew is of great biological and hydrological significance, as is explained later in this chapter. The role of atmospheric humidity in the processes of evapotranspiration and photosynthesis is discussed in Chapters 7 and 8, respectively, and its relation to frost incidence is described in Chapter 10.

5.2 PHYSICAL REVIEW

5.2.1 Dalton's Law of Partial Pressures

Dalton's law states that "the pressure P exerted by a mixture of ideal gases in a given volume V at absolute temperature T is equal to the sum of the pressures P_i, which would be exerted by each respective individual gas if it alone occupied volume V at the same temperature," so that

$$\sum P_i V = (\sum n_i) RT \tag{5.1}$$

where n_i is the number of moles of the gases and R is the universal gas constant with the value 8.314 KJ kg^{-1} mol^{-1} °K^{-1}.* Moist air behaves as a nearly ideal gas and obeys Dalton's law, so that

$$P = P_d + e_a \tag{5.2}$$

where P_d is the partial pressure of dry air and e_a is the partial pressure of water vapor in the mixture. Let X_w be the mole fraction of water vapor and let X_d be the mole fraction of dry air in a sample of moist air ($X_w + X_d = 1$). Then

$$e_a = X_w P \tag{5.3}$$

$$P_d = X_d P \tag{5.4}$$

5.2.2 Density of Moist Air

Consider a sample of moist air with pressure P, temperature T, volume V, and total mass m. The density ρ_a of the moist air is given by

$$\rho_a = \frac{m}{V} \tag{5.5}$$

---

* The derivations in this chapter follow Harrison (1965).

The mass m_w of the vapor alone is given by

$$m_w = \frac{e_a V M_w}{RT} \tag{5.6}$$

where M_w is the molecular weight of water (18.016 g mol^{-1}).
 The mass of the dry air in the mixture is given by

$$m_d = \frac{P_d V M_a}{RT} \tag{5.7}$$

where M_a is the molecular weight of dry air (28.966 g mol^{-1}). The total mass m of the moist air sample is

$$m = m_w + m_d = \frac{(e_a M_w + P_d M_a)V}{RT} \tag{5.8}$$

The ratio of the mole weight of water vapor and air (ϵ) is

$$\epsilon = \frac{M_w}{M_a} = \frac{18.016}{28.966} = 0.6220$$

so that $(1 - \epsilon) = 0.3780$. From (5.2), (5.5), and (5.8) the density of moist air ρ_a is

$$\rho_a = \frac{M_a}{R} \frac{(P - 0.3780 e_a)}{T} \tag{5.9}$$

From the basic relationships given above, other humidity relations having special applications in meteorology and physics are developed. These are given below.

5.3 MEASURES OF HUMIDITY

Specific humidity q is defined as the ratio of the mass of water vapor in a sample of moist air to the total mass of the sample. Thus, the specific humidity is

$$q = \frac{m_w}{m} = \frac{m_w}{m_w + m_d} \tag{5.10}$$

Specific humidity is related to the vapor pressure by

$$q = \frac{0.622 e_a}{P - 0.378 e_a} \tag{5.11}$$

The specific humidity can be expressed in units of kilograms of water vapor per kilogram of moist air.

170 ATMOSPHERIC HUMIDITY AND DEW

Absolute humidity, χ is defined as the ratio of the mass of water vapor to the total volume of moist air in which it is contained. Absolute humidity is given by

$$\chi = \frac{m_w}{V} \tag{5.12}$$

and can be related to the vapor pressure by

$$\chi = \frac{e_a M_w}{RT} \tag{5.13}$$

Absolute humidity is expressed in terms of kilograms of water vapor per cubic meter of moist air.

Mixing ratio W is the ratio of the mass of water vapor contained in a sample of moist air to the mass of dry air in the sample. The mixing ratio is given by

$$W = \frac{m_w}{m_d} \tag{5.14}$$

The mixing ratio is related to vapor pressure by

$$W = \frac{e_a M_w}{P_d M_a} = \frac{0.622 e_a}{P_d} \tag{5.15}$$

Since $P_d = P - e_a$ and $e_a \ll P$, the mixing ratio can usually be calculated from the total atmospheric pressure as

$$W \approx 0.622 \frac{e_a}{P} \tag{5.16}$$

Mixing ratio is usually expressed in terms of kilograms of water vapor per kilogram of dry air.

5.4 THE CONCEPT OF SATURATION

The kinetic theory of gases indicates that evaporation occurs when molecules in a liquid succeed in overcoming mutual attractive forces and escape from a free surface to form a vapor in the space above. Some molecules strike the surface and are recaptured until, ultimately, a state of dynamic equilibrium exists in which the exact number of escaping molecules is balanced by the number of recaptured molecules. At this point, the vapor is saturated. For a given temperature, there is a definite saturation vapor pressure or saturation vapor density.

Saturation vapor pressure increases exponentially with increasing temperature, as shown in Fig. 5.1. A numerical expression of the curve in Fig. 5.1 is

$$e_s = 0.61078 \exp\left(\frac{17.269 T}{T + 237.30}\right) \tag{5.17}$$

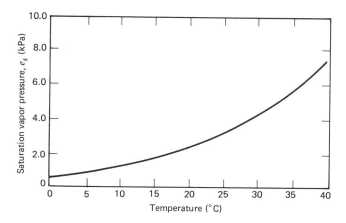

Fig. 5.1 Saturation vapor pressure e_s of water as a function of temperature.

where e_s is the saturation vapor pressure in kP$_a$ and T is the temperature in °C. For water below 0°C, the surface for reference must be ice or supercooled water rather than the liquid, and a different expression applies.

Saturated air is moist air whose vapor pressure is equal to the saturation vapor pressure. When vapor pressure is less than the saturation vapor pressure, air is unsaturated.

5.5 SATURATION-BASED MEASURES OF HUMIDITY

The saturation concept provides additional ways of characterizing water vapor relationships as follows:

The **relative humidity** RH is the ratio of actual to saturation vapor pressure at the same temperature. In percentage terms

$$RH = \frac{e_a}{e_s} \times 100 \qquad (5.18)$$

This is a frequently used parameter for the description of humidity conditions. However, since e_s is temperature dependent, relative humidity when expressed without the temperature is useful only in a qualitative way. Relative humidity should be used with caution in scientific work.

The **vapor pressure deficit** $e_s - e_a$ is the difference between the saturation vapor pressure (the total possible e_s at the ambient temperature) and the actual vapor pressure e_a. As such, it is a rough measure of the drying power of the air. Rates of evapotranspiration and transpiration are indicated by the magnitude of the vapor pressure deficit but cannot be directly predicted by that parameter (see Chapter 7).

The **dew point** T_d is defined as the temperature to which a given parcel of air must be cooled at constant pressure and constant water vapor content in order for saturation to occur. Since e_s is a function of temperature, there

172 ATMOSPHERIC HUMIDITY AND DEW

exists a temperature T_d (called the dew-point temperature) for which e_s is equal to e_a. T_d is independent of the air temperature as long as the air remains unsaturated. If the temperature drops below the dew point, moisture can be lost by condensation and T_d will, consequently, be lowered.

5.6 HUMIDITY STRUCTURE OF AIR

Water vapor is furnished to the air only from the evaporating surfaces of land and water. Water vapor transport into and through the layer of air adjacent to the ground is analogous to heat transport. Heat and water vapor are transferred to the bulk air primarily by convection or turbulent transport.

The net transport of heat is usually upward during the day and downward at night. The net vapor transport often follows this pattern with evaporation occurring by day and dew deposition by night. There exist, however, many climatic situations wherein evaporation continues throughout much or all of the night. Thus, frequently at night and sometimes during the day (see Chapter 7), we observe temperature inversions in conjunction with lapse profiles of vapor pressure. Typical midday profiles of vapor pressure and relative humidity over an irrigated crop are shown in Fig. 5.2.

Figure 5.3 shows mean daily pattern of vapor pressure over and within a dry bean crop in western Nebraska. Daytime vapor pressure reaches its peak near midday and is greatest within the canopy. Figures 5.3 and 5.4 also

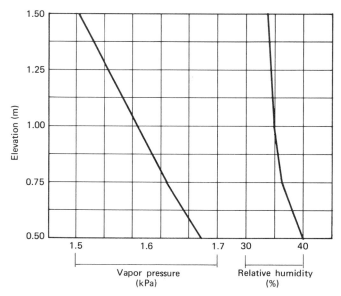

Fig. 5.2 Typical midday profiles of vapor pressure and relative humidity over irrigated soybeans during summer at Mead, Nebraska.

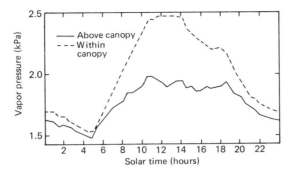

Fig. 5.3 Diurnal pattern of vapor pressure above and within the canopy of irrigated dry beans at Scottsbluff, Nebraska.

illustrate the fact that the peak in vapor pressure occurs between noon and early afternoon.

The daily waves of solar and net radiation peak at noon whereas temperature and evapotranspiration usually peak in early afternoon. When strong evaporative stress occurs in the early afternoon, as it does often in the Great Plains and other semiarid and arid regions, that stress is frequently sufficient to cause leaf stomates to close. Then the rate of transpiration decreases and, since turbulent transport of vapor away from the crop continues, the vapor pressure decreases. Relative humidity is greatest at night even though actual vapor pressure is least then. The minimum in relative humidity occurs in midafternoon because of the high temperature even though vapor pressure may reach its maximum then (Fig. 5.4).

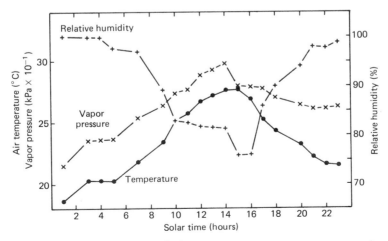

Fig. 5.4 Typical diurnal pattern of air temperature, vapor pressure and relative humidity over a soybean crop canopy. Data from Mead, Nebraska, mid-July.

174 ATMOSPHERIC HUMIDITY AND DEW

Data on atmospheric humidity are not as extensive as data on temperature, for example. Nonetheless, general information on the seasonal patterns of dew point temperature, relative humidity, and so on can be obtained from various sources. Gringorten et al. (1966) have prepared tables of atmospheric humidity (dew point and relative humidity) and probability statements for the Northern Hemisphere. Yao (1974) has examined voluminous records of surface relative humidity and finds that this parameter can be characterized statistically by a β distribution, that is, by an S-shaped curve with very few occurrences near 0 and 100%.

Court and Waco (1965) advise caution in use of statistics on mean daily relative humidity. When calculated as the midrange (maximum + minimum)/2, the value is biased because of the nonlinear dependence of relative humidity on atmospheric temperature.

5.7 PROFILES OF VAPOR PRESSURE

As has already been pointed out, the profiles of vapor pressure above the ground surface are almost always lapse (decreasing with elevation) except when condensation is occurring. The shape of the vapor pressure profile varies throughout the day. At dawn, it may be nearly isohumic (no change in vapor pressure with elevation). As the day progresses with increasing

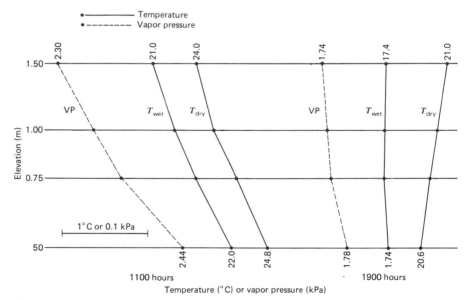

Fig. 5.5 Daytime and nighttime profiles of dry- and wet-bulb temperature and calculated water vapor pressure over an irrigated soybean crop in early July at Mead, Nebraska.

evaporation or transpiration, the vapor concentration near the surface increases. The effectiveness of vapor removal from the surface increases with increasing windiness and turbulence, so that by midday profiles showing a steep lapse in vapor pressure develop. By late afternoon or early evening, evaporative flux decreases as does turbulence and the humidity gradients again become moderate.

Figure 5.5 shows profiles of dry- and wet-bulb temperature measured with psychrometers (described below) and the profiles of vapor pressure calculated from these data. The data are shown for midday and early evening over a soybean crop grown at Mead, Nebraska. The profile of dry-bulb temperature goes from sharply lapse during midday to an inversion condition in early evening. Vapor pressure profiles remain lapse throughout the day although the gradients vary in intensity from time to time.

5.8 DEW

Dew is defined in the *Glossary of Meteorology* as "water condensed onto grass and other objects near the ground, the temperatures of which have fallen below the dew point of the surface air due to radiational cooling during the night, but are still above freezing."

Dew has great importance in agriculture and ecology. In certain arid regions of the world, dew may be the primary source of water on which local species of plants survive. In the summer climate of Vancouver, Canada, for example, Tuller and Chilton (1973) found that 12–14% of the normal monthly precipitation is provided by dew. Fritschen and Doraiswamy (1973) found that dew provides about 20% of the daily water consumption in a Douglas fir forest in Washington State.

Yet, the influence of dew on the physiology and water status of the plant may be more important than these percentages suggest. Heavy dew may suppress evaporation for a period of time. Dew may possibly be absorbed directly into plants and affect the internal water balance. Kerr and Beardsell (1975) noted that the internal water potential of *Paspalum* leaves in a pasture remained high until after dew was evaporated off the leaves. Canopy resistance remained low in the early morning in the presence of dew.

Dew duration is of major importance in the pathology of fungus disease. Certain pathogens, for example, can only penetrate stomates of leaves if they are enveloped in liquid water. Others require free water for a length of time in order for the spores to germinate (e.g., Blad et al., 1978; Weiss et al., 1980).

Wallin (1967) reviewed the climatological importance of dew. He summarized reports that show a range of nocturnal dewfall from 0.01 to 5 mm. Monteith (1957) did an analysis of possible dewfall amounts based on energy balance considerations. In his example sufficient air at relative humidity of 40% is available to deposit 1 g of water onto a soil or plant surface. This quantity of air and its vapor would have to be cooled by the equivalent of about 2100 J to reach its dew point and cause condensation. Condensation releases a further 2400 J g^{-1} of water condensed. To dissipate this quantity of energy (4500 J) by radiative cooling during the course of an average night (2.5 MJ is a reasonable nocturnal net radiative flux) would require a surface area of about 1.8×10^{-3} m^2. One gram of water (1 ml) will spread over this surface in a layer about 0.5 mm thick.

That Monteith's estimate is reasonable for the humid-maritime climate is shown in a number of reports. Burrage (1972) found nightly dewfall of 0.2–0.3 mm in England. Aslyng (1965) reports that in the humid, maritime climate of Copenhagen, Denmark, total dew deposition is only 51 mm yr^{-1}. Baier (1966) working in the semiarid interior of South Africa also found dewfall to total only about 12 mm yr^{-1}. Despite the small quantities of dew in both cases, plants are observed to be wet on many nights of the year (at least 20 nights per month in late summer and fall, according to Baier).

More favorable meteorological conditions might result in dewfalls 2–3 times as high as Monteith predicts but some very high rates reported in the literature are likely erroneous because of the difficulties inherent in dew measurement. There are also many observations reported in the literature of water on plant surfaces at night in conditions that cannot be conducive to dew formation. Noffsinger (1965) points out that other phenomena can be involved as well. Liquid water seen on plant surfaces may be rain, dew (the deposit of vapor condensed from the air on radiationally cold surfaces) or it may be **distillation** (the condensation of vapor originating from the soil), **water of guttation** (water expressed through the organs of translocation in leaves), or fog interception in which liquid water droplets are intercepted by the plant or ground surfaces. Distillation appears to explain the high frequency of wetted plants noted by Aslyng (1965), Baier (1966), and many others. Burrage (1972) concluded from a study of wet-bulb and dry-bulb temperature profiles in a wheat crop that condensation was caused by dew deposition above 0.6 m and to distillation below that height.

A relative humidity of 100% in the ambient air is not required for dewfall to begin. Monteith (1957) observed dewfall to occur while relative humidity was in the range of 91–99%. In that case, the leaves were colder than air by about 0.4–1.4°C. As long as the surface on which deposition takes place is colder than the dew point temperature of the contacting air, dew will form.

It is generally thought that dew formation requires clear skies so that surfaces cool rapidly and, also, that the air must be calm. Monteith (1957), however, considers that it is distillation of vapor from the air onto the soil and plants which occurs on calm nights. Distillation occurs primarily by molecular diffusion. Dew, according to Monteith, occurs, on the other hand, by turbulent transfer and is, in fact, negligible when winds are less than 0.5 m s^{-1}. Distillation rates are in the order of .01–.02 kg m^{-2} h^{-1}, but dew deposition is more rapid at 0.03–0.04 kg m^{-2} h^{-1}.

The period during which dew is deposited and its duration on leaf surfaces

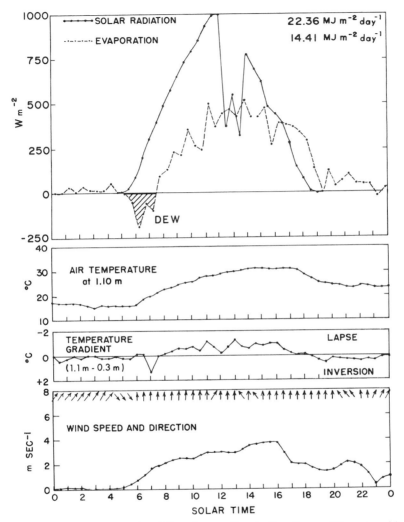

Fig. 5.6 Evaporation and dewfall on bare soil and climatic conditions in mid-August at Mead, Nebraska.

is also a matter of importance. In the late summer and autumn of interior South Africa, Baier (1966) found that plants may remain moisture-coated for 12–15 h from the onset of dew or distillation. Burrage (1972) found dew to last from 4–14 h on a wheat crop in England. Lomas (1965) found considerably longer dew duration to be common in the climatic conditions of Israel. As with quantity of dew, a wide range of values is given in the literature for dew duration and this range may result as much from the instrumental difficulties involved as from geographic and ecological differences.

178 ATMOSPHERIC HUMIDITY AND DEW

One case of dewfall measurement is presented here. Rosenberg (1969) measured dewfall on bare soil with precision weighing lysimeters at Mead, Nebraska, during the summer and fall. Lysimeters (if accurate) offer the best means of measuring actual dewfall. The lysimeter is a weighable block of soil in which plants can be grown (see Chapter 7). Whether plants are grown or not, dewfall can be ascertained since condensation of water released by the soil (or water of guttation, if plants are present) will cause no net change in the weight of the system. Thus, these phenomena are not measured and any change in weight of the system can be attributed to dewfall. A daily pattern of evaporation and dewfall typical of the region is shown in Fig. 5.6.

Since evaporation can continue during much of the night in the eastern Great Plains climate, dewfall was observed only during the early morning hours immediately after sunrise. The quantities measured during the growing season ranged from 0 to 0.52 mm, with the greatest amounts occurring during August when the air is most humid in this region.

5.9 INSTRUMENTATION FOR HUMIDITY MEASUREMENT

Water vapor content is measured with hygrometers. There are a number of types of hygrometers in common use. These may involve (1) the measurement of air temperatures; (2) the change of physical dimensions upon the absorption of moisture; (3) the condensation of a film of water; (4) the change in chemical or electrical properties upon absorption or adsorption of water vapor; (5) the absorption spectra of water vapor.

These general types are described and examples of instruments from categories (1) through (5) are given below:

1. The psychrometer consists of two ventilated thermometers. One of these measures air temperature directly, while the other, covered with a wick saturated in distilled water, measures a temperature lowered by an amount determined by the evaporative cooling caused by the sample air. Water is evaporated into the humid air flowing past the wet thermometer until equilibrium is reached. The temperature of the wet thermometer at this point is T_w, the **wet bulb temperature**. At equilibrium

$$e_a = e_s - AP(T - T_w) \qquad (5.19)$$

where T and T_w are the dry- and wet-bulb temperatures (°C), P is the air pressure, e_a is the actual vapor pressure, and e_s is the saturation vapor pressure at T_w. A is a constant of proportionality and is slightly dependent on T. A, according to Harrison (1965), has a value of

$$A = 6.6 \times 10^{-4} (1 + 1.15 \times 10^{-3} T_w) \qquad (5.20)$$

when the aspiration rate is sufficiently rapid, say, 3–4 m s^{-1} or more.

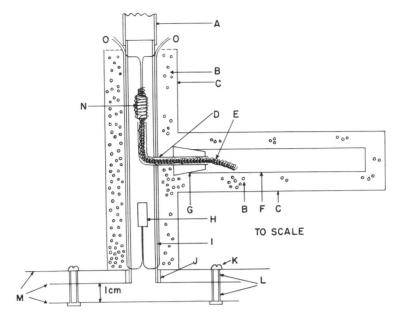

Fig. 5.7 Diagrammatic cross section of the aspirated psychrometer: A, plastic tube to aspiration manifold; B, styrofoam pipe cover; C, chrome-plated Mylar; D, plastic tube; E, cotton wick (shoestring); F, glass water reservoir; G, rubber stopper; H, dry bulb; I, Lucite tube; J, Lucite tube for friction fitting; K, plastic bolt; L, spacers; M, baffles; N, wet bulb; O, 30-gauge copper–constantan wires (after Rosenberg and Brown, 1974).

From a knowledge of wet- and dry-bulb temperatures in the bulk air, it is possible to derive an estimate of the vapor pressure, vapor pressure deficit, relative humidity, and dew point temperature.

One type of psychrometer, adapted for field micrometeorology by Rosenberg and Brown (1974), is shown in Fig. 5.7. Thermocouples embedded in Teflon plugs measure T and T_w. An assembly of the psychrometers is shown in Fig. 5.8. The psychrometers are aspirated and also shielded against radiation with aluminum foil or Mylar film. For use in the field, they are housed in a Styrofoam case that provides thermal insulation. As arranged, the assembly can be rotated to a horizontal position so that all sensors are at the same elevation. In this position the variations in output from the individual sensors can be measured and used as the basis for correcting or calibrating the instruments (see Rosenberg and Brown, 1974, for details).

Fine-wire thermocouples can be used for the measurement of rapid fluctuations in both temperature and humidity. One such instrument was developed by Tillman (1973) for use in eddy correlation studies of heat and vapor transport. The sensor consists of two 1-mil chromel–constantan thermocouples, one dry and one wetted by capillary flow of water through a

180 ATMOSPHERIC HUMIDITY AND DEW

Fig. 5.8 Assembly of "self-checking" thermocouple psychrometers for field use (after Rosenberg and Brown, 1974).

wick. The very small fluctuations in output signal are amplified at the sensor. Reference thermocouples are also included in the electronics. A set of fine-wire dry- and wet-bulb thermocouples used by Redford et al. (1980) are shown in Fig. 5.9.

Thermocouple psychrometers have also been "miniaturized" for placement within plant canopies so that profiles of temperature and humidity can be measured there. Stigter and Welgraven (1976) designed such a sensor using 0.2 mm (~8 mil) constantan–manganin thermocouples. A modified sensor of this kind is shown in Fig. 5.10. The sensor must be shielded from radiation but the fineness of the wires and the cotton wick help reduce the need for aspiration. Aspiration within the crop canopy would, likely, disrupt the temperature and humdity structure to be measured. To test the accuracy of the miniature thermocouple psychrometer, the instrument was compared with an Assmann psychrometer, which is ventilated at a constant 4 m s^{-1}. Errors no greater than 5% were found even when wind speeds within the canopy were as low as 0.1 to 0.2 m s^{-1}.

Thermocouple psychrometers are also used for determination of soil moisture content (potential) and water potential in plant tissue. In the former case, porous blocks containing the thermocouples are placed within the soil where they come into equilibrium with the soil air and its vapor content. In the latter case, samples of plant tissue are excised and placed in chambers and equilibrated at constant temperature until the vapor liberated from the tissue is in equilibrium with the water held in the tissue. Fine-wire psychrometers within these chambers measure the temperature of the air and the "wet bulb depression." The psychrometric cooling effect is achieved either by introduction of a small drop of water or by cooling the air to dew point with Peltier-type thermocouples. According to Rawlins (1966), the water potential Ψ in plant tissue or in soil can be determined by

$$\Psi = \frac{RT}{M_w} \ln \frac{e_a}{e_s} \qquad (5.21)$$

For additional detail on these applications of psychrometry, see Box (1965), Zollinger et al. (1966), and Valancogne and Daudet (1974). Savage et al. (1981) have developed a detailed procedure for calibration of these sensors. Because of their construction, they find the sensors require individual calibration.

2. Instruments that change their physical dimensions are many. The hair hygrometer is perhaps the most commonly used instrument of this type,

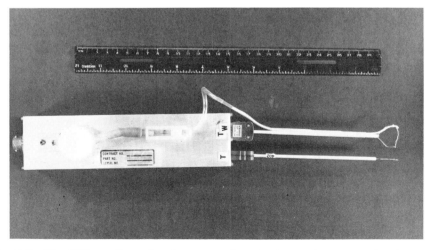

Fig. 5.9 The Tillman fine-wire thermocouples: (top) wet thermocouple; (bottom) dry thermocouple (from Redford et al., 1980).

182 ATMOSPHERIC HUMIDITY AND DEW

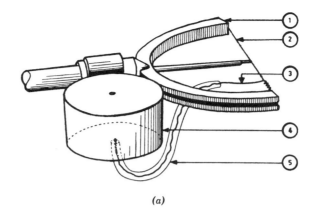

(a)

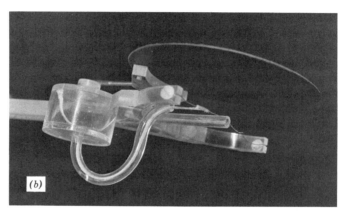

Fig. 5.10 The Wageningen Laboratory of Physics and Meteorology differential thermocouple psychrometer. (*a*) Schematic: 1, frame holding the thermoelectric wires and having the form of half a circle with an inner diameter of 85 mm; 2, constantan (inner) and manganin (outer) thermocouple wires (0.2 mm), welded together; 3, thick cotton lead wire bringing water to the small diameter cotton wire around the wet junction and neighboring wire parts; 4, Perspex container for the deionized water used; 5, Perspex sleeve guiding the cotton lead wire and filled with water (after Stigter and Welgraven, 1976). (*b*) Closeup view of the sensor.

although animal membranes and other materials are sometimes used. Human hair lengthens upon absorption of vapor from the air. Since changes in hair length are more linearly related to relative than to absolute humidity, hair can be arranged to activate a marker arm calibrated for relative humidity.

Hair hygrometers are difficult to maintain in calibration since the hairs tend to degenerate over time. These instruments have some peculiar properties to contend with as well. For example, hairs normally increase in length with increasing humidity, but a wet hair will shorten as it goes from saturated air to wetness. In this case, dew is indicated as 96% relative humidity.

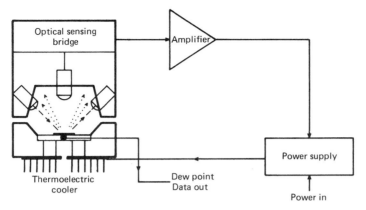

Fig. 5.11 Principle of operation of the dew-point hygrometer (courtesy of EG&G Environmental Equipment, Waltham, Massachusetts).

3. The condensation of a film of water is used in dew-point hygrometers, instruments that are meant to determine dew point temperatures (Fig. 5.11). In one common application, a mirrorlike surface is cooled slowly until the dew-point temperature of the ambient air is reached. At that temperature, a film of moisture is deposited on its cooled surface. A photocell detects the presence of dew and indicates the event on a time trace.

4. Chemical or electrical properties of certain systems can be altered upon the absorption of water vapor. This effect is used in the Dewcell and Dewprobe types of humidity sensors. These instruments use cloth or other materials impregnated with a hygroscopic chemical, such as lithium chloride, that takes up water if the ambient relative humidity is >11%. In the Dewprobe, the cloth is wrapped around a bobbin in contact with a bifilar winding of resistance wire. The system is powered. When the cloth is exposed to air containing water vapor, absorption occurs until the relative humidity of the air in contact with the bobbin is 11%. The electrical resistance of the absorbed chemical is decreased in proportion to the amount of water vapor absorbed. Increased electrical current flow will drive off some of the absorbed water vapor. When this has occurred, the current flow is greatly reduced because the electrical resistance has again become great. By these "overshooting" steps, an electrical equilibrium of the impregnated cloth and the air is achieved. The cavity temperature of the probe bobbin can be related to the dew point temperature of the ambient air. The chief disadvantage of the dew probe is its slow response to changes in vapor pressure and its limitation to ambient conditions of >11% relative humidity. Richards and Decker (1963) found that electrolytic resistance hygrometers also had very long time constants and were generally unreliable at humidities above 99%.

Adsorption of water vapor rather than **ab**sorption is the mechanism used in electro humidity sensors of another kind. A chemically treated styrene

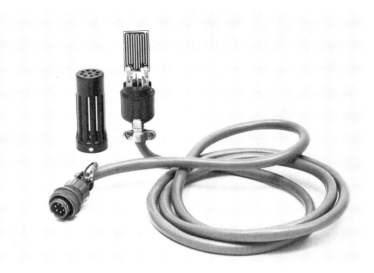

Fig. 5.12 A probe for sensing relative humidity and temperature (courtesy of Physical-Chemical Research Corp., New York, New York).

copolymer which has an electrically conducting surface layer is constructed on a thin plastic wafer. Changes in relative humidity cause the surface resistivity to vary. Such a sensor is shown in Fig. 5.12. The PCRC-HPB humidity probe includes a temperature sensor, as well, which provides automatic temperature compensation for the humidity sensor.

Other instruments based on chemical or electrical properties of materials have been proposed. Goltz et al. (1970), for example, suggested a barium fluoride film humidity sensor for use in eddy correlation measurements of evapotranspiration (see Chapter 7) because of its very rapid response to changing ambient humidity.

5. The absorption spectrum of water vapor provides the mechanism by which an infra red gas analyzer (IRGA) can be used to measure the water vapor content of air. A schematic of an IRGA is shown in Fig. 5.13. A special winding of metallic wire is heated to a temperature at which it emits strongly in the infrared (ir) range. The ir beam is "chopped" in order to pulse energy through each of two tubes in the analyzer. A standard gas with 0 or a known concentration of vapor is sealed into the reference cell. Air for analysis is moved through the sample cell. The absorption of ir radiation in the cells will vary with the concentration of the absorptive gas (in this case water vapor) in the sample stream. Less energy will reach the detector on the sample side than on the reference side if the concentration of water vapor is greater in the sample gas. The detector contains another gas that absorbs ir radiation. The side of the detector receiving less radiation (say, the sample side) will develop a lower pressure than will the reference side of the de-

tector. Differences of pressure that develop in the detector are a measure of differences in water vapor concentration in the gas stream.

IRGA systems are very accurate and can be made to respond rapidly. These systems can be used to monitor water vapor content in a continuous stream of air or in any number of air streams sampled in a predetermined sequence. IRGAs may also be designed to measure the difference between two unknown concentrations of water vapor in separate air streams flowing simultaneously through the cells. Resolution is normally improved when this type of differential analysis is made than when a wide range of absolute concentrations is measured with the use of a known reference gas. The IRGA principle is also used in the measurement of CO_2 concentrations and gradients in photosynthesis research and can be used to measure the concentration of other polar molecules such as CO.

Water vapor, because of its dielectric properties, absorbs microwave radiation (wavelength range of 1 mm to 1 m). Microwave refractometers used in meteorology are based on the principle that the resonant frequency of a cavity depends on the dielectric constant of its contents. A sealed cavity and a cavity in which air is brought for analysis are compared for difference

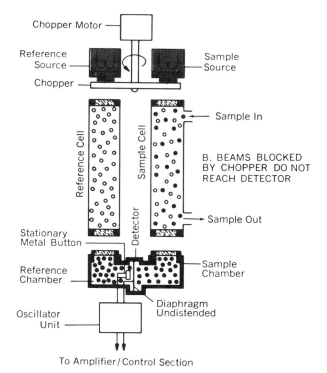

Fig. 5.13 Schematic of an Infra Red Gas Analyzer (IRGA). See the text for an explanation of operating principles (courtesy of Beckman Instruments, Fullerton, California).

Fig. 5.14 A Lyman-alpha hygrometer used at the Agricultural Meteorology Laboratory, Mead, Nebraska.

in frequency, which is almost linearly related to change in refractive index of the perforated cavity's contents. Microwave refractometers have been used on aircraft platforms sampling the troposphere, but have not been used extensively in micrometeorology. McGavin et al. (1971) reported on one microwave refractometer used in conjunction with a vertical sonic anemometer to measure evaporation near the ground by eddy-correlation.

Lyman-alpha (α) radiation is also used for rapid response measurements of water vapor concentration and fluctuations near the ground. Buck (1973, 1976) describes the function of a Lyman-α hygrometer in which hydrogen atoms and molecules are excited by a high voltage current and emit radiation in the far ultraviolet (121.56 nm). This wave-length of radiation is strongly absorbed by water vapor. The emitter is spaced a few centimeters from an absorbing chamber protected by a MgFl window that cuts off radiation outside of the 115–132 nm band. Nitrous oxide is ionized by the Lyman-α radiation entering the chamber. The flux density of that radiation is determined by the concentration of water vapor in the path between emitting and receiving chambers. Response time of the Lyman-α hygrometer is 12 ms. Thus, the instrument is capable of measuring rapid fluctuations in water vapor concentration which makes it suitable for use in eddy correlation studies.

When used in conjunction with a drag anemometer in eddy correlation studies, the Lyman-α gave accurate estimates of water vapor flux. A Lyman-α hygrometer in field use (similar to that described in Redford et al., 1980) is shown in Fig. 5.14.

5.10 INSTRUMENTS FOR MEASUREMENT OF DEW

Dew quantity is best measured with precision lysimeters, since errors due to distillation are avoided. Lysimeters, however, are extremely expensive and scarce. There are a number of sensitive weighing devices that can measure condensed water which forms on their surfaces. These, however, do not distinguish between dew and distillation. One simple and useful instrument for estimating the quantity of dewfall is the "Duvduvani dew plate," a polished wooden plate with surface characteristics that make it cool more or less like plant or soil materials. The patterns of droplets that form on the plates overnight are compared with a set of photographs of condensation patterns. This sensor, too, cannot resolve the question of whether the condensation stems from dew or distillation.

Dew duration is easier to measure than dew quantity and may be of greater significance, at least with respect to disease occurrence. Animal membranes which stretch when wet can be used to move a marker arm onto a moving clock-driven drum. The trace made represents the time of dew persistence. Lomas (1965) developed a simple recorder in which a clock driven rotating plate is in constant contact with a marker pencil. Only when the surface is wet, however, is a trace actually made on the plate.

Another principle of measurement involves the use of papers impregnated with salt to separate electrodes in a circuit. When the paper becomes wet enough, the resistance is decreased and the resultant current flow can be monitored. A printed circuit board has been used for the same purpose by Davis and Hughes (1970). An improved version of their duration meter involved the use of alternating current to avoid polarization of the sensing surface. Gillespie and Kidd (1978) used an electrical grid to sense the duration of moisture retention on mock leaves. In a comparison with observed moisture retention on onion foliage, a very light gray mock leaf held moisture for a maximum of 27 minutes longer in sunny conditions. Depending on weather conditions, errors of -20 to $+27$ min were found. Other colors on the mock leaves led to errors as great as two hours. Angle of sensor deployment also influenced the accuracy of this type of dew duration recorder.

Häckell (1980) used a network of exposed wires clamped onto broad, firm leaves, stalks, or heads of cereal plants to signal "yes–no" whether the surface is wet. Weiss and Lukens (1981) have developed another type of dew duration meter than involves use of an electrical grid and a water absorbent cotton cloth, the latter to simulate a leaf.

REFERENCES

Aslyng, H. C. 1965. Rain, snow and dew measurements. *Acta Agric. Scandinavica* **15**:275–283.

Baier, W. 1966. Studies on dew formation under semi-arid conditions. *Agric. Meteorol.* **3**:103–112.

Blad, B. L., J. R. Steadman, and A. Weiss. 1978. Canopy structure and irrigation influence on white mold disease and microclimate of dry edible beans. *Ecol. Epidemiol.* **68:**1431–1437.

Box, J. E. 1965. Design and calibration of a thermocouple psychrometer which uses the Peltier effect. *Humidity and Moisture* (A. Wexler, ed.), *Meas. Contr. Sci. Ind.* **1:**110–121.

Buck, A. L. 1973. Development of an improved Lyman-alpha hygrometer. *Atmos. Tech.* **2:**43–46.

Buck, A. L. 1976. The variable path Lyman-alpha hygrometer and its operating characteristics. *Bull. Am. Meteorol. Soc.* **57:**1113–1118.

Burrage, S. W. 1972. Dew on wheat. *Agric. Meteorol.* **10:**3–12.

Court, A., and D. Waco. 1965. Means and mid-ranges of relative humidity. *Mon. Weather Rev.* **93:**517–522.

Davis, D. R., and J. E. Hughes. 1970. A new approach to recording the wetting parameter by the use of electrical resistance sensors. *Plant Disease Reporter* **54:**474–479.

Fritschen, L. J., and P. Doraiswamy. 1973. Dew: An addition to the hydrologic balance of Douglas fir. *Water Resour. Res.* **9:**891–894.

Gillespie, T. J., and G. E. Kidd. 1978. Sensing duration of leaf moisture retention using electrical impedance grids. *Can. J. Plant Sci.* **58:**179–187.

Goltz, S. M., C. B. Tanner, G. W. Thurtell, and F. E. Jones. 1970. Evaporation measurements by an eddy correlation method. *Water Resour. Res.* **6:**440–446.

Gringorten, I. I., H. A. Samela, I. Solomon, and J. Sharp. 1966. Atmospheric humidity atlas–Northern Hemisphere. Air Force Surveys in Geophysics No. 186, AFCRL-66-621, August, 1966.

Häckel, H. 1980. New developments of an electrical method for direct measurement of the wetness-duration on plants. *Agric. Meteorol.* **22:**113–119.

Hadley, N. F. 1979. Wax secretion and color phases of the desert Tenebrionid beetle Cryptoglossa verrucosa (Le Conte). *Science* **203:**367–369.

Harrison, L. D. 1965. Fundamental concepts and definitions relating to humidity. *Humidity and Moisture* (A. Wexler, ed.), *Meas. Contr. Sci. Ind.* **3:**3–70.

Hoffman, G. J. 1973. Humidity effects on yield and water relations of nine crops. *Trans. ASAE* **16:**164–167.

Kerr, J. P., and M. F. Beardsell. 1975. Effect of dew on leaf water potentials and crop resistances in a Paspalum pasture. *Agron. J.* **67:**596–599.

Lomas, J. 1965. Note on dew duration recorders under semiarid conditions. *Agric. Meteorol.* **2:**351–359.

McGavin, R. E., P. B. Uhlenhopp, and B. R. Bean. 1971. Microwave evapotron. *Water Resour. Res.* **7:**424–428.

Monteith, J. L. 1957. Dew. *Q. J. Roy. Meteorol. Soc.* **83:**322–341.

Noffsinger, T. L. 1965. Survey of techniques for measuring dew. *Humidity and Moisture* (A. Wexler, ed.), *Meas. Contr. Sci. Ind.* **2:**523–531.

Rawlins, S. L. 1966. Theory of thermocouple psychrometers used to measure water potential in soil and plant samples. *Agric. Meteorol.* **3:**293–310.

Redford, T. G., S. B. Verma, and N. J. Rosenberg. 1980. Humidity fluctuations over a vegetated surface measured with a Lyman-alpha hygrometer and a fine-wire thermocouple psychrometer. *J. Appl. Meteorol.* **19:**860–867.

Richards, L. A., and D. L. Decker. 1963. Difficulties with electrolytic resistance hygrometers at high humidity. *Soil Sci. Soc. Am. Proc.* **27:**481.

Rosenberg, N. J. 1969. Evaporation and condensation on bare soil under irrigation in the east central Great Plains. *Agron. J.* **61**:557–561.

Rosenberg, N. J., and K. W. Brown. 1974. 'Self-checking' psychrometer system for gradient and profile determinations near the ground. *Agric. Meteorol.* **13**:215–226.

Savage, M. J., A. Cass, and J. M. de Jager. 1981. Calibration of thermocouple hygrometers. *Irrig. Sci.* **2**:113–125.

Stigter, C. J., and A. D. Welgraven. 1976. An improved radiation protected differential thermocouple psychrometer for crop environment. *Arch. Met. Geophys. Bioklimatol. B*, **24**:177–187.

Tillman, J. 1973. Wet and dry bulb thermocouple psychrometry. *Atmos. Tech.* **2**:77.

Tuller, S. E., and R. Chilton. 1973. The role of dew in the seasonal moisture balance of a summer-dry climate. *Agric. Meteorol.* **11**:135–142.

Valancogne, C., and F. A. Daudet. 1974. Adaptation of the Spanner psychrometer technique for the measurement of water potential in soils under field conditions. Problem raised by temperature. Ann. Agron. **25**:733–751.

Wallin, J. R. 1967. Agrometeorological aspects of dew. *Agric. Meteorol.* **4**:85–102.

Weiss, A., L. E. Hipps, B. L. Blad, and J. R. Steadman. 1980. Comparison of within-canopy microclimate and white mold disease (*Sclerotinia sclerotiorum*) development in dry edible beans as influenced by canopy structure and irrigation. *Agric. Meteorol.* **22**:11–21.

Weiss, A., and D. L. Lukens. 1981. Electronic circuit for detecting leaf wetness and comparison of two sensors. *Plant Dis.* **65**:41–43.

Yao, A. Y. M. 1974. A statistical model for the surface relative humidity. *J. Appl. Meteorol.* **13**:17–21.

Zollinger, W. D., G. S. Campbell, and S. A. Taylor. 1966. A comparison of water potential measurements made using two types of thermocouple psychrometers. *Soil Sci.* **102**:231–239.

CHAPTER 6 – MODIFICATION OF THE SOIL TEMPERATURE AND MOISTURE REGIMES

6.1 INTRODUCTION

In Chapter 2 the physics of soil heat flux and the temperature distribution in the soil mantle were discussed. The biological importance of soil temperature is stressed in this chapter and methods to make soil temperatures optimal for plant growth are explored. But what is optimum?

In the life cycle of the plant, temperature of the soil at germination is critical. Some examples follow: Corn seeds will not germinate if the soil temperature is lower than about 10°C (Alessi and Power, 1971). Lindstrom et al. (1976) indicate that the time from planting to emergence of winter wheat increases linearly as temperature decreases below 25°C. The range of temperatures used in their study was 5–35°C. The minimum water potential required for germination increases with increasing soil temperature so that if water is limiting, a lower temperature may be better for rapid germination.

Walker (1969) demonstrated that minor temperature changes in soil can effect dramatic changes in growth and nutritional behavior of maize seedlings. He found that calcium and boron uptake were uniquely related to soil temperature. According to Dalton and Gardner (1978) the osmotic permeability coefficient of root membranes and the rate constant for active nutrient uptake are also directly affected by temperature of the soil. Mederski and Jones (1963) found that heating a field soil in Ohio to increase temperature from 3 to 8°C before planting and at seedling emergence increased the uptake of nitrogen and potassium by 25% and of phosphorus by 100% (determined 30 days after emergence) with a consequent increase in final yield as well. Kaspar et al. (1981) have found that the pattern of distribution in soybean is affected by soil temperature. In cold soil the roots spread more laterally lining up almost parallel to the surface. In warmer soil the roots penetrate more deeply which is advantageous with respect to nutrient and water uptake.

Kuo and Boersma (1971) found in laboratory experiments that soybean nitrogen fixation responds to root temperature. The optimum temperature for maximal N fixation and net photosynthesis was about 27°C. Water uptake (transpiration) was also affected by root temperature, the maximum occurring at 30°C. Kramer (1969) presents many examples that illustrate the fact that transpiration can be impeded by low soil temperature. Temperature affects the viscosity of water, one factor contributing to the reduction in transpiration at low temperature.

Soil respiration is controlled by soil temperature in a manner described by (2.7). The breakdown of dead organic materials in the soil and the respiratory release of CO_2 by living roots depends not only on temperature but on the availability of moisture in the soil, as well. The range of soil respiration rates varies very widely. For example, Peterson and Billings (1975) found emission of CO_2 from a tundra soil in Alaska to be about 125 mg m^{-2} h^{-1}. Mogensen (1977) found respiration in a ryegrass field in Denmark to be about 900 mg m^{-2} h^{-1}.

Yield in a wide range of species is affected by soil temperature. Cannell et al. (1963) report that the greatest aboveground growth of tomato occurred with controlled soil temperature near 20°C (range 12–36°C) in conjunction with a high soil moisture potential. When soil temperature was increased from 12 to 20°C dry weight of snap beans increased (Mack et al., 1966). In Sitka spruce Turner and Jarvis (1975) found rates of photosynthesis for both cold-hardened and unhardened shoots to be greatly reduced by soil temperatures lower than -0.5°C. Photosynthesis for plants exposed to light increased rapidly and maintained high rates for 4–6 h (or as long as the illumination lasted). Shoots in colder soil, however, did not reach as high rates of net photosynthesis upon illumination, and rates decreased from the maxima achieved earlier after about 2 h. After 6 h rates of net photosynthesis in unhardened shoots growing in soils colder than -4.5°C were essentially zero.

Thus, we see that the influence of soil temperature on plant growth is multifaceted. Soil temperature affects the rate of biological reactions, for example, enzymatic activity and respiration in living roots. The activity of soil microorganisms and the decomposition of dead organic matter are also controlled by soil temperature. Thus, indirectly, soil temperature controls the composition of the soil atmosphere.

Soil temperature effects on plant growth are interactive with soil moisture status and also with the nutrient status of the soil. Hence, it is difficult to propose a specific optimal temperature for each type of plant at each stage of its growth. For many species, however, a generally good idea of the temperature optima has emerged through growth chamber and field research and through analysis of ecological settings.

In the following pages we will examine ways to manipulate the soil temperature regime.

6.2 SLOPE AND ASPECT

Both **slope** and **aspect** (direction) affect soil temperature and soil heat flux since capture of radiation is determined in part by these factors. Slope affects the receipts of radiation per unit area of absorbing surface because of the

192 MODIFYING TEMPERATURE AND MOISTURE IN SOIL

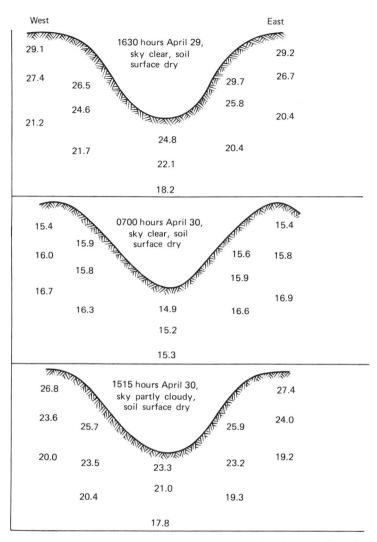

Fig. 6.1 Soil temperature (°C) at depths of 25, 75, and 150 mm below the surface, late April at Ames, Iowa (after Shaw and Buchele, 1957).

"cosine effect." The more normal the sun's rays are to the surface, the greater the flux density of the radiation impinging on the surface. Aspect affects the heat flux in an obvious way. Those slopes that face the sun capture more energy than those that face away from it.

The importance of slope and aspect varies with latitude. Differences in slope and aspect are more critical in the receipts of direct beam solar radiation in the high latitudes than in the tropics where the sun remains high throughout the year. However, because of low sun angle the ratio of diffuse to direct beam radiation is large in the high latitudes. This effect reduces the importance of slope and aspect because the sky is a nearly isotropic source of diffuse radiation that is transmitted in nearly equal intensity in all directions. Slope and aspect are more important in summer than in winter because the low sun angles in the winter season increase the proportion of diffuse radiation. Whatever the season, slope and aspect are of lesser importance in cloudy than in clear conditions for the same reason. By using Lambert's cosine law as a guideline, it is possible to design methods that beneficially alter the soil temperature regime. Ridging and shaping of agricultural fields are examples of such methods.

In a now classic experiment, Shaw and Buchele (1957) measured temperatures at three depths in a number of positions in a north–south running furrow (Fig. 6.1). Temperature differences in spring are important since the earlier planting allows a longer growing season and, generally, yields are improved. Temperatures were lowest in the furrow. The crests and slopes alternated in highest temperatures, depending on the time of day. Generally, the west-facing slope was warmer than the opposite slope. This is because the east-facing slope is illuminated at a time of day when the soil is most cool. Evaporation of dew at this time also tends to keep temperatures low. Figure 6.2 shows the patterns of temperature in the ridge–furrow system in May and again in July. Regardless of the season, the slopes remain the warmest parts of the system during the day.

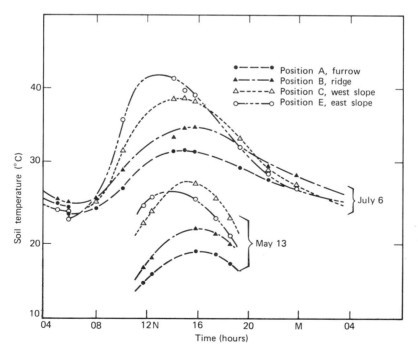

Fig. 6.2 Soil temperature at the 75 mm depth in a ridge–furrow profile, spring and summer at Ames, Iowa (after Shaw and Buchele, 1957).

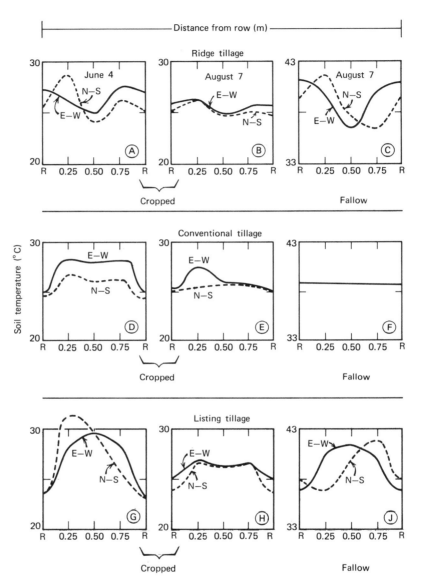

Fig. 6.3 Distribution of soil temperatures across the corn-row spacing at the 100 mm depth at 1600 hours on June 4 and August 7. Graphs A, B, D, E, G, and H are for cropped plots; graphs C, F, and J are for fallow plots. Each graph gives the temperature distribution for rows running north and south and running east and west (after Burrows, 1963).

In a similar but more detailed study, Burrows (1963) measured soil temperatures in corn rows planted in furrows made by listing, in ridges, and in conventional flat microtopography. Early in the season, it was found, soil temperatures were highest in the ridge planted rows, particularly during late afternoon. Soil temperatures were lowest in the corn rows planted in listed furrows. Temperatures in the flat microtopography were intermediate. The experiment was replicated in north–south and east–west running rows. Row direction had little effect on soil temperature. By August neither position of planting nor row orientation had significant influence on soil temperature because the heavy crop cover fully shaded the ground.

Figure 6.3 from Burrow's studies illustrates the effects on soil temperature of cropping, row orientation, and type of tillage. The data are for late afternoon at a depth of 0.1 m.

The developed crop (August 7) clearly damps all differences due to row orientation and tillage practice that existed on June 4. In the listed topography the row is coolest whether cropped or not. East–west rows in listed fallow are warmest on the ridge top. In north–south rows the west-facing slope is warmest. Under conventional tillage the rows tend to be cooled slightly by the plants in early June. In ridge tillage, east–west rows have a symmetrical temperature pattern, whereas the north–south rows seem warmest to the east and coolest slightly to the west of the rows. Burrow's studies show that for rapid emergence and early growth east–west ridge planting was best, followed by north–south ridge, east–west conventional tillage, and north–south conventional. Listing in any direction was found least effective.

The emergence of grain sorghum and corn is sensitive to temperature. Adams (1967) used mulches and bed configuration as means of hastening germination and improving stands for these crops in the black land region of North Texas. The beds were sloped for ridge planting. One bed had a 20° south-facing slope. In both cases, temperatures were increased at the depth of 70 mm as compared to conventional flat tillage. Emergence time for grain sorghum decreased by 1 day per 0.9°C temperature increase in the seedbed.

Cultivation may affect plant performance through mechanisms other than soil temperature change. Cartee and Hanks (1974), for example, found that seed and dry matter yield of pinto beans in Utah are affected by the type and timing of ridging and cultivation practiced early in the season. Ridging before emergence produced greater yield and total dry matter than did planting on the ridge or postemergence ridging. The results were correlated with soil moisture content above the 0.15 m depth and with the incidence of root rot infection. No direct correlation with soil temperature was found.

6.3 MULCHING

We define **mulching** as the application or creation of any soil cover that constitutes a barrier to the transfer of heat or vapor. Many types of mulches have been used in agriculture. Some examples of mulches made with in situ materials are the following: (1) the dust mulch, created by finely pulverizing the upper layer of soil to interrupt continuity of capillaries and, hence, to create a barrier to vapor flow; (2) weed or trash mulches, created by cutting weeds and other residual crop materials and allowing these to dry in a layer over the soil surface; (3) stubble mulch, created by permitting residues of small grain crops to remain standing in the field so as to increase surface roughness and reduce soil blowing; (4) straw mulch, created by combines blowing out small-grain straw over fields at harvest time.

Some examples of manufactured or "imported" mulching materials used for moisture conservation and/or soil temperature modification include paper of various textures and colors, aluminum foil, gravel, coal, cinders, petroleum by-products and black, transparent, and white opaque plastics of various kinds.

The principles of soil mulching are nicely illustrated in a classic work by Waggoner et al. (1960). They applied black plastic, translucent plastic, aluminum foil, paper, and hay mulches to soils and measured the resulting alterations in surface energy balance and in soil and air temperature profiles. Their results are shown in Figs. 6.4 and 6.5.

Flux density (W m^{-2})

	Open	Black	Paper	Hay
R_n	642	712	433	607
H	−362	−635	−349	−489
LE	−195	0	−42	−84
S	−85	−77	−42	−35

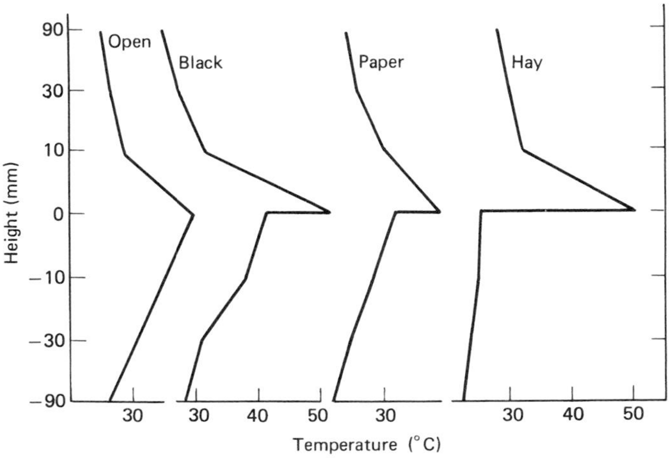

Fig. 6.4 Midday temperature profiles and energy balance in mid-June at Hamden, Connecticut (after Waggoner et al., 1960).

	Open	Flux density (W m^{-2}) Black	Translucent	Aluminum
R_n	447	503	447	279
H	−279	−453	−328	−265
LE	−70	0	0	0
S	−98	−50	−119	−14

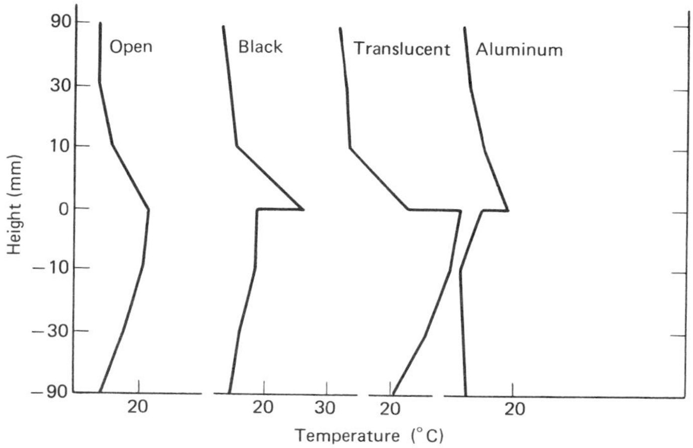

Fig. 6.5 Midday temperature profiles and energy balance in mid-October at Hamden, Connecticut (after Waggoner et al., 1960).

Total hemispherical radiation (shortwave and longwave) was about 1400 W m^{-2} during the day shown in Fig. 6.4. Black plastic reduced the outgoing radiation as evidenced by the greater net radiation, whereas paper, especially, and hay increased it. All mulches reduced the quantity of energy consumed in evaporation by blocking the transport of vapor out of the soil. Hay was least effective in this regard. Sensible heat generated at the surface was increased by the black plastic and the hay. The temperature of the hay and black plastic surfaces was high. Black plastic led to an increase in soil temperature whereas hay and paper had a cooling effect and reduced the amount of heat penetrating into the soil.

On a day of less intense solar radiation, black plastic, translucent plastic, and aluminum foil mulches were compared with an unmulched soil (Fig. 6.5). Outgoing radiation was sharply increased by the highly reflectant aluminum. The three materials effectively shut off vapor transport so that no energy was consumed by evaporation. The surface under aluminum was cooled relative to the unmulched soil, but it was warmed by the plastic. Aluminum caused a net cooling of the soil, whereas the translucent plastic led to marked warming. The translucent plastic apparently created a green-

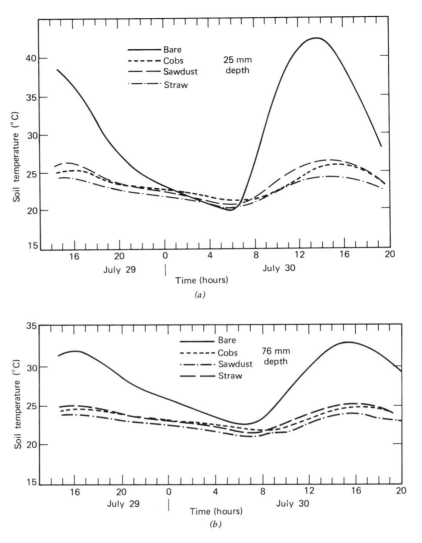

Fig. 6.6 (*a*) Soil temperature at the 25 mm depth as influenced by type of mulch. Sky clear, ground surface dry. Ames, Iowa (after Denison et al., 1953). (*b*) Same as (*a*) for the 75 mm depth.

houselike condition. Visible light penetrated the plastic, but condensation of vapor on the underside trapped infrared radiation within. Thus, the plastic acted in a manner similar to glass in a greenhouse.

Waggoner et al. (1960) summarized the influence of the various mulching materials as follows: Black plastic absorbs much radiation but transmits little. The mulch is heated greatly but transmits little energy downward by conduction because of the intermediate layer of still air. Translucent plastic transmits much of the incident radiation through in the visible range, but it permits little of the longwave radiation back out because of vapor condensation on the underside. Aluminum transmits no radiation but remains cool because of its high reflection. The soil below remains cool. Paper reflects much radiation and transmits little. Hay absorbs radiation like soil but transmits little. The hay gets hot but the soil stays cool.

Ekern (1967) in Hawaii found that black plastic mulch reduced water use by pineapple grown in lysimeters. Growth was also increased by one-third in agreement with the known response of pineapple to winter soil temperature, which was increased by 1.6°C by the mulch.

As shown above in Figs. 6.4 and 6.5 various types of mulching materials can affect the energy balance at the soil surface. Another important influence of mulching materials is demonstrated in Fig. 6.6a,b from Denison et al. (1953). Here it is seen that, acting as thermal insulators, certain opaque mulches of vegetative origin can damp the diurnal wave of soil temperature considerably. Barkley et al. (1965) also used straw, sawdust, and wood fiber cellulose, as well as a plastic emulsion, as aids in turf establishment and found that these materials, too, moderate soil temperature.

Specific uses of clear plastic mulch have been reported for sorghum and corn (Adams, 1965, 1967, 1970). The mulches consistently raised soil temperature. Miller (1968) "mulched" a sweet corn plot with water-filled plastic bags. By absorbing solar radiation the water bags increased soil temperature sufficiently to hasten harvest by 9 days. Chalk soils in the Champagne district of France warm very slowly in spring. Clear plastic was found more effective than opaque plastic in warming these soils (Ballif and Dutil, 1975). Andrew et al. (1976) noted that clear plastic is helpful in sweet corn production in Wisconsin, advancing silking and harvest by about 6 days. As one would expect, the effect is most important in cool years.

Petroleum substances have also been used as mulches to raise soil temperature and reduce evaporation. Qashu and Evans (1967) used a black granular by-product of petroleum refining. This was placed in strips of varying width and configuration on a flat surface soil in Arizona. Temperature beneath the mulch was increased as was the temperature gradient indicating heat movement into the soil. Evaporative losses were suppressed by the physical presence of the mulch.

Kowsar et al. (1969) applied a petroleum mulch to soil in Oregon and found an increase of 5°C in maximum soil temperature at the 10 mm depth. Untreated soil lost water rapidly from the upper 40 mm of soil. Treated soil lost water in the upper 10 mm but gained water below this zone, apparently because of thermally induced liquid and vapor flow (see Chapter 3). Since this is the depth in which seeds are normally placed the vapor redistribution effect can be quite important.

Phipps and Cochrane (1975) in England sprayed 0.15–0.20 m bands of bitumen mulch 1–2 mm in thickness over corn that had already been seeded for forage. The result during one season was an increase in temperature of about 4°C at the 0.1 m depth. Yields were increased in treated areas. In a second trial during a dry year, Phipps and Cochrane (1977) found that the bitumen mulch increased maximum and mean soil temperature by 0.7 and 1.3°C in May and June but that neither emergence, silking dates, yield, nor dry matter was affected by the treatment. They assume that positive effects of the mulch were masked by the drought.

Gravel mulches have been used in a number of instances. Gravel mulch was found, in a laboratory study by Benoit and Kirkham (1963), to be more effective in suppressing evaporation than either a dust mulch or a mulch of ground corncobs. Of the three, dust mulch was least effective. Fairbourn (1973) conducted laboratory and field tests using gravel as a mulch. He found that gravel conserved more water and raised soil temperatures to a greater degree than did a vegetative mulch. Black and white gravels were used on plots of corn, sorghum, tomatoes, and soybeans and yield improvements occurred in all but the soybeans.

As a possible substitute for gravel, which must be brought into the field, Fairbourn and Gardner (1975) manufactured water repellent clods of local soil. These materials reduced evaporation from bare soil by significant amounts. The synthesized clods lasted for about 1 yr under field conditions. Fairbourn (1974) also used pieces of coal ranging in size from 10 to 30 mm in diameter as a mulch to reduce evaporation and increase soil temperature.

Moody et al. (1963) found that 6.73 t ha^{-1} of straw on the soil surface temporarily retarded early season growth of corn as compared to plots where the same amount of straw was plowed under. However, overall growth and final yields were superior on the mulched plots. The early season retardation was due to lower soil temperature but later growth and yield increase were due to a more favorable moisture status.

Sorghum germination and growth were affected by variable amounts of straw mulch applied in early spring in northwest Texas. Unger (1978) reports that, whereas temperature increases were delayed by straw applied at rates of 1, 2, 4, 8, and 12 t ha^{-1}, optimum levels were reached before normal planting times. Soil temperature remained low but later, when moisture became the limiting factor, plants on high rates of mulch grew best. Mulching reduced the mean, maximum, and minimum and the standard deviation of soil temperature.

Straw mulching has received considerable attention in the Great Plains region of the United States and Canada. Greb (1966) found that soil water evaporation losses in Colorado were reduced linearly by increased surface application of wheat straw up to 90% soil cover. Over a 20-day period with mean maximum temperature of 22°C, soil water losses were reduced from 11.4 mm without straw to 5.8 mm with 3.36 t ha^{-1}. Straw reduced maximum daily soil temperature at 0.25 m depth by about 1.5°C for each t ha^{-1} applied.

Adams et al. (1976) tested the combined effects of row spacing and straw mulching on the rate of soil drying in a sorghum field in Texas. Figure 6.7 shows the influence of mulch rate (oat straw) on first stage evaporation (i.e.,

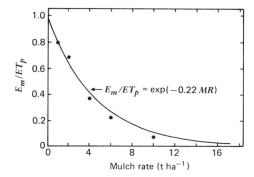

Fig. 6.7 Comparison of effect of mulch rate (MR) on first stage evaporation at Temple, Texas (after Adams et al., 1976). Reprinted from *Soil Science Society of America Journal*, Volume **40**, page 441, 1976, by permission of the Soil Science Society of America.

when the soil surface is wet and evaporation is comparable to that from a free water surface). The ordinate shows evaporation with mulch (E_m) as a fraction of the potential evaporation from a wet surface (ET_p). The first increments of straw are the most effective but increments up to 10 t ha^{-1} decrease evaporation rate still further.

6.4 ARTIFICIAL HEATING OF THE SOIL

When the first edition of this book was written (1972–1974), it was not totally unreasonable to assume that artificial means of altering soil temperature might be economically justified. Research had already demonstrated that grass might be established and maintained out of season by means of electrical heating systems installed in the soil (e.g., Ledeboer et al., 1971). In Fig. 6.8, the results of electrical heating of the soil with polyvinyl chloride (PVC)-jacketed heating cables, buried 0.15 and 0.30 m apart in a silty loam are shown. Attempts were made to keep the soil near 8–10°C throughout the winter season in Rhode Island. Protective screen covers over the soil were used in some plots. Temperature changes near the cables were directly related to power input. Near the surface, weather changes had the greatest effect on the soil temperature. In a similar study, McBee et al. (1968) used underground wires to keep Bermuda grass and St. Augustine grass green during winter in Texas. In this study, varietal differences in response to the heating were noted.

An alternative to soil heating by electrical power is the use of waste industrial heat for that purpose. This may also provide certain environmental benefits. As an example, consider that the conversion of fossil fuels into electrical energy is a very inefficient process in thermodynamic terms. Coal-

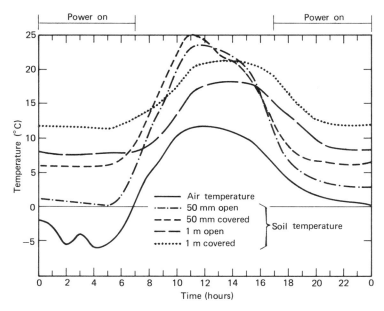

Fig. 6.8 Soil temperatures at two depths with and without protective cover and ambient air temperature in an area where soil was heated with 108 Wm^{-2}. Kingston, Rhode Island (after Ledeboer et al., 1971).

fired power plants are only about 50% efficient in converting energy stored in the fossil fuel to electrical energy. Nuclear power plants are only about 40% efficient. Thus, very large quantities of heat are produced that must be removed from the plants and dissipated into the environment. Water is the coolant most generally used in power plants.

Many chemical and manufacturing processes also produce large quantities of heat that must be disposed of. Water enters the plant at a low temperature and is discharged at a temperature that is usually 5–10°C higher. In some cases, water is drawn from a river or large lake and discharged back into it at a higher temperature. In other cases, the warm water is ponded and cools slowly by evaporation and sensible heat transport to the air. Forced draft and free draft evaporative cooling towers are also commonly used to dissipate the heat. In each of the cases mentioned above, the thermal energy is treated as a waste product to be disposed of at the least cost.

A number of schemes for using power plant or industrial waste heat have been proposed for this purpose. One of the most complete is reported by DeWalle (1974) and involves plans for an Agropower–Waste Water System as a means of land disposal of waste heat and waste water. The general scheme is shown in Fig. 6.9. A power plant supplies heated water for circulation in buried pipes underlying agricultural fields. The city powered by the plant produces waste water which, after treatment, is used to irrigate the heated fields.

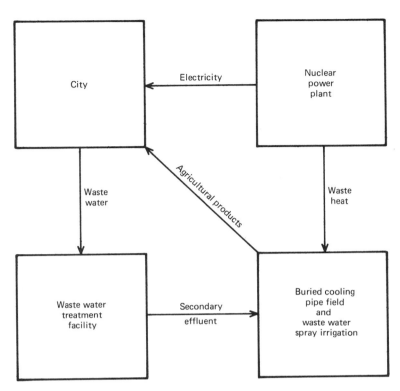

Fig. 6.9 Diagram of a proposed Agropower–Waste Water System (after DeWalle, 1974).

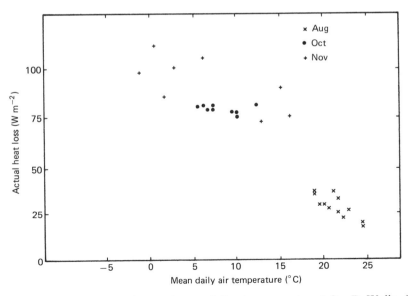

Fig. 6.10 Actual soil heat loss and mean daily air temperature (after DeWalle, 1974).

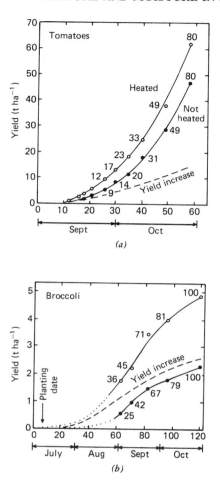

Fig. 6.11 (a) Yield and yield increase of ripe tomatoes. (b) Yield and yield increase of broccoli. The numbers by each data point indicate the accumulated yield in percent of final yield (after Rykbost et al., 1975b). Reprinted from *Agronomy Journal*, Volume 67, page 741, 1975, by permission of the American Society of Agronomy.

For the greatest effectiveness, the designers proposed that the pipes be buried in sandy soils and that these soils be left moist so far as possible in order to maintain a high thermal conductivity that would lead to the greatest heat dissipation. Actual heat losses measured in a field plot were related to mean daily air temperature as is shown in Fig. 6.10 (from DeWalle, 1974). The efficiency of the system as a means of dissipating waste heat decreased markedly from winter to summer. In another test of this concept Baille and Mermier (1980) used waste water circulating in underground pipes from a nuclear power plant in France to warm soil. Water at 15°C increased temperature at the surface by an average of 2°C and by a constant 3°C at a depth of 0.2 m.

Rykbost et al. (1976) studied the physics of heating soil with underground heating cable to simulate the use of waste condenser heat. Increases in temperature with respect to unheated soil were greatest with increasing depth so that at 2.2 m the temperature was 10 and 13°C greater in summer and winter, respectively. Dissipation of energy to the atmosphere was about 8 and 21 W m^{-2} in summer and winter, respectively. For the dissipation of heat from power plants by these means, very large areas of land would be required. The heat sources would, also, need to be placed near the surface.

Rykbost et al. (1975a,b) were able to show that the soil warming of about 10°C in Oregon could lead to yield increases of 19–50%, varying from year to year, for a number of agronomic crops and 19–100% for a wide range of vegetable crops. In the latter case, the greatest advantages occurred from soil warming early in the growing season, although growth rates continued to be most rapid on heated soil throughout the season. Cumulative yields of tomatoes and broccoli on heated and unheated soils are shown in Fig. 6.11a and b.

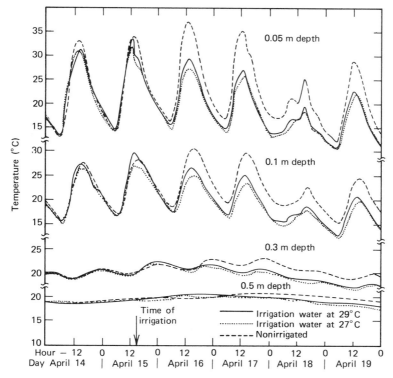

Fig. 6.12 Diurnal variation in soil temperatures at 0.05, 0.10, 0.30, and 0.50 m depths in irrigated and nonirrigated soil (after Wierenga et al., 1971).

Despite the physical problems associated with soil heating by these means, an economic analysis of the Rykbost et al. studies by Schmisseur et al. (1975) indicated that high value crops could be produced profitably with soil warming using waste heat in cooling water from power plants. The limitation (as of 1975) was the high capital cost of installation of the heat dispensing systems.

An alternative way of dissipating waste industrial heat may be by irrigating soil directly with heated water. A number of studies have shown that the temperature change in soil due to adding warm (or cold) water is small and quite transitory. For example, Wierenga et al. (1971) applied cold irrigation water to soils at a number of sites. Figure 6.12 is typical of their results and shows a temperature effect due to the water on the order of 2–3°C. The effect lasted for less than a day.

Wierenga and his co-workers consider that the biological effects of warm water irrigation would not be detrimental but the anticipated agricultural benefits would be small and would not justify any expenses over and above the cost of the water itself.

From the above, it seems evident that soil can provide an enormous sink for industrial waste heat. Evaporative cooling at the surface would be expected to dissipate the largest portion of the excess heat contained in the cooling water. Because of its great heat capacity, soil temperatures will not be markedly changed by heat contained in water at the normal temperatures of industrial coolants. A more cost effective and agronomically useful technique may lie in systems designed to deliver waste industrial heat to soils by means of heat exchange systems in which warm water is circulated, returning cool to the power plant site. Such systems may be economically viable since the yields of economically high value crops can be improved in temperate regions by means of soil heating.

REFERENCES

Adams, J. E. 1965. Effect of mulches on soil temperature and grain sorghum development. *Agron. J.* **57**:471–474.

Adams, J. E. 1967. Effect of mulches and bed configuration. I. Early-season soil temperature and emergence of grain sorghum and corn. *Agron. J.* **59**:595–599.

Adams, J. E. 1970. Effect of mulches and bed configuration. II. Soil temperature and growth and yield responses of grain sorghum and corn. *Agron. J.* **62**:785–790.

Adams, J. E., G. F. Arkin, and J. T. Ritchie. 1976. Influence of row spacing and straw mulch on first stage drying. *Soil Sci. Soc. Am. J.* **40**:434–442.

Alessi, J., and J. F. Power. 1971. Corn emergence in relation to soil temperature and seeding depth. *Agron. J.* **63**:717–719.

Andrew, R. H., D. A. Schlough, and G. H. Tenpas. 1976. Some relationships of a plastic mulch to sweet corn maturity. *Agron. J.* **68**:422–425.

Baille, A., and M. Mermier. 1980. Utilisation des eaux de rejet en agriculture. II. Modifications microclimatiques dues a l'irrigation a l'eau tiede. *Agric. Meteorol.* **22**:85–91.

Ballif, J. L., and P. Dutil. 1975. The warming of chalk soils by plastic films. Thermic measurements and evaluations. *Ann. Agron.* **26**:159–167.

Barkley, D. G., R. E. Blaser, and R. E. Schmidt. 1965. Effect of mulches on microclimate and turf establishment. *Agron. J.* **57**:189–192.

Benoit, G. R., and D. Kirkham. 1963. The effect of soil surface conditions on evaporation of soil water. *Soil Sci. Soc. Am. Proc.* **27**:495–498.

Burrows, W. C. 1963. Characterization of soil temperature distribution from various tillage-induced microreliefs. *Soil Sci. Soc. Am. Proc.* **27**:350–353.

Cannell, G. H., F. T. Bingham, J. C. Lingle, and M. J. Garber. 1963. Yield and nutrient composition of tomatoes in relation to soil temperature, moisture and phosphorus levels. *Soil Sci. Soc. Am. Proc.* **27**:560–565.

Cartee, R. L., and R. J. Hanks. 1974. Effect of ridging and early-season cultivation on bean yield. *Agron. J.* **66**:632–635.

Dalton, F. N., and W. R. Gardner. 1978. Temperature dependence of water uptake by plant roots. *Agron. J.* **70**:404–406.

Denison, E. L., R. H. Shaw, and B. F. Vance. 1953. Effect of summer mulches on yields of everbearing strawberries, soil temperature and soil moisture. *Iowa State College J. Sci.* **28**:167–175.

DeWalle, D. R., ed. 1974. An agro-power-waste water complex for land disposal of waste heat and waste water. Institute for Research on Land and Water Resources. The Pennsylvania State Univ. Res. Pub. No. 86, 195 pp.

Ekern, P. C. 1967. Soil moisture and soil temperature changes with the use of black vapor-barrier mulch and their influence on pineapple (*Ananas comosus* (L.) Merr.) growth in Hawaii. *Soil Sci. Soc. Am. Proc.* **31**:270–275.

Fairbourn, M. L. 1973. Effect of gravel mulch on crop yields. *Agron. J.* **65**:925–928.

Fairbourn, M. L. 1974. Effect of coal mulch on crop yields. *Agron. J.* **66**:785–789.

Fairbourn, M. L., and H. R. Gardner. 1975. Water-repellent soil clods and pellets as mulch. *Agron. J.* **67**:377–380.

Greb, B. W. 1966. Effect of surface-applied wheat straw on soil water losses by solar distillation. *Soil Sci. Soc. Am. Proc.* **30**:786–788.

Kaspar, T. C., D. G. Wooley, and H. M. Taylor. 1981. Temperature effect on the inclination of lateral roots of soybeans. *Agron. J.* **73**:383–385.

Kramer, P. J. 1969. *Plant and Soil Water Relationships.* McGraw-Hill, New York.

Kowsar, A., L. Boersma, and G. D. Jarman. 1969. Effects of petroleum mulch on soil water content and soil temperature. *Soil Sci. Soc. Am. Proc.* **33**:783–786.

Kuo, T., and L. Boersma. 1971. Soil water suction and root temperature effects on nitrogen fixation in soybeans. *Agron. J.* **63**:901–904.

Ledeboer, F. B., C. G. McKiel, and C. R. Skogley. 1971. Soil heating studies with cool season turfgrass I–IV. *Agron. J.* **63**:677–691.

Lindstrom, M. J., R. I. Papendick, and F. E. Koehler. 1976. A model to predict winter wheat emergence as affected by soil temperature, water potential and depth of planting. *Agron. J.* **68**:137–141.

McBee, G. G., W. E. McCure, and K. R. Beerwinkle. 1968. Effect of soil heating on winter growth and appearance of Bermuda grass and St. Augustine grass. *Agron. J.* **60**:228–231.

Mack, H. J., S. C. Fang, and S. B. Apple, Jr. 1966. Response of snap beans (*Phaseolous vulgaris* L.) to soil temperature and phosphorus fertilization on five western Oregon soils. *Soil Sci. Soc. Am. Proc.* **30**:236–239.

Mederski, H. J., and J. B. Jones, Jr. 1963. Effect of soil temperature on corn plant development and yield: I. Studies with a corn hybrid. *Soil Sci. Soc. Am. Proc.* **27**:186–189.

Miller, D. E. 1968. Emergence and development of sweet corn as influenced by various soil mulches. *Agron. J.* **60**:369–371.

Mogensen, V. O. 1977. Field measurements of dark respiration rates of roots and aerial parts in Italian ryegrass and barley. *J. Appl. Ecol.* **14**:243–252.

Moody, J. E., J. N. Jones, Jr., and J. H. Lillard. 1963. Influence of straw mulch on soil moisture, soil temperature and the growth of corn. *Soil Sci. Soc. Am. Proc.* **27**:700–703.

Peterson, K. M., and W. D. Billings. 1975. Carbon dioxide flux from tundra soils and vegetation as related to temperature at Barrow, Alaska. *Am. Midl. Nat.* **94**:88–98.

Phipps, R. H., and J. Cochrane. 1975. The production of forage maize and the effect of bitumen mulch on soil temperature. *Agric. Meteorol.* **14**:399–404.

Phipps, R. H., and J. Cochrane. 1977. A note on the effect of bitumen mulch on soil temperature and forage maize production. *Agric. Meteorol.* **17**:397–399.

Qashu, H. K., and D. D. Evans. 1967. Effect of black granular mulch on soil temperature, water content and crusting. *Soil Sci. Soc. Am. Proc.* **31**:429–435.

Rykbost, K. A., L. Boersma, H. J. Mack, and W. E. Schmisseur. 1975a. Yield response to soil warming: Agronomic crops. *Agron. J.* **67**:733–738.

Rykbost, K. A., L. Boersma, H. J. Mack, and W. E. Schmisseur. 1975b. Yield response to soil warming: Vegetable crops. *Agron. J.* **67**:738–743.

Rykbost, K. A., L. Boersma, and G. D. Jarman. 1976. Soil temperature increases induced by subsurface line heat sources. *Agron. J.* **68**:94–99.

Shaw, R. H., and W. F. Buchele. 1957. The effect of the shape of the soil surface profile on soil temperature and moisture. *Iowa State College J. Sci.* **32**:95–104.

Schmisseur, W. E., L. Boersma, and K. A. Rykbost. 1975. Yield response to soil warming: Economic feasibility. *Agron. J.* **67**:794–798.

Turner, N. C., and P. G. Jarvis. 1975. Photosynthesis in Sitka spruce (*Picea sitchensis* (*Bong.*) *Carr.*) IV. Response to soil temperature. *J. Appl. Ecol.* **12**:561–575.

Unger, P. W. 1978. Straw mulch effects on soil temperatures and sorghum germination and growth. *Agron. J.* **70**:858–864.

Walker, J. M. 1969. One-degree increments in soil temperatures affect maize seedling behavior. *Soil Sci. Soc. Am. Proc.* **33**:729–736.

Waggoner, P. E., P. M. Miller, and H. C. DeRoo. 1960. Plastic mulching-principles and benefits. Bull. No. 634. Conn. Agric. Exp. Stn., New Haven.

Wierenga, P. J., R. M. Hagan, and E. J. Gregory. 1971. Effects of irrigation water temperature on soil temperature. *Agron. J.* **63**:33–36.

CHAPTER 7–EVAPORATION AND EVAPOTRANSPIRATION

7.1 INTRODUCTION

7.1.1 SOME IMPORTANT PROPERTIES AND FUNCTIONS OF WATER IN PLANTS

Water has many unique properties including the ability to exist in the solid, liquid, or vapor phase within the range of temperatures that occur on earth. It is possible for all three phases to exist together but it is the change from one phase to another that is of interest to us, particularly the change from liquid to vapor. The phase change from ice to water requires about 0.34 MJ kg^{-1} of energy whereas the change from liquid to vapor requires approximately 2.45 MJ kg^{-1} at 20°C. The energy required for this latter change is called the latent heat of vaporization L.

Plant growth and productivity are directly related to the availability of water but only about 1% of the water taken up by plants is actually involved in metabolic activity. Water is a raw material in the photosynthetic fixation of CO_2 in which sugars are formed. It is the solvent in which the constituents of plant cells are transported or translocated. Many plant biochemical reactions require that cytoplasm be hydrated. The maintenance of plant turgor depends on adequate hydraulic pressure in leaves and thus on an adequate water supply. Most of the water used by plants, however, passes through the plants and is vaporized into the air.

7.1.2 Historical Background and Reviews

The scientific study of evaporation and transpiration processes has been under way for many years. A historical review in Rosenberg et al. (1968) traces "research" back to Aristotle who concluded in the fourth century BC that "wind is more influential in evaporation than the sun." They suggest that modern study began with Dalton who, in the late eighteenth century, theorized that evaporation from a surface must be a consequence of the combined influence of the wind, atmospheric moisture content, and characteristics of the surface. Further historical progress in evapotranspiration theory and measurement is shown in the forthcoming sections. The reader is advised to refer also to the well-documented reviews of McIlroy (1957), Monteith (1964), Thornthwaite and Hare (1965), Slatyer et al. (1970), Federer (1975), and Blad (1983).

7.1.3 Definitions

Evaporation is defined in the *Glossary of Meteorology* as "the physical process by which a liquid or solid is transferred to the gaseous state." The evaporation of water into the atmosphere occurs from water bodies such as oceans, lakes, and rivers, from swamps, from soils, and from wet vegetation. **Sublimation** is the direct transition of a solid to the vapor phase, and vice versa. It can be an important phenomenon in the water and energy balance of the cold regions.

Most water evaporated at plant surfaces is water that has passed through the plant, entering at the roots, passing through the vascular tissue to the leaves or other organs and exiting into the surrounding air primarily via stomates, but also at times through the cuticle. Evaporation of water that has passed through the plant is called **transpiration**. Direct evaporation from the soil E_s and transpiration T occur simultaneously in nature, and there is no easy way to distinguish the water vapor produced by the two processes. Therefore, the term **evapotranspiration** ET is used to describe the total process of water transfer into the atmosphere from vegetated land surfaces.

Equilibrium evapotranspiration ET_{eq} as defined by Slatyer and McIlroy (1961) is

$$ET_{eq} = \left(\frac{s}{s + \gamma}\right)(R_n + S) \qquad (7.1)$$

where s is the slope of the saturation vapor pressure curve at the mean wet bulb temperature of the air, γ is the psychrometric constant ($\gamma = PC_p/L\epsilon$) where P is the atmospheric pressure, C_p is the specific heat of air at constant pressure, and $\epsilon = M_w/M_a$ where M_w and M_a are the mole weights of water vapor and air, respectively, R_n is the net radiation, and S is the soil heat flux. ET_{eq} defines the minimum possible evaporation rate from a moist surface and depends only on the temperature and available energy (Davies, 1972). Certain evapotranspiration models discussed later in this chapter are based on ET_{eq}.

Another concept widely used in the study of evaporation and evapotranspiration is that of potential evapotranspiration ET_p. In 1948 both Thornthwaite and Penman put forth their definitions of "potential evapotranspiration." Since then the concept has had major influence on geographic studies of world climate, on general hydrologic research, and on attempts to predict water needs in dryland and irrigation agriculture. For example, the quantity, precipitation $-ET_p$ is frequently used, especially by geographers, as an index of aridity. ET_p is often included as a variable in regression equations for estimating crop yields (e.g., Williams, 1972).

The concept of potential evapotranspiration has been so widely accepted that a detailed discussion of its meaning is warranted. The following synthesis of definitions (our own) should be helpful: **Potential evapotranspiration** is the evaporation from an extended surface of a short green crop which fully shades the ground, exerts little or negligible resistance to the flow of water, and is always well supplied with water. Potential evapotranspiration cannot exceed free water evaporation under the same weather conditions.

7.1.4 Expansion on Potential Evapotranspiration

The condition that ET_p cannot exceed free water evaporation under the same weather conditions probably applies well in humid regions. Pruitt and Lourence (1968), for example, reported that the amount of water used by fescue and ryegrass at Davis, California, is about 80% of that evaporated from evaporation pans, except when winds are strong and the air is hot and dry. Under these conditions plants use, relatively, even less water apparently because of an increase in the stomatal resistance.

In the Great Plains and other more arid regions, well-watered crops that exert little canopy resistance can consume more energy and transpire more water than is evaporated from free water surfaces. A case in point occurred during a period of strong regional advection of sensible heat into eastern Nebraska. Daily evaporation from three class A evaporation pans with land exposures and three with lake exposures was smaller than evapotranspiration measured with precision weighing lysimeters in an irrigated alfalfa field (Rosenberg and Powers, 1970).

Rosenberg and Powers (1970) also compared soybean evapotranspiration to pan evaporation. On the 20 days uninterrupted by rain or irrigation, soybean water use averaged 8.33 mm day^{-1} compared to the average evaporation of 6.39 mm day^{-1}, or about 23% less.

We see from the foregoing that free water evaporation need not always indicate the maximum evapotranspiration in subhumid and arid regions as it does, apparently, in humid regions.

We know that actual ET differs from ET_p under most circumstances because actual water use depends on meteorological, plant, and soil factors and on the availability of water. It may help to look at the way in which Bouchet and Robelin (1969) attempted to deal, conceptually, with these factors, particularly those stemming from the plant and soil. They consider, as is shown in Fig. 7.1, that ET_p is entirely controlled by climate and varies as a function of season. ET_m is actual maximum evapotranspiration, a special case of the actual evapotranspiration, ET_r. These quantities are ordered in magnitude according to $ET_r \leq ET_m \leq ET_p$.

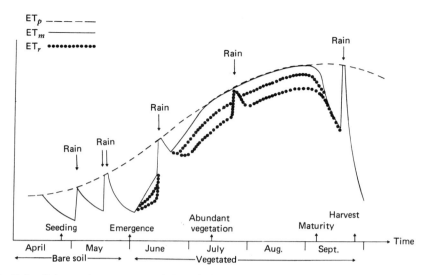

Fig. 7.1 Schematic treatment of the relations between potential ET_p, maximal ET_m, and actual ET_r evapotranspiration (after Bouchet and Robelin, 1969).

ET_m is defined in a given climate by the development of the plant and its physiology when it is well supplied with water. In essence, $ET_p - ET_m$ is a measure of the canopy resistance. ET_r results when poor soil water supply causes a reduction in water available to the plant. The ratio ET_r/ET_p may be expressed as a function of soil water potential (Eagleman and Decker, 1965) and has been used as an index of water supply in relation to water demand (e.g., Baier, 1969; Seguin, 1975; Yao, 1969, 1974).

7.2 IMPORTANCE OF EVAPORATION AND TRANSPIRATION

7.2.1 Evaporation and Transpiration in the Hydrologic Cycle

Because of the critical importance of water to plants and animals alike, we are concerned with the role of evapotranspiration in the hydrologic cycle. The change of phase liquid ⇌ vapor provides a major mechanism for the redistribution of energy within an ecosystem and throughout the atmosphere.

It is estimated that 97.5% of the world's water exists in the oceans and that 2% is locked into the Arctic and Antarctic ice caps and snow fields. A very small amount of the total water on earth, then, is available to support terrestrial life. This relatively small amount of water cycles between the atmosphere, the continents, and the oceans. Of the water which falls on the continental United States, about 30% is believed to run off into drainage systems, eventually reaching the oceans, or to percolate to depths where it enters groundwater storage. The remainder returns to the atmosphere by direct evaporation or by transpiration.

In the Great Plains of the United States, as well as in other relatively dry regions of the world, 90% or more of the precipitation is evaporated. Maxwell (1965) estimated that 113.6×10^{10} m^3 of water are returned daily to the atmosphere by evaporation from the continental United States. A 1% reduction for one day only would provide the yearly water needs of a city of 2,000,000 people.

Many schemes have been proposed to alter the hydrologic cycle: artificial rainmaking, seawater distillation, alteration of albedo (e.g., blackening of the polar ice pack to increase melting; blackening of coastal sands to induce lifting of air to condensation levels) in order to augment water supply. One idea that recently received considerable attention is that of towing icebergs (probably from the Antarctic region) to regions where melted water can be put to use in irrigation (Herman, 1977).

Notwithstanding the exotic possibilities described above, manipulation of the hydrologic cycle to augment the water supply may, perhaps, because of the immense quantities of water involved, be most easily and effectively accomplished by manipulation of the evapotranspiration process.

7.2.2 Evaporation and Transpiration in Plants and Soils

Plants require very large amounts of water. For example, about 1000 kg of water are required to produce 1 kg of wheat (Krogman and Hobbs, 1976). Every day an actively growing plant uses 5–10 times as much water as it can hold at one time.

Probably the most important physical effect of transpiration is the cooling that occurs at the transpiring surface. Because large quantities of energy are required in the change of phase from liquid to vapor, evaporation provides a very efficient mechanism for the dissipation of heat.

Reduced transpiration can easily result in an increase of 2–3°C in plant temperature with increases perhaps as great as 10°C under extreme conditions (Poljakoff-Mayber and Gale, 1972). Blad et al. (1981) measured canopy temperature differences between well-watered and stressed corn as great as 12°C.

Whereas meteorological or soil factors may limit plant productivity in a specific region, it seems safe to say that, on a worldwide scale, water availability is the factor most critical in determining plant survival, development, and ultimate productivity. Crop yield is directly related to the availability of soil moisture during the course of a growing season. A rich literature supports this view but only a few examples are given here.

De Wit (1958) provided one of the first analytical approaches to the question of the relationship of yield and water use:

$$Y/T = m/E_0 \tag{7.2}$$

where Y is the total dry matter production, T is the total transpiration during the growing season, m is a constant governed by plant species, and E_0 is the seasonal mean daily free water evaporation. This relationship worked well for arid climates but in humid climates a better description was

$$Y/T = n \qquad (7.3)$$

where n is a constant.

Bierhuizen and Slatyer (1965) suggested that differences in de Wit's expressions for humid and arid regions can be accounted for by differences in vapor pressure deficits of the different climates. They proposed that

$$Y/T = k_1/(e_s - e_a) \qquad (7.4)$$

where k_1 is a crop-specific constant and $e_s - e_a$ is the vapor pressure deficit of the air. This approach was also used by Tanner (1981). Tanner showed that year to year variability in potato yield as a function of transpiration was essentially eliminated by dividing the transpiration by the vapor pressure deficit (Fig.7.2).

Dale and Shaw (1965) found that the yield of corn in Iowa was directly related to the number of days on which plants are free of moisture stress. Musick et al. (1976) showed that soil water depletion in the top 1.20 m of a clay loam soil typical of the Texas Panhandle accounted for between 57 and 99% of the variability in the yields of irrigated sorghum, wheat, and soybeans. Their findings in a drier than normal season are shown in Fig. 7.3.

That the degree to which yields are reduced is a function of the growth stage at which water stress occurs has been shown by Howell and Hiler (1975) for sorghum, Gardner et al. (1981) for corn, and Choudhury and Kumar (1980) for wheat. Yields are generally most severely reduced if water stress occurs after the vegetative growth period.

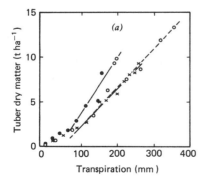 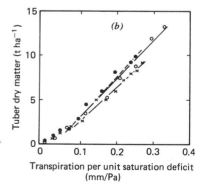

Fig. 7.2 Relation of dry matter accumulation in potato tubers to (*a*) transpiration and (*b*) transpiration/$(e_s - e_a)$ (after Tanner, 1981). Reproduced from *Agronomy Journal,* Volume **73**, 1981, pages 59–64, by permission of the American Society of Agronomy.

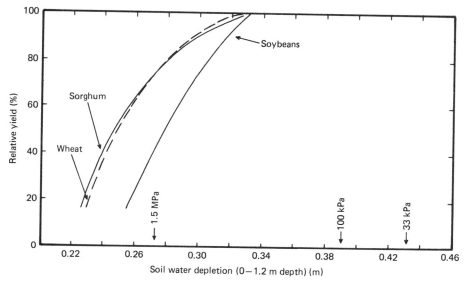

Fig. 7.3 Relationship of soil water depletion in the 0 to 1.2 m depth to relative yield of sorghum, wheat, and soybeans in drier than normal seasons (after Musick et al., 1976, *Trans. ASAE* **19**:489–493).

In an interesting study of yield–transpiration relationships, Blanchet et al. (1977) found that the grain yields and the production of total dry matter in 13 cultivars of soybeans were linearly related to the amount of water consumed (Fig. 7.4). Retta and Hanks (1980) reported similar results for corn and alfalfa. They also showed that the specific relationship between

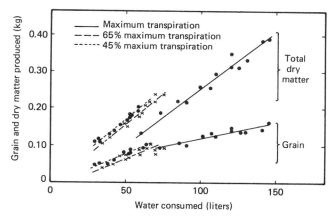

Fig. 7.4 Production of total dry matter and of grain as a function of water transpired (after Blanchet et al., 1977, *Annales Agronomique* **28**:261–275).

ET and dry matter production or yield was influenced by the choice of cultivar.

Plants exposed to nearly 100% relative humidity show a reduction in growth which has been interpreted to mean that the translocation of plant products is interrupted when transpiration ceases. Transpiration is also involved in the transport of essential nutrients from the soil to plant tissue where the nutrients enter into the photosynthetic processes. Water containing dissolved nutrients is absorbed through the roots and moved through the plant in the stream of water that is ultimately transpired. Thus it is not unexpected that a relationship would be found among fertilizer concentration, transpiration rate, and crop yield (Stutler et al., 1981). Salts in the soil also move in response to evaporative demands. Sinha and Singh (1974), for example, showed that salt concentration near corn roots increased with increasing rate of transpiration.

7.2.3. Evapotranspiration in Irrigation Management

Irrigation provides an important degree of stability to food production in many nations. In the United States more than 24.3 million hectares are now irrigated. The ability to accurately estimate evapotranspiration and from these estimates to determine the status of the soil water supply is of considerable importance for properly scheduling irrigations.*

Underirrigation or ill-timed irrigations result in lower than maximal yields. Excessive irrigation is also a cause of concern. Many farmers tend to apply excess water; some overirrigate by 50–100%. Overirrigation wastes fuel, degrades soil quality, depletes groundwater supplies more rapidly than necessary (thus increasing pump lifts and further increasing fuel requirements), may cause leaching of essential plant nutrients, and can reduce crop yield.

Irrigation can be scheduled rationally field by field on the basis of certain knowledge: soil type, moisture-holding capacity, antecedent soil moisture content, crop type, stage of development and water requirement at each stage; irrigation system characteristics and efficiency; and, not least, cumulative ET since the last irrigation or rainfall event. This knowledge allows an irrigator to know when to initiate an irrigation and how much water must be supplied to recharge the soil's root zone. The irrigation should be scheduled sufficiently early to avoid adverse effects on the crop and as late as possible to permit efficient irrigation. In general, it is best to have an irrigation schedule that maximizes the length of time between irrigations for this allows greater opportunity for use of natural rainfall (Cull et al., 1981). Wright and Jensen (1978) describe one of the more modern, computer-aided approaches for scheduling irrigation.

---

* For an analysis of the economic benefits of irrigation scheduling, see National Academy of Sciences (1980). See also Fischbach and Thompson (1979).

7.3 SOIL, PLANT, AND CLIMATIC INFLUENCES ON EVAPOTRANSPIRATION

7.3.1 Water Availability

So long as there is adequate water, evapotranspiration will proceed at the maximum possible rate depending only on the amount of energy available and the control exerted by vegetation, if any. When the soil surface dries or when soil water becomes limiting the rate of ET will decrease. Thus, the amount of water at the soil surface and within the plant root zone is very important in the evapotranspiration process.

Soil Evaporation. Water evaporates from the soil surface at rates comparable to evaporation from a free water surface so long as the surface is wet and the soil is not shaded by plants or mulches (Adams et al., 1976). Evaporation E_s from a wet soil surface occurs during what has been called "first stage drying." "Second stage drying" begins when the soil surface becomes visibly dry. This generally occurs 1–5 days after irrigation or precipitation. During the initial portion of second stage drying soil evaporation rates are controlled by hydraulic properties, which determine the rate at which water will move through the soil and to the soil surface. During the latter portion of this stage most of the water evaporated at the soil surface moves through the soil in the form of vapor.

During second stage drying E_s decreases approximately as the square root of the time elapsed (Ritchie, 1972; Gardner, 1974). Second stage drying continues until adsorptive forces at the soil particle–liquid interfaces exert control over the evaporation rate. Then "third stage drying" begins (Lemon, 1956; Idso et al., 1974). During this third stage evaporation consumes less than 5% of the energy absorbed at the soil surface (Massee and Cary, 1978).

Methods for calculating E_s during the various stages of soil drying are discussed by Fuchs et al. (1969), Ritchie (1972), Jackson et al. (1976), and van Bavel and Hillel (1976).

Soil Water for Plants. The contribution of soil evaporation to total evapotranspiration decreases as plant cover increases. Thus with increasing plant cover the ratio ET/ET_p depends mainly on the soil water status and on the ability of plant roots to extract the available water. Certainly, as soil water becomes less available to plants transpiration decreases. There are other cases, generally near midday under a high radiation and heat load, when, even with adequate water, plants cannot extract water from the soil at a rate sufficient to meet the evaporative demand (Lang and Gardner, 1970). When such a situation occurs stomatal diffusion resistance will increase and the transpiration rate will decrease.

"Well supplied with water" suggests no soil-imposed restriction to reduce the meteorologically determined rates of water use. Obviously, in rain-fed

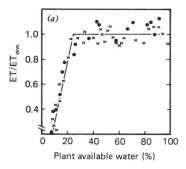 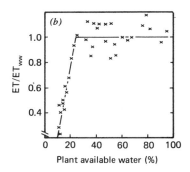

Fig. 7.5 Dependence of actual evapotranspiration ET and evapotranspiration ET_{ww} of a well watered treatment on plant available water for (*a*) wheat and (*b*) soybeans (after Meyer and Green, 1981).

agriculture, plants are not always well supplied with water. This can be the case in irrigation agriculture as well, inadvertently or by intention, for when the water supply becomes limited it is more important to get the greatest yield per unit of water expended than per unit of land used. We know that strategic irrigation at certain stages of crop growth may lead to great increases in yield over no irrigation and that, when irrigation water is limited, the best (most economic) results can be obtained by spreading the water over a larger land area than by intensively irrigating a small area.

Marlatt et al. (1961) and others have shown that, with decreasing soil moisture availability, the rate of ET is reduced below the potential. Van Bavel (1967) suggested that transpiration in alfalfa begins to diminish after a soil water potential of about -0.4 MPa is reached. He cites other works in which this break point ranges from -0.02 to -1.0 MPa for corn and cotton, respectively.

Ritchie et al. (1972) and Ritchie and Jordan (1972) have shown that ET by sorghum and cotton is conditioned by soil moisture status after a critical level of water extraction has occurred. Before the critical moisture level is reached, the ratio of the latent heat flux* LE to net radiation R_n is about 1. Thereafter, the ratio drops rapidly until evapotranspiration virtually ceases. Similar relations for wheat and soybeans are reported by Meyer and Green (1981) (Fig. 7.5).

Due primarily to capillary flow of water through the soil, plants are capable of extracting water from below the root zone. Stone et al. (1973) observed conditions during a 31-day study in which 60 mm of the water used by sorghum came from the soil below the root zone. Reicosky et al. (1977) found that during a drought period the total upward flux of water from below the root zone accounted for 34% of the total ET. Although the total amount

---

* LE is computed as the product of the latent heat of vaporization L and the water vapor flux E.

of water moving upward was small it was sufficient to keep the crop alive until rainfall occurred.

Water supplied to growing vegetation from shallow water tables can also be important and should be accounted for when scheduling irrigations. Failure to do so can lead to deleterious effects on crops (Wallender et al., 1979).

Dew and Intercepted Water. In addition to water supplied from the soil, dew and intercepted rainfall or irrigation will be evaporated from plant surfaces. The importance of dew is discussed in Chapter 5. The contribution of dew as a source of water for evapotranspiration can be considerable depending on climatic conditions and the type of vegetation.

Intercepted water evaporates at potential rates and can lead to greatly increased evapotranspiration. Stewart (1977) found that, at the same level of radiation, the average rate of evaporation from intercepted water was 3 times the average rate of transpiration from a pine forest. McNaughton and Black (1973) observed 20% higher ET from a Douglas fir forest wetted by intercepted water than from a nonwetted canopy. Fritschen et al. (1977) and Gash and Stewart (1977) made measurements in coniferous forests showing that 25–40% of the seasonal ET came from the evaporation of intercepted water. Calder (1976) reported that most of the water evaporated from a spruce forest in central Wales was intercepted water. Thus, in many forests dew and/or intercepted water contribute significantly to the total water lost by evapotranspiration. It should, however, be noted that the evaporation of the free-standing water on plant surfaces reduces transpiration and thus helps to conserve soil water.

7.3.2 Plant Factors

Internal Plant Resistance to Water Flow. For plants to transpire at potential rates requires that they behave passively, acting as wicks for the transport of water from soil to air. Penman and Schofield (1951) recognized that this assumption did not hold in all circumstances and attempted to adjust for stomatal influence on transpiration.

Considerable effort has been devoted to understanding the manner in which stomates control transpiration. Hansen (1974a), for example, has shown that transpiration in Italian ryegrass is a curvilinear function of stomatal resistance r_s (Fig. 7.6a) or a linear function of stomatal conductance $1/r_s$ (Fig. 7.6b).

A number of environmental factors including leaf temperature, light, leaf water potential, and, probably, vapor pressure deficit affect stomatal resistance. For example, the stomatal resistance of Sitka spruce is very sensitive to increases in dryness of the air and to needle temperature as is shown in Fig. 7.7.

Rosenberg (1969a) shows an interesting effect of cold weather on the

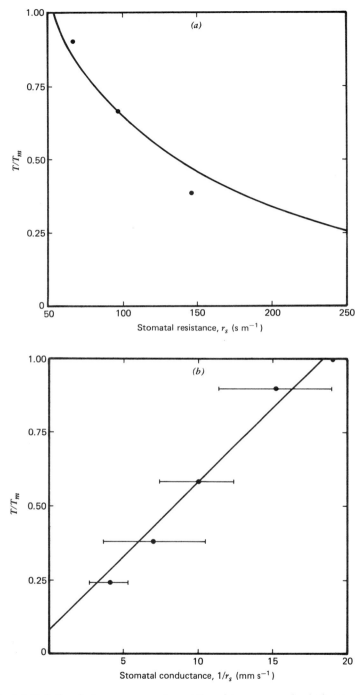

Fig. 7.6 (a) Relation between mean stomatal resistance r_s and relative transpiration rate T/T_m, and (b) relation between mean stomatal conductance $1/r_s$ and T/T_m. Bars in (b) indicate standard deviation of the means (after Hansen, 1974a).

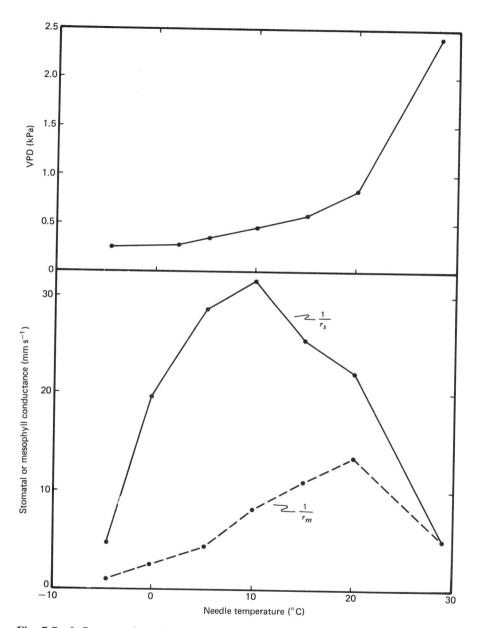

Fig. 7.7 Influence of needle temperature on stomatal conductance $1/r_s$ and mesophyll conductance $1/r_m$ of a shoot of *Picea sitchensis*. Leaf-air vapor pressure difference VPD at various temperatures is also shown (after Nielsen and Jarvis, 1975).

222 EVAPORATION AND EVAPOTRANSPIRATION

resistance to vapor diffusion exerted by an alfalfa crop (Fig. 7.8a,b). Evapotranspiration was greater than R_n on April 21, but after a cold night ET on April 22 was reduced, even though R_n was much higher. A major increase in canopy resistance caused by the cold night is suggested.

Light, particularly photosynthetically active radiation (PAR), has a great effect on r_s at visible light flux densities below about 200 W m^{-2}. Above that value r_s decreases very slowly with increasing irradiance. An idealized curve of the effect of light on r_s is shown in Fig. 7.9.

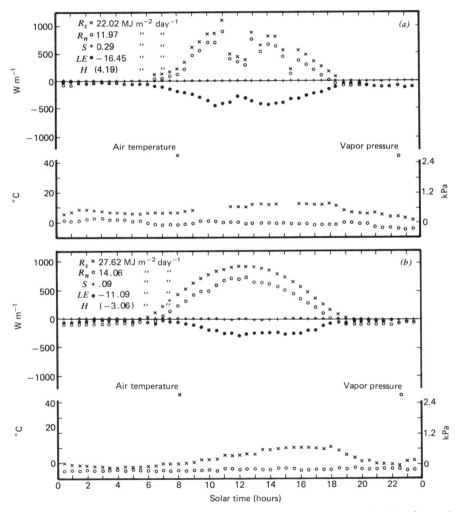

Fig. 7.8 Energy balance with lysimetrically measured evapotranspiration from alfalfa at Mead, Nebraska, on (a) April 21 and (b) April 22. Air temperature and vapor pressure measured at 1.00 m (after Rosenberg, 1969a).

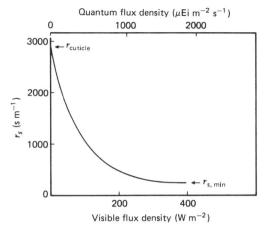

Fig. 7.9 General shape of the dependence of stomatal resistance r_s on photosynthetically active radiation (after Norman, 1979, with permission of the American Society of Agricultural Engineers). The stomatal resistance with the stomates fully open, $r_{s,\,min}$, and with stomates closed, $r_{cuticle}$, are also shown.

A further complication is identified by Teare and Kanemasu (1972) who found that r_s is not uniform in the canopies of either sorghum or soybean. Generally, r_s is greater in the lower portion of the canopy. The position of minimum resistance within the canopy also varies with time of day. This may be explained, at least in part, by the influence on r_s of the different light regimes found within the crop canopies.

Differences between species in diffusive resistance to water vapor transport are, evidently, considerable. Alfalfa, Sudan grass, and soybeans exert very little canopy resistance to evapotranspiration when well supplied with water (van Bavel et al., 1962; van Bavel, 1967; van Bavel and Ehrler, 1968; Blad and Rosenberg, 1974a,b). However, sorghum has been shown to use less water than soybeans perhaps because of its greater stomatal resistance (Teare et al., 1973). There is evidence of significant stomatal influence on ET rates in sugar beets (Brown and Rosenberg, 1970), dry and snap beans (Rosenberg, 1966; Rosenberg et al., 1967), pastures (Blad and Rosenberg, 1974a), and orange trees (van Bavel et al., 1967).

When the rate of water intake to plants cannot equal the rate of water loss stomates close either partially or completely, thereby increasing r_s. The specific response of plant stomates to water stress is conditioned by the prior history of the plant, that is, whether the plant has been adequately supplied with water or, instead, has experienced periods of water stress. The stomatal response to water stress appears to change throughout the plant's life cycle. For example, Ackerson and Krieg (1977) showed that stomates of corn and sorghum responded to water stress during the vegetative growth stage but were insensitive during the reproductive stage. During the reproductive stage leaf resistances were also minimal.

It should be clear from the foregoing that plants rarely behave as the "passive wicks" so useful in theory. Plant internal resistance, whether it originates in the root tissues in contact with soil water films, in the water transporting structures, in the mesophyll, in the stomatal guard cells, or in the cuticle exerts a major influence on the transpiration process. Lee (1967) stimulated a spirited argument by suggesting a major role for stomatal regulation of evapotranspiration and, thus, a major role in the hydrologic cycle of earth. A review of these arguments and rebuttals by van Bavel and by Idso make interesting reading.

Commonly stomates are distributed unequally on the top and bottom of plant leaves. In some plants stomates are found on only one side (hypostomatous) whereas other plants may have an equal number of stomates on both sides of the leaf (amphistomatous). To determine the effective leaf stomatal resistance r_s, the stomatal resistance of the top (adaxial) r_{st} and the bottom (abaxial) r_{sb} sides are considered to be connected in parallel. When there are unequal numbers of stomates on the top and bottom, the effective stomatal resistance can be expressed as

$$r_s = \frac{r_{sb} r_{st}}{r_{sb} + r_{st}} \tag{7.5}$$

For amphistomatous leaves, the expression is

$$r_s = \frac{r_{sb}}{2} = \frac{r_{st}}{2} \tag{7.6}$$

and for hypostomatous leaves,

$$r_s = r_{sb} \quad \text{or} \quad r_{st} \tag{7.7}$$

Various procedures for averaging r_s values from individual leaves are used to calculate the total canopy resistance r_c (Black et al., 1970; Brun et al., 1973; Szeicz et al., 1973). In one case r_c is calculated as the mean of r_s values of all leaves and in another it is the mean of resistances for various leaf layers considered as parallel resistors. A better estimate of r_c can be obtained by calculating the average r_c for the sunlit and shaded leaves separately. r_c can then be calculated as the mean of the two values weighted by sunlit and shaded leaf area index (Norman, 1982).

Because it is somewhat difficult to calculate r_c from r_s values, r_c is sometimes estimated from more easily measured variables. Verma and Rosenberg (1977) expressed r_c in terms of R_n. This simplification applies only to well watered crops. Callander and Woodhead (1981) found r_c to be a function of R_n and vapor pressure deficit. Singh and Szeicz (1980) developed an equation for overall canopy resistance as a function of solar irradiance and soil moisture content, and Russell (1980) derived r_c as a function of boundary layer resistance r_a and the ratio ET/ET$_p$.

The Influence of Crop Cover. Row crops do not normally shade the ground fully except at advanced stages in their development, and broadcast crops such as alfalfa do not fully shade the ground for some time after periodic cuttings. We know that water use may continue to increase with increasing leaf area even when leaf area great enough to fully shade the ground has been achieved.

Before full crop closure is achieved, row direction can influence evapotranspiration. For example, Hicks (1973) observed eddy fluxes in a vineyard at two stages of growth and noted that evapotranspiration increased as the wind swung away from flow along the crop rows.

El Nadi (1974) found that the ratio of ET to pan evaporation increased as cotton and *Dolichos* (hyancinth bean) plants grew to cover the rows. Kristensen (1974) obtained data showing that the ratio of ET to ET_p approaches unity at a leaf area index LAI of about 3 for barley, sugar beets, and grass (Fig. 7.10).

Leaf area or amount of ground covered determines the ratio of direct evaporation from the soil to transpiration from the plants. Brun et al. (1972) found that the proportion of water lost as transpiration in soybean and sorghum fields was closely correlated to LAI. Transpiration was approximately 50% of the total evapotranspiration at an LAI of 2 and increased to 95% at an LAI of 4. Shading is an obvious factor controlling this ratio. Less obvious perhaps but important, nonetheless, are changes in crop roughness with consequent changes in the value of the aerodynamic parameters z_0 and d. Ritchie and Burnett (1971) related transpiration to ET_p as a function of

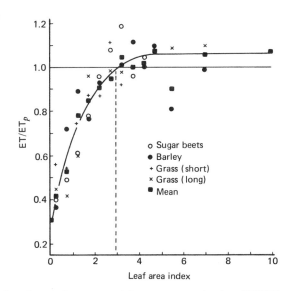

Fig. 7.10 Ratio of actual to potential evapotranspiration ET/ET_p as a function of the leaf area index (after Kristensen, 1974, *Nordic Hydrology* **5**:173–182).

LAI (Fig. 7.11) as follows:

$$T = ET_p(-0.21 + 0.70 LAI^{1/2}) \quad \text{for} \quad 0.1 \leq LAI \leq 2.7 \quad (7.8)$$

There are at least two possible reasons for the nonlinear relationship of Fig. 7.11: (1) there is less competition for radiation per unit leaf area when the plants are small and (2) a large fraction of incident net radiation is repartitioned at the dry soil surface between plant rows, increasing canopy temperature and, consequently, transpiration. With increasing crop growth the latter effect, particularly, diminishes in importance.

Evapotranspiration from one of several native grassland areas in the southwestern United States was observed by Ritchie et al. (1976) and is shown in Fig. 7.12. During midyear ET/ET_p is related to the increase in leaf area index of the grass sward. A rapid drop off in this ratio prior to cutting is the result of soil moisture shortage.

The evidence seems conclusive that transpiration in most mesophytic crop plants and other mesophytic vegetation well supplied with water increases with leaf area to an LAI of about 3. This relationship may not hold for xerophytes, however. Gindel (1971) showed, for example, that three *Eucalyptus* species that were stressed for water transpired more rapidly when the water supply was adequate than did plants that had suffered no previous water stress. This was true even though the plants subjected to prior stress

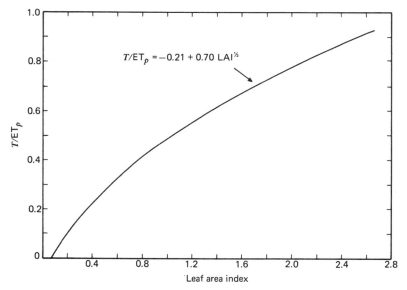

Fig. 7.11 Ratio of transpiration to potential evapotranspiration T/ET_p as a function of leaf area index when soil water is not limited (after Ritchie and Burnett, 1971). Reproduced from *Agronomy Journal*, Volume **63**, 1971, pages 56–62, by permission of the American Society of Agronomy.

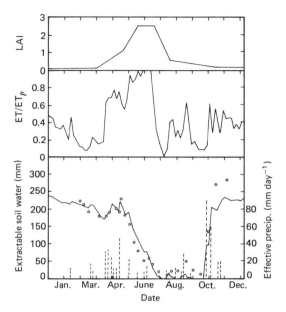

Fig. 7.12 Seasonal variation of leaf area index LAI, ratio of calculated evapotranspiration to potential evapotranspiration ET/ET_p, calculated extractable soil water (continuous curve), measured soil water (open squares), and effective precipitation (rainfall minus runoff) (after Ritchie et al., 1976).

had 30–40% less leaf area. This work suggests that biological control can be very important when plants grow under extreme climatic conditions.

Influence of Plant Height. In general, the taller the crop, the more aerodynamically rough it will be and its extraction of energy from the ambient air will be increased. Stanhill (1965) and El Nadi and Hudson (1965) have shown that the quantities of water evaporated by alfalfa increase as the crop increases in height. Tall crops, especially in areas of strong sensible heat advection, are able to extract more sensible heat from the air than are short crops. Slabbers (1977) stresses the importance of taking crop roughness into account when calculating ET.

The consumption of water by phreatophytic vegetation is of great significance, especially in arid areas. Van Hylckama (1966) estimates an annual consumption by phreatophytes of over 29×10^9 m^3 of water in the southwestern United States. The height of many species of phreatophytes (e.g., salt cedar, cottonwood) enhances the extraction of energy from the atmosphere and contributes to the strong water use by such plants.

Denmead (1969) found greater evapotranspiration in an Australian pine forest than that in an adjacent wheat field. Although many factors contributed to this result, height of the vegetation was one of the most important.

Influence of Plant Morphology. We now know that the type of leaf will influence transpiration and that, all things being equal, broad-leafed plants will transpire more than will the grasses (Fritschen, 1965; Blad and Rosenberg, 1974a). The size of the leaf may be important since larger leaves tend to be less efficient in dissipating heat through convective transfer and will, therefore, have more energy available for evaporation than will smaller leaves.

Leaves may be coated with waxes, cutin, or other materials, which act as vapor barriers and reduce the loss of water from plant surfaces. Plants having such leaves are commonly found in the more arid regions of the world. In the very arid regions plants are also found (primarily cacti and other succulents), which open their stomates only at night. The entry and capture of CO_2 for photosynthesis occurs, then, at a time when the atmospheric demand for water vapor is lowest (Tinus, 1974) (see Chapter 8 for further details).

Other morphological features, including pubescence, color, leaf shape, and presence of awns and other specialized structures may influence the amount of water used by plants. This may occur through alteration of the plant energy balance, through modification of the leaf and canopy boundary layer resistance to energy and mass exchange, or even, perhaps, through an influence on the way in which dew is deposited on plant surfaces. Manipulation of plant architectural features for physical effects is discussed in Chapter 11.

7.3.3 Meteorological Conditions

Weather plays a major role in determining rates of ET and ET in turn may affect climate (Shukla and Mintz, 1982). Water for ET is provided by precipitation and water vapor is transported away from the crop in the bulk air. The great quantities of energy consumed in evapotranspiration are supplied almost entirely from two sources: radiant energy and energy from air which is warmer than the crop. Both sources of energy are traceable to solar radiation.

The relationship between energy sources and energy consumers is summarized in the energy balance equation

$$R_n + H + S + LE + PS + M = 0 \qquad (7.9)$$

where R_n is net radiation, S is soil heat flux, H is sensible heat flux, LE is latent heat flux, and P and M represent photosynthesis and miscellaneous energy exchanges, respectively. LE is the major energy consumer where water is present and R_n is the major energy supplier, except in special instances when H is also a significant source of energy.

Net Radiation. It has been clearly demonstrated that R_n is the major source of energy for evapotranspiration. In fact, in humid regions daily R_n is a good measure of LE when potential evaporation conditions prevail (Ritchie, 1971; Parmele and McGuiness, 1974). In humid regions R_n generally sets the upper limit on the amount of energy consumed as LE. Lemon et al. (1971), for example, report that R_n in the eastern United States in summertime is proportioned as follows: LE (40–90%), H (10–60%), and S (5–10%).

Differences in cultural practices or morphological differences in plants have been shown to alter the energy balance over plant canopies. Water use rates were found to be lower in 0.3 m than in 0.9 m east–west-oriented rows of peanuts or than in north–south-oriented rows. This effect was attributed to low R_n in the former (Chin Choy et al., 1977). When the albedo of a surface is increased R_n is decreased and ET rates will be reduced correspondingly. It was with this in mind that we reflectorized soybean canopies to increase their albedo and found that decreased R_n resulted in decreased ET (Baradas et al., 1976; Lemeur and Rosenberg, 1976; see Chapter 11 for additional details).

Sensible Heat Advection. **Advection** is defined in the *Glossary of Meteorology* as "the process of transport of an atmospheric property solely by the mass motion of the atmosphere. . . ." We will use the term **advection** as synonomous with the transport of energy or mass in the horizontal plane in the downwind direction. We are most concerned with the advection of sensible heat from one field to another or from one region to another, namely, heat that is used to evaporate water.

Tanner (1957) suggested a terminology to describe advection effects. When advected energy is drawn from the air, an **oasis effect** prevails. When air is horizontally transferred from areas where sensible heat has been generated through crops, a **clothesline effect** exists. "Clotheslines" typify small plots and are particularly disturbing when agronomic treatments create height or density differences.

The terms **local** and **global, regional,** or **large-scale advection** have been used to identify sources of preconditioned air. In this regard, Slatyer and McIlroy (1961) commented on the global scale of advective transport. They suggest that even in moist homogeneous areas large-scale or upper-level advection due to movement of weather systems may lead to latent heat consumption exceeding average net radiation received. This may arise because of passage across the evaporating surface of relatively warm, dry air masses generated over large relatively hot, dry surfaces elsewhere. "Certain regions, even though quite well provided with incoming radiant energy, will act as long-term sinks of sensible heat, owing to the general pattern of circulation over them . . . provided, of course, that sufficient moisture is available for the amount of evaporation involved."

In nature advection is the normal rather than the abnormal situation. Only when the surface under consideration is identical in its characteristics of color, roughness, moisture availability, and so on, with an infinite surface upwind will nonadvective conditions prevail. Many have attempted to define fetch requirements for micrometeorological research, but so long as upwind fetch is finite only locally generated advective affects will be eliminated. Air masses conditioned in very distant regions and climates continue to move over surfaces of different characteristics and their effects on local micrometeorological processes remain observable.

The presence of sensible heat advection may be inferred by reference to (7.9). If $LE > (R_n + S)$, sensible heat has been drawn from the air and consumed in evaporation. This effect (consumption rather than generation of sensible heat) has often been considered prima facie evidence of advection and, except in a few cases, correctly so.

The origin of sensible heat for advection to other localities depends on surface conditions. Generally, the availability of water determines the partitioning of energy among sensible, latent, and soil heat fluxes. This is well illustrated in work of Fritschen and van Bavel (1962) at Phoenix, Arizona (Fig 7.13).

When the soil was moist almost all of the energy supplied by R_n was consumed as latent heat. Small quantities of energy were distributed to the soil and sensible heat flux from early morning until about 1500 hours on the third day. By the fifth day, with a dry soil surface, the consumption of net radiation as latent heat was greatly reduced and sensible heat was generated. This heat, if transported over a wetter field, would then be available for consumption in the evapotranspiration process.

. Heat generated at a dry soil surface can also be consumed through increased transpiration by the plants adjacent to the dry soil. This source of sensible heat is known as **within-row advection** (Hanks et al., 1971). Within-row advection can be important as shown by Kanemasu and Arkin (1974) who reported that ET from wide rows (0.91 m) in sorghum was about 10% greater than from narrow rows (0.46 m). This was attributed to increased transpiration in the wider rows due to the consumption of sensible heat generated at the soil surface.

Local clothesline effects are well illustrated by results of van Bavel et al. (1962), shown in Fig. 7.14. Sudan grass growing in a small field near Phoeinx, Arizona, used 9.76 mm of water on July 23. The crop was cut except for plants in a set of precision lysimeters. On a day of almost identical weather conditions, 14.66 mm of water was consumed by the isolated plants. In order for this large quantity of water to be transpired, considerably more energy was consumed than was avalable from the net radiation. The added energy was sensible heat advected from the surroundings, including the Sudan grass stubble. Sellers (1965) has argued that some of the additional transpiration in this experiment could also have been due to the increased penetration of radation into the sides of the exposed Sudan grass.

Since "infinite fetch" does not truly exist, advection is a factor to be

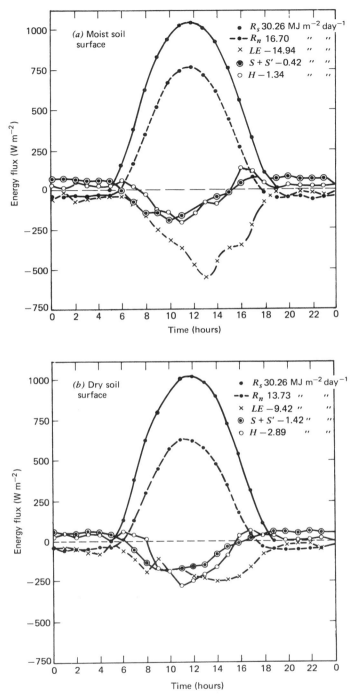

Fig. 7.13 Hourly distribution of energy balance components over a wetted soil surface at Phoenix, Arizona, on (a) the third day after irrigation and (b) the fifth day. S' is the change in heat storage in the soil (after Fritschen and van Bavel, 1962).

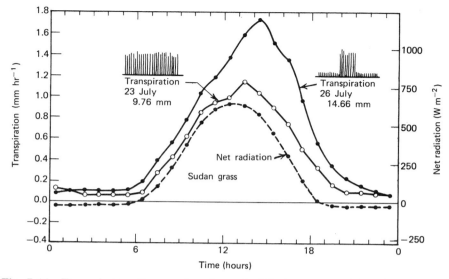

Fig. 7.14 Transpiration from a closed stand of Sudan grass at Phoenix, Arizona, on July 23 and from an isolated 1 m² plot of the same Sudan grass 3 days later. Average net radiation for the 2 days which differed by only 5% is shown (after van Bavel et al., 1962).

considered in any analysis of evapotranspiration. For our convenience the sources of advected energy will be classified as regional and local.

Regional advection or the "oasis effect" has been demonstrated in experiments conducted on a wide range of crops grown at a number of locations throughout the world. Rijks (1971) working in the Gezira region of the Sudan found evapotranspiration from actively growing cotton was often 1.8 times greater than the amount of net radiant energy available to the crop. In southern Idaho, a well-watered aerodynamically rough alfalfa crop providing full cover consumed 11.0 mm of water on a single day and averaged 8.8 mm during a peak 10-day period (Wright and Jensen, 1972). These quantities were in excess of R_n. Other documented cases of a significant sensible heat advection contributing to ET are given by van Bavel (1967), Rosenberg (1969a,b), Blad and Rosenberg (1974a), and Rosenberg and Verma (1978) for alfalfa; by Rosenberg (1972) for soybeans; and by Pruitt and Lourence (1968) for ryegrass.

The advection of sensible heat is also of major importance in determining the regional water balance and the moisture stress imposed on crops grown in large parts of the Soviet Union where the so-called Sukhovey winds prevail with varying frequency (Dzerdzeevski, 1957; Lydolph, 1964). Regional sensible heat advection is a major component of the energy balance in regions of Australia (Stern, 1967; Slatyer and McIlroy, 1961) as well as the Great Plains of North America (Rosenberg, 1969a,b; 1972).

A demonstration of the great hydrologic significance of sensible heat advection during an extended drought is found in a report of Rosenberg and Verma (1978). ET from irrigated alfalfa at Mead, Nebraska, throughout the course of an extended drought in 1976 ranged from 4.75 to 14.22 mm day^{-1} and exceeded 10 mm day^{-1} on one-third of the days studied (see Fig. 7.15 for examples of 2 days). R_n provided energy sufficient for no more than 7 mm day^{-1} of ET. On *each and every day* during that season, LE exceeded $(R_n + S)$.

Local advection occurs when wind blows across a surface that is discontinuous in temperature, humidity, or roughness, as from a dry field to an adjacent wet field. Rider et al. (1963), Dyer and Crawford (1965), Goltz and Pruitt (1970), and others have shown that local advection results in increased ET immediately downwind from a leading edge. As the distance from the leading edge (the fetch) increases, the influence of local advection on ET decreases until, finally, horizontal homogeneity is established.

Brakke et al. (1978) conducted a study at Mead, Nebraska, to determine the relative magnitudes of the local and regional components of sensible heat advection. The impact of the sensible heat on evapotranspirtion on a day in late June is shown in Fig. 7.16. The greatest LE fluxes occur near the leading edge (the border between the irrigated and dry fields). A rapid reduction in LE occurs with distance into the field (between A and B). Further decreases may be small, especially during midday when fluxes are great. Note, however, that even at the position furthest from the leading edge of the field LE far exceeds $(R_n + S)$.

Figure 7.17 is a schematic representation of the energy balance as it changes with distance downwind of the leading edge. Latent heat flux is greatest at the leading edge because of the energy contribution from local advection H_{loc}. The influence of the local advection diminishes rapidly with distance into the field since the source of H_{loc} is a limited area upwind. Where equilibrium latent heat flux (no further change with distance) occurs, *LE still exceeds* $(R_n + S)$. That difference is the regional contribution of sensible heat H_{reg}.

During the season of this experiment sensible heat advection contributed 15–50% of the daily total energy consumed in evapotranspiration at the location furthest downwind. At the upwind station (location A in Fig. 7.16) local advection supplied 1–14% of the energy consumed by LE. This study suggests that, in the Great Plains region, regional advection can be a major source of energy for evapotranspiration.

Wind. The wind often plays an important role in the evapotranspiration process. Strong winds enhance turbulence, thereby reducing the boundary layer resistance and, consequently, facilitating the movement of vapor laden

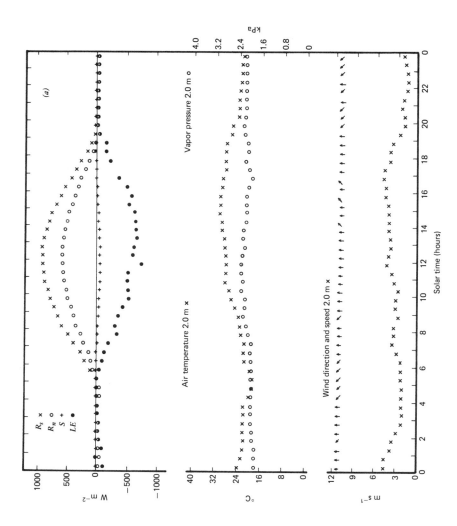

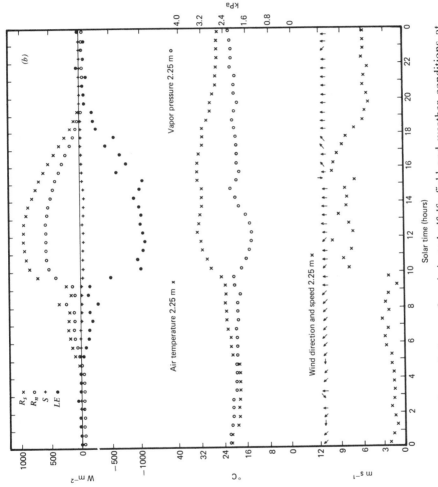

Fig. 7.15 Energy balance of an irrigated alfalfa field and weather conditions at Mead, Nebraska on 2 days during an extended drought: (a) June 10 and (b) June 11 (after Rosenberg, N. J., and Verma, S. B., 1978, *J. Appl. Meteorol.* **17**:934–941, American Meteorological Society).

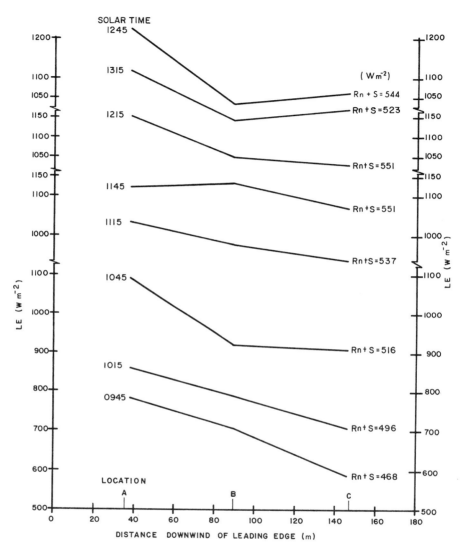

Fig. 7.16 Variation in latent heat flux *LE* over irrigated alfalfa at Mead, Nebraska. Locations *A*, *B*, and *C* are 38, 80, and 146 m downwind from the leading edge, respectively (after Brakke, T. W., Verma, S. B., and Rosenberg, N. J., 1978, *J. Appl. Meteorol.* **17**:955–963, American Meteorological Society).

air into the drier atmosphere. Winds also serve as a means of transporting sensible heat from dry surroundings into wet fields. Thus, as Brakke et al. (1978) observed, the contribution of regional sensible heat advection is greatest on days of strong wind.

One effect of reduced wind speed (as can be achieved through use of windbreaks) is reduced evapotranspiration from wet surfaces (Skidmore and Hagan, 1970). When surfaces are not wet, however, decreasing the wind speed may act either to increase or decrease evapotranspiration. The effect will depend on the plant resistance to diffusion of water vapor (Monteith, 1964; van Bavel, 1966) or on the relative humidity of the air (Linacre, 1964). Seginer (1971) modeled the effects of wind on evapotranspiration. He calculated increased or decreased evaporation with increased wind speed depending on atmospheric and surface conditions.

Humidity. Evaporation of small droplets of water, evaporation from wet soil or water surfaces, and transpiration are all influenced by the vapor content of the nearby air. If the air is saturated evaporation will not occur; if not, evaporation can proceed. As a general rule, evaporation or evapotranspiration increases in response to an increasing difference between the vapor pressure at the evaporating surface and the vapor pressure of the air. Showalter (1971) has shown that, for the case of small water droplets evaporating into the air, the evaporative capacity of the air E_c can be defined by the wet-bulb depression $(T_a - T_w)$ (a measure of the humidity of the air) and the atmospheric pressure P. His relationship is given as

$$E_c = PA(T_a - T_w) \qquad (7.10)$$

where A is a constant used in the psychrometric equation (5.20).

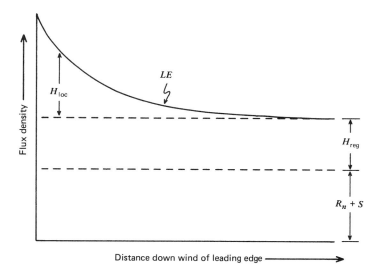

Fig. 7.17 Variations in energy balance with distance from the leading edge of an irrigated field. Energy balance components include latent heat flux LE, regional sensible heat advection H_{reg}, local sensible heat advection H_{loc}, net radiation R_n, and soil heat flux S (after Brakke, T. W., Verma, S. B., and Rosenberg, N. J., 1978, *J. Appl. Meteorol.* **17**:955–963, American Meteorological Society).

The water potential of air Ψ_a, a measure of the evaporative demand of the air, is given by

$$\Psi_a = \frac{RT_a}{V_w} \ln \frac{e_a}{e_s} = \frac{RT_a}{V_w} \ln \text{RH} \qquad (7.11)$$

where R is the gas constant, T_a is the air temperature in °K, V_w is the volume occupied by 1 mole of water vapor, e_a is the vapor pressure of the air, e_s is the saturation vapor pressure of the air at the given air temperature, and RH is the relative humidity (e_a/e_s) of the air. Ψ_a is always zero or negative in sign; the greater the absolute value of Ψ_a, the stronger the evaporative demand.

Air is seldom at 100% relative humidity, especially during daytime so that the magnitude of Ψ_a generally exceeds plant water potential by several MPa. Thus, the atmospheric water potential is almost always conducive to transpiration.

The evaporation of free water from bodies of water and from soil or plant surfaces will always increase as the humidity of the air decreases. Transpiration also increases as the air becomes drier as is shown in Fig. 7.18. There are some cases, however, when low atmospheric humidity can lead to stomatal closure thereby decreasing ET rates (Lange et al., 1971). Under very dry conditions with closed stomates transpiration rates may actually increase with increasing humidity, because cuticular resistance can be inversely related to the relative humidity of the surrounding air (Moreshet, 1970). Under such conditions, however, transpiration rates are very low, so that the amounts of water involved are not great.

Temperature. The temperature of the air and/or that of the evaporating surface exerts a major influence on evapotranspiration. In general the higher the temperature, whether of air or evaporating surface, the greater will be the rate of evaporation. Because of the strong dependence of evaporation

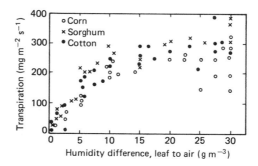

Fig. 7.18 Relation of transpiration to the difference in absolute humidity between the leaf and air. Data are from two irrigation treatments after initiation of reproductive growth (after Ackerson and Krieg, 1977).

on temperature and because temperature is a good integrator of several environmental variables, many models for predicting ET use temperature as a major input.

Temperature influences evapotranspiration in the following four ways.

The amount of water vapor that air can hold increases exponentially with increasing temperature (Chapter 5). Thus, as the surface temperature increases the vapor pressure at the evaporating surface increases as does the vapor pressure deficit between the surface and nearby air. This holds true so long as the water supply is adequate to nearly saturate air at the evaporating surface. Because air can hold more vapor as its temperature increases the vapor pressure deficit between air and the evaporating surface becomes larger and evaporative demand is increased as air is warmed.

Warm dry air may supply energy to an evaporating surface. The rate of evaporation is dependent on the amount of heat transferred; therefore, the warmer the air the stronger the temperature gradient and the higher the rate of evapotranspiration. On the other hand, if it is the evaporating surface that is warmed less sensible heat will be extracted from the air and evaporation will decrease. This occurs when the water supply is inadequate to meet the evaporative demand or when the surface is modified in a manner that reduces the dissipation of radiation (Chapter 11). Paradoxically, perhaps, under certain conditions warm air moving over a cool surface may actually suppress evaporation and cause condensation to occur. This may happen in the spring when, for example, warm humid air passes over a cool lake. Sensible heat exchange between the water and the air causes the air to be cooled to its dew point, which may be above that of the water surface. Then vapor condenses from the air to the surface of the lake.

Another, although perhaps minor, effect of increasing temperature at the evaporating surface results from the fact that less energy is required to evaporate warm than cool water. For example, values of L at 20 and 35°C are 2.44 and 2.40 MJ kg^{-1}, respectively. Thus, for the same energy input more water will be evaporated at the higher temperature.

Temperature may also influence evaporation through its influence on stomatal aperture. Stomatal reactions to temperature may be somewhat variable due to interactions between temperature and other environmental factors (Meidner and Mansfield, 1968). In general, however, stomatal aperture increases with rising temperature (Hofstra and Hesketh, 1969).

7.4 SOIL–PLANT–ATMOSPHERE CONTINUUM

Plants function like water pumps moving water from the soil and into the atmosphere in response to differences in water potential in soil, plant, and air. The hydraulic system that conveys water from the soil through the plant and into the air acts as a true continuum, for pressure changes at the roots are faithfully manifested at the leaves (Nulsen et al., 1977). The movement of water through the soil–plant–atmosphere system can be explained by the laws and principles of thermodynamics, some of which will be presented here. For a more extensive discussion see Leyton (1978).

The movement of water through the soil–plant–atmosphere continuum can be regarded as occurring along a gradient of decreasing water potential Ψ. Typical values of Ψ for an actively transpiring corn plant in air at 30°C and relative humidity of 50% when the soil water is at field capacity are shown in Table 7.1. Further examples of water potentials in air, leaf, root and soil for varying atmospheric and soil water conditions are presented by Reicosky and Lambert (1978). By far, the biggest drop in Ψ occurs between the leaves and the air, corresponding to the pathway of water vapor through the stomates. Water potential gradients in the soil are normally small compared with those through the roots (Reicosky and Lambert, 1978).

The process of water movement through the plant may be better understood, if we regard the flow of water as analogous to the flow of an electric current. According to Ohm's law, current = voltage (driving force)/resistance. In the water transport process, the driving force is the water potential difference $\Delta\Psi$ between two parts of the system working against a resistance r in that same segment. Under steady-state conditions, the flow through each segment is equal and the potential gradients and resistances are related as follows:

$$\text{flow rate} = \frac{\Psi_g - \Psi_r}{r_g} = \frac{\Psi_r - \Psi_l}{r_r + r_x} = \frac{\Psi_l - \Psi_a}{r_s + r_a} \quad (7.12)$$

where Ψ_g, Ψ_r, Ψ_l, and Ψ_a are water potentials of the soil, root, leaf, and air, respectively, and r_g is the resistance through the soil to the root surface, r_r is the resistance across the root, r_x is the xylem resistance, r_s is the stomatal resistance, and r_a is the aerial boundary layer resistance. The resistance to flow in the vapor phase is the greatest of all (Cowan and Milthorpe, 1968). To overcome this strong resistance, a large drop in water potential between

Table 7.1 Approximate water potential values for a transpiring corn plant[a]

Component	Water potential (MPa)
Soil	−0.01
Roots	−0.15
Leaves	−0.80
Air	−80.0

[a] Soil water is at field capacity, relative humidity is 50%, and air temperature is 30°C. Data are from Neumann et al. (1974) and Leyton (1978).

the air and leaf is required and does, in fact, occur (Table 7.1). At low wind speed r_a limits transpiration whereas at high wind speeds r_s becomes the dominating factor.

In studies of liquid water flow through plants, the quasi-steady-state model of van den Honert (1948) provides an acceptable approximation. This model, a simplification of (7.12), is given by

$$T = \frac{\Psi_g - \Psi_l}{r_g + r_p} \qquad (7.13)$$

where r_p is the total resistance ($r_p = r_r + r_x$) in the plant (exclusive of the stomatal resistance which involves a phase change of water from the liquid to the vapor state). The total plant resistance is composed of resistance to water flow through the roots, through the stem, and to the sites within the leaves where it is converted into vapor. Boyer (1974) has shown that most of the change in r_p that occurs with varying transpiration rate is due to changes in resistance within the leaves.

Theoretical and field studies have been conducted to evaluate the relative importance of the root resistance versus the total plant resistance. The approach taken by Gardner (1960) assumes that r_g is negligible at high values of Ψ so that r_p can be calculated from (7.13). The total resistance r_p is then held constant so that any increase in the sum of the resistances ($r_g + r_p$) as Ψ_g decreases is ascribed to r_g. Using this approach, r_g is much larger than r_p. Newman (1969) assumed that r_p is much greater than r_g except when effective root densities or soil water potentials are very low. Newman's conclusions were substantiated exerimentally by Hansen (1974a,b) and Molz (1975).

There are limitations to the resistance approach for describing water flow in the soil–plant–atmosphere system; nevertheless, the approach is useful for conceptualizing water flow through the soil and plant and into the atmosphere. One problem is the demonstration that, at least for some species, transpiration rates may increase several fold with no increase in the driving force (leaf water potential) (Baars, 1973; Kaufmann and Hall, 1974). For this to occur, there must be a reduction in the resistance to liquid flow through the plant.

The flow of water through a plant is a dynamic rather than a steady-state process, but the flux is typically much greater than temporal changes in the quantity of water stored in the plant. Diurnal changes in water flow can, therefore, be approximated by a sequence of steady-state conditions even under variable environmental conditions (Cowan and Milthorpe, 1968).

7.5 ESTIMATION OF EVAPORATION AND EVAPOTRANSPIRATION

The need for expanded food production has led to dramatic increases in irrigated land area in the subhumid to arid regions of the world. Agriculture faces increased competition for water by industries, municipalities, and other groups. This growing demand for water and the increased costs of energy to move it make it imperative that water management procedures be improved. The accurate assessment of evapotranspiration is essential if this is to be done.

Since Dalton first introduced the mass transport equation numerous methods for estimating or measuring ET have been developed. Certain of these methods are accurate and reliable; others provide only a rough approximation. In the sections that follow many of the most commonly used techniques for estimating E and ET are described.

7.5.1 Hydrologic or Water Balance Method

Essentially all water on the earth is held in the oceans, lakes, and ice or snow fields. Only about 0.5% of the total water on earth is involved at any given time in the hydrologic cycle. Nevertheless, because essentially all water which falls on land surfaces is eventually returned to the atmosphere by evaporation and transpiration, it is important to understand the water balance of these surfaces. The hydrologic approach for estimating ET is widely used and, in some cases, provides the most practical approach. The water balance may be written as

$$PI + \Delta SW \pm RO - D - ET = 0 \quad (7.14)$$

where PI is the precipitation and/or irrigation, RO is the runoff, D is the percolation or deep drainage, and ΔSW is the change in content of water stored in soil. Inputs to the system are positive and outputs are negative.

Equation 7.14 can be applied to any scale, ranging from continental land masses (Malhotra and Brock, 1970) and hydrologic catchments (Baumgartner, 1970; Woolhiser et al., 1970; Lettau and Baradas, 1973) to small fields or even individual plants. In most water balance studies, all elements in (7.14) other than ET are measured or otherwise estimated and ET is calculated as the residual.

The hydrologic approach has been used extensively to gather data for water-planning purposes. The accuracy of this method depends on the accuracy with which rainfall, runoff, and ΔSW are measured. The errors in measurement of these parameters are sometimes significant. The assumptions concerning deep percolation also require qualification. van Bavel et al. (1968), Wight (1971), and Wilcox and Sly (1974) have shown that measurements may not truly reflect the quantities of water that either percolate or evaporate from the soil because large quantities of vapor and liquid flow in both directions as a consequence of thermal and hydraulic gradients. The magnitude of thermally induced flow is discussed in detail in Chapter 2.

Davidson et al. (1969) measured the soil water flux that occurs at various depths with normal and suppressed surface evaporation in order to develop better estimates of percolation or deep drainage D for use in estimating ET in hydrologic studies. When D is properly accounted for, ET can be estimated from changes in ΔSW that occur during the period between rains or irrigations. In other instances, primarily for water balance studies over entire watersheds, ΔSW is assumed to be zero, PI and RO are measured, and D is neglected or otherwise accounted for. In this case, ET = PI − RO. This approach is generally used for periods of time during which it is reasonable to assume that the soil water content is essentially unchanged.

The advantages of the soil water balance approach as compared to aboveground measurements of water vapor flux are, according to Slatyer (1968), the ease of data processing and the integration by the soil water reservoir of extraction rates between observations. The disadvantages are the low level of measurement accuracy and the difficulties of assessing ET during rainy periods. Even with careful measurements, it is difficult to detect soil water changes with an accuracy better than ±2 mm of water. Errors associated with the water balance approach invalidate it for estimating ET on a daily basis. If drainage can be adequately accounted for, the method should be acceptable for 2- or 3-day intervals. When it is applied to large areas, the major problem is not the method itself, but the lack of good spatial averaging of inputs and outputs due to the variation in rainfall over large areas and the lack of homogeneity in the topography and soils that underlie them.

Weeks and Sorey (1973) proposed a method for the estimation of ET from groundwater by nonmeteorological means. They analyzed water level data from observation wells installed in five-point arrays and used finite-difference approximations of the differential equation to describe the groundwater flux.

7.5.2 Climatological Methods

Air Temperature-Based Formulas. In certain regions of the world, meteorological and climatological data may be quite limited. Models based almost solely on air temperature may be used in such cases to provide estimates of ET. If estimates are made for periods of several weeks or a month, reasonable first approximations are possible even with the defects inherent in these models (Hashemi and Habibian, 1979). Some of the more common temperature-based models are described below.

Thornthwaite Method. Thornthwaite (1948) described the biological and physical importance of evapotranspiration in climatic classification. As a consequence of his efforts he developed an equation for estimating ET_p. The Thornthwaite method for estimating monthly ET_p in mm may be written as

$$ET_p = 16 \left(\frac{l_1}{12}\right) \left(\frac{N}{30}\right) \left(\frac{10T_a}{I}\right)^{a_1} \tag{7.15}$$

where l_1 is actual day length (h), N is the number of days in a month, T_a is the mean monthly air temperature (°C), and a_1 is defined as

$$a_1 = 6.75 \times 10^{-7} I^3 - 7.71 \times 10^{-5} I^2 + 1.79 \times 10^{-2} I + 0.49 \quad (7.16)$$

where I is a heat index derived from the sum of 12 monthly index values, i, obtained from

$$i = \left(\frac{T_a}{5}\right)^{1.514} \quad (7.17)$$

Nomograms and tables have been prepared which greatly simplify these calculations (Thornthwaite and Mather, 1955; Palmer and Havens, 1958).

Certain shortcomings are inherent in the method. Calculated ET is underestimated at the time of annual maximum radiation reception during summer and is consequently out of phase in fall as well. Furthermore, application of the Thornthwaite concept to short time periods often leads to serious errors because short-term mean temperature is not a suitable measure of incoming radiation (Pelton et al., 1960). The success of the method on a long-term basis is explained as being due to the fact that both temperature and ET are similar functions of net radiation and are, therefore, autocorrelated when the periods considered are long.

Blaney–Criddle Method. Blaney and Criddle (1950) developed a method for estimating actual evapotranspiration or, as they have termed it, "consumptive use." The consumptive use c_u on a monthly basis is

$$c_u = k_m f \quad (7.18)$$

where k_m is an emprically derived monthly consumptive use coefficient (a function of the type of crop), and f is a monthly consumptive use factor, $0.01(1.8T_a + 32)p$, where T_a is the monthly mean temperature (°C) and p is the monthly percentage of total annual daylight hours. The total consumptive use for the season C_U is

$$C_U = \sum c_u = \sum k_m f \quad (7.19)$$

The method is relatively easy to use and the needed data are readily available. It has been used extensively, especially in the western United States, with results accurate enough for many practical applications.

Hargreaves Method. Hargreaves (1974) developed a method to estimate ET_p which emphasizes simplicity and requires a minimum amount of climatic data. This method can be written as

$$ET_p = MF(1.8T_a + 32)CH \quad (7.20)$$

where ET_p is in mm month^{-1}, MF is a monthly latitude-dependent factor, which is given in a table, T_a is the mean monthly temperature (°C), and CH is a correction factor for relative humidity RH to be used only when mean daily relative humidity values exceed 64%. CH can be calculated from the following equation:

$$CH = 0.166(100 - RH)^{1/2} \qquad (7.21)$$

For mean daily RH of 64% or less CH is unity. Hargreaves tested his method against lysimeter measurement of ET at several locations around the world and developed regression equations relating actual ET to ET_p in different climatic regimes.

Linacre Method. Linacre (1977) proposed a method that requires knowledge of the elevation, latitude, daily maximum and minimum temperature, and mean dew-point temperature of the site. The equation for estimating ET in mm is

$$ET = \frac{700 T_m/(100 - l) + 15(T_a - T_d)}{(80 - T_a)} \qquad (7.22)$$

where $T_m = T_a + 0.006z$, z is the elevation (m), T_a is the mean temperature (°C), l is the latitude (degrees), and T_d is the mean dew-point temperature (°C). Linacre found that values of ET given by (7.22) typically differed from measured values by 0.3 mm day^{-1} on an annual basis to 1.7 mm day^{-1} on a daily basis.

Solar Radiation Formulas. Experimental evidence (Tanner and Lemon, 1962) suggests that most of the energy for evapotranspiration,* even in arid regions, is derived from solar radiation. Thus, ET is correlated with solar radiation and, as shown by Aslyng (1974), ET_p is linearly and strongly dependent on R_s (Fig. 7.19). The dependence of ET on solar radiation changes as climate and surface conditions change with season of the year. Certain methods based on solar radiation also involve a temperature term.

Regression Methods. The relationship of ET_p and R_s has been established empirically and can be described by a simple linear regression of the form

$$ET_p = aR_s + b \qquad (7.23)$$

where a and b are empirical constants that change with location and season. Stanhill (1961) and Tanner (1967) propose values for a and b. Such regession models are simple to use but, because of their highly empirical nature, have only a very limited range of applicability. The lack of adequate instrumental networks to provide data on R_s also limits use of such models.

---

* In arid regions advection supplies a large portion of the energy in subhumid and semi arid regions (see Section 7.3.3, Sensible Heat Advection). In such cases, although the advected sensible heat is derived from solar radiation, it is unlikely that ET_p will be linearly related to R_s.

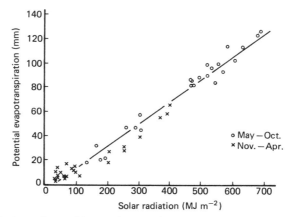

Fig. 7.19 Relation of monthly total potential evapotranspiration to monthly totals of solar radiation (after Aslyng, 1974).

Makkink Method. Makkink (1957) proposed the following regression type formula for estimating ET_p in mm day^{-1} from solar radiation measurements:

$$ET_p = R_s \left(\frac{s}{s + \gamma}\right) + 0.12 \qquad (7.24)$$

where the energy content of R_s is converted to equivalent units of water evaporated. Makkink's formula has given good results in cold wet climates but has not been found satisfactory in arid regions.

Jensen–Haise Method. Jensen and Haise (1963) summarized data collected in arid regions of the western United States and developed the following equation:

$$ET_p = R_s(0.025T_a + 0.08) \qquad (7.25)$$

where T_a is the mean daily air temperature in °C, R_s is the daily total solar radiation in units equivalent to mm of water, and ET_p is in mm day^{-1}. This model was tested in Nebraska against lysimetric measurements with the results shown in Fig. 7.20. The model tended to seriously underestimate ET under advective conditions but gave good results under nonadvective conditions. Jensen et al. (1970) proposed a modification of (7.25) but this modified method tended to underestimate ET under advection even more seriously (Table 7.2). The Jensen–Haise method and others based on similar principles are described and discussed by Linacre (1967).

Solar Thermal Unit Method. Caprio (1974) proposed a method to estimate ET based on a solar thermal unit STU concept. STUs are defined as the product of the mean daily temperature (°C) minus a threshold temperature of -0.55°C (31°F) times the daily total solar radiation in J m^{-2}. In SI units

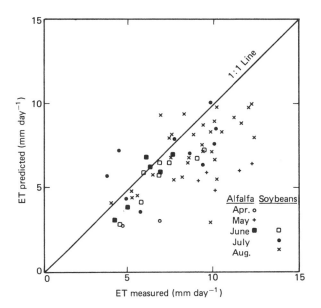

Fig. 7.20 Comparison of the Jensen and Haise (1963) method predictions with lysimetric measurements of evapotranspiration by well-watered soybeans and alfalfa at Mead, Nebraska.

with ET_p in mm day^{-1} the STU method can be expressed as

$$ET_p = 6.1 \times 10^{-9} R_s (1.8 T_a + 1) \qquad (7.26)$$

The method has not, as yet, been widely tested. It would appear, however, that this method would be subject to the same limitations that affect other solar radiation and temperature based empirical methods.

Solar and Thermal Radiation Method. An empirical method developed by Idso et al. (1975, 1977) is based on the supposition that LE is proportional

Table 7.2 Measured evapotranspiration under potential conditions, and estimations of potential evapotranspiration by the Jensen–Haise methods of 1963 and 1970

				Estimated	
Crop	Season	Days	Measured lysimetrically (mm day^{-1})	Jensen and Haise (1963) (mm day^{-1})	Jensen et al. (1970) (mm day^{-1})
Alfalfa	1967	13	7.65	5.13	4.80
Soybean	1969	31	7.47	6.25	5.64
Soybean	1970	20	9.22	7.57	6.78

to R_n when potential evaporation conditions prevail. Their equation is

$$ET_p = 4.11 \times 10^{-7}[R_{sw\downarrow} - R_{sw\uparrow} + 1.56 \times (R_{lw\downarrow} - R_{lw\uparrow}) + 6.53 \times 10^6] \qquad (7.27)$$

where ET_p is the 24-h evaporation rate in mm day^{-1}, 4.11×10^{-7} is a conversion factor for changing energy units from J m^{-2} to mm of water, and 1.56 and 6.53×10^6 are empirical coefficients. $R_{sw\downarrow}$, $R_{sw\uparrow}$, $R_{lw\downarrow}$, and $R_{lw\uparrow}$ are 24-h totals in J m^{-2} of incoming solar, relfected solar, incoming thermal radiation from the atmosphere, and outgoing thermal radiation from the surface, respectively. The method was developed for use in remote sensing. Inputs are daily solar radiation, albedo of the moist surface, and maximum and minimum surface and air temperatures. $R_{lw\downarrow}$ is calculated from the Idso–Jackson equation (Idso and Jackson, 1969) and $R_{lw\uparrow}$ is calculated from the Stefan–Boltzmann equation using the average daily maximum and minimum surface temperature. The model calculations were in good agreement with lysimetric measurements of evaporation from bare soil, water surfaces, and crops. Theoretical criticisms of the method are presented by Kalma et al. (1977) and McKeon and Rose (1977).

Combination Formulas. Penman (1948) was among the first to develop a method considering the factors of both energy supply and turbulent transport of water vapor away from an evaporating surface. In "combination models" LE is calculated as the residual in the energy balance equation with sensible heat flux estimated by means of an aerodynamic equation. One form of the combination equation is

$$LE = -\left[R_n + S + \rho_a C_p \frac{(T_a - T_s)}{r_a}\right] \qquad (7.28)$$

where ρ_a is the density of air, T_a is the air temperature, T_s is the surface temperature, and r_a is the boundary layer resistance. Verma et al. (1976), Blad and Rosenberg (1976), and Heilman and Kanemasu (1976) have shown the method to provide reliable estimates of LE fluxes when surface temperature is measured directly by means of ir thermometry. They found this true under both advective and nonadvective conditions and on both a short-period and a daily basis.

Jackson et al. (1977) made some assumptions to simplify (7.28) so that LE is determined as a function of R_n and $(T_a - T_s)$ only. The availability of remote sensing techniques allows this approach to be used in estimating LE over large areas; however, the equation contains an empirically derived constant that requires local calibration.

When T_s cannot be measured directly $(T_a - T_s)$ can be eliminated by application of the Clausius–Clapyron equation to give

$$LE = -\left[\frac{s}{s + \gamma}(R_n + S) + \frac{\rho_a C_p}{(s + \gamma)} \frac{(e_{ss} - e_a)}{r_a}\right] \qquad (7.29)$$

For a detailed description of the transformations involved see Kanemasu et al. (1979). The combination methods of Penman, van Bavel, and Slatyer and McIlroy, discussed below, are derived from (7.29).

Penman Method. The Penman method is the most popular and widely used combination model for estimating ET_p. Tests of this model continue more than 30 years after its introduction, attesting to its importance in agricultural and hydrological studies. The method, first presented by Penman (1948), was designed to estimate evaporation from open water surfaces and may be written as

$$E_o = \frac{sR_{no} + \gamma E_a}{(s + \gamma)} \quad (7.30)$$

where R_{no} is the net radiation over open water and E_a is given by an equation of the form

$$E_a = f(U)(e_s - e_a) \quad (7.31)$$

When e_s and e_a are expressed in mb and U is the windrun in km day^{-1} at 2 m, $f(U)$ (according to Doorenbos and Pruitt, 1975) becomes

$$f(U) = 0.27(1 + U/100) \quad (7.32)$$

Stigter (1980) reviewed other equations for estimating $f(U)$. Based on the approach of Thom and Oliver (1977) Stigter concludes that

$$f(U) = 0.37(1 + U/160) \quad (7.33)$$

gives results similar to (7.32) and that both equations are acceptable as generalized wind functions.

A major reason for the widespread popularity of the Penman method is that it requires that meteorological measurements be made at only one level above the surface. Thus, it is easier to apply than either the energy balance or aerodynamic equations from which it was derived. Other than R_n and S the required weather parameters are measured routinely at most weather observatories. R_n, not generally measured, can be estimated using procedures proposed by Penman (1948) or Linacre (1968, 1969). S is generally neglected in applying the Penman equation.

To deal with water use by vegetation, Penman related ET_p to E_o by the following expression:

$$ET_p = f_1 E_o \quad (7.34)$$

where f_1 is an empirical factor with values of 0.8 in summer and 0.6 in winter. Although these values were determined for England they are, according to Monteith (1973), valid to within about ±15% for all temperate climates. Thom and Oliver (1977) suggest that (7.30) will yield ET_p directly and that use of (7.34) is unnecessary when $(R_n + S)$ is measured over the vegetation itself rather than over some hypothetical water surface, that is, $(R_n + S)$ is used rather than R_{no}.

250   EVAPORATION AND EVAPOTRANSPIRATION

The Penman method was not intended for use in the presence of advection. Thus it often fails under such conditions (Slatyer et al., 1970). Under conditions of strong sensible heat advection, Rosenberg (1969a,b) found that the Penman method consistently underestimated ET from bare soil and from alfalfa (Fig. 7.21). More recently Cull et al. (1981) found that a modified form of the Penman equation using a wind function derived by Wright and Jensen for a semiarid climate worked quite well when regional advection was significant.

Equation 7.30 estimates only E_o [ET_p if $(R_n + S)$ is measured over vegetation]. Other approaches are used to estimate actual ET by incorporating plant and soil water factors into the Penman method (and other methods for estimating ET_p). Penman developed a drying curve to account for the influence of limited water on actual ET (Penman, 1949). Kristensen and Jensen (1975) used an approach in which ET_p is reduced according to vegetation density, soil water content, and rainfall distribution. A common and widely used practice for obtaining ET for use in irrigation scheduling is to multiply ET_p values by an empirically derived crop coefficient (e.g., see Doorenbos and Pruitt, 1975; Burt et al., 1981). However, crop coefficients are limited to specific crops and change during the course of the growing season. A

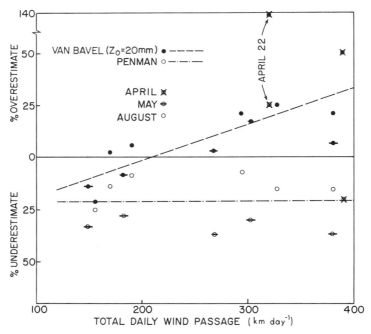

Fig. 7.21 Effects of windiness on the deviation of Penman and van Bavel method estimates of potential evapotranspiration from measured evapotranspiration at Mead, Nebraska. See Section 7.3.2, Internal Plant Resistance to Water Flow, for discussion of April 22 points (after Rosenberg, 1969a).

very good discussion and step by step instructions for use of the Penman method to estimate crop water requirements is given by Doorenbos and Pruitt (1975).

Penman Method Modified by Monteith. Monteith (1963, 1964) introduced resistance terms into the method of Penman and arrived at the following equation for *LE* from surfaces with either optimal or limited water supply:

$$LE = -\frac{s(R_n + S) + \rho_a C_p (e_s - e_a)/r_a}{(s + \gamma)[(r_a + r_c)/r_a]} \quad (7.35)$$

where all terms have been previously defined. This method has been successfully used to estimate ET from crops (Slabbers, 1977) and from forests (Calder, 1977). This model, of course, requires data on r_a and r_c (not readily available), so that the Monteith–Penman model has been limited to use in research applications, primarily.

Bailey and Davies (1981) present an approach to obtaining a bulk stomatal resistance for crops by solving for r_c in (7.35) and measuring the other quantities. Values obtained in this manner compared favorably with r_c values calculated from stomatal diffusion measurements. Using independently determined bulk stomatal resistance values in (7.35) they obtained ET estimates from soybeans that agreed well with Bowen ratio–energy balance method estimates.

Van Bavel Method. Following the method of Penman, van Bavel (1966) derived the following expression:

$$LE = -\frac{s(R_n + S) + \gamma L B_v (e_s - e_a)}{s + \gamma} \quad (7.36)$$

B_v is defined as

$$B_v = \frac{\rho_a \epsilon k^2}{P} \frac{U}{[\ln(z/z_0)]^2} \quad (7.37)$$

where U is the wind speed at a given height z, z_0 is a surface roughness parameter, and k is von Karman's constant. van Bavel considered this model an improvement over that of Penman since it involves no empirical constants or functions. He obtained excellent agreement between measured *LE* rates and those calculated with (7.36) for open water, wet bare soil, and well-watered alfalfa. However, Rosenberg (1969a) and Evans (1971) found the van Bavel method to be very sensitive to windiness and heavily dependent on the value assigned to z_0 (Fig. 7.21). The model underestimated in calm conditions and overestimated in strong winds.

Slatyer and McIlroy Method. The combination method given by Slatyer and McIlroy (1961) is similar to the Penman method except for use of the wet-bulb depression in place of the vapor pressure deficit. This model may be expressed as

$$ET_p = \frac{s}{s + \gamma}(R_n + S) - \rho_a C_p \frac{D_0 - D_z}{r_a} \tag{7.38}$$

where D_0 is the wet-bulb depression at the surface and D_z is the wet-bulb depression at height z. From this generalized equation two special cases can be derived (Davies, 1972). If the surface is well supplied with water so that the air next to the surface is saturated, $D_0 = 0$ and

$$ET_p = \frac{s}{s + \gamma}(R_n + S) + \frac{\rho_a C_p}{r_a} D_z \tag{7.39}$$

The second case describes the situation where D_z goes to zero. Under these conditions the equilibrium evaporation defined in (7.1) prevails.

Priestley–Taylor Model. Priestley and Taylor (1972) showed that, in the absence of advection, ET_p is directly related to the equilibrium evaporation, that is,

$$ET_p = \alpha \frac{s}{s + \gamma}(R_n + S) \tag{7.40}$$

Because α, which may be considered to be the ratio ET_p/ET_{eq}, is an empirically derived constant, the model is semiempirical in nature. It may also be considered a simplified form of the Penman equation in which the additive γE_a term in (7.30) is replaced by the constant α.

Using sets of reliable data from diverse well-watered surfaces, Priestley and Taylor obtained values for α between 1.08 and 1.34 with an overall mean of 1.26. Davies and Allen (1973) and Stewart and Rouse (1977) indicate that values of α vary slightly, depending on the air temperature, but are close to 1.26 for temperatures between 15 and 30°C. Barton (1979) suggests that α is dependent on the nature of the surface and that for some surfaces (e.g., forests) α is nearer unity. Barton also suggests, as do Davies and Allen (1973), that α is a function of the soil moisture at the surface and that (7.40) can be made more general if α is modified to account for changes due to the influence of soil moisture. Williams et al. (1978), however, were unable to find any consistent relationship between α and soil moisture under nonpotential conditions. Marsh et al. (1981) suggested an equation relating α to soil moisture at a high Arctic site. They concluded that the relation of α to moisture was site specific.

Thompson (1975) verified the value of $\alpha = 1.26$ for wet surfaces. He was also able to show that α decreased as the ratio LE/R_n decreased. Tanner and Jury (1976) found that, when 24-h values of R_n were used in (7.40), α was 1.35 ± 0.10, depending somewhat on the crop and local climate. If daytime R_n values alone were used, α was smaller.

ESTIMATION OF EVAPORATION AND EVAPOTRANSPIRATION

Jury and Tanner (1975) showed that α increased with advection and suggested a procedure for adapting the Priestley–Taylor method to advective conditions. Kanemasu et al. (1976) found that ET_p was grossly underestimated in advective conditions even with the Jury–Tanner correction. On the other hand, Shouse et al. (1980) found that the advection-modified Priestley–Taylor method was adequate under advective conditions whereas the unmodified Priestley–Taylor model, even with local calibration of α, did not give satisfactory results (Fig. 7.22). Williams and Stout (1981) also found that the Jury and Tanner modification led to accurate estimates of ET_p for alfalfa immediately following irrigation.

The Priestley–Taylor method is most reliable in humid areas. It has not yet been adequately tested in arid regions. Since the model utilizes meteorological data that are readily available, it is likely to be widely employed for estimating ET_p. With modification, the model may be made reliable for estimating actual ET, although much research is still needed to evaluate its performance under nonpotential conditions.

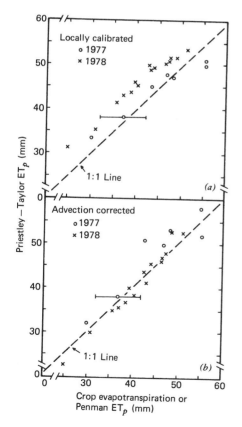

Fig. 7.22 Weekly estimates of potential evapotranspiration ET_p, using (a) the locally calibrated Priestley–Taylor method and (b) an advection modified Priestley–Taylor method. These estimates are compared with ET measured with a lysimeter in 1977 (○) and with ET_p calculated by the Penman method in 1978 (×) (after Shouse et al., 1980). Reproduced from *Agronomy Journal*, Volume 72, 1980, pages 994–998, by permission of the American Society of Agronomy.

7.5.3 Micrometeorological Methods

Mass Transport Methods. Dalton (about 1800) is credited with a general formula that predicts evaporation as a function of vapor pressure:

$$E_0 = C(e_0 - e_a) \tag{7.41}$$

where C is an empirically determined constant involving some function of windiness, e_0 is the vapor pressure at the surface, and e_a is the actual vapor pressure in the air at some point above the surface. This apparently simple method is not easily applied because of the difficulty in determining e_0. Only if the surface is saturated (e.g., a lake) can one make the assumption that $e_s = e_{ss}$ at surface temperature T_s.

Some variations of the Daltonian equation have been proposed. Rohwer (1931) gives

$$E_0 = (0.44 + 0.118U)(e_{ss} - e_a) \tag{7.42}$$

where U is wind speed. This relationship was derived from pan evaporation data at elevations above 1500 m in Colorado. Penman (1948) proposed

$$E_0 = 0.40(e_{ss} - e_a)(1 + 0.17U_2) \tag{7.43}$$

where U_2 is wind measured 2 m above ground surface.

Harbeck (1962) developed a slightly different equation for estimating evaporation from reservoirs:

$$E_0 = NU_2(e_{ss} - e_a) \tag{7.44}$$

where N is a coefficient related to the reservoir surface area.

The mass transport method, despite some problems, offers the advantage of simplicity in calculation, once the empirical constants have been developed. Improvements in the empirical constants, such as those of Brutsaert and Yu (1968), will continue to make the method attractive for estimating E_0 from lakes or reservoirs. This method has also been used to estimate evaporation from bare soils (Conaway and van Bavel, 1967; Ripple et al., 1970) and evapotranspiration from vegetated surfaces (Pruitt and Aston, 1963; Blad and Rosenberg, 1976).

There are some inherent difficulties in the method for estimating ET. To obtain e_{ss} for vegetative surfaces from measurements of T_s, it must be assumed that air spaces (substomatal cavities) within the leaf are at the temperature of the leaf and at, or very near, a relative humidity of 100%.

Estimates of water vapor fluxes from vegetative surface by the mass transport model become less accurate as crops experience water stress. This occurs because, as the temperature of the vegetative surface becomes elevated, e_{ss} estimated from these elevated temperatures will be overestimated. In spite of such difficulties, the model works well, when C is locally determined, in advective and in nonadvective cases when water is not limiting (Blad and Rosenberg, 1976).

Daltonian equations have been and will continue to be used widely because of their relatively simple data requirements. Measurements errors in the input parameters (e.g., temperature, relative humidity, wind speed) must be considered if the reliability of the final estimates are to be appraised realistically. Hage (1975) provides a useful analysis of the errors involved in Daltonian equations.

Aerodynamic Method. Theories, principles, and procedures involved in the aerodynamic methods are discussed in Chapter 4. Thornthwaite and Holzman (1942) were among the first modern micrometeorologists to apply the aerodynamic approach to measurement of ET. They proposed a relationship involving the gradients of specific humidity q and the logarithmic wind profile. Their expression, given here without derivation, is

$$E = \rho_a k^2 \frac{(q_2 - q_1)(U_2 - U_1)}{\ln(z_2/z_1)^2} \qquad (7.45)$$

Over a rough cropped surface $z - d$ is substituted for z. An error analysis of this method is given by Thompson and Pinker (1981).

Following Thornthwaite and Holzman's work, many others (e.g., Pasquill, 1950; Munn, 1961; Pruitt, 1963; Lumley and Panofsky, 1964; Oke, 1970; Businger et al., 1971; Dyer, 1974) have proposed stability-corrected aerodynamic methods for estimating the flux of vapor. Aerodynamic methods require stringently accurate observations of wind speed and specific humidity or vapor pressure at a number of heights above the surface, as well as temperature measurement to permit stability corrections to be made. Webb (1965), Pierson and Jackman (1975), and Kanemasu et al. (1979) provide reviews that are good sources of further information on aerodynamic models.

Because of its origins in classical fluid dynamics theory, aerodynamic methods have been popular with scientists. With recent refinements these models yield sufficiently accurate results. However, the methods have not reached a degree of development that makes them applicable for routine use, for example, in irrigation scheduling. On the other hand, adaptations of the aerodynamic method have been used to estimate evapotranspiration on a geographic scale appropriate for use in global climatic models (e.g., Yu, 1977; Manabe and Wetherald, 1980).

Bowen Ratio–Energy Balance Method. Bowen (1926) introduced a relationship between LE and H known as the Bowen ratio β. This is defined by

$$\beta = \frac{H}{LE} = \frac{PC_p}{L\epsilon}\left(\frac{K_h}{K_w}\right)\frac{\partial T/\partial z}{\partial e/\partial z} = \gamma \frac{K_h}{K_w}\frac{\partial T/\partial z}{\partial e/\partial z} \qquad (7.46)$$

This relationship is generally simplified by assuming that the turbulent exchange coefficient for heat transport K_h = the exchange coefficient for water vapor transport K_w and that $(\partial T/\partial z)/(\partial e/\partial z) \approx \Delta T/\Delta e$ where $\Delta T = T_2 - T_1$, and $\Delta e = e_2 - e_1$. Equation 7.46 then becomes

$$\beta \approx \gamma \frac{\Delta T}{\Delta e} \tag{7.47}$$

A simplified form of the energy balance equation (7.9) at the earth's surface can be written as

$$R_n + S + LE + H = 0 \tag{7.9}$$

From (7.46), $H = \beta LE$. Substitution into (7.9) and solution for LE yields

$$LE = -\left(\frac{R_n + S}{1 + \beta}\right) = -\left[\frac{R_n + S}{1 + \gamma \frac{\Delta T}{\Delta e}}\right] \tag{7.48}$$

Equation 7.48 is the so-called "Bowen Ratio–Energy Balance" (BREB) method of estimating LE.

Tanner (1960), Pruitt and Lourence (1968), and Denmead and McIlroy (1970) found BREB estimates of LE in good agreement with lysimeter measurements in nonadvective conditions. Lang (1973) offered a method of modifying the Bowen ratio method for use in locally advective conditions. Blad and Rosenberg (1974b) reported that the BREB method underestimated LE under conditions of regional sensible heat advection (Fig. 7.23). They suggested that the underestimation is due to an inequality in K_h and K_w. Later studies by Verma et al. (1978) have shown that $K_h > K_w$ when conditions of regional sensible heat advection prevail. Estimates of LE with the Bowen ratio method in regions of sensible heat advection will be improved by using actual K_h/K_w ratios rather than the assumption that $K_h/K_w = 1$.

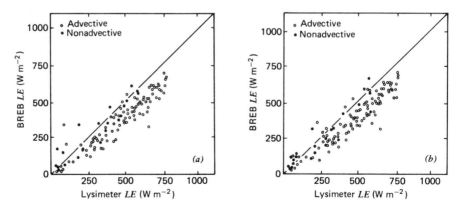

Fig. 7.23 Relation between latent heat flux LE calculated with the Bowen Ratio–Energy Balance method BREB and lysimetrically measured LE. Temperature and vapor pressure data for BREB calculations were from (a) 1.00 to 1.25 m level and (b) 1.25 to 1.50 m level (after Blad, B. L., and Rosenberg, N. J., 1974b, *J. Appl. Meteorol.* 13:227–236, American Meteorological Society).

The Bowen ratio method has certain distinct advantages for field measurement of LE. Instrumental requirements are relatively simple and rapid response instruments are not needed since mean fluxes over 15 min to 1 h are adequate for most purposes. It is important, however, that sensors used to measure ΔT and Δe be adjusted, alternated, or otherwise controlled to minimize errors. This is required since the ΔT and Δe gradients are usually quite small.

Fuchs and Tanner (1970) provide an analysis of errors in Bowen ratio calculations. Fuchs (1973a) suggests that daily estimates of LE cannot be made directly from daily average $(R_n + S)$ and daily average β. Instead, integration of 30–60-min estimates of LE over the day will yield the best results. McIlroy (1971) and Dilley (1974) describe an instrument that combines sensing and computing functions into a simple package for making continuous measurement of evapotranspiration based on the Bowen ratio method.

Resistance Methods. Earlier it was shown that the transport of sensible heat from surface to air proceeds at a rate directly proportional to the temperature gradient and inversely proportional to the aerial resistance r_a to heat transfer:

$$H = \rho_a C_p \frac{T_a - T_s}{r_a} \tag{3.4}$$

Similarly, it can be shown that the transport of vapor is directly proportional to the gradient in vapor pressure from the evaporating surface to the air and inversely proportional to aerial and stomatal resistance to the transport of water molecules.

Unlike sensible heat, which originates at the surface, the source of water vapor in the transpiration process is the substomatal cavity of the leaf. In that cavity air is saturated, or nearly so, unless the plant is under severe water stress or is desiccated. The vapor must diffuse through the stomatal openings controlled by the guard cells. Therefore, a second resistance must be considered in transpiration from a leaf: the stomatal resistance r_s or, if from a plant, the canopy resistance r_c.

Monteith (1963) proposed a method for estimating LE based on the mass transport (7.41). In his model C is replaced by a term that includes r_a and r_c. Equation 7.41 now becomes

$$LE = \frac{\rho_a L \epsilon}{P} \frac{e_a - e_{ss}}{r_a + r_c} = \frac{\rho_a C_p}{\gamma} \frac{e_a - e_{ss}}{r_a + r_c} \tag{7.49}$$

To use this method one may estimate e_{ss} from data on the surface temperature T_s, measure e_a, estimate r_a from wind speed measurements, and calculate r_c from data on stomatal resistance (see Section 7.3.2, Internal Plant Resistance to Water Flow).

The total resistance to vapor transport is considerably greater than is the resistance to sensible heat transport. Use of the resistance analogy to estimate both sensible and latent heat flux is subject to some criticism. Tanner (1963) and Phillip (1966) suggested that the sources and sinks for sensible and latent heat in a crop canopy are unlikely to be identical and, therefore, that the same resistance values cannot be used.

Brown and Rosenberg (1973) proposed a somewhat more complicated resistance model than that of Monteith's which eliminates the need to obtain values for e_{ss}. This model can be used to evaluate the influence of changing weather and plant factors on LE. Their model, which requires iterative solutions, is

$$LE = \left\{ \frac{f[(R_n + S + LE)r_a/\rho_a C_p + T_a] - e_a}{r_a + r_c} \right\} \frac{\rho_a C_p}{\gamma} \quad (7.50)$$

Verma and Rosenberg (1977) evaluated the performance of this model. They also simplified the model by the use of functional expressions for r_a and r_c. They report that the model predictions of hourly and daily values of LE generally agreed to within 10–15% of LE measured by lysimeters.

Eddy Correlation Methods. Water vapor is transported in the vertical direction by the upward and downward motion of small parcels of air, called eddies. Swinbank (1951) proposed the eddy correlation method to estimate the vertical flux of water vapor. The derivation of the eddy correlation equation for the transport of water vapor is given in Chapter 4. For a horizontal surface and with adequate upwind fetch the vertical transport of water vapor can be determined from

$$E = \frac{\epsilon}{P} \rho_a \overline{w' e'_a} \quad (7.51)$$

where w' and e'_a are instantaneous departures from the mean vertical wind speed and mean vapor pressure, respectively.

Theoretically, the eddy correlation technique should provide very accurate estimates of evapotranspiration. The major difficulties lie in the instrumentation and data collection as discussed in detail in Chapter 4.

7.6 MEASUREMENT OF EVAPOTRANSPIRATION

7.6.1 Lysimeters

Lysimeters are large blocks of soil isolated from, but as identical as possible with, surrounding soils. According to Pelton (1961), lysimeters were first used to study the percolation of water through soils. It was not until the twentieth century that modifications in lysimeter design were made to permit the study of evapotranspiration. Different types of lysimeters range widely in the accuracy with which changes in the soil water content are detected. The most accurate lysimeters can detect losses as small as 0.01 mm of water and can be used to accurately detect ET rates over time periods shorter than 1 h (measurements over 5–10-min periods have been reported). Other lysimeters are only sensitive enough to detect changes that occur over a day or longer.

Lysimeters provide the only direct measure of water flux from a vegetative surface. As such, they provide a standard against which other methods can be tested and calibrated. To provide reliable measurement of ET, lysimeters should meet the following criteria (Pelton, 1961):

1. They should be constructed so that moisture relationships inside the lysimeter correspond closely to those of soils under natural conditions.
2. The lysimeters should be sufficiently deep to extend well below the plant root zone or should use a tensioning device at the bottom of the soil column to maintain moisture at or near the same level as surrounding areas.
3. Lysimeters should be managed in exactly the same manner as the surrounding area.
4. The ratio of the wall surface area to the enclosed lysimeter area should be small to avoid small-scale advection from the uncropped surface.

Reviews of various kinds of lysimeters are given by Harrold (1966) and Rosenberg et al. (1968). Some of the more common types of lysimeters are:

Potential Evapotranspirometers. Potential evapotranspirometers are lysimeters in which the water table is held constant in a mass of soil so that evaporation may proceed at a potential rate. By continually monitoring the water added to the system and the amount of drainage, it is possible to estimate the rate of ET. Van Hylckama (1968) describes probable reasons for diurnal fluctuations in water levels and suggests some corrective procedures that would improve the accuracy of water use measurements made with evapotranspirometers. Potential evapotranspirometers should not be used to provide estimates for periods shorter than 1 day.

Potential evapotranspirometers were initially developed by Thornthwaite (Wilm, 1946). Reichman et al. (1979) describe the construction and operation of evapotranspirometers that are large enough to adequately sample most crops and which allow for conventional farming operation over them. At the other end of the size scale Tomar and O'Toole (1980) describe a small potential evapotranspirometer (microlysimeter).

Floating Lysimeters. Floating lysimeters are instruments that float on liquids such as water, oil, or heavy liquids such as solutions of zinc chloride. In floating lysimeters, the displacement of the fluid by the lysimeter bin is measured, generally with a manometer and its weight is determined using the Archimedean principle.

King et al. (1956) developed one of the first floating-type lysimeters. Their lysimeter floated on water and used a modified water stage recorder to detect water level changes. This lysimeter required large air spaces in the soil-filled bin to maintain buoyancy. McMillan and Paul (1961) described a lysimeter that consisted of a large soil container floating in a solution of zinc chloride. The use of heavy liquids such as the solution of zinc chloride eliminates the need for maintaining large air spaces in the lysimeter bins.

A variant of the floating lysimeter is one in which the weight of the soil column is supported by liquid filled bags. The pressure exerted on the bags, a function of the total lysimeter weight, is then measured with manometers (Glover and Forsgate, 1962; Hanks and Shawcroft, 1965).

Floating-type lysimeters equipped with manometers should not be used to obtain estimates of ET for periods shorter than 1 day and, even then, care must be exercised to account for errors due to temperature fluctuations. For periods as long as 1 week, however, such lysimeters may provide sufficiently accurate measurements of ET (Fulton and Findlay, 1966). Lourence and Goddard (1967) describe a floating lysimeter equipped with linear variable differential transducers to continuously record weight changes of the floating tank. This lysimeter has a resolution of 0.025 mm of water. With this system, reliable estimates of ET for periods of less than 1 day have been obtained.

Weighing Lysimeters. A simple weighing lysimeter can be made by filling a small container with soil and burying it in the ground. The container can be removed from the soil periodically and weighed on a scale. Such lysimeters have been used to study evapotranspiration from suburban lawns, as for example by Suckling (1980) and Danielson et al. (1981). Heatherly et al. (1980) describe a relatively inexpensive weighing lysimeter, which utilizes a hanging crane scale to measure weight changes. This particular lysimeter requires no power source and can be operated in isolated areas.

Very large lysimeters, such as those described by Harrold and Dreibelbis (1958) and Pruitt and Angus (1960), use mechanical scales installed below the ground. A large monolith (undisturbed block of soil) lysimeter (4 m in diameter and 1 m deep) is described by Sammis (1981). This lysimeter has a resolution of 0.19 mm of water and has been used to measure ET from a creosote bush (Sammis and Gay, 1979).

Very accurate estimates of ET can be obtained from precision weighing lysimeters. van Bavel and Meyers (1962) developed a very good precision weighing lysimeter. Improvements in the design of this lysimeter were suggested by Rosenberg and Brown (1970). A schematic of the modified lysimeter is shown in Fig. 7.24. The soil bin, which is 1 m^2 × 1.6 m deep, weighs about 3 tons. Weight changes, detected with a strain gage load cell, permit the measurement of ET with an accuracy approaching 0.01 mm of water. A

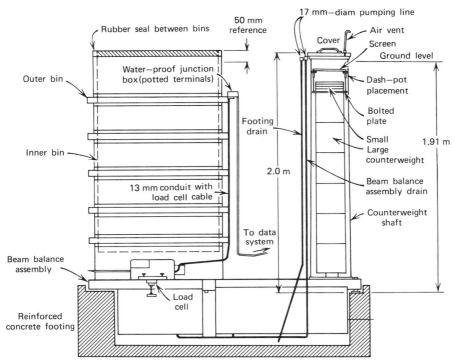

Fig. 7.24 Modified van Bavel–Myers lysimeter (after Rosenberg and Brown, 1970).

precision weighing lysimeter with dimensions of 1.8 × 1.8 × 1.2 m deep is described by Ritchie and Burnett (1968). Weight changes are detected with a strain gage load cell with a resolution of 0.025 mm.

Parton et al. (1981) describe the use of a large monlith lysimeter designed by Armijo et al. (1971) to measure ET from grasslands. This lysimeter contains an undisturbed soil core in a soil bin of 3.04 m diameter with a depth of 1.22 m. An electronic load cell is used to detect weight changes with a resolution equivalent to 0.025 mm of water.

7.6.2 Evaporimeters

Evaporation Pans. Pans of many shapes and sizes have been used to measure free water evaporation. Pans are inexpensive, relatively easy to maintain and simple to operate. However, care must be taken in relating evaporation from pans to actual ET, especially in arid climates. Because of the smaller aerodynamic roughness of water surfaces relative to vegetative surfaces, the former will extract less sensible heat energy from the passing air. Thus pans may, at times, show less evaporation than ET from vegetation.

Rosenberg and Powers (1970) observed, for example, that during a 5-day period in May, ET measured lysimetrically from alfalfa was 51.9 mm while evaporation pans (U.S. Weather Bureau Class A type) lost only 45.7 mm. In contrast, Pruitt and Lourence (1968) observed that the ratio of ET from fescue and ryegrass to pan evaporation E_{pan} decreased with strong hot and dry winds. This was probably due to increased stomatal resistance of the vegetation under stress conditions.

In humid regions pans may give realistic estimates of ET_p. Crop water use in such climates is usually about 60–90% of E_{pan}. Pruitt and Lourence (1968) reported that evapotranspiration from grasses was about 80% of that from evaporation pans except when winds were strong and the air was dry and hot. Stewart et al. (1977) found that the ratio of ET from well-watered corn to that of E_{pan} varied as a function of the stage of growth although for much of the year the ratio of ET/E_{pan} was about 0.9. Tan and Fulton (1980) found similar results for corn but different ratios for potatoes and tomatoes. The ratio of ET to E_{pan} is commonly called the crop coefficient K_1 (K_1 is also often expressed as the ratio of ET to ET_p). Thus empirically derived values of K_1 coupled with pan evaporation data make it possible to estimate ET from crops.

The exposure of an evaporation pan is very important. Fuchs (1973b) reported that the ratio of E_{pan} to calculated ET_p varied from 0.6 to 2.0, depending on the pan exposure. The smaller values generally occurred when the pan was partially shaded or was placed within an irrigated field. Because animals may drink from a pan or debris may fall into it, screens are sometimes used. When this is done evaporation is reduced. Campbell and Phene (1976), for example, found evaporation from screened pans to be reduced by about 13%. Kristensen (1979) compared evaporation from pans with 0, 9, and 40% of the pan surface screened to ET_p calculated with the Penman method. In all cases pan evaporation was less than ET_p with the greatest difference in the most densely screened pan. The 40% screen has been used in Denmark for several years as a standard method for estimating ET_p. Kristensen suggests reducing the screen density to even less than 9%.

Water level changes in pans are commonly measured with a micrometer gauge. This may be good enough for obtaining daily values but for accurate short period measurements of pan evaporation other ways of detecting the water level changes are required. Bloemen (1978) reviews various methods for monitoring water level changes and suggests a system which operates on the constant water level principle. Ambrus et al. (1981) propose a constant water level system that continuously logs evaporation with an accuracy of ±0.1 mm.

Evaporation Brush. Palland (1979) describes an "evaporation brush" in which evaporation takes place through vegetable fibers anchored in its base. The brush is placed in a tray in which water is kept at a constant level 20 mm above the tray bottom. The amount of water required to maintain a constant water level indicates the amount of evaporation. Daily evaporation from the brush agreed well with daily ET from meadow grass as calculated with a modified Penman formula.

7.6.3 Atmometers

Various types of atmometers are described by Livingston (1935), Carder (1960), Pelton and Korven (1969), and Dilley and Helmond (1973). Atmometers are, essentially, porous ceramic or paper evaporating surfaces and are generally colored either black or white. The evaporating surface is continuously supplied with water. Atmometers are, thus, passive devices, which expose a moist porous surface to the atmosphere. They respond relatively quickly to environmental influences, particularly radiation, wind speed, and the dryness of the air. Atmometers are especially sensitive to wind so that care should be exercised in their placement and exposure.

Atmometers are relatively inexpensive instruments for obtaining estimates of ET_p. Stanhill (1961) found that weekly and monthly evaporation rates from a Piche atmometer were linearly related with measured ET_p. However, correlation coefficients were only about 0.69. Dilley and Helmond (1973) found that estimates of ET_p from atmometers were in good agreement with ET from weighing lysimeters whereas Baier (1971) noted that atmometer estimates of ET_p agreed well with Penman method estimates. Pelton and Korven (1969), using empirical data on the relationship of atmometer evaporation to actual ET, reported that black Bellani plate and black porous disk atmometers could be used to estimate daily ET of alfalfa to an accuracy of about 1.5 mm day^{-1}. Wilcox and Sly (1974) found that a second-degree polynomial equation was required to accurately estimate ET_p from atmometer measurements. It would seem, then, that properly cared for and calibrated atmometers can be used to obtain reasonable estimates of ET_p.

7.6.4 Chamber Techniques

It is possible to estimate the amount of water evaporated from plants by enclosing them in a chamber and measuring the air flow rate and the concentration of water vapor entering and exiting the chamber. The change in concentration of water vapor multiplied by the volume of air passing through the chamber is equal to the amount of water lost by the vegetation and soil within the chamber. Because the placement of a chamber can drastically disturb the plant environment, data should be collected as soon as possible after the chamber is put in place. Even then, chamber data must be interpreted with care.

Peters et al. (1974) describe a chamber that travels along a track for measuring ET from crops. The chamber can be stopped at various locations and

264 EVAPORATION AND EVAPOTRANSPIRATION

Fig. 7.25 Tractor mounted portable field chamber for measurement of evapotranspiration (photograph courtesy of Dr. Don Reicosky, Soil Scientist, North Central Soil Conservation Laboratory, Agricultural Research Service, USDA, Morris, Minnesota).

a set of measurements made within about 20 s. An inexpensive portable chamber which can be mounted on the front of a tractor for making rapid measurements of ET in the field was designed by Reicosky and Peters (1977) (Fig. 7.25). They report that errors in ET estimates made with this chamber can be kept to within 11–19% so long as dry- and wet-bulb temperature of the air can be measured with an accuracy of ±0.1°C both before and 1 min after the chamber is put in place. They concluded that accuracy and rapidity of measurement combined with portability and relatively low cost make the chamber technique useful for evaluating the influence of soil and water management practices on plant water use.

The "ultimate" chamber may have been that used by Sebenik and Thomas (1967) who measured ET from a tree by enclosing it in a large plastic tent.

7.6.5 Sap Flow Techniques

The transpiration rate of a plant can be monitored by evaluating the rate at which sap flows through the conducting tissue. There are several ways to determine the rate of sap movement in a plant. Heat (Bloodworth et al.,

1955; Swanson, 1972; Balek and Pavlik, 1977), injected tritiated water (Kline et al., 1970), and radioactive phosphorous (Owston et al., 1972) have been used as tracers. The transpiration rate is calculated from the time lapse between tracer passage of two points in the stem. The sap flow techniques are somewhat specialized methods, applicable primarily to the measurement of water flow through individual plants, especially larger plants such as trees. Therefore, they are not often used for measuring ET of crops.

7.7 SEPARATION OF EVAPORATION AND TRANSPIRATION

Evaporation and transpiration are complex processes driven by the atmospheric demand for water. When plants cover only a small portion of the soil surface, evapotranspiration is dominated by evaporation from the soil. Transpiration becomes increasingly important as the amount of leaf area increases. When ground crop cover becomes complete, transpiration accounts for most of the water lost as evapotranspiration. It is sometimes desirable to measure or model evaporation and transpiration separately.

7.7.1 Direct Measurement

Separation of the evaporation and transpiration components can be accomplished with two lysimeters. These must be treated identically in terms of water input, soil conditions, crop type, and vegetative cover. A cover, impermeable to water and water vapor is placed over one lysimeter with the plants protruding through. This cover restricts the soil evaporation E_s so that the only water lost from the lysimeter is that which is transpired by the crop. The other lysimeter is left uncovered so that the water loss is total ET. The contribution of E_s is obtained by difference.

It is obvious that certain errors may occur with this approach. For example, the cover may change the surface energy balance, generating sensible heat and augmenting transpiration of the covered lysimeter. This will lead to underestimation of the E_s component of the uncovered lysimeter. The cover may also cause a change in temperature of the surface soil which could affect the rate of water movement to the plant. The suppression of evaporation from the covered lysimeter may, when water becomes limiting, make more water available for plant transpiration than might otherwise have been available. These and other problems should be considered in planning studies to directly separate soil evaporation and transpiration.

Another, and perhaps more accurate, approach to separating E_s and T when soil surfaces are kept relatively wet is the use of small shallow lysimeters filled with soil which are placed under the plant canopy. E_s, thus determined, can then be subtracted from ET and T obtained by difference. Arkin et al. (1974) have used this approach.

7.7.2 Models

Ritchie (1972) and Tanner and Jury (1976) propose somewhat similar approaches for separating E_s and T. Only the Ritchie approach will be described here. E_s is calculated in the constant rate and the falling rate stages. In the constant rate stage the supply of energy reaching the soil limits the rate of E_s. Hydraulic properties of the soil limit the rate of water flow to the surface and thus control the transition from the constant rate stage to the falling rate stage.

When plants begin to cover the soil, E_s will differ from that of a bare soil because of shading, lowered wind speed, and reduced vapor pressure deficits. Ritchie's equation to estimate E_s from the soil in the constant rate stage is

$$E_s = \frac{s}{s + \gamma} R_{ns} = \frac{s}{s + \gamma} R_n \exp(-0.4 \text{LAI}) \qquad (7.52)$$

where R_{ns} is the net radiation at the soil surface. When (7.52) is applied to the soil surfaces partially covered by mulch, the exponent becomes $-0.4(\text{LAI} + 2.5M)$ where M is the fraction of the soil surface covered by the mulch. Ritchie (1975) suggests that E_s estimates calculated with (7.52) can be improved by multiplying the right-hand side of the equation by an empirically derived Priestley–Taylor term (7.40) where

$$\alpha = 0.92 + 0.4 \frac{R_{ns}}{R_n} \qquad (7.53)$$

During the falling rate stage, E_s becomes more dependent on the hydraulic properties of the soil and less dependent on the available energy. Black et al. (1969) have shown that the cumulative evaporation from the soil can be expressed as

$$\sum E_s = \alpha_s t^{1/2} \qquad (7.54)$$

where α_s is a parameter dependent on soil hydraulic properties. α_s can be evaluated for individual soils from data on cumulative evaporation during a single drying cycle. t is the time (in days) from the onset of the falling rate stage.

The equation developed by Ritchie and Burnett (1971) for estimating transpiration as a function of leaf area index (see Section 7.3.2, The Influence of Crop Cover) when soil water is not limited is

$$T = \text{ET}_p(-0.21 + 0.70 \text{LAI}^{1/2}) \qquad \text{for } 0.1 \leq \text{LAI} \leq 2.7 \qquad (7.8)$$

Ritchie suggests using either the Penman method (7.30) or the Priestley–Taylor method (7.40) to calculate ET_p. For crop canopies with LAI greater than about 2.7, T is obtained from

$$T = \text{ET}_p - E_s \qquad \text{for LAI} \geq 2.7 \qquad (7.55)$$

Ritchie (1975) and Ritchie et al. (1976) suggest that soil water reaches a critical threshold value when the amount of extractable water (SW_e) remaining in the root zone of the soil is approximately 25% of the total extractable water (SW_t). The following equations for calculating T are suggested for the case when soil water is limiting:

$$T = 4(-0.21 + 0.70 \text{LAI}^{1/2})\text{ET}_p \frac{SW_e}{SW_t} \quad \text{for LAI} < 2.7 \quad (7.56)$$

and

$$T = 4\text{ET}_p \frac{SW_e}{SW_t} \quad \text{for LAI} > 2.7 \quad (7.57)$$

The models have been used with reasonable success by Ritchie and his co-workers to estimate ET from agricultural crops and small watersheds. The major problem with these models has been an apparent overestimate of transpiration with wet soils and when LAI is large. Other problems may arise when the model is applied in areas where sensible heat advection is common because of the difficulty of obtaining reliable estimates of ET_p under advective conditions.

7.8 APPLICATION OF EVAPOTRANSPIRATION METHODS TO SPECIAL SITUATIONS

Our interest in evapotranspiration is not confined to agricultural crops and other more regular surfaces but extends to understanding and measuring evaporation from natural surfaces such as forests, ice and snow, swamps, ponds, lakes, and oceans. Most principles and methods previously discussed will apply but in some cases they may fail or require adjustment.

7.8.1 Forest Evapotranspiration

Considerable effort has been devoted to understanding and measuring or estimating ET from forests of many types. A review of some of the theories and problems of various methods is given by Federer (1970). Forests differ from short vegetation in many ways that relate to evapotranspiration. They are aerodynamically rough. Because of the size of trees and the distribution of foliage throughout their canopies, significant quantities of sensible heat are exchanged between the air and leaves so that even in humid climates sensible heat advection can be an important source of energy for transpiration in forests (Shuttleworth and Calder, 1979). The relatively large heat capacity in the canopy volume can also cause errors in short period estimates of ET (Black and McNaughton, 1972). Interception of precipitation by forest canopies can be significant and in wetter climates evaporation of intercepted water can be twice that transpired (Calder, 1978; Pearce and Rowe, 1979).

Dew can also be a major source of water and has been shown to contribute as much as 20% of the total evapotranspiration from a forest (Fritschen and Doraiswamy, 1973).

A relatively high percentage of the water that reaches the soil surface should be available to the trees because most forest floors have a cover of litter that acts as a mulch to reduce soil evaporation.

Lysimeters have been used successfully by Gay and Fritschen (1979) to measure ET from salt cedar, from Douglas fir by Fritschen et al. (1977), and from spruce by Calder (1978). As might be expected, there are some special problems associated with the use of lysimeters in which large trees are growing. Fritschen et al. (1977) give ways to deal with the problems of sample size, root pruning, irrigation, drainage, locking of limbs, wind, and rainfall interception. They also discuss the problem of extrapolating data from a single tree to an area of forest.

Calder (1977) and Gash and Stewart (1977) used Monteith's modification of the Penman model (7.35) to obtain ET estimates from spruce and pine forests that agree well with measured values. Gash and Stewart (1977) found little difference between ET from the Scots pine of Thetford Forest and ET_p from short grass plentifully supplied with water. Pearce and Rowe (1979), on the other hand, cite studies that show that forests require more water than do crops.

Shuttleworth and Calder (1979) evaluated the Priestley–Taylor equation for estimating forest ET. They conclude that, because of advection and other factors unique to tall vegetation, the Priestley–Taylor method is of limited value and they warn against its indiscriminate use. They do, however, suggest modifying the model by adjusting α and including an interception term. With this modification Priestley–Taylor estimates agreed to within 20% of long-term total ET measurements in two forests.

In spite of the problem of small vertical temperature and humidity gradients above forest canopies the Bowen ratio technique has been used with considerable success (e.g., McNaughton and Black, 1973; McNeil and Shuttleworth, 1975; Black, 1979). Eddy correlation methods have likewise been used with reasonable success to determine energy fluxes in forests. Spittlehouse and Black (1979) suggested a method that combines energy balance and eddy correlation measurements. The eddy correlation approach (4.25) was used to obtain the sensible heat flux H. LE was then calculated by combining H with other measured terms in (7.9). When wind speeds were moderate their method proved ET estimates in good agreement with BREB estimates. With low wind speed the method failed because of problems associated with the stalling of the mechanical anemometers.

7.8.2 Evaporation from Ice and Snow

The study of evaporation from snow and ice is of considerable importance in the water balance of temperate and arctic zones. Significant water losses can occur from snow and ice through sublimation when air temperatures are below freezing. Branton et al. (1971), for example, found sublimation gains on 6 days and losses on 31 days during a 115-day period at Palmer, Alaska. During this period 30–70 mm of water was lost. Palm and Tveitereid (1979) found that thermal convection occurs within layers of dry snow with strong vertical temperature gradients and large air permeability (i.e., old snow). This thermal convection increases the flux of water vapor through and out of the snow and alters its density and insulating power.

McKay and Thurtell (1978) observed no direct relationship between R_n and the fluxes of H or LE during the course of a day. Maximum snow evaporation, as determined by an eddy correlation technique, occurred after warm moist air masses were replaced by cold dry air masses. During such periods LE exceeded R_n. They concluded that the Bowen ratio method fails because snow stores large quantities of energy. In spring, when snow begins to melt, most of the radiant energy and sensible heat available is consumed in melting the snow and only a small amount (3%, in a study by DeWalle and Meiman, 1971) is used in evaporation.

7.8.3 Evaporation from Swamps

Linacre et al. (1970) reviewed investigations relating evapotranspiration from swamps to evaporation from lakes or pans. Evaporation has been measured with a tank placed within a swamp (Blaney and Ewing, 1946), estimated by means of the Bowen ratio method (Rijks, 1969; Munro, 1979), the mass transfer equation (Eisenlohr, 1967), and the eddy correlation method (Linacre et al., 1970). Early studies indicated that swamps lose water at higher rates than do lakes and ponds (Blaney and Ewing, 1946). However, most later studies (Rijks, 1969; Linacre et al., 1970) suggest that, except immediately after rains, the loss of water from swamps is significantly lower than that from open bodies of water. Linacre et al. (1970) attributes lower rates of ET in swamps to the higher albedo of the reedy vegetation sheltering the water surface and the internal resistance of the reeds to water vapor flux. Based on theoretical and experimental evidence, Idso (1981) concluded that over extensive surfaces swamp evaporation will be less than open water evaporation and only for young and vigorous vegetation will evaporation from the swamp approach that lost from the open water.

7.8.4 Evaporation from Water Bodies

Practically all methods used to estimate ET from vegetative surfaces have been applied to estimating E_0 from bodies of water. The two methods most commonly applied are (1) a water budget method in which the lake level, precipitation, inflows and outflows are measured and E_0 is calculated as the residual (e.g., Shih, 1980) and (2) the mass transport approach. There appear to be many forms of the mass transport equation used for estimating evaporation from bodies of water. The apparent differences are due to the numerous approaches used in evaluating the mass transfer coefficient C in (7.41). Generally, the value for C is determined empirically in one of the following ways: (1) by dividing the result of an evaporation estimate made by some other means by the term $(e_{ss} - e_a)$ or (2) by linear regression of a water level change in a reservoir ΔWL against the term $C(e_{ss} - e_a)$. The regression takes the form

$$\Delta\text{WL} = C(e_{ss} - e_a) \pm D \tag{7.58}$$

where D is the average seepage loss from the reservoir (Lakshman, 1972).

Many estimates of lake evaporation have been based on the mass transfer formulations from the classic Lake Hefner study (U.S. Geological Survey, 1954) but more recently other approaches have been used to determine C. These include the use of wind profile data to define C as a function of the size and shape of the water body as well as a function of turbulent boundary layer parameters. This latter approach has been successfully applied to water bodies ranging from less than 0.5 ha in size to large lakes (Lakshman, 1972). In another approach, due to Quinn (1979), C is given as a function of the wind speed gradient, the potential temperature gradient and the Monin–Obukhov length. This latter method has been applied to estimating evaporation from the Great Lakes. Results for Lake Ontario suggest that the Lake Hefner mass transfer equation may overestimate large lake evaporation by about 20%. Expressions for C have also been developed from standard heat transfer expressions (Goodling et al., 1976) and from the principles of fluid mechanics (Weisman, 1975).

Our growing ability to measure water surface temperatures with aircraft and satellite-borne sensors assures that mass transport models will continue to be used for estimating evaporation from ponds and lakes.

The Penman and Priestley–Taylor models were developed to estimate evaporation from saturated surfaces and should be applicable to estimating evaporation from bodies of water. One major problem in applying these approaches to lakes lies in the practical difficulty of evaluating lake heat storage. For evaporation estimates over long time periods, the storage term can be neglected (Stewart and Rouse, 1977). For shorter periods measurements of water temperature profiles and soil heat flux contributions at the lake bottom are required (Bill et al., 1980). In spite of this difficulty the Priestley–Taylor method has been shown by Stewart and Rouse (1977) to work reasonably well for small shallow lakes and by deBruin and Keijman (1979) for a large shallow lake. In the latter study it was shown, however, that α varies with time of day and with season. deBruin (1978) combined the Priestley–Taylor and Penman approaches to eliminate the $(R_n + S)$ term. The resulting equation estimates evaporation from air temperature, saturation deficit, and wind speed at 2 m. For periods of 10 days or longer, this model gave acceptable results as compared to energy balance measurements.

DeBruin and Keijman (1979) describe the use of the Bowen ratio method for estimating E_0. The calculation of β was based on the temperature and vapor pressure difference between the water surface and air at 2 m. Hicks (1977) suggested a method for obtaining β based solely on the water surface temperature. Bill et al. (1980) compared the Hicks model with eddy correlation measurements and found it inadequate for estimating lake evaporation on a short term basis. They suggested modifications to improve the Bowen ratio method. Keijman (1974) simplified the Bowen ratio method for estimating lake evaporation. His simplified approach requires measurements of T_a, relative humidity, wind speed, and sunshine or solar radiation.

Another approach to estimating lake evaporation is that of Morton (1979) who developed a climatological model requiring monthly observations of temperature, humidity, and solar radiation. This model, which has been used to produce maps of lake evaporation over large areas, provides reasonably accurate estimates for shallow lakes on a monthly basis and for any type of lake on an annual basis.

An attempt at more direct measurement of lake evaporation has been made by Jarvinen (1978). He measured evaporation using floating pans located in the middle of four Finnish lakes. The floating pan worked satisfactorily except when high winds caused the water to become choppy.

Estimating evaporation from the oceans is more difficult than estimating evaporation from lakes, even large lakes. In theory, several evaporation models described previously should be usable for this purpose but the paucity of data, the vast size of the oceans, the currents, and the upwelling and downwelling all combine to make estimation of ocean evaporation a formidable task. However, because of the dynamic role of ocean evaporation in the hydrologic cycle, and in the generation of global weather systems, reliable estimates are required to improve our short- and long-term weather forecasts and as inputs to models for predicting possible climatic changes due to natural events or anthropogenically caused changes in the environment.

An example of such a model is that of Manabe and Wetherald (1980) who have predicted possible climatic changes due to the increasing atmospheric CO_2 concentration. That model uses an aerodynamic approach described in Manabe (1969) to estimate evaporation. Because of its fundamental importance, increased emphasis is likely to be placed on measuring and modeling evaporation from the oceans.

REFERENCES

Ackerson, R. C., and D. R. Krieg. 1977. Stomatal and nonstomatal regulation of water use in cotton, corn and sorghum. *Plant Physiol.* **60**:850–853.

Adams, J. E., G. F. Arkin, and J. T. Ritchie. 1976. Influence of row spacing and straw mulch on first stage drying. *Soil Sci. Soc. Am. J.* **40**:436–442.

Ambrus, L., E. Antal, and H. A. Karsai. 1981. New electronic evaporation and rain measuring equipment. *Agric. Meteorol.* **25**:35–43.

Arkin, G. F., J. T. Ritchie, and J. E. Adams. 1974. A method for measuring first-stage soil water evaporation in the field. *Soil Sci. Soc. Am. Proc.* **38**:951–954.

Armijo, J. D., G. A. Twitchell, R. D. Burman, and J. R. Nunn. 1971. A large, undisturbed, weighing lysimeter for grassland studies. ASAE Paper No. 71-582, pp. 1–16.

Aslyng, H. C. 1974. Evapotranspiration and plant production directly related to global radiation. *Nordic Hydrol.* **5**:247–256.

Baars, H. D. 1973. Controlled environment studies of the effects of variable atmospheric water stress on photosynthesis, transpiration and water status of *Zea mays* L. and other species. *Plant Response to Climatic Factors*, Proc. Uppsala Symp. UNESCO, pp. 249–258.

Baier, W. 1969. Concepts of soil moisture availability and their effect on soil moisture estimates from a meteorological budget. *Agric. Meteorol.* **6**:165–178.

Baier, W. 1971. Evaluation of latent evaporation estimates and their conversion to potential evaporation. *Can. J. Plant Sci.* **51**:255–266.

Bailey, W. G., and J. A. Davies. 1981. Bulk stomatal resistance control on evaporation. *Boundary-Layer Meteorol.* **20**:401–415.

Balek, J., and O. Pavlik. 1977. Sap stream velocity as an indicator of the transpirational process. *J. Hydrol.* **34**:193–200.

Baradas, M. W., B. L. Blad, and N. J. Rosenberg. 1976. Reflectant induced modification of soybean canopy radiation balance. V. Longwave radiation balance. *Agron. J.* **68**:848–852.

Barton, I. J. 1979. A parameterization of the evaporation from nonsaturated surfaces. *J. Appl. Meteorol.* **18**:43–47.

Baumgartner, A. 1970. Water and energy balances of different vegetation covers. *World Water Balance*, Proc. Reading Symp. Int. Assoc. Scientific Hydrology, pp. 57–65.

Bierhuizen, J. F., and R. O. Slatyer. 1965. Effect of atmospheric concentration of water vapor and CO_2 in determining transpiration-photosynthesis relationships of cotton leaves. *Agric. Meteorol.* **2**:259–270.

Bill, R. J., Jr., A. F. Cook, L. H. Allen, Jr., and J. F. Bartholic. 1980. Predicting fluxes of latent and sensible heat of lakes from surface water temperatures. *J. Geophys. Res.* **85**:507–512.

Black, T. A., W. R. Gardner, and G. W. Thurtell. 1969. The prediction of evaporation, drainage and soil water storage from a bare soil. *Soil Sci. Soc. Am. Proc.* **33**:655–660.

Black, T. A., C. B. Tanner, and W. R. Gardner. 1970. Evapotranspiration from a snap bean crop. *Agron. J.* **62**:66–69.

Black, T. A., and K. G. McNaughton. 1972. Average Bowen ratio methods of calculating evapotranspiration applied to a Douglas fir forest. *Boundary-Layer Meteorol.* **2**:466–475.

Black, T. A. 1979. Evapotranspiration from Douglas fir stands exposed to soil water deficits. *Water Resour. Res.* **15**:164–170.

Blad, B. L., and N. J. Rosenberg. 1974a. Evapotranspiration by subirrigated alfalfa and pasture in the east central Great Plains. *Agron. J.* **66**:248–252.

Blad, B. L., and N. J. Rosenberg. 1974b. Lysimetric calibration of the Bowen ratio-energy balance method for evapotranspiration estimation in the central Great Plains. *J. Appl. Meteorol.* **13**:227–236.

Blad, B. L., and N. J. Rosenberg. 1976. Evaluation of resistance and mass transport evapotranspiration models requiring canopy temperature data. *Agron. J.* **68**:764–769.

Blad, B. L., B. R. Gardner, D. G. Watts, and N. J. Rosenberg. 1981. Remote sensing of crop moisture status. *Remote Sensing Q.* **3**:4–20.

Blad, B. L. 1983. Atmospheric demand for water. *Plant Water Relations* (I. D. Teare, ed.), Wiley, New York, pp. 1–44.

Blanchet, R., N. Gelfi, and M. Bosc. 1977. Relations entre consommation d'eau et production chez divers types varietaup de soja (*Glycine max* L. Merr.). *Ann. Agron.* **28**:261–275.

Blaney, H. F., and P. A. Ewing. 1946. Irrigation practices and consumptive use of water in Salinas Valley, California. U.S. Dept. Agric., Los Angeles, California. 93 pp.

Blaney, H. F., and W. D. Criddle. 1950. Determining water requirements in irrigated areas from climatological and irrigation data. USDA Soil Conservation Service Tech. Paper No. 96. 48 pp.

Bloemen, G. W. 1978. A high-accuracy recording pan-evaporimeter and some of its possibilities. Tech. Bull. 107, *J. Hydrol.* **39**:159–173.

Bloodworth, J. E., J. B. Page, and W. R. Cowley. 1955. A thermoelectric method for determining the rate of water movement in plants. *Soil Sci. Soc. Am. Proc.* **19**:411–414.

Bouchet, R. J., and M. Robelin. 1969. Evapotranspiration potentielle et reele-domain d'utilisation-portee pratique. *Bull. Tech. Inform.* **238**:1–9.

Bowen, I. S. 1926. The ratio of heat losses by conduction and by evaporation from any water surface. *Phys. Rev.* **27**:779–787.

Boyer, J. S. 1974. Water transport in plants: Mechanisms of apparent changes in resistance during absorption. *Planta (Berlin)* **117**:187–207.

Brakke, T. W., S. B. Verma, and N. J. Rosenberg. 1978. Local and regional components of sensible heat advection. *J. Appl. Meteorol.* **17**:955–963.

Branton, C. I., L. D. Allen, and J. E. Newman. 1971. Some agricultural implications of winter sublimation gains and losses at Palmer, Alaska. *Agric. Meteorol.* **10**:301–310.

Brown, K. W., and N. J. Rosenberg. 1970. Effect of windbreaks and soil water potential on stomatal diffusion resistance and photosynthetic rate of sugar beets (*Beta vulgaris*). *Agron J.* **62**:4–8.

Brown, K. W., and N. J. Rosenberg. 1973. A resistance model to predict evapotranspiration and its application to a sugar beet field. *Agron. J.* **65**:341–347.

Brun, L. J., E. T. Kanemasu, and W. L. Powers. 1972. Evapotranspiration from soybean and sorghum felds. *Agron. J.* **64**:145–148.

Brun, L. J., E. T. Kanemasu, and W. L. Powers. 1973. Estimating transpiration resistance. *Agron. J.* **65**:326–328.

Brutsaert, W., and S. L. Yu. 1968. Mass transfer aspects of pan evaporation. *J. Appl. Meteorol.* **7**:563–566.

Burt, J. E., J. T. Hayes, P. A. O'Rourke, W. H. Terjung, and P. E. Todhunter. 1981. A parametric crop water use model. *Water Resour. Res.* **17**:1095–1108.

Businger, J. A., J. C. Wyngaard, Y. Izumi, and E. F. Bradley. 1971. Flux-profile relationships in the atmospheric surface layer. *J. Atmos. Sci.* **28**:181–189.

Calder, I. R. 1976. The measurement of water losses from a forested area using a natural lysimeter. *J. Hydrol.* **30**:311–325.

Calder, I. R. 1977. A model of transpiration and interception loss from spruce forest in Plynlimon, Central Wales. *J. Hydrol.* **33**:247–265.

Calder, I. R. 1978. Transpiration observations from a spruce forest and comparisons with predictions from an evaporation model. *J. Hydrol.* **38**:33–47.

Callander, B. A., and T. Woodhead. 1981. Canopy conductance of estate tea in Kenya. *Agric. Meteorol.* **23**:151–167.

Campbell, R. B., and C. J. Phene. 1976. Estimating potential evapotranspiration from screened pan evaporation. *Agric. Meteorol.* **16**:343–352.

Caprio, J. M. 1974. The solar thermal unit concept in problems related to plant development and potential evapotranspiration. *Phenology and Seasonality Modeling* (H. Leith, ed.), Springer-Verlag, New York, pp. 353–364.

Carder, A. C. 1960. Atmometer assemblies, a comparison. *Can. J. Plant Sci.* **40**:700–706.

Chin Choy, E. W., J. F. Stone, and J. E. Garton. 1977. Row spacing and direction effects on water uptake characteristics of peanuts. *Soil Sci. Soc. Am. J.* **41**:428–432.

Choudhury, P. N., and V. Kumar. 1980. The sensitivity of growth and yield of dwarf wheat to water stress at three growth stages. *Irrig. Sci.* **1**:223–231.

Conaway, J., and C. H. M. van Bavel. 1967. Evaporation from a wet soil surface calculated from radiometrically determined surface temperatures. *J. Appl. Meteorol.* **6**:650–655.

Cowan, I. R., and F. L. Milthorpe. 1968. Plant factors influencing the water status of plant tissues. *Water Deficits and Plant Growth* (T. T. Kozlowski, ed.), Academic Press, New York, Vol. 1, pp. 137–193.

Cull, P. O., R. C. G. Smith, and K. McCaffery. 1981. Irrigation scheduling of cotton in a climate with uncertain rainfall. *Irrig. Sci.* **2**:141–154.

Dale, R. F., and R. H. Shaw. 1965. Effect on corn yields of moisture stress and stand at two fertility levels. *Agron. J.* **57**:475–479.

Danielson, R. E., C. M. Feldhake, and W. E. Hart. 1981. Urban lawn irrigation and management practices for water saving with minimum effect on lawn quality. Colorado Water Resources Institute, OWRT Project No. A-043-COLO, 120 pp.

Davidson, J. M., L. R. Stone, D. R. Nielsen, and M. E. Larue. 1969. Field measurement and use of soil water properties. *Water Resour. Res.* **5**:1312–1321.

Davies, J. A. 1972. Actual, potential and equilibrium evaporation for a beanfield in southern Ontario. *Agric. Meteorol.* **10**:331–348.

Davies, J. A., and C. D. Allen. 1973. Equilibrium, potential and actual evaporation from cropped surfaces in southern Ontario. *J. Appl. Meteorol.* **12**:649–657.

deBruin, H. A. R. 1978. A simple model for shallow lake evaporation. *J. Appl. Meteorol.* **17**:1132–1134.

deBruin, H. A. R., and J. Q. Keijman. 1979. The Priestley-Taylor evaporation model applied to a large, shallow lake in the Netherlands. *J. Appl. Meteorol.* **18**:898–903.

Denmead, O. T. 1969. Comparative micrometeorology of a wheat field and a forest of Pinus Radiata. *Agric. Meteorol.* **6**:357–371.

Denmead, O. T., and I. C. McIlroy. 1970. Measurements of non-potential evaporation from wheat. *Agric. Meteorol.* **7**:285–302.

DeWalle, D. R., and J. R. Meiman. 1971. Energy exchange and late season snowmelt in a small opening in Colorado subalpine forest. *Water Resour. Res.* **7**:184–188.

deWit, C. T. 1958. *Transpiration and Crop Yields*. Wageningen, The Netherlands. Versi-Landbouwk. Onderz 646. 88 pp. (Inst. of Biological and Chemical Research on Field Crops and Herbage).

Dilley, A. C., and I. Helmond. 1973. The estimation of net radiation and potential evapotranspiration using atmometer measurements. *Agric. Meteorol.* **12**:1–11.

Dilley, A. C. 1974. An energy partition evaporation recorder. CSIRO Aust. Div. Atmos. Phys. Tech. Paper No. 24, pp. 1–25.

Doorenbos, J., and W. O. Pruitt. 1975. Crop water requirements. Irrigation and Drainage Paper, FAO, Rome. 179 pp.

Dyer, A. J., and T. V. Crawford. 1965. Observations of the modification of the microclimate at a leading edge. *Q. J. Roy. Meteorol. Soc.* **91**:345–348.

Dyer, A. J. 1974. A review of flux-profile relationship. *Boundary-Layer Meteorol.* **7**:363–372.

Dzerdzeevski, B. L., ed. 1957. Sukhovies and Drought Control. Akad. Nauk. SSSR, Inst. Geogr. (Israel Program for Scientific Translations, Jerusalem, 1963, OTS-63-11140).

Eagleman, J. R., and W. L. Decker. 1965. The role of soil moisture in evapotranspiration. *Agron. J.* **57**:626–629.

Eisenlohr, W. S., Jr. 1967. Measuring evapotranspiration for vegetation-filled prairie potholes in North Dakota. *Water Resour. Bull.* **3**:59–65.

El Nadi, A. H., and J. P. Hudson. 1965. Effects of crop height on evaporation from lucerne and wheat grown in lysimeters under advective conditions in the Sudan. *Exp. Agric.* **1**:289–298.

El Nadi, A. H. 1974. The significance of leaf area in evapotranspiration. *Ann. Bot.* **38**:607–611.

Evans, G. N. 1971. Evaporation from rice at Griffith, New South Wales. *Agric. Meteorol.* **8**:117–127.

Federer, C. A. 1970. Measuring forest evapotranspiration—theory and problems. USDA Forest Service Res. Paper NE-165. 25 pp.

Federer, C. A. 1975. Evapotranspiration. *Rev. Geophys. Space Phys.* **13**:442–445, 487–494.

Fischbach, P. E., and T. L. Thompson. 1979. Irrigation scheduling—Technology transfer program using AGNET computer system and other tools. Trans. ASAE and CSAE, Winnipeg, Canada, Summer Meeting. 11 pp.

Fritschen, L. J., and C. H. M. van Bavel. 1962. Energy balance components of evaporating surfaces in arid lands. *J. Geophys. Res.* **67**:5179–5185.

Fritschen, L. J. 1965. Evapotranspiration rates of field crops determined by the Bowen-ratio method. *Agron. J.* **58**:339–342.

Fritschen, L. J., and P. Doraiswamy. 1973. Dew: An addition to the hydrologic balance of Douglas fir. *Water Resour. Res.* **9**:891–894.

Fritschen, L. J., J. Hsia, and P. Doraiswamy. 1977. Evapotranspiration of a Douglas fir determined with a weighing lysimeter. *Water Resour. Res.* **13**:145–148.

Fuchs, M., C. B. Tanner, G. W. Thurtell, and T. A. Black. 1969. Evaporation from drying surfaces by the combination method. *Agron. J.* **61**:22–26.

Fuchs, M., and C. B. Tanner. 1970. Error analysis of Bowen ratio measured by psychrometry. *Agric. Meteorol.* **7**:329–334.

Fuchs, M. 1973a. Water transfer from the soil and the vegetation to the atmosphere. *Arid Zone Irrigation* (B. Yaron, E. Danfors, and Y. Vaadia, eds.), Springer-Verlag, New York, pp. 143–152.

Fuchs, M. 1973b. The estimation of evapotranspiration. *Arid Zone Irrigation* (B. Yaron, E. Danfors, and Y. Vaadia, eds.), Springer-Verlag, New York, pp. 241–247.

Fulton, J. M., and W. I. Findlay. 1966. Reproducibility of evaporation measurements from floating lysimeters. *Can. J. Plant Sci.* **46**:685–686.

Gardner, H. R. 1974. Prediction of water loss from a fallow field soil based on soil water flow theory. *Soil Sci. Soc. Am. Proc.* **38**:379–382.

Gardner, B. R., B. L. Blad, R. E. Maurer, and D. G. Watts. 1981. Relationship between crop temperature and the physiological and phenological development of differentially irrigated corn. *Agron. J.* **73**:743–747.

Gardner, W. R. 1960. Dynamic aspects of water availability to plants. *Soil Sci.* **89**:63–73.

Gash, J. H. C., and J. B. Stewart. 1977. The evaporation from Thetford Forest during 1975. *J. Hydrol.* **35**:385–396.

Gay, L. W., and L. J. Fritschen. 1979. An energy budget analysis of water use by saltcedar. *Water Resour. Res.* **15**:1589–1592.

Gindel, I. 1971. Transpiration in three eucalyptus species as a function of solar energy, soil moisture and leaf area. *Physiol. Plant.* **24**:143–149.

Glover, J., and S. A. Forsgate. 1962. Measurement of evapotranspiration from large tanks of soil. *Nature (London)* **195**:1330.

Goltz, S. M., and W. O. Pruitt. 1970. Spatial and temporal variations of evapotranspiration downwind from the leading edge of a dry fallow field. Div. Tech. Rep. ECOM68-G10-1. Dept. of Water Sci. and Engineering, Univ. of California–Davis. 95 pp.

Goodling, J. S., B. L. Sill, and W. J. McCabe. 1976. An evaporation equation for an open body of water exposed to the atmosphere. *Water Resour. Bull.* **12**:843–853.

Hage, K. D. 1975. Averaging errors in monthly evaporation estimates. *Water Resour. Res.* **11**:359–361.

Hanks, R. J., and R. W. Shawcroft. 1965. An economical lysimeter for evapotranspiration studies. *Agron. J.* **57**:634–636.

Hanks, R. J., L. H. Allen, and H. R. Gardner. 1971. Advection and evapotranspiration of wide-row sorghum in the central Great Plains. *Agron. J.* **63**:520–527.

Hansen, G. K. 1974a. Resistance to water flow in soil and plants, plant water status, stomatal resistance and transpiration of Italian ryegrass as influenced by transpiration demand and soil water depletion. *Acta Agric. Scand.* **24**:84–92.

Hansen, G. K. 1974b. Resistance to water transport in soil and young wheat plants. *Acta Agric. Scand.* **24**:37–48.

Harbeck, G. E., Jr. 1962. A practical field technique for measuring reservoir evaporation utilizing mass-transfer theory. U.S. Geol. Surv. Paper 272-E:101–105.

Hargreaves, G. H. 1974. Estimation of potential and crop evapotranspiration. *Trans. ASAE* **17**:701–704.

Harrold, L. L., and F. R. Dreibelbis. 1958. Evaluation of agricultural hydrology by monolith lysimeters (1944–1955). U.S. Dept. Agric. Tech. Bull. No. 1179. 166 pp.

Harrold, L. L. 1966. Measuring evapotranspiration by lysimetry. Evapotranspiration and its role in water resources management. *Proc. Am. Soc. Agric. Eng.*, pp. 28–33.

Hashemi, F., and M. T. Habibian. 1979. Limitations of temperature-based methods in estimating crop evapotranspiration in arid-zone agricultural development projects. *Agric. Meteorol.* **20**:237–247.

Heatherly, L. G., B. L. McMichael, and L. H. Ginn. 1980. A weighing lysimeter for use in isolated field areas. *Agron. J.* **72**:845–847.

Heilman, J. L., and E. T. Kanemasu. 1976. An evaluation of a resistance form of the energy balance to estimate evapotranspiration. *Agron. J.* **68**:607–612.

Herman, C. 1977. Icebergs might supply water to arid regions. *Irrigation Age*, Nov.–Dec., p. 42.

Hicks, B. B. 1973. Eddy fluxes over a vineyard. *Agric. Meteorol.* **12**:203–215.

Hicks, B. B., and G. D. Hess. 1977. On the Bowen ratio and surface temperature at sea. *J. Phys. Oceanogr.* **7**:141–145.

Hofstra, G., and J. D. Hesketh. 1969. The effect of temperature on stomatal aperture in different species. *Can. J. Bot.* **47**:1307–1310.

Howell, T. A., and E. A. Hiler. 1975. Optimization of water use efficiency under high frequency irrigation—I. Evapotranspiration and yield relationship. *Trans. ASAE* **5**:873–878.

Idso, S. B., and R. D. Jackson. 1969. Thermal radiation from the atmosphere. *J. Geophys. Res.* **74**:5397–5403.

Idso, S. B., R. J. Reginato, R. D. Jackson, B. A. Kimball, and F. S. Nakayama. 1974. The three stages of drying of a field soil. *Soil Sci. Soc. Am. Proc.* **38**:831–837.

Idso, S. B., R. D. Jackson, and R. J. Reginato. 1975. Estimating evaporation: A technique adaptable to remote sensing. *Science* **189**:991–992.

Idso, S. B., R. J. Reginato, and R. D. Jackson. 1977. An equation for potential evaporation from soil, water and crop surfaces adaptable to use by remote sensing. *Geophys. Res. Letters* **4**:187–188.

Idso, S. B. 1981. Relative rates of evaporative water losses from open and vegetation covered water bodies. *Water Res. Bull.* **17**:46–48.

Jackson, R. D., S. B. Idso, and R. J. Reginato. 1976. Calculation of evaporation rates during the transition from energy-limiting to soil-limiting phases using albedo data. *Water Resour. Res.* **12**:23–26.

Jackson, R. D., R. J. Reginato, and S. B. Idso. 1977. Wheat canopy temperature: A practical tool for evaluating water requirements. *Water Resour. Res.* **13**:651–656.

Jarvinen, J. 1978. Estimating lake evaporation with floating evaporimeters and with water budget. *Nordic Hydrol.* **9**:121–130.

Jensen, M. E., and H. R. Haise. 1963. Estimating evapotranspiration from solar radiation. *J. Irrig. Drainage Div. ASCE* **89**:15–41.

Jensen, M. E., D. C. N. Robb, and C. E. Franzoy. 1970. Scheduling irrigation using climate-crop soil data. *J. Irrig. Drainage Div. ASCE* **96**:25–38.

Jury, W. A., and C. B. Tanner. 1975. Advective modification of the Priestley and Taylor evapotranspiration formula. *Agron. J.* **67**:840–842.

Kalma, J. D., P. M. Fleming, and G. F. Byrne. 1977. Estimating evaporation: Difficulties of applicability in different environments. *Science* **196**:1354–1355.

Kanemasu, E. T., and G. F. Arkin. 1974. Radiant energy and light environment of crops. *Agric. Meteorol.* **14**:211–225.

Kanemasu, E. T., L. R. Stone, and W. L. Powers. 1976. Evapotranspiration model tested for soybean and sorghum. *Agron. J.* **68**:569–573.

Kanemasu, E. T., M. L. Wesely, B. B. Hicks, and J. L. Heilman. 1979. Techniques for calculating energy and mass fluxes. *Modification of the Aerial Environment of Crops* (B. J. Barfield and J. F. Gerber, eds.). *Am. Soc. Agric. Eng. Monogr.* **2**:156–182.

Kaufmann, M. R., and A. E. Hall. 1974. Plant water balance—its relationship to atmospheric and edaphic conditions. *Agric. Meteorol.* **14**:85–98.

Keijman, J. Q. 1974. The estimation of the energy balance of a lake from simple weather data. *Boundary-Layer Meteorol.* **7**:399–407.

King, K. M., C. B. Tanner, and V. E. Suomi. 1956. A floating lysimeter and its evaporation recorder. *Trans. Am. Geophys. Union* **37**:738–742.

Kline, J. R., J. R. Martin, C. F. Jordan, and J. J. Kovanda. 1970. Measurement of transpiration in tropical trees with tritiated water. *Ecology* **51**:1068–1073.

Kristensen, K. J. 1974. Actual evapotranspiration in relation to leaf area. *Nordic Hydrol.* **5**:173–182.

Kristensen, K. J., and S. E. Jensen. 1975. A model for estimating actual evapotranspiration from potential evapotranspiration. *Nordic Hydrol.* **6**:170–188.

Kristensen, K. J. 1979. A comparison of some methods for estimation of potential evaporation. *Nordic Hydrol.* **10**:239–250.

Krogman, K. K., and E. H. Hobbs. 1976. Scheduling irrigation to meet crop demands. Publ. 1590, Agriculture Canada. 18 pp.

Lakshman, G. 1972. An aerodynamic formula to compute evaporation from open water surfaces. *J. Hydrol.* **15**:209–225.

Lang, A. R. G., and W. R. Gardner. 1970. Limitation to water flux from soils to plants. *Agron. J.* **62**:693–695.

Lang, A. R. G. 1973. Measurement of evapotranspiration in the presence of advection by means of a modified energy balance procedure. *Agric. Meteorol.* **12**:75–81.

Lange, O. L., R. Losch, E. D. Schulze, and L. Kappen. 1971. Response of stomata to changes in humidity. *Planta* **100**:76–86.

Lee, R. 1967. The importance of transpiration control by stomata. *Water Resour. Res.* **3**:737–752; Comments and replies in *Water Resour. Res.* **4**(1968):665–669, 1387–1390.

Lemeur, R., and N. J. Rosenberg. 1976. Reflectant induced modification of soybean canopy radiation balance. III. A comparison of the effectiveness of Celite and kaolinite reflectants. *Agron. J.* **68**:30–35.

Lemon, E. R. 1956. The potentialities for decreasing soil moisture evaporation loss. *Soil Sci. Soc. Am. Proc.* **20**:120–125.

Lemon, E., D. W. Stewart, and R. W. Shawcroft. 1971. The sun's work in a cornfield. *Science* **174**:371–378.

Lettau, H. H., and M. W. Baradas. 1973. Evapotranspiration climatonomy II: Refinement of parameterization exemplified by application to the Mabacan River watershed. *Mon. Weather Rev.* **101**:636–649.

Leyton, L. 1978. Some thermodynamic concepts of water movement. *Fluid Behavior in Biological Systems.* Oxford Univ. Press, New York. 235 pp.

Linacre, E. T. 1964. Calculations of the transpiration rate and temperature of a leaf. *Arch. Meteorol. Geophys. Bioklimatol.* **B13**:391–399.

Linacre, E. T. 1967. Climate and the evaporation from crops. *J. Irrig. Drainage Div. ASCE* **93**:61–79.

Linacre, E. T. 1968. Estimating the net radiation flux. *Agric. Meteorol.* **5**:49–63.

Linacre, E. T. 1969. Net radiation to various surfaces. *J. Appl. Ecol.* **6**:61–75.

Linacre, E. T., B. B. Hicks, G. R. Sainty, and G. Grauze. 1970. The evaporation from a swamp. *Agric. Meteorol.* **7**:375–386.

Linacre, E. T. 1977. A simple formula for estimating evapotranspiration rates in various climates, using temperature data alone. *Agric. Meteorol.* **18**:409–424.

Livingston, B. E. 1935. Atmometers of porous porcelain and paper. *Ecology* **16**:438–472.

Lourence, F. J., and W. B. Goddard. 1967. A water-level measuring system for determining evapotranspiration rates from a floating lysimeter. *J. Appl. Meteorol.* **6**:489–492.

Lumley, J. L., and H. A. Panofsky. 1964. *The Structure of Atmospheric Turbulence.* Wiley-Interscience, New York. 239 pp.

Lydolph, P. E. 1964. The Russian Sukhovey. *Ann. Assoc. Am. Geogr.* **54**:291–309.

McIlroy, I. C. 1957. The measurement of natural evaporation. *J. Australian Inst. Agric. Sci.* **23**:4–17.

McIlroy, I. C. 1971. An instrument for continuous recording of natural evaporation. *Agric. Meteorol.* **9**:93–100.

McKay, D. C., and G. W. Thurtell. 1978. Measurements of the energy fluxes involved in the energy budget of a snow cover. *J. Appl. Meteorol.* **17**:339–349.

McKeon, G. M., and C. W. Rose. 1977. Estimating evaporation: Difficulties of applicability in different environments. *Science* **196**:1355–1356.

McMillan, W. D., and H. A. Paul. 1961. Floating lysimeter. *Agric. Engr.* **42**:498–499.

McNaughton, K. G., and T. A. Black. 1973. A study of evapotranspiration from a Douglas fir forest using the energy balance approach. *Water Resour. Res.* **9**:1579–1590.

McNeil, D. A., and W. J. Shuttleworth. 1975. Comparative measurements of the energy fluxes over a pine forest. *Boundary-Layer Meteorol.* **9**:297–313.

Makkink, G. F. 1957. Ekzameno de la formula de Penman. *Neth. J. Agric. Sci.* **5**:290–305.

Malhotra, G. P., and P. Brock. 1970. Hydrologic budget of North America and subregions formulated using atmospheric vapor flux data. *World Water Balance.* Proc. of Reading Symp. Int. Assoc. Scientific Hydrology. pp. 501–524.

Manabe, S. 1969. Climate and ocean circulation: I. The atmospheric circulation and the hydrology of the earth's surface. *Mon. Weather Rev.* **97**:739–774.

Manabe, S., and R. T. Wetherald. 1980. On the distribution of climate change resulting from an increase in CO_2 content of the atmosphere. *J. Atmos. Sci.* **37**:99–118.

Marlatt, W. E., A. V. Havens, N. A. Willits, and G. D. Brill. 1961. A comparison of computed and measured soil moisture under snap beans. *J. Geophys. Res.* **66**:535–541.

Marsh, P., W. R. Rouse, and M. K. Woo. 1981. Evaporation at a high Arctic site. *J. Appl. Meteorol.* **20**:713–716.

Massee, T. W., and J. W. Cary. 1978. Potential for reducing evaporation during summer fallow. *J. Soil Water Conserv.* **33**:126–129.

Maxwell, J. C. 1965. Will there be enough water? *Am. Scientist* **53**:97–103.

Meidner, H., and T. A. Mansfield. 1968. In *Physiology of Stomata*. McGraw-Hill, London. 179 pp.

Meyer, W. S., and G. C. Green. 1981. Plant indicators of wheat and soybean crop water stress. *Irrig. Sci.* **2**:167–176.

Molz, F. J. 1975. Potential distributions in the soil-root system. *Agron. J.* **67**:726–729.

Monteith, J. L. 1963. Gas exchange in plant communities. *Environmental Control of Plant Growth* (L. T. Evans, ed.), Academic Press, New York, pp. 95–112.

Monteith, J. L. 1964. Evaporation and environment. The State and Movement of Water in Living Organisms. *19th Symp. Soc. Exp. Biol.*, Academic Press, New York, pp. 205–234.

Monteith, J. L. 1973. *Principles of Environmental Physics*. Edward Arnold, London. 241 pp.

Moreshet, S. 1970. Effect of environmental factors on cuticular transpiration resistance. *Plant Physiol.* **46**:815–818.

Morton, F. I. 1979. Climatological estimates of lake evaporation. *Water Resour. Res.* **15**:64–76.

Munn, R. E. 1961. Energy budget and mass transfer theories of evaporation. *Proc. 2nd Hydrol. Symp. (Toronto)*, pp. 8–30. Cat R32-361/2, Queens Printer, Ottawa.

Munro, D. S. 1979. Daytime energy exchange and evaporation from a wooded swamp. *Water Resour. Res.* **15**:1259–1265.

Musick, J. T., L. L. New, and D. A. Dusek. 1976. Soil water depletion–yield relationships of irrigated sorghum, wheat and soybeans. *Trans. ASAE* **19**:489–493.

National Academy of Sciences/National Research Council. 1980. Irrigation demonstration project. A Strategy for the National Climate Program. Report of the Workshop to Review the Preliminary National Climate Program Plan. National Academy of Sciences, Washington, D.C. Appendix B, pp. 56–57.

Neumann, H. H., G. W. Thurtell, and K. R. Stevenson. 1974. In situ measurements of leaf water potential and resistance to water flow in corn, soybean and sunflower at several transpiration rates. *Can. J. Plant Sci.* **54**:175–184.

Newman, E. J. 1969. Resistance to water flow in soil and plant. I. Soil resistance in relation to amounts of root: Theoretical estimates. *J. Appl. Ecol.* **6**:1–12.

Neilson, R. E., and P. G. Jarvis. 1975. Photosynthesis in Sitka spruce (*Picea sitchensis* (Bong.) Carr.). VI. Response of stomata to temperature. *J. Appl. Ecol.* **12**:879–891.

Norman, J. M. 1979. Modeling the complete crop canopy. *Modification of the Aerial Environment of Plants* (B. J. Barfield and J. F. Gerber, eds.). *Am. Soc. Agric. Engr. Monogr.* **2**:249–277.

Norman, J. M. 1982. Simulation of microclimate. *Biometeorology in Integrated Pest Management* (J. L. Hatfield and I. J. Thomason, eds.), Academic Press, New York, pp. 62–99.

Nulsen, R. A., G. W. Thurtell, and K. R. Stevenson. 1977. Response of leaf water potential to pressure changes at the root surface of corn plants. *Agron. J.* **69**:951–954.

Oke, T. R. 1970. Turbulent transport near the ground in stable conditions. *J. Appl. Meteorol.* **8**:778–786.

Owston, P. W., J. L. Smith, and H. G. Halverson. 1972. Seasonal water movement in tree stems. *Forest Sci.* **18**:266–272.

Palland, C. L. 1979. The "evaporation brush" an evaporimeter for measuring the potential evaporation of meadow grass. *J. Hydrol.* **41**:363–369.

Palm, E., and M. Tveitereid. 1979. On heat and mass flux through dry snow. *J. Geophys. Res.* **84**:745–749.

Palmer, W. C., and A. V. Havens. 1958. A graphical technique for determining evapotranspiration by the Thornthwaite method. *Mon. Weather Rev.* **86**:123–128.

Parmele, L. H., and L. J. McGuiness. 1974. Comparisons of measured and estimated daily potential evapotranspiration in a humid region. *J. Hydrol.* **22**:239–251.

Parton, W. J., W. K. Lauenroth, and F. M. Smith. 1981. Water loss from a shortgrass steppe. *Agric. Meteorol.* **24**:97–110.

Pasquill, F. 1950. Some further considerations of the measurements and indirect evaluation of natural evaporation. *Q. J. Roy. Meteorol. Soc.* **76**:287–301.

Pearce, A. J., and L. K. Rowe. 1979. Forest management effects on interception, evaporation and water yield. New Zealand Forest Service, #1293, O.D.C. No. 116, *J. Hydrol.* **18**:73–87.

Pelton, W. L., K. M. King, and C. B. Tanner. 1960. An evaluation of the Thornthwaite method for determining potential evapotranspiration. *Agron. J.* **52**:387–395.

Pelton, W. L. 1961. The use of lysimetric methods to measure evapotranspiration. *Proc. 2nd Hydrology Symp. (Toronto)*, pp. 106–134.

Pelton, W. L., and H. C. Korven. 1969. Evapotranspiration estimates from stomates and pans. *Can. J. Plant Sci.* **49**:615–621.

Penman, H. L. 1948. Natural evapotranspiration from open water, bare soil and grass. *Proc. R. Soc. London Ser. A.* **193**:120–145.

Penman, H. L. 1949. The dependence of transpiration on weather and soil conditions. *J. Soil Sci.* **1**:74–89.

Penman, H. L., and R. K. Schofield. 1951. Some physical aspects of assimilation and transpiration. *Symp. Soc. Exp. Biol.* **5**:115–129.

Peters, D. B., B. F. Clough, R. A. Graves, and G. R. Stahl. 1974. Measurement of dark respiration, evaporation and photosynthesis in field plots. *Agron. J.* **66**:460–462.

Phillip, J. R. 1966. Plant water relations: Some physical aspects. *Annu. Rev. Plant Physiol.* **17**:245–268.

Pierson, F. W., and A. P. Jackman. 1975. An investigation of the predictive ability of several evaporation equations. *J. Appl. Meteorol.* **14**:477–487.

Poljakoff-Mayber, A., and J. Gale. 1972. Physiological basis and practical problems of reducing transpiration. *Water Deficits and Plant Growth* (T. T. Kozlowski, ed.), Academic Press, New York, pp. 277–306.

Priestley, C. H. B., and R. J. Taylor. 1972. On the assessment of surface heat flux and evaporation using large-scale parameters. *Mon. Weather Rev.* **100**:81–92.

Pruitt, W. O., and D. E. Angus. 1960. Large weighing lysimeter for measuring evapotranspiration. *Trans. Am. Soc. Agric. Engr.* **3**:13–18.

Pruitt, W. O. 1963. Application of several energy balance and aerodynamic evaporation equations under a wide range of stability. Final Report to USAEPG on Contract No. DA-36-039-SC-80334. Univ. of California-Davis. pp. 107–124.

Pruitt, W. O., and M. J. Aston. 1963. Atmospheric and surface factors affecting evapotranspiration. Final Report to USAEPG on Contract No. DA-36-039-SC-80334. Univ. of California–Davis, pp. 69–105.

Pruitt, W. O., and F. J. Lourence. 1968. Correlation of climatological data with water requirement of crops. Dept. of Water Sci. and Engr. Paper No. 9001. Univ. of California–Davis. 59 pp.

Quinn, F. H. 1979. An improved aerodynamic evaporation technique for large lakes with application to the international field year for the Great Lakes. *Water Resour. Res.* **15**:935–940.

Reichman, G. A., T. J. Doering, L. C. Benz, and R. F. Follett. 1979. Construction and performance of large automatic (nonweighing) lysimeters. *Trans. ASAE* **22**:1343–1346.

Reicosky, D. C., and D. P. Peters. 1977. A portable chamber for rapid evapotranspiration measurements on field plots. *Agron. J.* **69**:729–732.

Reicosky, D. C., C. W. Doty, and R. B. Campbell. 1977. Evapotranspiration and soil water movement beneath the root zone of irrigated and non-irrigated millet (*Panicum miliaceum*). *Soil Sci.* **124**:95–101.

Reicosky, D. C., and J. R. Lambert. 1978. Field measured and simulated corn leaf water potential. *Soil Sci. Soc. Am. J.* **42**:221–228.

Retta, A., and R. J. Hanks. 1980. Corn and alfalfa production as influenced by limited irrigation. *Irrig. Sci.* **1**:135–147.

Rider, N. E., J. R. Philip, and E. F. Bradley. 1963. The horizontal transport of heat and moisture—A micrometeorological study. *Q. J. Roy. Meteorol. Soc.* **89**:507–530.

Rijks, D. A. 1969. Evaporation from a papyrus swamp. *Q. J. Roy. Meteorol. Soc.* **95**:643–649.

Rijks, D. A. 1971. Water use by irrigated cotton in Sudan: III. Bowen ratios and advective energy. *J. Appl. Ecol.* **8**:643–663.

Ripple, C. D., J. Rubin, and T. E. A. van Hylckama. 1970. Estimating steady-state evaporation rates from bare soils under conditions of high water table. U.S. Geol. Sur. Open-file Rept. Water Res. Div., Menlo Park, California, 62 pp.

Ritchie, J. T., and E. Burnett. 1968. A precision weighing lysimeter for row crop water use studies. *Agron. J.* **60**:545–549.

Ritchie, J. T. 1971. Dryland evaporative flux in a subhumid climate: I. Micrometeorological influences. *Agron. J.* **63**:51–55.

Ritchie, J. T., and E. Burnett. 1971. Dryland evaporative flux in a subhumid climate: II. Plant influences. *Agron. J.* **63**:56–62.

Ritchie, J. T. 1972. Model for predicting evapotranspiration from a row crop with incomplete cover. *Water Resour. Res.* **8**:1204–1213.

Ritchie, J. T., E. Burnett, and R. C. Henderson. 1972. Dryland evaporative flux in a subhumid climate: III. Soil water influence. *Agron. J.* **64**:168–173.

Ritchie, J. T., and W. R. Jordan. 1972. Dryland evaporative flux in a subhumid climate. IV. Relation to plant water status. *Agron. J.* **64**:173–176.

Ritchie, J. T. 1975. Evaluating irrigation needs for southeastern U.S.A. Contribution of Irrigation and Drainage to World Food Supply. Proc. Irrigation and Drainage Div. Specialty Conf., Biloxi, MS. Aug. 14–16. Am. Soc. Civil Engr., New York, pp. 262–279.

Ritchie, J. T., E. D. Rhoades, and C. W. Richardson. 1976. Calculating evaporation from native grassland watersheds. *Trans. ASAE* **19**:1098–1103.

Rohwer, C. 1931. Evaporation from free water surfaces. *USDA Tech. Bull.* **217:**1–96.
Rosenberg, N. J. 1966. Microclimate, air mixing and physiological regulation of transpiration as influenced by wind shelter in an irrigated bean field. *Agric. Meteorol.* **3:**197–224.
Rosenberg, N. J., D. W. Lecher, and R. E. Neild. 1967. Wind shelter induced growth and physiological responses in irrigated snap beans. *Proc. Am. Soc. Horticul. Sci.* **90:**169–179.
Rosenberg, N. J., H. E. Hart, and K. W. Brown. 1968. Evapotranspiration—Review of Research. Nebr. Agr. Exp. Station Misc. Bull. No. 20. 80 pp.
Rosenberg, N. J. 1969a. Seasonal patterns in evapotranspiration by irrigated alfalfa in the central Great Plains. *Agron. J.* **61:**879–886.
Rosenberg, N. J. 1969b. Advective contribution of energy utilized in evapotranspiration by alfalfa in the east central Great Plains. *Agric. Meteorol.* **6:**179–184.
Rosenberg, N. J., and W. Powers. 1970. Potential for evapotranspiration and its manipulation in the Plains region. Proc. of the Symp. on Evapotranspiration in the Great Plains. Great Plains Agr. Council Publ. No. 50. pp. 275–300.
Rosenberg, N. J., and K. W. Brown. 1970. Improvements in the van Bavel-Meyers automatic weighing lysimeter. *Water Resour. Res.* **6:**1227–1229.
Rosenberg, N. J. 1972. Frequency of potential evapotranspiration rates in the central Great Plains. *J. Irrig. Drainage Div. ASCE* **98:**203–206.
Rosenberg, N. J., and S. B. Verma. 1978. Extreme evapotranspiration by irrigated alfalfa: A consequence of the 1976 midwestern drought. *J. Appl. Meteorol.* **17:**934–941.
Russell, G. 1980. Crop evaporation, surface resistance and soil water status. *Agric. Meteorol.* **21:**213–226.
Sammis, T. W., and L. W. Gay. 1979. Evapotranspiration from an arid zone plant community. *J. Arid. Environ.* **2:**313–321.
Sammis, T. W. 1981. Lysimeter for measuring arid-zone evapotranspiration. *J. Hydrol.* **49:**385–394.
Sebenik, P. G., and J. L. Thomas. 1967. Water consumption by phreatophytes. *Prog. Agric. Ariz.* **19:**10–11.
Seginer, I. 1971. Wind effect on the evaporation rate. *J. Appl. Meteorol.* **10:**215–220.
Seguin, B. 1975. Influence de l'evapotranspiration regionale sur la mesure locale d'evapotranspiration potentielle. *Agric. Meteorol.* **15:**355–370.
Sellers, W. D. 1965. *Physical Climatology*. Univ. of Chicago Press, Chicago. 272 pp.
Shih, S. F. 1980. Water budget computation in Lake Okeechobee. *Florida Sci.* **43:**84–92.
Shouse, P., W. A. Jury, and L. H. Stolzy. 1980. Use of deterministic and empirical models to predict potential evapotranspiration in an advective environment. *Agron. J.* **72:**994–998.
Showalter, A. K. 1971. Evaporative capacity of unsaturated air. *Water Resour. Res.* **7:**688–691.
Shukla, J., and Y. Mintz. 1982. Influence of land-surface evapotranspiration on the earth's climate. *Science* **215:**1498–1500.
Shuttleworth, W. J., and I. R. Calder. 1979. Has the Priestley-Taylor equation any relevance to forest evaporation? *J. Appl. Meteorol.* **18:**639–646.

Singh, B., and G. Szeicz. 1980. Predicting the canopy resistance of a mixed hardwood forest. *Agric. Meteorol.* **21**:49–58.

Sinha, B. K., and N. T. Singh. 1974. Effect of transpiration rate on salt accumulation around corn roots in a saline soil. *Agron. J.* **66**:557–560.

Skidmore, E. L., and L. J. Hagan. 1970. Evaporation in sheltered areas as influenced by windbreak porosity. *Agric. Meteorol.* **7**:363–374.

Slabbers, P. J. 1977. Surface roughness of crops and potential evapotranspiration. *J. Hydrol.* **34**:181–191.

Slatyer, R. O., and I. C. McIlroy. 1961. *Practical Microclimatology*. CSIRO, Australia and UNESCO.

Slatyer, R. O. 1968. The use of soil water balance relationships in agroclimatology. Agroclimatological Methods, *UNESCO Natl. Resour. Res.* **7**:73–87.

Slatyer, R. O., C. E. Hounam, K. C. Leverington, and W. C. Swinbank. 1970. Estimating evapotranspiration—An evaluation of techniques. Australian Water Resources Council, Dept. of National Development, Hydrological Series **5**:3–23.

Spittlehouse, D. L., and T. A. Black. 1979. Determination of forest evapotranspiration using Bowen ratio and eddy correlation measurements. *J. Appl. Meteorol.* **18**:647–653.

Stanhill, G. 1961. A comparison of methods of calculating potential evapotranspiration from climatic data. *Israel J. Agric. Res.* **11**:159–171.

Stanhill, G. 1965. The concept of potential evapotranspiration in arid zone agriculture. *Proc. Montpellier Symp.* UNESCO. pp. 109–117.

Stern, W. R. 1967. Seasonal evapotranspiration of irrigated cotton in a low latitude environment. *Australian J. Agric. Res.* **18**:259–269.

Stewart, J. B. 1977. Evaporation from the wet canopy of a pine forest. *Water Resour. Res.* **13**:915–921.

Stewart, J. I., R. M. Hagan, W. O. Pruitt, R. E. Danielsen, W. T. Franklin, R. J. Hands, J. P. Riley, and E. B. Jackson. 1977. Optimizing crop production through control of water and salinity levels in the soil. PRWG 151-1, Utah Water Res. Lab., Utah State Univ., Logan. 191 pp.

Stewart, R. B., and W. R. Rouse. 1977. Substantiation of the Priestley and Taylor parameter $\alpha = 1.26$ for potential evaporation in high latitudes. *J. Appl. Meteorol.* **16**:649–650.

Stigter, C. J. 1980. Assessment of the quality of generalized wind functions in Penman's equations. *J. Hydrol.* **45**:321–331.

Stone, L. R., M. L. Horton, and T. C. Olson. 1973. Water loss from an irrigated sorghum field: I. Water flux within and below the root zone. *Agron. J.* **65**:492–495.

Stutler, R. K., D. W. James, T. M. Fullerton, R. F. Wells, and E. R. Shipe. 1981. Corn yield functions of irrigation and nitrogen in Central America. *Irrig. Sci.* **2**:79–88.

Suckling, P. W. 1980. The energy balance microclimate of a suburban lawn. *J. Appl. Meteorol.* **19**:606–608.

Swanson, R. H. 1972. Water transpired by trees is indicated by heat pulse velocity. *Agric. Meteorol.* **10**:277–281.

Swinbank, W. C. 1951. The measurement of vertical transfers of heat and water vapor by eddies in the lower atmosphere. *J. Meteorol.* **8**:135–145.

Szeicz, G., C. H. M. van Bavel, and S. Takami. 1973. Stomatal factors in the water use and dry matter production by sorghum. *Agric. Meteorol.* **12**:361–389.

Tan, C. S., and J. M. Fulton. 1980. Ratio between evapotranspiration of irrigation crops from floating lysimeters and class A pan evaporation. *Can. J. Plant Sci.* **60**:197–201.

Tanner, C. B. 1957. Factors affecting evaporation from plants and soils. *J. Soil Water Conserv.* **12**:221–227.

Tanner, C. B. 1960. Energy balance approach to evapotranspiration from crops. *Soil Sci. Soc. Am. Proc.* **24**:1–9.

Tanner, C. B., and E. R. Lemon. 1962. Radiant energy utilized in evapotranspiration. *Agron. J.* **54**:207–212.

Tanner, C. B. 1963. Energy relations in plant communities. *Environmental Control of Plant Growth* (L. T. Evans, ed.), Academic Press, New York, pp. 141–148.

Tanner, C. B. 1967. Measurement of evapotranspiration. *Irrigation of Agricultural Lands* (R. M. Hagen, M. R. Haise, and T. W. Edminster, eds.), Am. Soc. Agron., Madison, Wisc., pp. 534–574.

Tanner, C. B., and W. A. Jury. 1976. Estimating evaporation and transpiration from a row crop during incomplete cover. *Agron. J.* **68**:239–243.

Tanner, C. B. 1981. Transpiration efficiency of potato. *Agron. J.* **73**:59–64.

Teare, I. D., and E. T. Kanemasu. 1972. Stomatal-diffusion resistance and water potential of soybean and sorghum leaves. *New Phytol.* **71**:805–810.

Teare, I. D., E. T. Kanemasu, W. L. Powers, and H. S. Jacobs. 1973. Water use efficiency and its relation to crop canopy area, stomatal regulation, and root distribution. *Agron. J.* **65**:207–211.

Thom, A. S., and H. R. Oliver. 1977. On Penman's equation for estimating regional evaporation. *Q. J. Roy. Meteorol. Soc.* **103**:345–357.

Thompson, J. R. 1975. Energy budgets for three small plots—substantiation of Priestley and Taylor's large scale evaporation parameter. *J. Appl. Meteorol.* **14**:1399–1401.

Thompson, O. E., and R. T. Pinker, 1981. An error analysis of the Thornthwaite-Holzman equations for estimating sensible and latent heat fluxes over crop and forest canopies. *J. Appl. Meteorol.* **20**:250–254.

Thornthwaite, C. W., and B. Holzman. 1942. Measurement of evaporation from land and water surface. *USDA Tech. Bull.* **817**:1–75.

Thornwaite, C. W. 1948. An approach toward a rational classification of climate. *Geogr. Rev.* **38**:55–94.

Thornthwaite, C. W., and J. R. Mather. 1955. The water balance. *Climatology* **8**:1–104.

Thornthwaite, C. W., and F. K. Hare. 1965. The loss of water to the air. *Meteorol. Monogr.* **6**:163–180. P. E. Waggoner, ed. Am. Meteorol. Soc., Boston.

Tinus, R. W. 1974. Impact of the CO_2 requirement on plant water use. *Agric. Meteorol.* **14**:99–112.

Tomar, V. S., and J. C. O'Toole. 1980. Design and testing of a microlysimeter for wetland rice. *Agron. J.* **72**:689–692.

U.S. Geological Survey. 1954. Water loss investigations, Lake Hefner studies. Prof. Paper 269. Washington, D.C. 158 pp.

van Bavel, C. H. M., L. J. Fritschen, and W. E. Reeves. 1962. Transpiration by Sudan grass as an externally controlled process. *Science* **141**:269–270.

van Bavel, C. H. M., and L. E. Meyers. 1962. An automatic weighing lysimeter. *Agric. Eng.* **43**:580–583, 587–588.

van Bavel, C. H. M. 1966. Potential evaporation: The combination concept and its experimental verification. *Water Resour. Res.* **2**:455–467.

van Bavel, C. H. M. 1967. Changes in canopy resistance to water loss from alfalfa induced by soil water depletion. *Agric. Meteorol.* **4**:165–176.

van Bavel, C. H. M., J. E. Newman, and R. H. Hilgeman. 1967. Climate and estimated water use by an orange orchard. *Agric. Meteorol.* **4**:27–37.

van Bavel, C. H. M., K. J. Brust, and G. B. Stirk. 1968. Hydraulic properties of a clay loam soil and the field measurement of water uptake by roots. II. The water balance of the root zone. *Soil Sci. Soc. Am. Proc.* **32**:317–321.

van Bavel, C. H. M., and W. L. Ehrler. 1968. Water loss from a sorghum field and stomatal control. *Agron. J.* **60**:84–86.

van Bavel, C. H. M., and D. I. Hillel. 1976. Calculating potential and actual evaporation from a bare soil surface by simulation of concurrent flow of water and heat. *Agric. Meteorol.* **17**:453–476.

van den Honert. 1948. Water transport in plants as a catenary process. *Discuss. Faraday Soc.* **3**:146–153.

van Hylckama, T. E. A. 1966. Effect of soil salinity on the loss of water from vegetated and fallow soil. *Proc. Wageningen Symp. June 1966*, pp. 635–644.

van Hylckama, T. E. A. 1968. Water level fluctuation in evapotranspirometers. *Water Resour. Res.* **4**:761–768.

Verma, S. B., N. J. Rosenberg, B. L. Blad, and M. W. Baradas. 1976. Resistance-energy balance method for predicting evapotranspiration: Determination of boundary layer resistance and evaluation of error effects. *Agron. J.* **68**:776–782.

Verma, S. B., and N. J. Rosenberg. 1977. The Brown-Rosenberg resistance model of crop evapotranspiration modified tests in an irrigated sorghum field. *Agron. J.* **69**:332–335.

Verma, S. B., N. J. Rosenberg, and B. L. Blad. 1978. Turbulent exchange coefficients for sensible heat and water vapor under advective conditions. *J. Appl. Meteorol.* **17**:330–338.

Wallender, W. W., D. W. Grimes, D. W. Henderson, and L. K. Stromberg. 1979. Estimating the contribution of a perched water table to the seasonal evapotranspiration of cotton. *Agron. J.* **71**:1056–1060.

Webb, E. K. 1965. Aerial microclimate. *Agricultural Meteorology* (P. E. Waggoner, ed.), Chapter 2. *Meteorol. Mongr.* **6**(No. 28):27–58. Am. Meteorol. Soc., Boston.

Weeks, E. P., and M. L. Sorey. 1973. Use of finite-difference arrays of observation wells to estimate evapotranspiration from ground water in the Arkansas River Valley, Colorado. Water Supply Paper 2029-C. U.S. Geol. Survey, Washington, D.C. 27 pp.

Weisman, R. N. 1975. Comparison of warm water evaporation equations. *J. Hydraulics Div. ASCE* **101**:1303–1313.

Wight, J. R. 1971. Comparison of lysimeter and neutron scatter techniques for measuring evapotranspiration from semiarid rangelands. *J. Range Management* **24**:390–393.

Wilcox, J. C., and W. K. Sly. 1974. Ratio between evapotranspiration from lysimeters and evaporation from small evaporimeters using 2- and 3-hour periods of measurement. *Can. J. Plant Sci.* **54**:559–564.

Williams, G. D. V. 1972. Geographical variations in yield-weather relationships over a large wheat growing region. *Agric. Meteorol.* **9**:265–283.

Williams, R. J., K. Boersma, and A. L. van Ryswyk. 1978. Equilibrium and actual evapotranspiration from a very dry vegetated surface. *J. Appl. Meteorol.* **17**:1827–1832.

Williams, R. J., and D. G. Stout. 1981. Evapotranspiration and leaf water status of alfalfa growing under advective conditions. *Can. J. Plant Sci.* **61**:601–607.

Wilm, H. G. 1946. Report of the committee on evaporation and transpiration, 1945–1946. *Trans. Am. Geophys. Union* **27**:720–725.

Woolhiser, D. A., R. E. Smith, and C. L. Hanson. 1970. Evaporation components of watershed models for the Great Plains. Evapotranspiration in the Great Plains. Great Plains Agric. Council Publ. No. 50. pp. 111–136.

Wright, J. L., and M. E. Jensen. 1972. Peak water requirements of crops in southern Idaho. *J. Irrig. Drainage Div. Proc. ASCE* **98**:193–201.

Wright, J. L., and M. E. Jensen. 1978. Development and evaluation of evapotranspiration models for irrigation scheduling. *Trans. ASAE* **21**:88–91, 96.

Yao, A. Y. M. 1969. The R index for plant water requirement. *Agric. Meteorol.* **6**:259–273.

Yao, A. Y. M. 1974. Agricultural potential estimated from the ratio of actual to potential evapotranspiration. *Agric. Meteorol.* **13**:405–417.

Yu, T. W. 1977. Parameterization of surface evaporation rate for use in numerical modeling. *J. Appl. Meteorol.* **16**:393–400.

CHAPTER 8 – FIELD PHOTOSYNTHESIS, RESPIRATION, AND THE CARBON BALANCE

8.1 INTRODUCTION AND DEFINITIONS

Green plants through the process of **photosynthesis** extract solar energy and convert it into a form usable by other levels in the food chain. In this process chlorophyll-containing plants produce primary sugars by the light-intermediated reaction:

$$6CO_2 + 12H_2O \xrightarrow[\text{chlorophyll}]{\text{light } 0.469 \text{ MJ (mole } CO_2)^{-1}} C_6H_{12}O_6 + 6H_2O + 6O_2$$

(8.1)

Primary sugars produced in this process form the raw material from which the more complex sugars, starches, cellulose, and other plant constituents are subsequently synthesized. Photosynthesis provides the materials for plant growth and the development of synthesizing and storage organs such as leaves, stems, tubers, and fruits. The energy of the sun is captured through photosynthesis and is stored for shorter or longer periods in such diverse materials as the easily converted sugars and the fossil fuels that must be mined and purposefully combusted in order to release that energy.

Carbon dioxide fixation proceeds by the pathway shown in (8.1), producing simple sugars from the raw materials carbon dioxide and water. The processes of growth and the synthesis of more complex compounds require the input of energy. **Respiration,** the oxidation of the carbon compounds produced in photosynthesis, releases energy for the execution of the varied biological and chemical work in the plant and provides carbon "skeletons" that become the basis of other organic compounds. Thus, respiration is the inverse of photosynthesis although the chemical pathways of the two processes are very different. Chlorophyll is not involved in respiration.*

Photosynthesis begins at sunrise, but the rates are low and it is usually sometime later before the fixation of carbon dioxide exceeds the rate of

---

* Those desiring more detailed explanations of the physiological and biochemical processes involved in photosynthesis should consult a text such as that of Salisbury and Ross (1978).

respiratory release. The irradiance level at which photosynthesis and respiration are equal is known as the **light compensation point**.

Plant leaves normally show an increasing rate of photosynthesis with increasing irradiance above the compensation point. This increase is in the form of a rectangular hyperbola or "diminishing returns curve." The level at which the photosynthetic rate becomes independent of irradiance is called the **light saturation point** (usually expressed as a flux density). Normally light saturation for a whole plant or a field or community of plants occurs at a flux density somewhat higher than for illuminated isolated leaves. This is so since plants in a community or leaves of the same plant tend to mutually shade one another. The concepts of light compensation point and light saturation point are illustrated in Fig. 8.1.

The green plants can be classified into three major groups on the basis of their photosynthetic mechanisms: these are the C_4, C_3, and CAM plants. **C_4 plants** utilize the C_4-dicarboxylic acid chemical pathway for photosynthesis. C_4 species are generally the tropical grasses, for example, corn, sorghum, millet, and sugar cane.

C_3 plants utilize a photosynthetic pathway involving a three-carbon intermediate product. The C_3 group includes virtually all other species: small grains (wheat, barley); leguminous species (alfalfa, soybean, and many others). A list of major C_3 and C_4 species is given in Table 8.1. Note that a single genus (e.g., *Atriplex*) may include both C_3 and C_4 species.

A third, but relatively minor, group of plants accomplish photosynthesis through crassulacean acid metabolism (CAM). These plants maintain open stomates at night during which time they fix CO_2 in the form of organic acids. During daytime, the stored CO_2 is reduced photosynthetically. Many of the xerophytic desert species are of the CAM group. Pineapple is one of the few cultivated CAM plants.

All plants consume, by respiration, some portion of the photosynthate they produce. Respiration proceeds in both C_3 and C_4 species by an essen-

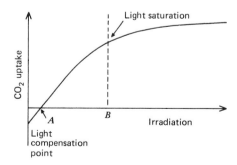

Fig. 8.1 Leaf photosynthetic rate as a function of irradiance (from Goudriaan, J., and Atjay, G. L., The possible effects of increased CO_2 on photosynthesis. *The Global Carbon Cycle* (*SCOPE 13*) (B. Bolin, E. T. Degens, S. Kempe, and P. Ketner, eds.). Copyright 1979. Reprinted by permission of John Wiley and Sons, Ltd.)

Table 8.1 Some common C_3 and C_4 plants[a]

C_3	C_4
Rice (*Oryza sativa* L.)	Purple lovegrass (*Eragrostis spectabilis* (Pursh) Steud.)
Alta fescue (*Festuca arundinacea* Schreb.)	Rhodes grass (*Chloris gayana* Kunth)
Bluegrass (*Poa pratensis* L.)	Pearl millet (*Pennisetum glaucum* (L.) R. Br.)
Oats (*Avena sativa* L.)	
Crested wheatgrass (*Agropyron desertorum* Fisch. Schult.)	Smooth cordgrass (*Spartina alterniflora* Loisel)
Barley (*Hordeum vulgare* L.)	Sugar cane (*Saccharum officinarum* L.)
Rye (*Secale cereale* L.)	Sorghum (*Sorghum bicolor* (L.) Moench)
Wheat (*Triticum aestivum* L.)	
Sugar beet (*Beta vulgaris* L.)	Maize (*Zea mays* L.)
Spearscale (*Atriplex patula* L.)	Red orache (*Atriplex rosea* L.)
Bean (*Phaseolus vulgaris* L.)	
Soybean (*Glycine max* (L.) Merr.)	
Alfalfa (*Medicago sativa* L.)	
Cotton (*Gossypium hirsutum* L.)	
Potato (*Solanum tuberosum* L.)	
Tomato (*Lycopersicon esculentum* Mill.)	
Sunflower (*Helianthus annuus* L.)	
Walnut (*Juglans regia* L.)	

[a] We thank Professor R. C. Lommasson, School of Life Sciences, University of Nebraska–Lincoln for help in assembling this table.

tially identical biochemical pathway throughout the day and night. However, the C_3 plants have an additional respiratory mechanism that is controlled by light and the availability of oxygen. The respiratory mechanism common to C_3 and C_4 plants is called **dark respiration** since it occurs regardless of light. The additional respiratory mechanism of C_3 plants is called **photorespiration** and occurs only during daytime.

Characteristics of C_3 and C_4 plants are given in Table 8.2 (from Goudriaan and Ajtay, 1979). At the light compensation point, the internal CO_2 con-

Table 8.2 Some characteristics of C_3 and C_4 plants

	C_3	C_4
CO_2 assimilation rate in high light	2–4 g CO_2/m^2 h	4–7 g CO_2/m^2 h
Temperature optimum	20–25°C	30–35°C
CO_2 compensation point in high light	50 ppm	10 ppm
Photorespiration	Present	Not present

SOURCE: Goudriaan and Ajtay (1979).

centration is considerably greater within the leaves of C_3 plants, due to the rapid release of CO_2 in photorespiration. **The CO_2 compensation point,** which is the CO_2 concentration inside a leaf under strong irradiance, is considerably greater in the C_3 than in the C_4 plants. This fact has implications on the differential response of C_3 and C_4 plants to the presence of high concentrations of CO_2 in the ambient air (see Chapter 11).

8.2 GROSS AND APPARENT PHOTOSYNTHESIS

In the field CO_2 is provided to the crop from three sources: the air above, the soil and roots below, and the crop itself. The source of CO_2 in the latter two cases is respiration. Thus, it is convenient to define two additional measures of photosynthesis, the **gross photosynthesis** gPS and the **apparent photosynthesis,** aPS. Figure 8.2 represents the fluxes of CO_2 schematically.

Gross photosynthesis is given by

$$gPS = F_c + R_c + R_g + R_r \qquad (8.2)$$

where F_c is the flux of CO_2 from the air above; R_c is the quantity of CO_2 released in dark and photorespiration (when present) so that

$$R_c = R_l + R_d \qquad (8.3)$$

where R_l is photorespiration and R_d is dark respiration; R_g is the quantity of CO_2 released by respiratory activity of soil flora and fauna; and R_r is CO_2 released by the living roots.

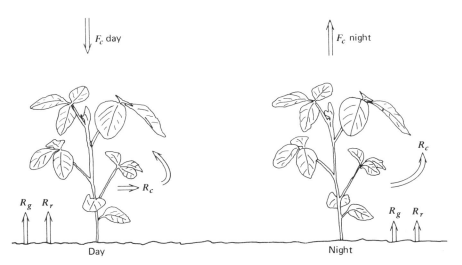

Fig. 8.2 Schematic representation of CO_2 fluxes between soil, plant, and atmosphere during the day and night.

292 FIELD PHOTOSYNTHESIS, RESPIRATION, AND CARBON BALANCE

Apparent photosynthesis is a term proposed by Shibles (1976) to replace the more familiar "net photosynthesis" and is given by

$$\text{aPS} = F_c + R_g \qquad (8.4)$$

Thus, gross photosynthesis during the day includes the recapture of CO_2 released by plant respiration. The apparent photosynthesis considers only the CO_2 delivered from the atmosphere and the soil. At night there is a loss of CO_2 from soil and crop to the atmosphere. The concepts of apparent and gross photosynthesis are discussed further in Section 8.6 after a treatment of the environmental factors that control these processes.

8.3 PHOTOSYNTHESIS AS A RESISTANCE PROCESS

Previously, we have established that transfer to and from vegetated surfaces can be described by a resistance analogy to Ohm's law. For sensible heat flux H and latent heat flux LE the resistance equations are

$$H = \frac{\rho_a C_p (T_a - T_s)}{r_a} \qquad (3.4)$$

and

$$LE = \frac{\rho_a L \epsilon}{P} \frac{e_a - e_{ss}}{r_a + r_c} \qquad (7.49)$$

These fluxes are seen to be directly proportional to the gradient of temperature or vapor pressure and inversely proportional to the resistances. Since sensible heat originates at the surface and moves in response to the temperature gradient from surface to air, the only resistance involved is the aerodynamic resistance r_a. Vapor originates within the substomatal cavities of plant leaves. Thus, the gradient of interest is that of vapor pressure between the ambient air and the substomatal cavity (assumed to be at or near 100% relative humidity, the saturation vapor pressure at leaf temperature). The resistances involved in this case are r_a and r_c. The latter changes as the stomates open and close.

As with water vapor in the transpiration process, carbon dioxide in photosynthesis must move through the air in contact with the leaf and through the stomatal opening. In order to reach the site of the photosynthetic reaction, carbon dioxide must also diffuse through the stomatal cavity and into the cells, penetrating cell walls and chloroplasts into the grana. The additional resistances can be considered, together, to constitute a mesophyll resistance r_m. Photosynthesis PS can then be described by

$$\text{PS} = \frac{C_a - C_{chl}}{r_a + r'_s + r_m} \qquad (8.5)^*$$

where the driving force is the carbon dioxide density difference between air and the chloroplast. Often it is assumed that C_{chl} is at or near zero, so that if the carbon dioxide density in air is known the gradient may be easily established. Brown (1969), however, has calculated that C_{chl} may actually range from about 32 to 144 ppm, depending on the species and its response to irradiation. In Table 8.2 the values of 10 and 50 ppm were proposed as typical CO_2 compensation points in C_4 and C_3 plants, respectively.

The range of the resistances r_a and r_c under normal ambient conditions are known. r_a may vary from near zero in very turbulent air to about 300–400 s m^{-1} in still air. Stomatal resistance r_c may vary from about 50 to 100 s m^{-1} when stomates are wide open to very large values when tightly closed. Mesophyll resistance is actually calculated by solution of (8.5) and may range very widely, depending on very complex biochemical factors. A range of about 100 s m^{-1} for corn and sunflowers to 1000 s m^{-1} for Bermuda grass was found by El Sharkawy and Hesketh (1965). These r_m values were calculated from the rates of photosynthesis of plants exposed to 300 ppm CO_2 and intense irradiance. It was assumed in these calculations that C_{chl} is zero. Mesophyll resistance can be greater under other, less favorable, growing conditions.

Knowledge of the aerial, stomatal, and mesophyll resistances provides an important tool for improving primary production. As shown in Chapters 9 and 11, means are available for manipulating or controlling the magnitude of the aerial and stomatal resistances. Future research may provide the means to alter the magnitude of mesophyll resistance as well.

8.4 ENVIRONMENTAL FACTORS AFFECTING PHOTOSYNTHESIS

8.4.1 Light

The various species of plants differ greatly with respect to the light dependency of their carbon dioxide fixation rates. Leaves of the C_4 species (e.g., corn, sorghum, sugar cane) show a virtually linear increase in carbon dioxide uptake rate with increasing level of irradiance. Leaves of C_3 species (e.g., soybeans, sugar beets, many range and forage species) are less productive and may become light saturated at levels of irradiance as low as one-fourth

---

* The stomatal resistances to the transport of water vapor and carbon dioxide differ because of the differing diffusivities of these molecules. The ratio of stomatal resistances for CO_2 and H_2O, r'_s/r_s, is 1.6 at 20°C.

the full sunlight of the mid-latitudes. Some woody species and shade plants are light saturated at even lower irradiance. These differences in response to light are illustrated in Fig. 8.3 (from Moss, 1965). Leaves of the shade and woody species are light saturated at about 350 W m^{-2} of incandescent light. The orchard grass group appears light saturated at about 700 W m^{-2}, whereas the corn leaves are still unsaturated at 1400 W m^{-2}.

8.4.2 Water

Water is an essential component in the photosynthetic reaction. Shortages of soil moisture or extreme dryness of the atmosphere create a water stress that affects the efficiency of the photosynthetic reaction in the plant. Boyer (1970) states that moisture stress affects photosynthesis through a number of mechanisms: by affecting the levels of metabolic intermediates; by inhibiting the photosynthetic electron transport system; by causing stomatal closure and by altering rates of respiration.

One very direct influence of water availability on photosynthesis is through the impact on stomatal aperture. As stomates close in response to stress, resistance to the diffusion of carbon dioxide into the leaves increases. Moss (1965) speculates on the influence of soil moisture stress and atmospheric evaporative demand on photosynthesis at varying levels of irradiance (Fig. 8.4). With increasing soil moisture stress (increasing dryness), the optimum photosynthetic rate is reached at lower irradiance. When soil moisture stress is low and with little atmospheric evaporative demand, photosynthesis continues to rise even at high irradiance. High atmospheric stress and, par-

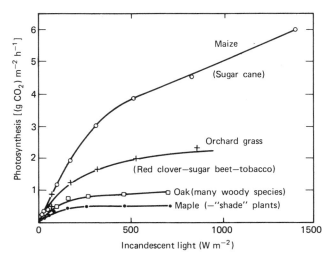

Fig. 8.3 Effect of light intensity (flux density) on photosynthesis by four groups of plants (after Moss, 1965).

ENVIRONMENTAL FACTORS AFFECTING PHOTOSYNTHESIS

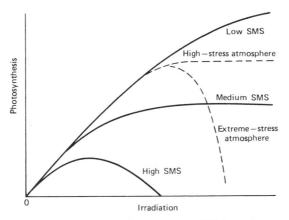

Fig. 8.4 Expected effects of soil moisture stress (SMS) and atmospheric stress of water on photosynthesis at varying levels of irradiation (after Moss, 1965).

ticularly, extreme atmospheric stress reduce photosynthesis, probably because rapid evaporation reduces turgor in the guard cells causing stomates to close.

Figure 8.5 (from Boyer, 1970) illustrates differences in the sensitivity of C_3 and C_4 species to water stress. Corn, a C_4 plant, shows a more or less constant decrease in apparent photosynthesis with leaf water potential decreasing to -1.6 MPa. Apparent photosynthesis in soybean, a C_3 plant, is nearly insensitive to water stress from -0.4 to -1.2 MPa and shows a sharp decline to -2.0 MPa. Note that, whereas corn photosynthesis is about 50% greater than that of soybean when water stress is minor, this advantage is lost at about -1.2 MPa.

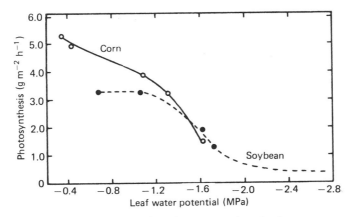

Fig. 8.5 Photosynthesis in corn and soybean at various leaf water potentials (after Boyer, 1970).

8.4.3 Temperature

The photosynthetic reaction is not strongly affected by ambient temperature in the normal range of plant's adaptation. Temperature does affect photosynthetic performance, but the effects may vary according to prior acclimation to hot or cold conditions. Responses to temperature in a number of desert species are, according to Bjorkman (1981), correlated with changes in the concentration of certain enzymes, especially RuP_2 carboxylase. Respiration, as shown in Section 2.8, is controlled quite directly by ambient temperature.

The C_4 plants have a generally greater photosynthetic potential under higher temperatures. Through controlled environment studies, Moss (1965) has shown that maize assimilates carbon dioxide more effectively as temperatures increase from 10 to 30°C (Fig. 8.6). An optimum temperature exists between 30 and 40°C, however.

Sugar production in sugar beets, on the other hand, is benefited by a decrease in temperature in the range of 15–29°C (Thomas and Hill, 1949). Baldocchi et al. (1981a) in a field study with alfalfa found the flux of CO_2 from air to crop to decrease with increasing temperature in the range 23–32°C (Fig. 8.7). In the field, this effect may be due not only to the direct influence of temperature on photosynthesis and crop respiration, but also

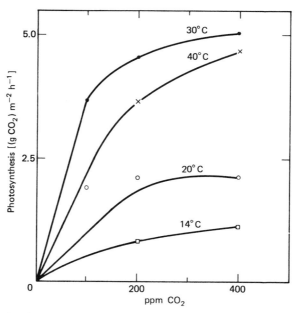

Fig. 8.6 Relation between photosynthesis of maize and carbon dioxide concentration at different temperatures (after Moss, 1965).

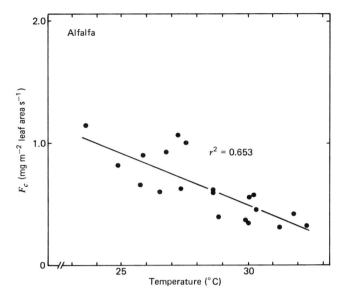

Fig. 8.7 CO_2 flux from air to an alfalfa crop as a function of air temperature at 1 m aboveground, Mead, Nebraska (after Baldocchi et al., 1981a).

to an increase in soil and root respiration that would cause a greater release of carbon dioxide from below. Such an increased release might diminish the need for CO_2 from the air above.

8.4.4 Carbon Dioxide Concentration

The fixation of carbon dioxide can generally be increased by increasing the ambient concentration of that gas. This phenomenon is demonstrated for maize in Fig. 8.6. The influence of CO_2 concentration on sugar beet photosynthesis is shown in Fig. 8.8. Although the individual curves in this figure differ considerably, all three sets of data agree in showing a linear increase in photosynthetic rate with increasing carbon dioxide concentration in the range 220–400 ppm. In the case of C_3 species, the increase in ambient CO_2 concentration may also act to suppress photorespiration since that process proceeds at a rate that depends on competition between oxygen molecules and carbon dioxide molecules for enzymatic sites (Chollet, 1977; Ehleringer and Bjorkman, 1977). The direct influence of CO_2 concentration on the photosynthetic rate of C_4 species is smaller than for C_3 species. As is seen in Fig. 8.6 (Moss et al., 1961) only a small increase in maize photosynthesis occurs with increasing CO_2 concentration in the range 200–400 ppm.

298 FIELD PHOTOSYNTHESIS, RESPIRATION, AND CARBON BALANCE

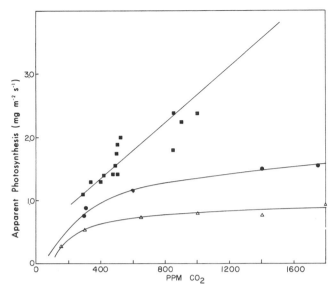

Fig. 8.8 Dependence of the net photosynthetic rate of sugar beets on the concentration of carbon dioxide in the ambient air. Solid squares are data for a field-grown crop (presumably under full sunlight, 840 W m^{-2}), and solid circles are for isolated leaves per unit leaf area at 840 W m^{-2} (after Thomas and Hill, 1949). The open triangles are for sugar beets in a growth chamber at 250 W m^{-2} (after Gaastra, 1959).

To capitalize on the response of C_3 plants, "fertilizing" the air with carbon dioxide in confined environments such as greenhouses has been practiced for some time in order to increase production of high value crops. Wittwer (1978) describes many such applications in the production of vegetables and ornamentals.

8.4.5 Wind and Turbulence

Wind and turbulence are among the environmental factors that influence photosynthesis. The resupply of carbon dioxide to levels at which it has been depleted by the actively photosynthesizing plant should generally be adequate whenever there is effective turbulent mixing. Figure 8.9 (from Baldocchi et al., 1981a) shows the interacting effects of windiness and irradiance on the flux of CO_2 to an alfalfa crop. F_c is seen to respond to an increase in the flux density of net radiation. However, the rates at any irradiance level increase with increasing windiness (described in the figure in terms of friction velocity u_*). This effect may be due to distortion of the canopy shape by the wind, thus facilitating the penetration of radiation to the lower, light-unsaturated leaves.

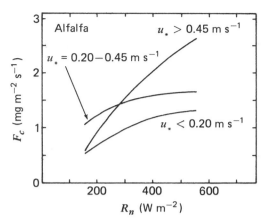

Fig. 8.9 Effects of net radiation and wind on photosynthetic rate of alfalfa (after Baldocchi et al., 1981a).

8.5 ENVIRONMENTAL INFLUENCES ON RESPIRATION

As was pointed out above, photosynthesis in the field consumes carbon dioxide supplied from the air above, from the soil and living roots below, and from the plants themselves. CO_2 emanating from the soil and from the plant is produced by respiration. Temperature and water supply are the primary environmental factors that control the rate of respiratory release.

8.5.1 Temperature

In Chapter 2 it is shown that respiration is a temperature-dependent process. The dependency is described by

$$R_T = R_0 Q_{10}^{(T-T_0)/10} \qquad (2.7)$$

where R_T is respiration rate at temperature T and R_0 is respiration rate at reference temperature T_0.

This expression can be applied to dark respiration R_d, and to the respiration that occurs in the soil which is due to the oxidation of humus, the metabolic activity of flora and fauna R_g, and the respiratory activity of the roots R_r.

That R_d responds to leaf temperature has been demonstrated in many laboratory studies. In Fig. 8.10 (from Regehr et al., 1975), the net photosynthesis, dark respiration, and transpiration are seen to respond to the temperature of *Populus deltoides* leaves. The decline in aPS above 30°C is due to the continuing increase in R_d as well as the decreasing leaf conductance.*

---
* Leaf conductance is the reciprocal of leaf stomatal resistance.

300 FIELD PHOTOSYNTHESIS, RESPIRATION, AND CARBON BALANCE

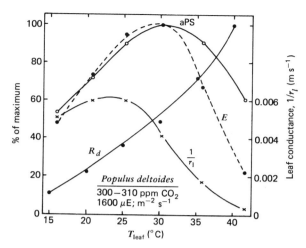

Fig. 8.10 Photosynthetic PS and transpiration E rates and leaf conductance $(1/r_l)$ in the light, and dark respiration rate (R_d) of single intact leaves at various temperatures (from Regehr et al., 1975).

Da Costa (1981) using field chambers has shown the exponential response of actual rates of respiration (per unit of leaf area) to increasing air temperature in a field of soybeans (Fig. 8.11). The system used in his studies is described in Section 8.8.6 of this chapter.

8.5.2 Soil Moisture

Availability of soil moisture affects respiration in the plant as well as in the soil. In their study of *Populus deltoides*, Regehr et al. (1975) stressed the plant by withholding water until an internal leaf water potential of about -1.8 MPa was reached (Fig. 8.12). Dark respiration decreased significantly in the range -0.4 to -0.8 MPa. This decreasing respiration rate was linked to the decrease in photosynthesis that occurred in the leaf as a result of water stress. Rates recovered rapidly after watering.

The influence of soil water availability on respiratory processes has also been demonstrated in the field by Da Costa (1981). In Fig. 8.13a, mean nocturnal release of CO_2 as a result of dark respiration in soybeans is shown for a number of nights during the growing season. In the early part of the season, soil moisture content was low. The respiratory rate appeared to be related to air temperature. In early August extensive rains filled the soil profile and the increase in respiration was more closely related to soil moisture availability through its general influence on photosynthesis.

The influence of moisture on respiratory activity in the soil is seen even more clearly in Fig. 8.13b. Temperature in the upper 30–100 mm of the soil was fairly uniform throughout the season. Respiration rose rapidly after the

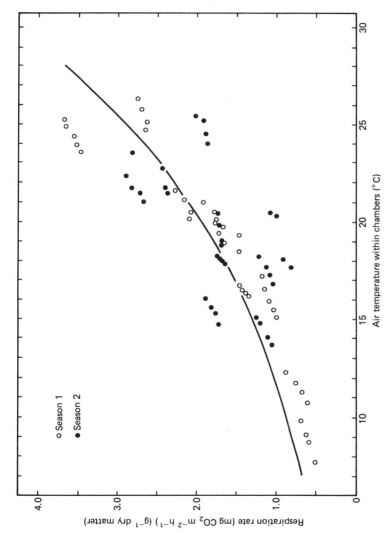

Fig. 8.11 Respiration rate from a full soybean crop (soil + roots + aerial parts) as a function of air temperature. Mead, Nebraska, 1979 and 1980. Rates are expressed on the basis of g dry matter of the aboveground portion of the crop (after Da Costa, 1981).

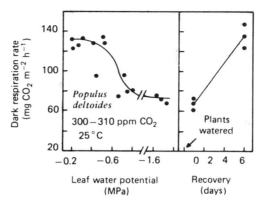

Fig. 8.12 Dark respiration rate in response to decreasing leaf water potential and for 6 days of recovery from drought (after Regehr et al., 1975).

rainfall but diminished as the season progressed. Part of the increased production of CO_2 in the soil can be related to the greater photosynthetic activity after rain; a part is also attributable to increased activity of microorganisms for which soil organic materials provide the substrate.

8.6 CARBON BALANCE IN THE FIELD

As is shown in Fig. 8.2 photosynthesis depends on CO_2 supplied from three sources: the air, the soil, and the plant itself. Equations 8.2 and 8.4 describe the gross and apparent photosynthesis. The former term denotes the total capture of CO_2 from all sources; the latter denotes only the actual gain since CO_2 released by respiration is merely recycled by the plant itself.

Thus, during the day the net capture of CO_2 by the crop can be given by the apparent photosynthesis as

$$\text{aPS} = F_{c\,(\text{day})} + R_g \tag{8.4a}$$

At night when photosynthesis has ceased there is a net release of CO_2 to the atmosphere. The sources of the CO_2 at night are the fluxes R_s and R_r, both normally diminished because of lower temperatures at night. The crop continues to respire but also at a diminished rate because of lower temperatures. In the C_3 species this decline is due, as well, to the cessation of photorespiration at night.

Thus, at night, the net release of CO_2 to the atmosphere can be represented by

$$F_{C\,(\text{noct})} = R_c + R_g + R_r \tag{8.4b}$$

The net daily gain or loss of carbon dioxide from a unit of land area can be represented by

$$F_{c\,(\text{net})} = F_{c\,(\text{day})} - F_{c\,(\text{noct})} \tag{8.6}$$

In the foregoing, the crop respiration term R_c has been used to represent the sum of the dark respiration R_d and photorespiration R_l (when present) (8.3). According to Biscoe et al. (1975) *gross photosynthesis might be determined in the field by adding an appropriate rate of respiration (including photorespiration) to the rate of apparent photosynthesis. Field determination of photorespiration is not practical since that process is coupled to photosynthesis and cannot be accurately measured if the environment about the leaf is altered.

In laboratory work, the sum of the apparent photosynthetic rate and the rate of respiration in the dark are regarded as a measure of gross photosynthesis. Biscoe et al. (1975) feel that a similar approach is permissible in the field so long as the effect of changing temperature is carefully accounted for. Since temperatures are lower at night and dark respiration is measured then, daylight estimates of R_d must be adjusted upward for the higher temperature. This approach does not account for photorespiration or for any depression in photosynthesis caused by accumulation of assimilates in the foliage. The total respiration over a day is the sum of the dark respiration over 24 h, $\sum R_d$, plus the sum of photorespiration during the day, $\sum R_l$.

In order to adjust respiration rates for temperature Biscoe et al. (1975) propose that the Q_{10} concept (2.7) be applied. For example, in their experiment with barley in England, the following conditions were assumed: a 14-h day with plant tissue temperature T_{ll}; a 10-h night (during which measurements of respiration R_d' are made) with tissue temperature T_{ld}; and a Q_{10} of 2. Thus, respiration rate in the light R_l' was calculated as

$$R_l = (14/10)R_d(2)^{(T_{ll} - T_{ld})/10} \tag{8.7}$$

Temperatures are measured for this purpose either throughout the plant canopy or at some reasonable level such as the height of maximum foliage. An arithmetic mean temperature for night and day can be applied for the purpose of calculation although, where a wide temperature range prevails during the course of 24 h, a reasonable R_d' and leaf temperature should be established and R_l' should be calculated for each hour on the basis of changing leaf temperature. Thus, by the Biscoe et al. (1975) analysis, the gross photosynthesis for a day can be estimated from

$$\text{gPS} = \sum_{\text{light}} (F_c + R_g) + 1.4 R_d'(2)^{(T_{ll} - T_{ld})/10} \tag{8.8}$$

Some examples of field carbon balance studies embodying these principles are given below. The carbon balance for an irrigated sugar beet field in western Nebraska was described by Brown and Rosenberg (1971). Carbon dioxide flux above the crop was estimated by using the Bowen ratio–energy balance method to provide an exchange coefficient for CO_2 transfer (see Section 8.8). Chambers were used to determine the soil contribution of carbon dioxide by respiration. The contribution of root respiration was estimated from laboratory data.

A typical pattern of carbon dioxide flux from the soil and apparent pho-

"*The term "gross photosynthesis" as Biscoe et al. have used it does not account for photorespiration. The notation R_d' and R_l' will be used to represent respiration in the dark and respiration in the light (excluding photorespiration)."

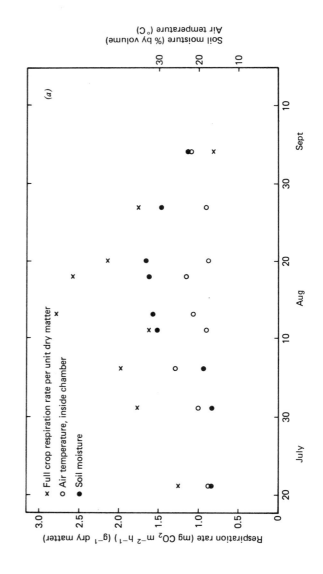

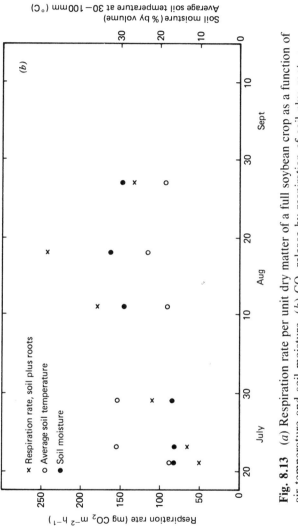

Fig. 8.13 (*a*) Respiration rate per unit dry matter of a full soybean crop as a function of air temperature and soil moisture. (*b*) CO_2 release by respiration of soil plus roots as a function of soil temperature and soil moisture in a soybean field. Mead, Nebraska, 1980 (after Da Costa, 1981).

tosynthetic fixation is given in Fig. 8.14. Soil emission of carbon dioxide follows a clear diurnal pattern which is in phase with soil temperature. The maximum photosynthetic rate measured was 1.26 mg m^{-2} s^{-1} and occurred at about 1300 hours. Generally, the carbon dioxide flux from the air increased during the first hours of daylight and leveled off thereafter. Typical flux rates during midday were about 0.8 mg m^{-2} s^{-1}.

In such a crop as sugar beets, which becomes light saturated at about two-thirds of the full solar flux density (~550 W m^{-2}) in the Great Plains, the rates of downward flux of carbon dioxide should not be expected to be

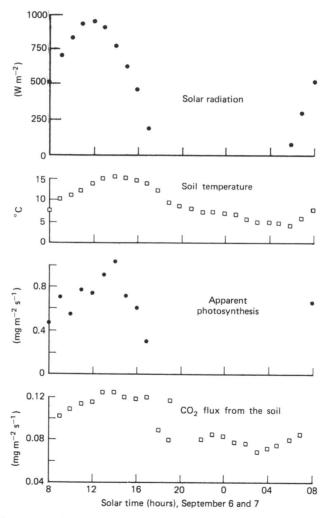

Fig. 8.14 Apparent photosynthesis by sugar beets, related to the solar radiation flux density, soil temperature, and carbon dioxide flux from the soil. Early September at Scottsbluff, Nebraska (after Brown and Rosenberg, 1971).

Table 8.3 Diurnal apparent photosynthesis in sugar beets, calculated from carbon dioxide flux components on a clear day[a]

	CO_2 (g m^{-2})
Period of flux from the atmosphere (daytime, 12 h)	
CO_2 flux from the air ($F_{c\ (day)}$)	27.8
CO_2 flux from the soil ($R_g + R_r$)	4.4
Net flux to the above-soil portion of the plant (F_c day $+ R_r + R_g$)	32.2
Estimated root respiration (R_r)	-2.0
Apparent photosynthesis during the daylight hours	30.2
Period of flux to the atmosphere (night, 12 h)	
CO_2 flux to the air ($F_{c(noct)}$)	-8.3
CO_2 flux from soil ($R_g + R_r$)	3.2
Net flux from the above-soil portion of the plant (R_d)	-5.1
Estimated root respiration (R_r)	-2.0
Apparent respiration flux during the night ($R_d + R_r$)	-7.1
Net carbon dioxide exchange (24 h total) ($F_{c(net)}$)	23.1

[a] After Brown and Rosenberg (1971). All values are given in grams of equivalent dry matter.

in phase with solar radiation. Variations will be due to other external and internal controlling factors, such as windiness, temperature, and stomatal closure.

A typical daily carbon dioxide balance for the sugar beet crop is given in Table 8.3 in which CO_2 flux values have been adjusted to their equivalent in dry matter. The net exchange of carbon dioxide (aPS) during the 12 h of daylight was about 30.2 g m^{-2}, composed of the downward and upward fluxes of carbon dioxide minus the root respiration during this period. During the night, root respiration continues. The carbon dioxide emitted by the soil and by the plants is discharged to the air for a net loss during the 12-h dark period of 7.1 g m^{-2}. A typical daily balance shows an apparent photosynthesis of 23.1 g m^{-2}.

In the study described above Brown and Rosenberg (1971) did not attempt to measure net flux of CO_2 to the atmosphere at night since the Bowen ratio–energy balance technique fails then. Biscoe et al. (1975) in their study of the carbon budget of a barley stand attempted to overcome this difficulty by using the aerodynamic method during the night. A complete carbon balance for the barley crop over its growing season was assembled in this most complete of available studies. The components of that balance are shown in Fig. 8.15. The greater magnitude of R_l than of R_d is apparent throughout

308 FIELD PHOTOSYNTHESIS, RESPIRATION, AND CARBON BALANCE

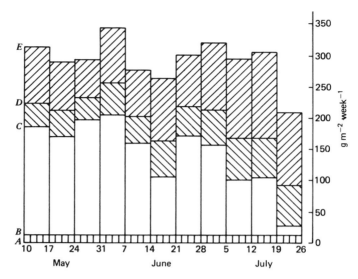

Fig. 8.15 Weekly amounts of soil microorganism respiration (*AB*), CO_2 uptake by the crop from the atmosphere (*BD*), net CO_2 fixation by the crop (*AC*), crop respiration in the dark (*CD*), crop respiration in the light (*DE*), and gross photosynthesis (*AE*) for the period early May to late July in the English Midlands (after Biscoe et al., 1975).

the season, as is the increasing size of the total plant respiration with respect to photosynthetic fixation as the season ended. For additional reviews of carbon balance in crops, see the contributions of Brown, Saugier, Denmead, and Uchijima in Monteith (1976).

8.7 RADIANT ENERGY CONVERSION IN PHOTOSYNTHESIS

In order for the photosynthetic reaction to occur, light must be available and impinging on the leaves and/or other photosynthetically active organs of the plant. Solar radiation supplies the energy required in the process. However, only light in the waveband from about 400 to 700 nm is effective in photosynthesis (photosynthetically active radiation, PAR). Of the total energy in the solar spectrum, only about 40–50% is contained in this waveband. The reduction of 1 mole of carbon dioxide to form carbohydrate requires about 0.469 MJ. One mole of photons (an Einstein, Ei) of light in the visible wavelengths has an energy content of about 0.172 MJ. It appears, then, that about three quanta of visible light should be needed to reduce one carbon dioxide molecule. A number of physiological experiments have shown, however, that the actual **quantum requirement** is between 8 and 12. Lemon (1967) reports the quantum requirement for corn grown in the field to be about 16.

Loomis and Williams (1963) on the basis of **photosynthetic efficiencies** (the fraction of radiant energy in either the PAR or the total solar radiation fixed in a biochemical reaction) estimated the theoretical maximum photosynthetic productivity on the basis of the following assumptions: in an area receiving 20.9 MJ m^{-2} of solar radiation in a day, the flux density of visible radiation (400–700 nm) would be about 2.06 µEi J^{-1} of available radiation; the quantum requirement of photosynthesis is assumed to be about 10 for the 400–700 nm range; respiration losses equal 33% of the gross photosynthetic production.

From this they calculate the potential production of 71 g (CH_2O) m^{-2} day^{-1} or 3.3 µg [CH_2O] J^{-1} total incident radiation. This level of productivity represents storage in carbon compounds with an energy content of 15.66 kJ g^{-1} or about 5.3% of the total radiant energy on a day of 20.9 MJ m^{-2} irradiation. The energy stored in photosynthate would be equivalent to about 12% of the PAR.

Chang (1968), on the other hand, calculated the upper limit of solar energy on the basis of the following assumptions: there is a 25% loss of the solar radiation impinging on the crop by reflection and transmission (Chang says 15%); 41% of the total solar spectrum is in the visible wavelengths; the quantum efficiency of energy conversion is only about 20%. Thus, [0.20 × (1 − 0.25) × 0.41] or 5% of the total solar radiation is convertible to carbohydrate.

Photosynthetic efficiencies approaching even 6% of solar irradiance have not yet been achieved in crop plants. Lemon (1969) points out that the annual dry matter production of a crop of corn, for example, is equivalent only to the energy supplied by the sun during a 1- or 2-day period at most. He considers that the best farming practices in use today yield photosynthetic efficiencies not greater than 1% of the total seasonal incident solar radiation.

Data on apparent photosynthesis as a function of solar radiation for a single day of corn growth are presented by Lemon (1967). Observations of photosynthetic fixation of CO_2 were made simultaneously in the field by means of an aerodynamic method and with an enclosed chamber system. The slope of the line in Fig. 8.16 represents a photosynthetic efficiency of 5%. The aerodynamic method indicates what that level of efficiency was achieved throughout most of the day except at the highest and lowest irradiances. The chamber technique generally indicated a lower photosynthetic efficiency, possibly an experimental artifact.

The relatively low photosynthetic efficiency reported above does not apply to all plants. Mooney et al. (1976) have shown, for example, that the desert plant *Camissonia claviformis,* a winter annual found in Death Valley, California, fixes absorbed noon sunlight (400–700 nm waveband) into chemical energy with an efficiency of 8.5%, about 80% of that theoretically possible. *Camissonia* is a C_3 plant; its ability to utilize such high irradiances is due apparently to a high stomatal conductance to carbon dioxide and to certain biochemical factors such as high levels of soluble protein and ribulose-1,5-diphosphate carboxylase/oxygenase.

The response of *Camissonia* to increasing irradiance is compared in Fig.

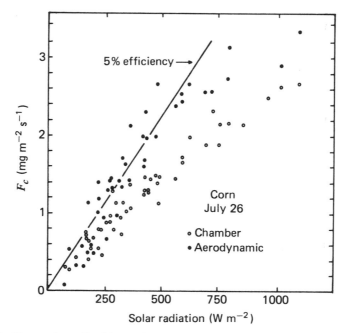

Fig. 8.16 Net carbon dioxide exchange rates in a cornfield as a function of total incident shortwave radiation. Late July at Ellis Hollow, New York (after Lemon, 1967).

8.17 to that of sugar beet and to that theoretically possible. Observations were made on *Camissonia* at 30°C and 320 ppm CO_2. The *Camissonia* response is shown as a function of irradiance "incident" on the leaf. The response is also plotted in terms of irradiance absorbed by reducing values by 19% to account for reflection and transmission. The sugar beet, a more typical C_3 plant, captures less than half as much CO_2 as does *Camissonia* under strong irradiance (see Mooney et al., 1976, for further details).

Biscoe et al. (1975) calculated the actual photosynthetic efficiency of a barley crop in the capture of solar radiation in terms of the proportion of radiant energy in the waveband 400–700 nm converted to chemical energy by the crop. In Fig. 8.18 data are presented on percent efficiency calculated from gross photosynthesis, apparent photosynthesis, and weakly intercepted PAR. It was found that in the early season (following emergence of the ears) maximum efficiency was observed on cloudy days when insolation was less than 6 MJ m^{-2} (about half full sunlight in the climate of the British Midlands). On mostly clear days the photosynthetic system of barley was saturated in bright sunshine at an efficiency of 4–6% PAR. Later, as the plants matured, photosynthetically active tissue diminished and so did the rate of photosynthesis in bright light. On very dull days maximal efficiency approaching 10% of PAR was achieved.

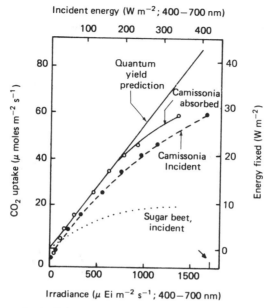

Fig. 8.17 Photosynthetic light response of *Camissonia claviformis* and sugar beet (*Beta vulgaris*) (see text and Mooney et al., 1976, for details; Copyright 1976 by the American Association for the Advancement of Science).

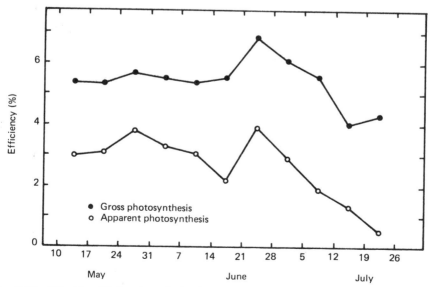

Fig. 8.18 Weekly efficiencies of a barley crop for the period early May to late July in the English Midlands calculated from gross photosynthesis and apparent photosynthesis and weekly intercepted PAR (from Biscoe et al., 1975).

By way of summary, we may say that the discrepancy between the actual energy conversion achieved in photosynthesis and the theoretical potential fixation is great. This point is emphasized further in Table 8.4 which is discussed in the section that follows. Methods to narrow this gap must be developed if a sustained food and fiber production for future generations is to be possible. That a potential exists to do so is shown by Mooney et al. (1976) for *Camissonia claviformis* discussed above. Adapting plants to the stresses of their environment may provide another way (see Chapters 9, 10, and 11).

8.8 WATER USE EFFICIENCY

Water use efficiency is the ratio of dry matter produced by photosynthesis to the water consumed in evapotranspiration. Lemon (1969) surveyed actual water use efficiencies in modern day agricultural systems (Table 8.4). For these calculations he assumed that 60% of the solar energy is consumed in evapotranspiration.

Lemon's table shows that the best intensive farming results in conversion of 1% of the solar radiation and produces 0.7–1.2 kg of dry matter per ton of water used. Subsistence farming may produce a solar energy fixation of

Table 8.4 Achievements and possibilities for photosynthetic and water use efficiencies[a]

System	Photosynthesis efficiency[b] (%)	Water use efficiency[c] (kg t^{-1})[d]
Subsistence farming		
Average	0.04–0.1	0.04–0.12
Best	0.08–0.2	0.10–0.24
Ranch farming		
Average	0.1–0.2	0.12–0.24
Best	0.2–0.4	0.24–0.48
Intensive farming		
Average	0.25–0.35	0.30–0.42
Best	0.6–1.0	0.72–1.2
Experimental		
Season	0.8–1.5	0.96–1.8
Weeks	1.5	1.8
Days	2–4	2.4–4.8
Theoretical upper limit	8–10	9.6–12.0

[a] After Lemon (1969).
[b] Incident total solar energy base.
[c] Assumes a 60% conversion of solar energy to latent heat.
[d] kg dry matter per ton of water.

0.1–0.2% and dry matter production of less than 0.2 kg per ton of water consumed. Even in the best experimental circumstances and under excellent growing conditions, the best 1-day photosynthetic and water use efficiencies are only in the order of 2–4% of the solar energy conversion and 2.4–4.8 kg dry matter per ton of water used. Lemon finds it theoretically possible to fix 8–10% of the solar radiation and to decrease water use so that as much as 9.6–12 kg of dry matter can be produced per ton of water consumed.

Thus, it appears theoretically possible to increase the photosynthetic efficiency of the best actual farming today 8- to 10-fold. The greater production of dry matter need not require the consumption of additional water so that an 8- to 10-fold improvement in water use efficiency is also theoretically possible. It seems likely, in fact, that improved plant growth would reduce somewhat the amount of water consumed by direct evaporation from the soil surface and further improve the water use efficiency.

The water use efficiency can also be usefully defined in agrometeorological experimentation in terms of the **CO_2–water vapor flux ratio** CWFR. CWFR is the ratio of the mass fluxes of carbon dioxide F_c and water vapor E and can be calculated for the entire day, the daylight period, or any portion of it:

$$\text{CWFR} = \frac{F_c}{E} \qquad (8.9)$$

F_c is, of course, the exchange of CO_2 between the crop and the atmosphere and as such neglects the contributions of CO_2 from root and soil respiration and from the plant itself. Nonetheless, in the field many interesting effects can be illustrated by calculation of CWFR.

Baldocchi et al. (1981a) have shown that CWFR of alfalfa grown under irrigation in eastern Nebraska is most sensitive to variations in net radiation and sensible heat advection. R_n provides energy that drives both the photosynthetic and evapotranspiration processes; advection of sensible heat serves only to provide additional energy for evapotranspiration. These effects, illustrated for alfalfa in Fig. 8.19, were also found to hold for soybeans (Baldocchi et al., 1981b).

Other environmental effects on CWFR in soybeans were reported by Baldocchi et al. (1981c). Three typical days are illustrated in Fig. 8.20a,b,c. August 4 was a clear day with strong sensible heat advection. Evapotranspiration rate was very high on this day. F_c rose rapidly in the morning but fell sharply as temperatures rose beyond the optimum for soybeans. August 16 was a cloudy day. Net radiation was partitioned between sensible and latent heat and evapotranspiration was low. Peak values of F_c were lower than on August 4 but remained at a plateau level through much of the day. September 3 was a clear day. Air temperatures remained below 30°C and sensible heat advection was minor. Latent heat flux was moderate. F_c rates reached high midday values and declined slowly as the afternoon progressed.

CWFR patterns for each of the three days in Fig. 8.20a–c are shown in

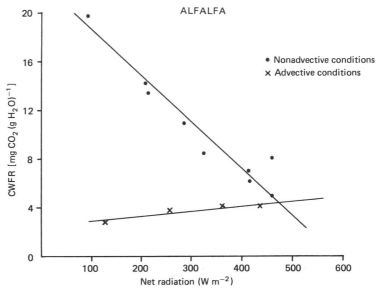

Fig. 8.19 CO_2 water flux ratio of alfalfa (CWFR) as a function of net radiation in nonadvective and advective conditions. Late summer at Mead, Nebraska (after Baldocchi et al., 1981a).

Fig. 8.21. The greatest water use efficiency, it is seen, occurred in cloudy weather (August 16); the lowest efficiency occurred when sensible heat advection caused strong evapotranspiration and high temperatures reduced photosynthesis (August 4).

There are other ways to approach the characterization of water use efficiency. Tanner (1981) expresses water use efficiency in terms of the ratio Y/T. Y is either the total dry matter production or marketable dry matter and T is the transpiration. Direct evaporation from soil, he holds, affects physiological processes only indirectly, for example, by hastening water depletion or by modifying the local microclimate, whereas transpiration is directly related to plant processes. Tanner tested an approach suggested by Bierhuizen and Slatyer (1965) that the water use efficiency could be represented by an expression of the form

$$Y/T = k_1/(e_s - e_a) \qquad (7.4)$$

where k_1 is a physico/physiologically based constant and $(e_s - e_a)$ is the average daytime water vapor pressure deficit. Measurements of potato tuber growth provide a value of $k_1 = 6.5 \pm 0.7$ Pa for tubers and total dry matter. This number sets the upper limit of water use efficiency, so defined since ET is greater than T. The particular merit of this approach is its definition of water use efficiency in terms of a physical factor known to strongly affect it, namely, the vapor pressure deficit.

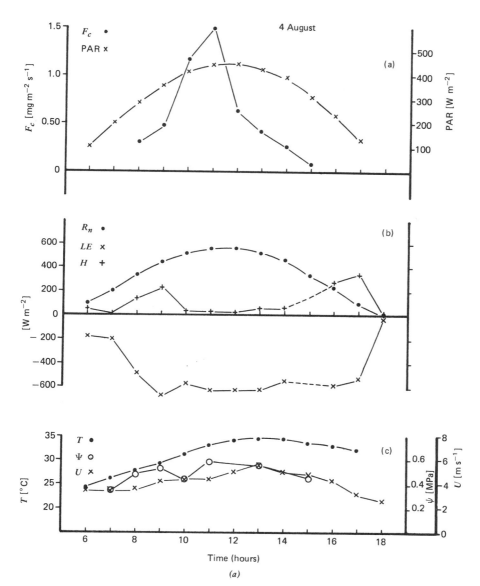

Fig. 8.20a Energy and mass fluxes and environmental conditions over a soybean canopy. August 4, 1979 at Mead, Nebraska. (a) diurnal course of canopy CO_2 flux F_c, and photosynthetically active radiation, PAR. (b) diurnal course of net radiation R_n, latent heat flux LE, and sensible heat flux H. (c) diurnal course of air temperature T at 0.7 m, and wind speed U at 1.5 m, and plant water potential Ψ (after Baldocchi et al., 1981b). Reprinted from *Agronomy Journal*, Volume **73**, pages 707, 708, and 709, 1981, by permission of the American Society of Agronomy.

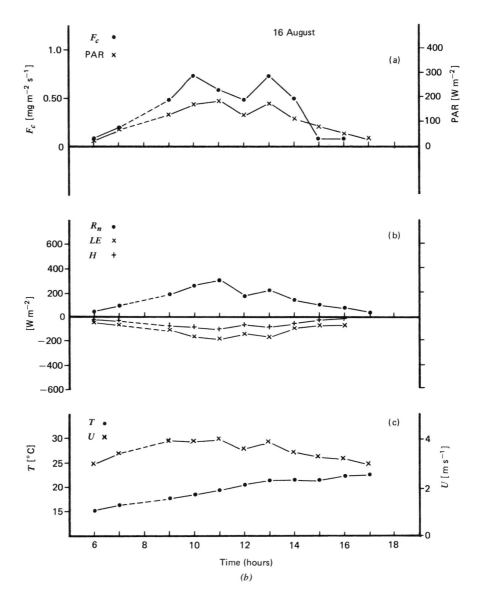

Fig. 8.20b Same as (*a*) for August 16, 1979.

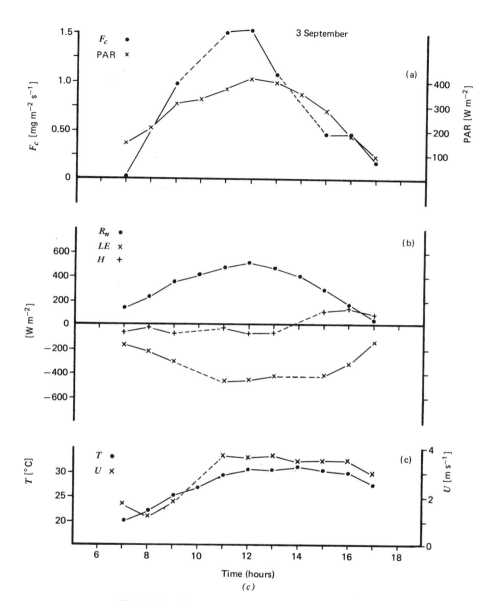

Fig. 8.20c Same as (*a*) for September 3, 1979.

318 FIELD PHOTOSYNTHESIS, RESPIRATION, AND CARBON BALANCE

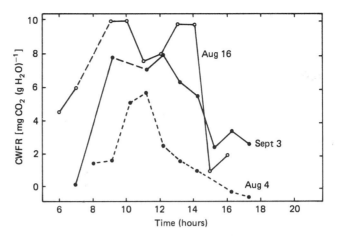

Fig. 8.21 Diurnal course of CO_2–water flux ratio CWFR for a soybean crop at Mead, Nebraska, on the 3 days represented in Fig. 8.20a–c (redrawn from data of Baldocchi et al., 1981b).

8.9 MEASUREMENT OF PHOTOSYNTHESIS IN THE FIELD

8.9.1 Micrometeorological Methods

The flux of carbon dioxide can be treated as a turbulent diffusion phenomenon analogous to others we have dealt with

$$F_c = K_c \frac{\partial \rho_c}{\partial z} \qquad (8.10)$$

where K_c is the exchange coefficient for CO_2 and $\partial \rho_c/\partial z$ is the gradient of the CO_2 density. Gradients of carbon dioxide density may be determined by infrared gas analysis or other techniques. The determination of an exchange coefficient for carbon dioxide, however, is difficult and requires that certain assumptions be made in each of the methods presented below. For an analysis of the errors involved in the first three micrometeorological methods described below, see Verma and Rosenberg (1973, 1975).

Flux-Gradient Methods: Aerodynamic Theory. Following the procedure described in Chapter 4, (8.10) can be rewritten as

$$F_c = k^2 z^2 \left(\frac{\partial U}{\partial z}\right) \left(\frac{\partial \rho_c}{\partial z}\right) \left(\frac{K_c}{K_m} \phi_m^{-2}\right) \qquad (8.11)$$

K_c is generally assumed to be equal to K_h and K_w. Therefore, F_c can be calculated by employing the wind speed, air temperature, and CO_2 density gradient data in (8.11) and the stability correction formulas described in Chapter 4 [(4.14) and (4.16)].

Early attempts to use an aerodynamic approach for estimating photosynthesis were made by Thornthwaite and Holzman (1942). Inoue et al. (1958) and Lemon (1960) did pioneering work in estimating photosynthesis in corn by this method. More recently Verma and Rosenberg (1976, 1981) have estimated the flux of CO_2 between 16 and 5.6 m above the ground at Mead, Nebraska, using the flux gradient method (8.11). By so doing, they were able to determine the annual patterns of source and sink strength for CO_2 of a typical agricultural region. Houghton and Woodwell (1980) used the technique to study the exchange of CO_2 between a salt marsh and the atmosphere. Baumgartner (1969) applied the aerodynamic approach to estimating the exchange of CO_2 in a spruce forest. There are many other examples of the use of this method for estimating F_c above various kinds of vegetated surfaces.

Bowen Ratio–Energy Balance Methods. In this method K_c is assumed to be equal to K_h or K_w. Therefore from (7.46) and (7.48), we can write

$$K_c = K_w = K_h = -\frac{R_n + S}{\rho_a \left(C_p \frac{\partial T}{\partial z} + \frac{M_w/M_a}{P} L \frac{\partial e}{\partial z} \right)} \qquad (8.12)$$

Thus, a by-product of sensible and latent heat measurement is an estimate of the value of K_c, which, with carbon dioxide gradient measurements, can be used to estimate F_c, as in (8.10). The detailed studies of Brown and Rosenberg (1971), Biscoe et al. (1975), and Baldocchi et al. (1981b) have made use of the Bowen ratio method to estimate F_c for sugar beets, barley, and soybeans, respectively.

Lysimetric Method. An alternative method for determining the exchange coefficient is through the use of precision weighing lysimeters. Lysimetry provides a means of accurately recording the flux of water vapor over periods as short as 10–15 min. Humidity gradients are measured simultaneously over the crop with psychrometers or hygrometers. The magnitude of the exchange coefficient K_w is determined from

$$K_w = \frac{LE_{(lys)}}{\rho_a \frac{M_w/M_a}{P} L \frac{\partial e}{\partial z}} \qquad (8.13)$$

where $LE_{(lys)}$ is the latent heat equivalent of the evapotranspiration measured by lysimeter. K_w is assumed equal to K_c (e.g., Brown and Rosenberg, 1971; Denmead and McIlroy, 1970; Baldocchi et al., 1981a). Again, as a by-product of evapotranspiration measurement, we may estimate F_c if only the carbon dioxide concentrations are measured concurrently.

Eddy Correlation. The principle of eddy correlation discussed in Chapter 4 can be easily applied to the determination of the vertical carbon dioxide flux above crop canopies:

$$F_c = \overline{w'\rho'_c} \qquad (8.14)$$

where w is the vertical velocity and ρ_c is the CO_2 concentration density.* Holmes and Carson (1964) designed an eddy correlation device that coupled a vertical anemometer to a rapid-response infrared gas analysis system (IRGA) for carbon dioxide. The analyzer was placed about 2 m above the ground in the field and sampled air directly on the site. The carbon dioxide concentration was recorded during the period when a vertical velocity anemometer turned in one direction (e.g., representing an updraft). The vertical wind speed was recorded simultaneously. When the direction of the vertical wind speed changed, the IRGA sampled air from the downcoming eddy and recorded its concentration.

Desjardins and Lemon (1974) measured turbulent CO_2 fluxes over corn by using a propeller anemometer with a modified infrared CO_2 gas analyzer in an eddy correlation technique similar to that proposed by Holmes and Carson (1964). Garratt (1975) and Wesely and Hicks (1975) were critical of this approach as, apparently, the effects of slow sensor response time were not correctly evaluated and the fluxes were substantially underestimated.

Although vertical wind speed sensors, adequate for eddy correlation measurements near the ground, have been developed in recent years (see Chapter 4), a suitable, fully tested fast response CO_2 sensor is not yet commercially available. Ohtaki and Seo (1976) and Seo and Ohtaki (1976) developed a "double-beam" CO_2 fluctuation analyzer that operates by taking the light from an infrared source and dividing it into two beams that are optically filtered to provide wavelengths of 3.9 and 4.3 μm. CO_2 is largely transparent at the former wavelength and absorbs very strongly in the latter wavelength. Jones and Smith (1977) reported preliminary measurements of eddy fluxes of CO_2 made in Nova Scotia with a similar sensor. Desjardins et al. (1982) also report an aircraft mounted sensor with which they have monitored surface carbon dioxide exchange over a cornfield, a forest, and a lake in southern Canada.

Bingham et al. (1978) reported the development of a miniature, rapid response CO_2 sensor also based on the differential absorption of CO_2 at the wavelengths 3.9 and 4.3 μm. This sensor is intended for use in aircraft and on a tower as well. A field model of this sensor is shown in Fig. 8.22.

The difficulties experienced in development of eddy correlation instrumentation have led to some ingenuous contraptions. For example, Desjardins (1972) developed an eddy accumulation device for F_c estimation. In his system, the rate and direction of rotation of a vertical anemometer determine the quantity of air drawn for mixing into one of two sample storage bags. Each upward or downward eddy contributes a volume of air proportional to its size, and the mean concentration times the vertical wind speed for

---

* See Section 4.6 for an explanation of the eddy correlation notation.

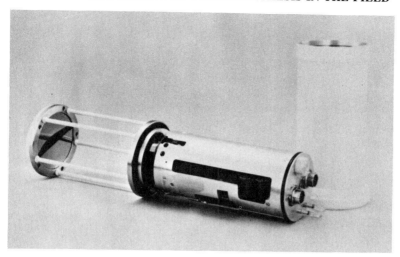

Fig. 8.22 A rapid response CO_2 sensor for measuring fluctuations in CO_2 concentration (courtesy of Dr. Gail Bingham, Lawrence Livermore National Laboratories, University of California).

each direction, when added algebraically, gives the net flux of the carbon dioxide. The method is not widely used.

Corrections in CO_2 Flux Measurements. As was indicated in Chapter 4 recent investigations (e.g., Smith and Jones, 1979; Webb et al., 1980; Webb, 1982) reveal that turbulent fluxes of CO_2 are influenced by density variations due to the flux of heat and/or water vapor. To account for such density effects the CO_2 flux should be corrected by the following equations developed by Webb et al. (1980):

$$F_c = F_{c\,(raw)} + (1.684 \times 10^{-3}H + 0.325 \times 10^{-3}LE) \quad (8.15)$$

where H and LE are in W m^{-2} and F_c is in mg m^{-2} s^{-1} (a mean CO_2 content of 330 ppm by volume relative to dry air was assumed in this derivation). The above equation applies to flux computations based on eddy correlation or mean gradient measurements made in unmodified, *in situ* air. If the mean gradient of CO_2 concentration is measured in air brought to a common temperature without predrying, then a correction arising only from water vapor flux is required and the following equation applies:

$$F_c = \left[\left(\frac{P}{P_I}\right)\left(\frac{T_I}{T_a}\right)^{-1}\right]\left[F_{c\,(raw)} + \left(\frac{\rho_c}{\rho_a\epsilon}\right)\left(1 + \frac{\sigma}{\epsilon}\right)E\right] \quad (8.16)$$

where P (kPa) and T_a (°K) are the ambient pressure and temperature, P_I (kPa) and T_I (°K) are the pressure and air temperature inside the CO_2 analyzer cells, ϵ is the ratio of mole weights of water vapor and dry air, σ is the ratio of the densities of water vapor and dry air, ρ_c (g m^{-3}) is the partial density of CO_2, ρ_a (g m^{-3}) is the density of moist air, and E is the water vapor flux (F_c and E have the same units).

No corrections, however, are needed if the air samples from different elevations are predried and brought to a common temperature before mean CO_2 concentrations are measured. For further details see Webb et al. (1980) and Webb (1982).

8.9.2 Nonmeteorological Methods

Nonmeteorological methods for determining the rates of field photosynthesis have been proposed by Thomas and Hill (1949), Musgrave and Moss (1961), and Baker and Musgrave (1964). These experimenters used plastic chambers to enclose small portions of a field in order to measure the exchange of carbon dioxide between the air entering and the air exiting the chamber. By measuring the CO_2 concentration of the incoming and outgoing air and the rate of air flow, the net exchange of carbon dioxide can be calculated.

These chambers, it has been found, can drastically disturb the plant's environment and, thus, cannot be used for long periods of time. To achieve a reasonable spacial average of a field would require very large numbers of chambers. The results of chamber observations, therefore, have often been of limited applicability to the general field situation. However, there are many experimental circumstances, such as in forests or wild lands where the surface cover is not uniform. Here chambers provide the only feasible method of determining both carbon dioxide and water vapor exchange. For a thorough review of chamber methods, see Sestak et al. (1971).

There are two ways in which the difficulties created by large chambers can be minimized. The climate within the chamber can be controlled by "air-conditioning" systems. Although this is difficult to do with large chambers, specially designed cuvette systems are helpful. Schulze et al. (1972) describe the use of temperature- and humidity-controlled cuvettes for measurement of net photosynthesis and transpiration. They calculated instrumental and recording errors with this system of 4.2 and 8.3% for net photosynthesis and transpiration, respectively. It would be useful to establish whether or not the presence of a chamber, no matter how well it air-conditions the plant, can induce changes of even greater magnitude in the rates of photosynthesis and transpiration.

Mooney et al. (1976) also used a sophisticated cuvette system for their field studies of *Camissonia claviformis* and other desert plants in which temperature, humidity, and CO_2 concentration were controlled. Sullivan et al. (1978) have developed a relatively simple leaf chamber which need be in place for only a very short time (Fig. 8.23). As the small lucite chamber is clamped about a selected leaf, a syringe extracts a sample of air for determination of CO_2 concentration; after 20 or 30 s another sample of air is extracted with another syringe. The change in concentration of CO_2 between

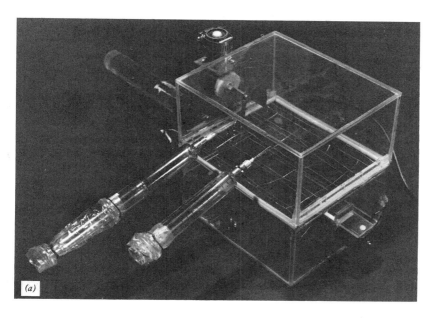

Fig. 8.23. A chamber for rapid measurement of photosynthesis in the field: (*a*) closeup view; (*b*) in field use. (Courtesy of Dr. Charles Y. Sullivan, U.S. Dept. of Agriculture, Agricultural Research Service and University of Nebraska–Lincoln).

sampling times is an indicator of net photosynthesis. A battery powered fan is included in the chamber to assure sufficiently good ventilation. Improvements in IRGAs, making them field portable, may eliminate the need for the use of syringes in such chambers.

8.10 MEASURING THE RESPIRATION COMPONENTS

The difficulties involved in chamber-type measurements of field photosynthesis are briefly noted above. Chambers do, however, provide a practical means for the measurement of soil and plant respiratory release of carbon dioxide. Musgrave and Moss (1961) and Moss et al. (1961) used a large plant chamber in the field to measure respiration from crops and soil. With the soil isolated by plastic sealed around the corn plants, the chamber was used to measure plant photosynthesis and respiration alone. With no seal they measured total respiration and, by difference, were able to calculate the soil respiration term.

Monteith et al. (1964) measured the respiration from bare and cropped soil during the course of 1 yr. A large inverted glass tank whose edges were pressed below the soil surface was used for this purpose. The tank was shielded with aluminum foil to minimize heating. The carbon dioxide produced by the soil was determined with a gravimetric method involving absorption in soda lime.

Brown and Rosenberg (1971) measured soil respiration in a sugar beet field with small chambers. Atmospheric air was mixed in a 200-liter tank and blown through the chambers at a flow rate of 10 liters min^{-1}. The carbon dioxide concentration of the incoming and outgoing air was measured with an infrared gas analyzer.

Mogensen and Rosenberg (1972) describe a chamber system for simultaneous measurement of the plant, soil, and root respiration. Figure 8.24 is a schematic of this system. Operation in the field is shown in Fig. 8.25. Three ventilated respiration chambers are used. One chamber is sufficiently large to accommodate the whole crop. The difference in carbon dioxide concentration of the entering and exiting air and the flow rates are measured. Respiration by crop, soil, and roots is the result in this case. One of the two smaller chambers is placed over bare soil from which aboveground portions of the plant have been freshly removed. The second small chamber is placed over soil in which the plants have been killed sometime earlier by spading. Edges of all the chambers are pressed into the soil to minimize the underground ventilation of the chamber. The tops of the chambers are removable so that more or less natural growth conditions prevail until just before the measurements begin. Measurements of the dark respiration R_d are made at night. Temperature is measured continually so that the Q_{10} of respiration can be established in order to estimate R_d during the daylight hours.

Air moved into the chambers is drawn from well above the crop and is

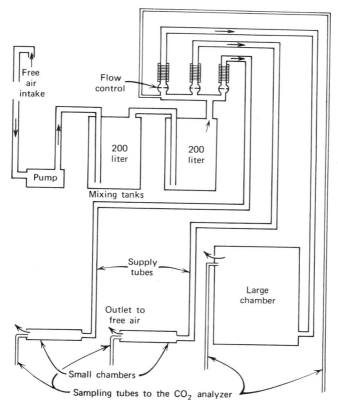

Fig. 8.24 Flow diagram of a field respiration measurement system (after Mogensen and Rosenberg, 1972).

thoroughly mixed. The concentration is measured at the outlet of the mixing tank and at the outlet of each of the respiration chambers. The flow rates of air through the chambers are controlled and known. The respiration rate is calculated from the flow rate and the increase of carbon dioxide in the air stream passing through the chambers.

The respiration rate of the above ground portion of the crop is calculated by subtracting the respiration rate for soil + root from that for soil + root + aboveground crop. The root respiration rate is calculated by subtracting the soil respiration from the adjusted soil + root respiration. This system was improved by Mogensen (1977). Da Costa (1981) has made further improvements by providing for measurement of pressure differences between the chamber and the outside air. As designed, it is also possible to use all chambers to provide replicated measurements of a single condition (e.g., soil + roots at a particular time).

Kanemasu et al. (1974) are critical of soil respiration measurements made with chambers of the type described above. Flux rates measured, they hold,

Fig. 8.25 An array of respiration chambers in the field. (Foreground) flow meters between mixing tanks and chambers; (right) large crop respiration chamber; (Background, left) two small chambers for measurement of soil and root respiration. Agricultural Meteorology Laboratory near Mead, Nebraska.

may be very much dependent on the magnitude of the pressure difference (positive in a forced air system; negative in a suction system) between the chamber and the outside air. Fluxes may be in error by as much as 50% when pressure differences are 2 Pa or greater. Mogensen (1977) reported a smaller pressure difference in practice and considered its effect insignificant. This question has not yet been satisfactorily resolved.

In a case where respiration chamber measurements were impractical, Woodwell and Dykeman (1966) determined the respiration from a forest by measuring carbon dioxide accumulation during temperature inversions. This was possible, in their view, because of the very limited transport of carbon dioxide out of the forest under such conditions, although it seems likely that, unless conditions were extremely calm, some turbulent transport out of the forest must have occurred.

REFERENCES

Baker, D. N., and R. B. Musgrave. 1964. Photosynthesis under field conditions. V. Further plant chamber studies of the effect of light on corn (*Zea mays*). *Crop Sci.* **4**:121–131.

Baldocchi, D. D., S. B. Verma, and N. J. Rosenberg. 1981a. Environmental effects

on the CO_2 flux and CO_2-water flux ratio of alfalfa. *Agric. Meteorol.* **24:**175–184.
Baldocchi, D. D., S. B. Verma, and N. J. Rosenberg. 1981b. Mass and energy exchanges of a soybean canopy under various environmental regimes. *Agron. J.* **73:**706–710.
Baldocchi, D. D., S. B. Verma, and N. J. Rosenberg. 1981c. Seasonal and diurnal variation in the CO_2 flux and CO_2-water flux ratio of alfalfa. *Agric. Meteorol.* **23:**231–244.
Baumgartner, A. 1969. Meteorological approach to the exchange of CO_2 between the atmosphere and vegetation, particularly forest stands. *Photosynthetica* **3:**127–149.
Bierhuizen, J. F., and R. O. Slatyer. 1965. Effect of atmospheric concentration of water vapor and CO_2 in determining transpiration-photosynthesis relationships of cotton leaves. *Agric. Meteorol.* **2:**259–270.
Bingham, G. E., C. H. Gillespie, and J. H. McQuaid. 1978. Development of a miniature, rapid-response carbon dioxide sensor. NSF Ecosystem Program, Project DEB77-16327.
Biscoe, P. V., R. K. Scott, and J. L. Monteith. 1975. Barley and its environment. III. Carbon budget of the stand. *J. Appl. Ecol.* **12:**269–291.
Bjorkman, O. 1981. The response of photosynthesis to temperature. *Plants and Their Atmospheric Environment* (J. Grace, E. D. Ford, and P. G. Jarvis, eds.), Blackwell, Oxford, Chapter 16, pp. 273–301.
Brown, K. W. 1969. A model of the photosynthesizing leaf. *Physiol. Plant* **22:**620–637.
Brown, K. W., and N. J. Rosenberg. 1971. Shelter-effects on microclimate, growth and water use by irrigated sugar beets in the Great Plains. *Agric. Meteorol.* **9:**241–263.
Brown, K. W. 1976. Sugar beet and potatoes. *Vegetation and the Atmosphere* (J. L. Monteith, ed.), Academic Press, London, pp. 65–86.
Boyer, J. S. 1970. Differing sensitivity of photosynthesis to low leaf water potentials in corn and soybean. *Plant Physiol.* **46:**236–239.
Chang, J. H. 1968. *Climate and Agriculture—An Ecological Survey,* Aldine, Chicago, p. 24.
Chollet, R. 1977. The biochemistry of photorespiration. *Trends Biochem. Sci.* **2:**155–159.
Da Costa, J. M. N. 1981. In Progress Rept. to the National Science Foundation on Grant ATM-7901017, S. B. Verma and N. J. Rosenberg, Principal Investigators.
Denmead, O. T., and I. C. McIlroy. 1970. Measurements of nonpotential evaporation from wheat. *Agric. Meteorol.* **7:**285–302.
Denmead, O. T. 1976. Temperate cereals. *Vegetation and the Atmosphere* (J. L. Monteith, ed.), Academic Press, London, pp. 1–32.
Desjardins, R. L. 1972. CO_2 flux measurements by eddy correlation methods. *Bull. Am. Meteorol. Soc.* **53:**1040. (Abstr.)
Desjardins, R. L., and E. R. Lemon. 1974. Limitations of an eddy-correlation technique for the determination of the carbon dioxide and sensible heat fluxes. *Boundary-Layer Meteorol.* **5:**475–488.
Desjardins, R. L., E. J. Brach, P. Alvo, and P. H. Scheupp. 1982. Aircraft monitoring of surface carbon dioxide exchange. *Science* **216:**733–735.
Ehleringer, J., and O. Bjorkman. 1977. Quantum yields for CO_2 uptake in C_3 and C_4 plants. *Plant Physiol.* **59:**86–90.

El Sharkawy, M., and J. Hesketh. 1965. Photosynthesis among species in relation to characteristics of leaf anatomy and CO_2 diffusion resistance. *Crop Sci.* **5:**517–521.
Gaastra, P. 1959. Photosynthesis of crop plants as influenced by light, carbon dioxide, temperature and stomatal diffusion resistance. *Meded. Landbouwhogeschool* (Wageningen, Netherlands) **59:**1–58.
Garratt, J. R. 1975. Limitations of the eddy-correlation technique for the determination of turbulent fluxes near the surface. *Boundary-Layer Meteorol.* **8:**255–259.
Goudriaan, J., and G. L. Ajtay. 1979. The possible effects of increased CO_2 on photosynthesis. *The Global Carbon Cycle (SCOPE 13)* (B. Bolin, E. T. Degens, S. Kempe, and P. Ketner, eds.), Wiley, New York, pp. 237–249.
Holmes, R. M., and H. W. Carson. 1964. Carbon dioxide flux in nature. Canada Dept. of Agriculture, Ottawa, Mimeo, 28 pp.
Houghton, R. A., and G. M. Woodwell. 1980. The flax pond ecosystem study: Exchanges between a salt marsh and the atmosphere. *Ecology* **6:**1434–1445.
Inoue, E., N. Tani, K. Imai, and S. Isobe. 1958. The aerodynamic measurement of photosynthesis over a nursery of rice plants. *J. Agric. Meteorol. Tokyo* **14:**45–53 (Japanese, English summary).
Jones, E. P., and S. D. Smith. 1977. A first measurement of sea-air CO_2 flux by eddy correlation. *J. Geophys. Res.* **82:**5990–5992.
Kanemasu, E. T., W. L. Powers, and J. W. Sij. 1974. Field chamber measurements of CO_2 flux from soil surface. *Soil Sci.* **118:**233–237.
Lemon, E. R. 1960. Photosynthesis under field conditions. II. An aerodynamic method for determining the turbulent carbon dioxide exchange between the atmosphere and a cornfield. *Agron. J.* **52:**697–703.
Lemon, E. R. 1967. Aerodynamic studies of CO_2 exchange between the atmosphere and the plant. *Harvesting the Sun: Photosynthesis in Plant Life* (A. San Petro, F. A. Greer, and T. J. Army, eds.), Academic Press, New York, pp. 117–137.
Lemon, E. R. 1969. Important microclimatic factors in soil-water-plant relationships. *Modifying the Soil and Water Environment for Approaching the Agricultural Potential of the Great Plains.* Great Plains Agr. Council Publ. No. 3, pp. 95–102.
Loomis, R. S., and W. A. Williams. 1963. Maximum crop productivity: An estimate. *Crop Sci.* **3:**67–72.
Mogensen, V. O., and N. J. Rosenberg. 1972. An improved method for measuring soil and root respiration. Simultaneous Determination of Short-Period Photosynthesis and Evapotranspiration, Final Report to NOAA on Grant No. E-293-68(G), pp. 52–58.
Mogensen, V. O. 1977. Field measurements of dark respiration rates of roots and aerial parts in Italian ryegrass and barley. *J. Appl. Ecol.* **14:**243–252.
Monteith, J. L., G. Szeicz, and K. Yabuki. 1964. Crop photosynthesis and the flux of carbon dioxide below the canopy. *J. Appl. Ecol.* **1:**321–337.
Monteith, J. L., ed. 1976. *Vegetation and the Atmosphere,* Vol. 2, *Case Studies.* Academic Press, London, 439 pp.
Mooney, H. A., J. Ehleringer, and J. A. Berry. 1976. High photosynthetic capacity of a winter annual in Death Valley. *Science* **194:**322–324.
Moss, D. N. 1965. Capture of radiant energy in plants. *Agricultural Meteorology* (P.

E. Waggoner, ed.), Chapter 5, *Meteorol. Monogr.* **6**(28):90–108. Am. Meteorol. Soc., Boston.

Moss, D. N., R. B. Musgrave, and E. R. Lemon. 1961. Photosynthesis under field conditions. III. Some effects of light, carbon dioxide, temperature and soil moisture on photosynthesis, respiration and transpiration of corn. *Crop Sci.* **1**:83–87.

Musgrave, R. B., and D. N. Moss. 1961. Photosynthesis under field conditions. I. A portable closed system for determining the rate of photosynthesis and respiration of corn. *Crop. Sci.* **1**:37–41.

Ohtaki, E., and T. Seo. 1976. Infrared device for measurement of carbon dioxide fluctuations under field conditions. II. Double beam system. *Ber. Ohara Inst. Landw. Biol. Okayama Univ.* **16**:183–190.

Regehr, D. L., F. A. Bazzaz, and W. R. Boggess. 1975. Photosynthesis, transpiration and leaf conductance of *Populus deltoides* in relation to flooding and drought. *Photosynthetica* **9**:52–61.

Salisbury, F. B., and C. W. Ross. 1978. *Plant Physiology*. Wadsworth, Belmont, California. 422 pp.

Saugier, B. 1976. Sunflower. *Vegetation and the Atmosphere* (J. L. Monteith, ed.), Academic Press, London, pp. 87–120.

Schulze, E. D., O. L. Lange, and G. Lembke. 1972. A digital registration system for net photosynthesis and transpiration measurements in the field and an associated analysis of errors. *Oecologia (Berlin)* **10**:151–166.

Seo, T., and E. Ohtaki. 1976. Infrared device for measurement of carbon dioxide fluctuations under field conditions. III. Adaptation to infrared hygrometry. *Ber. Ohara Inst. Landw. Biol. Okayama Univ.* **16**:191–198.

Sestak, Z., J. Catsky, and P. G. Jarvis, eds. 1971. *Plant Photosynthetic Production—Manual of Methods*. Junk, The Hague. 819 pp.

Shibles, R. M. 1976. Terminology pertaining to photosynthesis. *Crop Sci.* **16**:437–439.

Smith, S. D., and E. P. Jones. 1979. Dry-air boundary conditions for correction of eddy flux measurements. *Boundary-Layer Meteorol.* **17**:375–379.

Sullivan, C. Y., N. D. Clegg, and J. M. Bennett. 1978. A portable technique for measuring photosynthesis. Am. Assoc. Advan. Sci. Abstr., Washington, D.C., February, 1978, p. 124.

Tanner, C. B. 1981. Transpiration efficiency of potato. *Agron. J.* **73**:59–64.

Thomas, M. D., and G. R. Hill. 1949. Photosynthesis under field conditions. *Photosynthesis in Plants* (J. Frank and W. E. Loomis, eds.), Plant Physiol. Monogr. Iowa State Univ. Press, Ames, pp. 19–52.

Thornthwaite, C. W., and B. Holzman. 1942. Measurement of evaporation from land and water surfaces. U.S. Dept. Agric. Technical Bull. No. 817. 75 pp.

Uchijima, Z. 1976. Maize and rice. *Vegetation and the Atmosphere* (J. L. Monteith, ed.), Academic Press, London, pp. 22–64.

Verma, S. B., and N. J. Rosenberg. 1973. Systematized monitoring of CO_2 concentration and gradients in an agricultural region and implications with respect to air pollution. WMO/WHO Conf. on Observations and Measurement of Atmos. Pollution, Helsinki, Finland.

Verma, S. B., and N. J. Rosenberg. 1975. Accuracy of lysimetric, energy balance and stability-corrected aerodynamic methods of estimating above-canopy flux of CO_2. *Agron. J.* **67**:699–704.

Verma, S. B., and N. J. Rosenberg. 1976. Carbon dioxide concentration and flux in a large agricultural region of the Great Plains of North America. *J. Geophys. Res.* **81**:399–405.

Verma, S. B., and N. J. Rosenberg. 1981. Further measurements of carbon dioxide concentration and flux in a large agricultural region of the Great Plains of North America. *J. Geophys. Res.* **86**:3258–3261.

Webb, E. K., G. I. Pearman, and R. Leuning. 1980. Correction of flux measurements for density effects due to heat and water vapour transfer. *Q. J. Roy. Meteorol. Soc.* **106**:85–100.

Webb, E. K. 1982. On the correction of flux measurements for effects of heat and water vapour transfer. *Boundary-Layer Meteorol.* **23**:251–254.

Wesely, M. L., and B. B. Hicks. 1975. Comments on 'Limitations of an eddy-correlation technique for the determination of the carbon dioxide and sensible heat fluxes.' *Boundary-Layer Meteorol.* **9**:363–367.

Wittwer, S. H. 1978. Carbon dioxide fertilization of crop plants. *Crop Physiology* (U. S. Gupta, ed.), Oxford and IBH, New Delhi, India, pp. 310–333.

Woodwell, G. M. and W. R. Dykeman. 1966. Respiration of a forest measured by carbon dioxide accumulation during temperature inversions. *Science* **154**:1031–1034.

CHAPTER 9 – WINDBREAKS AND SHELTER EFFECT

9.1 INTRODUCTION

We have seen that the environment in which plants grow is not always the ideal or optimum for productivity or for comfort. Agriculturalists or horticulturalists, probably from primitive times, have attempted to find ways to protect their plantings from the hazards of the natural environment. Soil manipulation (Chapter 6) offers one good way to alter microclimate. Irrigation is another fairly obvious method of modifying the environment. Protection from excessively strong winds offers yet another beneficial method.

The use of windbreaks probably goes far back in the history of pastoral and agricultural civilizations. Wind problems have been of major importance in determining the characteristic agriculture in many regions of the world. In the lower Rhone Valley of France, for example, almost all agricultural enterprises require some degree of protection against the force of the mistral winds. This has led to the culture of small fields protected by a dense network of windbreaks (Fig. 9.1). In the Great Plains of the United States the planting of trees for wind shelter has been encouraged by legislation since early settlement times (Read, 1964). Windbreaks are also used for other specialized purposes. Heat consumption in greenhouses is reduced with windbreaks. Albrektsson et al. (1978) consider this technique especially well suited to flat, windy terrains. Windbreaks also provide an environment for wildlife (e.g., McClure, 1981).

We observe that grazing animals seek shelter from strong winds. This is a response to physical discomfort caused either by the chilling in cold wind, by desiccation in hot winds, or simply by the mechanical pressure on the animal. Bates and Phillips (1980) conclude that animals sheltering in the lee of a solid wooden windbreak during winter in Oregon are affected mostly by a reduction in wind chill. Lynch and Marshall (1969) have found that sheltered pasture land in New South Wales, Australia, produces more forage and that, if animal stocking rates are properly managed, this augmentation can be used to increase body weight gain by sheep.

Plants, too, are subject to damage caused by excessive chilling, high temperature, desiccation, and direct mechanical injury. Grace (1977) has provided an excellent monograph on the direct influences of wind on plant growth. **Windbreaks** (any structure that reduces wind speed) and **shelterbelts**

Fig. 9.1 Map of the Montfavet region in Provence, France, showing the intricate patterns of closely spaced windbreaks used in the region. The empty area is one cleared for experimental purposes (after Seguin and Gignoux, 1974).

(rows of trees planted for wind protection) can, by reducing these stresses, be profoundly beneficial to the growth of plants in their lee.

These generally established benefits have led to a considerable and widespread use of windbreaks. The drought years of the 1930s in North America and the serious wind erosion problems that followed prompted very extensive plantings of shelterbelts, particularly in the Great Plains. The Federal Shelterbelt Project of the 1930s and early 1940s led to the planting of thousands of kilometers of 6- to 8-row shelterbelts in the Great Plains states (Fig. 9.2). These shelterbelts were composed of a number of different tree species. It was prophesized by some at the time that a beneficial climatic change might follow.

As these shelterbelts grew to maturity in an era of rising land values and more moderate weather conditions, criticism arose as to the effectiveness and economy of the plantings. It became evident, too, that the tree shelterbelts compete with adjacent crops for soil nutrients and water and that the belts may shade the nearby crops sufficiently to reduce their production. Considerable follow-up research in the Great Plains (e.g., Read, 1958; Stoeckeler, 1962) has shown that, despite the reduction of yield near the windbreaks, the net yield per unit land area usually exceeds that in unsheltered adjacent fields. When the value of wood products from the shelterbelt is considered as well, the total economic return from the sheltered fields can be even greater.

INTRODUCTION

Despite the generally acknowledged positive benefits of windbreaks many have been removed in recent years. A report by the Comptroller General of the United States (GAO, 1975) indicates that increasing land values and incompatability of windbreaks with large scale farming operations is primarily responsible for this trend. Renovation of existing windbreaks and their protection is recommended in this report to the Congress of the United States, lest serious wind erosion occur. This danger is especially great during drought periods when, the report holds, other practices are less effective and may lead to a degradation of currently sheltered farmlands.

There are a number of good research reviews on the influence of windbreaks and shelterbelts. Jensen (1954) in Denmark and Caborn (1957) in Scotland prefaced books on their personal research with extensive literature reviews. United Nations agencies have sponsored more extensive reviews of windbreak influence (van der Linde, 1962, for FAO; van Eimern et al., 1964, for WMO). Guyot (1963), Rosenberg (1967, 1975, 1979), Marshall (1967), and Sturrock (1975) have also contributed additional reviews with emphasis on the mechanisms of windbreak effects on the microclimate and on plant growth.

There are very few published reports showing yield reduction caused by

Fig. 9.2 Pattern of field and farmstead windbreaks near Blair, Oklahoma (courtesy of Dr. David van Haverbeke, Rocky Mountain Forest and Range Expt. Station, Windbreak Research Laboratory, Lincoln, NE. Photo due to USDA–SCS, 1949).

windbreaks or shelterbelts, except in the zone immediately adjacent to the barrier. The van Eimern et al. (1964) review gives a thorough recounting of yield influences reported from all over the world. Grace (1977) augmented van Eimern et al. with a careful tabulation of yield results. With few exceptions these reports show shelter to be beneficial.

9.2 INTERRELATIONS OF WIND SHELTER, MOISTURE CONSERVATION, PLANT GROWTH, AND YIELD

Many think that the major influence of windbreaks on plant growth, particularly under dryland conditions, is due to the redistribution and conservation of soil water. In higher latitudes the windbreak can, if properly designed, aid in uniformly distributing snow and thus will improve the supply of soil moisture to crops in spring. By reducing the wind speed the direct evaporation of moisture from the soil is also reduced. Reference to the Penman and other Daltonian-type equations given in Chapter 7 shows that evaporation is directly related to wind speed so that significant water-saving effects should result from shelter.

Atmometers, evaporation pans, and wetted soils in isolated containers have been used to study the influence of wind shelter on evaporation from the soil. These devices measure the potential evaporation, which occurs with unrestricted availability of water at the evaporating surface. The results have been predictable: less wind, less evaporation. A useful rule of thumb relating potential evaporation to wind speed is that of Naegeli (van Eimern et al., 1964), who found that potential evaporation from flat moist containers is proportional to the square root of the wind speed.

Predicting the exact water conserving influence of windbreaks on bare soil under nonpotential conditions is more difficult. We have seen in Chapter 7 that the rate of drying in a wet soil will decrease after a few days of evaporation at the potential rate. After a period of time that depends on the prevailing evaporative demand, the soil texture, and the hydraulic conductivity, evaporation will essentially cease. In shelter the same soil, initially wetted to the same degree, will be subjected to less than potential evaporation rates. After some period of time evaporation will cease in the sheltered soil, as well, and both soils will be equally dry.

Nonetheless, the lower evaporation rate in sheltered soil may provide an important advantage in maintaining better conditions for seed germination (Rosenberg, 1966b). Aase and Siddoway (1976) found that the top 40 mm of soil in a field protected by perennial wheatgrass barriers remained wetter than a control area for about 3 days after irrigation. In fact, barriers of this kind nearly doubled soil moisture storage over winter according to Black and Siddoway (1976).

The effect of shelter on actual evapotranspiration is even more difficult to predict. For example, seeds that germinate rapidly because of a beneficial shelter effect grow into larger plants and ramify roots more quickly into the soil. The greater crop cover decreases the relative importance in shelter of direct evaporation from the soil. Assuming that transpiration is a function of leaf area alone, soil water in the sheltered area could be depleted more quickly since evaporation rate diminishes rapidly after a few days. This can lead to a more rapid development of soil moisture stress in shelter. Thus it is possible for the development of the sheltered plants to be checked while the unsheltered plants are less inhibited. Jensen (1954) in Denmark and Mastinskaja (cited by van Eimern et al., 1964) in the East Volga region of the USSR have shown that, in this way, the initial moisture advantage of sheltered plants can, eventually, be dissipated. Over a season during which some rain falls the more luxuriant crops grown in shelter may actually use more water than do plants grown in the open.

The relative proportion of water transpired to that evaporated may also be increased (Budyko, cited by van Eimern et al., 1964). Whether an increase in the ratio of transpiration to evaporation results in greater dry matter production or harvestable yield per unit of water used has been much discussed. An indirect answer to this question is to be found in the data of Zukovsky (cited by van Eimern et al., 1964). In one of the drier regions of the USSR it was observed that during wet years significant increases in the yield of grains, alfalfa, and pasture were obtained in shelter. During dry years, however, the percentage increases in yield were very great. Utilization of the limited water available for transpiration during these dry years is evidently more efficient in shelter.

In Saskatchewan, Staple and Lehane (1955) found the average production of wheat sheltered by a series of *Carragana* windbreaks to increase by 1.35 t ha^{-1} in the zone 15–25 h (where h is the height of the windbreak) downwind. In the zone 0–15 h during years of heavy snow yields were increased by 1.82 t ha^{-1}. In years of medium snowfall the yield increases were 1.61 and 1.48 t ha^{-1} in the 0–15 and 15–25 h zones, respectively. Pelton and Earl (1962), also in Saskatchewan, studied wheat yield and water use in a field sheltered by 2.4 m high snow fencing. They demonstrated that shelter results in increases of as much as 1 kg of wheat per mm of water consumed. The best results were obtained at 3 h to the east (predominantly leeward) of the barrier. Kaminski (1967) reported studies in Poland showing the yields of rye, barley, potato, and small beans are improved by shelter in the zone between 1 and 12 h of an old dense 36-m wide tree windbreak.

In a series of experiments done in Nebraska the yields of dry beans (Rosenberg, 1966a), snap beans (Rosenberg et al., 1967), and sugar beets (Rosenberg, 1966b; Brown and Rosenberg, 1972) were shown to have been either unaffected or increased in shelter with, except in one case, little or no increase in water use. Thus, crop production per unit of water consumption (the water use efficiency) was either improved or unaffected in shelter.

Detailed studies of wind shelter effects on growth and yield of soybeans have been reported by Radke and Burrows (1970) and Radke (1976). In western Minnesota, soybeans sheltered by double rows of corn were slightly, but significantly, taller than unsheltered plants. Yields of grain ranged from 28% more to 2% less in shelter, depending on the year. Overall the sheltered crops yielded 11.5% more grain. On the other hand, Frank et al. (1975, 1977a,b) found that soybeans in western North Dakota benefited from the shelter of a slat fence only with irrigation. Otherwise plants were sometimes found to suffer *greater* internal water stress in shelter. This may have been due to a greater plant size in shelter which occurs even when the soybean is grown without irrigation.

In the Nebraska studies, cited above, attempts were made to eliminate differences in soil moisture between sheltered and exposed sites by means of frequent irrigation. We hoped that, by so doing, the influence of shelter-induced microclimatic changes would be causal in determining plant growth. We learned, however, that even under ample irrigation, it is still the shelter influence on soil and plantwater relations that is the critical factor. Thus it should be apparent that an understanding and ability to predict the influence of windbreaks on the growth of plants that are sheltered is not an easy matter. In the section that follows we will attempt to systematically analyze the physical and physiological factors that lead to a shelter effect on plant growth. We must remember, as well, that shelter is not uniform in the field and that geometry, spacing, age, height, and density of the windbreak also affect its functioning.

9.3 WIND SPEED AND TURBULENCE IN SHELTER

The purpose of a windbreak is to reduce the force of the wind in the sheltered zone. To design a windbreak which will function efficiently in the field is a very difficult problem. Patterns of flow around barriers are very complex and difficult to define with precision. Plate (1971), for example, distinguished as many as seven separate flow zones with different aerodynamic behavior upwind and downwind of a wedge-shaped barrier. Windbreaks vary in effectiveness, depending on their height, porosity, and length. The higher the windbreak, the greater will be the distance of its downwind, as well as upwind, influence. As mentioned above, the length of the sheltered zone is normally described in terms of the parameter h, the height of the barrier.

Although estimates in the literature vary, the differing effectiveness of dense and porous wind barriers can be described in general terms from work by Caborn (1957) and Naegeli (cited by van Eimern et al., 1964). As is shown schematically in Fig. 9.3 a dense barrier may protect an area about 10–15 h downwind. By increasing the porosity to near 50%, the downwind influence can be increased to 20–25 h (Fig. 9.4). This increasing porosity permits penetration of the wind and prevents the turbulent return of air that has overtopped the barriers to the ground close by. Sturrock (1969) has observed an interesting relationship between porosity and distance of protective ef-

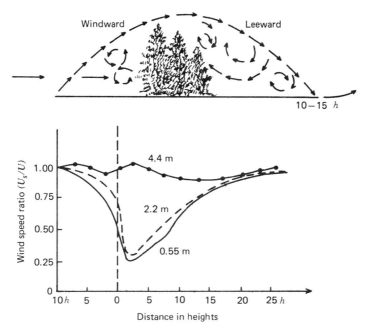

Fig. 9.3 Influence of a dense windbreak on the ratio of wind speed in shelter (U_s) and in the open (U) (from van Eimern et al., 1964).

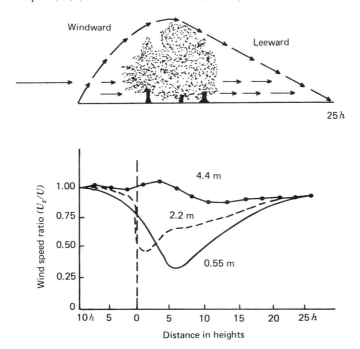

Fig. 9.4 Influence of a permeable windbreak on the ratio of wind speed in shelter (U_s) and in the open (U) (from van Eimern et al., 1964).

338 WINDBREAKS AND SHELTER EFFECT

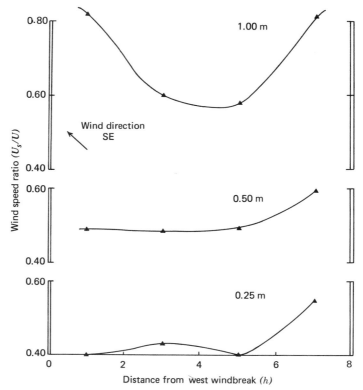

Fig. 9.5 Ratio of wind speeds at four locations over sheltered sugar beets to wind speeds at the same heights in the open. Scottsbluff, Nebraska (after Brown and Rosenberg, 1971).

fect. He measured wind profiles to the lee and windward of 10 types of windbreaks in New Zealand. The distance for initial recovery of wind speed was closely related to the minimum wind speed reached. In other words the lower the wind speed, the shorter the distance required for full speed recovery.

The longer the windbreak, the more constant is its influence. If a barrier is too short or if it has large gaps in it, jetting effects may actually increase, rather than reduce the wind speed and, consequently, damage to plants will be greater near the gaps. The effectiveness of the windbreak is also influenced by atmospheric thermal stability: the more unstable the air, the greater the protected distance downwind (Bean et al., 1975). Crop roughness, according to Marshall (1967), can also influence the extent of the downwind protection as can angle of incidence (Seginer, 1975).

It is generally agreed that for the best wind reduction and greatest downwind influence the windbreak should be most porous near the ground, where the wind speed is lowest. Ideally, the density of the barrier should increase logarithmically with height in accordance with the wind speed profile. In wind tunnel and field tests Moysey and McPherson (1966) found slat fence windbreaks with a porosity of 15–30% in the lower half to give better results than solid windbreaks. The shape of barrier openings was found unimportant.

Wind reduction is a function of location in the shelter as well as of the height above the plants. Figure 9.5 (from Brown and Rosenberg, 1971) shows the ratio of the mean wind speed U_s across the sheltered portion of a sugar beet field to the wind speed U in the open. The field was protected by 2 m tall corn plants in double rows spaced every 15 m. Windbreaks were planted on a north–south line. Near the windbreaks, wind speeds 1 m above the plants were barely affected, but in the center of the sheltered area wind speeds were reduced by more than 40%.

Windbreaks need not be fully developed or have achieved complete crown closure in order to generate a significant shelter effect. Miller et al. (1975) reported on a rapidly growing and highly permeable shelterbelt in Nebraska that was, after only 4 years, one-third as effective in slowing the wind as other fully grown windbreaks. That windbreak was composed of 2 rows: alternating cottonwood and eastern red cedar in one; alternating pairs of eastern red cedar and Scotch pine in the other. Wind speed reductions varied with thermal stability from about 25 to 40% at $2h$ to 15–25% at $8h$.

The reduction of wind speed and the reduction of turbulence by a windbreak are not uniquely related. Brown and Rosenberg (1971) describe patterns of wind speed and the degree of turbulent mixing that occur in shelter. Over a day the ratio of the wind speed in the corn-sheltered sugar beets to that in the open (U_s/U) ranged between 0.8 and 0.9. The ratio of exchange coefficients (K_s/K) determined by solution of the Bowen ratio energy balance (7.46) varied widely from a situation of a slightly greater turbulent exchange in shelter during the early morning to a 50% reduction at midafternoon.

To accomplish these measurements in a shelter where "fetch" was inadequate, instruments were placed very close to the top of the crop canopy. Psychrometers to measure ΔT and Δe were stacked one above the other in such a way that the Bowen ratio could be determined for a number of levels above the crop. Data showed that these measurements were representative of the newly developed local boundary layer in the shelter. The reduction in exchange coefficient provided a better explanation of the unique microclimate and plant growth in shelter than did the reduction in wind speed alone.

Miller et al. (1973) used lysimeters to provide an indirect measurement of the influence of shelter on the effectiveness of turbulent transport. In a field at Mead, Nebraska, a slat fence windbreak was moved frequently to shelter first one area including a precision weighing lysimeter, then another. The turbulent exchange coefficients were calculated from

$$K_w = \frac{LE_{(lys)}}{\rho_a[(M_w/M_a)/P]L(\partial e/\partial z)} \qquad (8.13)$$

where LE was measured lysimetrically and the vapor pressure gradient be-

340 WINDBREAKS AND SHELTER EFFECT

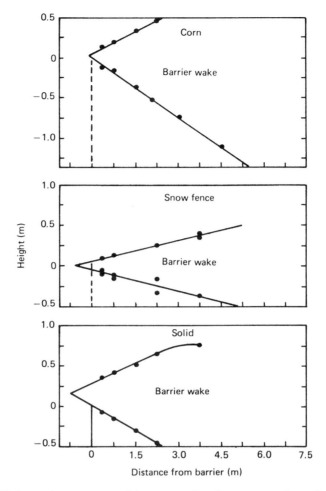

Fig. 9.6 Wake regions generated by a corn barrier, a snow fence barrier, and a solid board fence. The vertical lines represent the barriers and the *x*-axis represents the top of the soybean canopy (e.g., the solid barrier protrudes 0.5 m above the soybean crop). Between the wake limbs are regions of increased turbulent mixing (from Radke, 1976).

tween 1.25 and 2.00 m was measured psychrometrically. The exchange coefficient in shelter was reduced from 25 to 80% of that in the open field depending on weather and atmospheric stability conditions.

Direct measurements of the influence of shelter on wind structure and turbulence have become available through the use of rapid response instruments for measurement of multidimensional wind speeds (e.g., sonic anemometers, hot-wire and hot-film anemometers). Radke and Hagstrom (1974) used hot-film anemometers in the lee and upwind of four types of shelters in a soybean field. These were double rows of corn, double rows of sunflowers, a snow fence, and a solid board fence. The barriers were placed every 12 rows in the field (about 9.1 m apart). The wind structure and turbulence characteristics behind the solid fence differed from those for the porous barriers in that the wind went up and over rather than through the barrier. The porous barriers broke up the larger eddies into smaller ones, decreased wind velocity, and reduced the amount of turbulent energy at lower frequencies. The crop and snow fence barriers were similar with respect to wind speed reductions but the turbulent energy, frequency, and scale of turbulence were different in each case. Turbulent intensities were much higher over the first six rows in the lee of the solid windbreak because of the greatly reduced wind speeds there. Wind speed and turbulent energy were high midway between the solid barriers as a result of wind flowing over the top of the barriers.

Radke (1976) characterized the average wake regions of turbulent mixing generated by the top of the wind barriers described above (Fig. 9.6). The greater turbulent mixing in the wakes causes a higher potential evapotranspiration. For the solid barrier the wake intercepts the soybean canopy at a distance of only 2.50 m to the leeward.

Maki and Allen (1978) investigated the turbulent characteristics of an 11.1 m single line pine tree windbreak in Florida using a three-dimensional ultrasonic anemometer–thermometer. They made measurements at various elevations above ground at stations from 100 m windward to 100 m leeward of the windbreak. Downdrafts were found to prevail in the vertical profile, and the ratio of downdrafts to horizontal mean wind increased with height. Turbulent intensities were generally observed to decrease with height.

The study of single windbreaks, although instructive, is not always sufficient for understanding the effects of multiple windbreaks. Seguin and Gignoux (1974) present interesting information on the wind patterns that develop in and over a network of windbreaks near Avignon in southern France.

9.4 MICROCLIMATE IN SHELTER

The changes in wind speed and turbulence that occur as a result of windbreaks must affect the microclimate of the sheltered zone.

9.4.1 Radiation Balance

Solar and net radiation may be significantly reduced in the areas shaded by windbreaks. This effect has not been found to be of major importance in north–south oriented windbreak systems, since only small areas are shaded during the course of the day, especially during the growing season when the sun is high. On a full day basis, the difference in radiation balance between areas near and areas remote from the barrier may be entirely negligible. This follows, since an area shaded in the morning by a windbreak to the east will receive some additional radiation by reflection from the windbreak in the afternoon. East–west oriented windbreaks, on the other hand, may have a greater effect. Areas to the north, particularly during seasons when the sun is low, will be shaded for long periods. Areas to the south will be subject to reflection off the windbreak throughout the day.

Shading depends, of course, on the height of the barrier, on latitude, season, and time of day. Beyond 1 or 2 h the energetic effects of shading are probably unimportant although direct biological effects on the ecology of that zone cannot be discounted. Marshall (1967) suggests that whereas severe shading may suppress photosynthesis and dry matter production this effect can be offset by reduced evapotranspiration in the shaded zone.

In a very interesting work Guyot and Seguin (1978) report studies in Brittany on the changes that occur when networks of hedges ("bocage") are removed from the landscape in attempts to modernize the agriculture. They observed no significant changes in the net radiation as compared to an unaltered area since a slight increase in infrared loss is compensated by a similarly slight increase in absorption of solar radiation due to a reduction in the overall albedo of the region.

9.4.2 Air Temperature and Humidity

It is usually observed in clear weather that daytime air temperatures are greater in shelter than in open fields. This is due, apparently, to the reduction of turbulent mixing and the consequent reduction in the removal of sensible heat generated at the plant or soil surface. If evaporation is also suppressed in shelter, additional energy is available for sensible heat generation as well. When turbulent mixing is reduced, the aerial resistance r_a increases and the temperature gradients are intensified.

Temperature inversions normally develop at night in both sheltered and unsheltered areas; then the plant and soil surfaces become the sink for, rather than the source of, sensible heat. Windiness mixes the nocturnal inversion layer. The reduction of windiness and effectiveness of turbulent mixing in shelter means that temperature inversions will normally be more intense there. Unless total calm prevails, the air will generally be colder at night in shelter than in open fields. Kaminski (1968) found, however, that the incidence of frosts in Poland was reduced near a windbreak on both the windward and leeward sides. Between 4 and 16 h the incidence of frost was increased. Kaminski's paper gives no explanation for this phenomenon. The reduction of cooling near the windbreak could have been due to radiative exchange with the trees. Possibly the increased vapor content of the air in that zone may have reduced the rate of radiational cooling, as well.

Humidity and vapor pressure gradients are also increased in shelter. Tran-

Table 9.1 Mean daytime (0600–1800 hours) air temperature and water vapor pressure at 0.5 m in open plots and in plots of sugar beets sheltered by corn at Scottsbluff, Nebraska, 1966[a]

Date	Air temperature, T_a (°C)			Air vapor pressure, e_a (kPa)			Relative humidity (%)		
	Open	Shelter	Difference	Open	Shelter	Difference	Open	Shelter	Difference
Aug. 10	22.2	22.5	+0.3	2.06	2.50	+0.44	77	92	+15
11	22.4	23.6	+1.2	2.13	2.38	+0.25	79	82	+3
14	19.5	19.1	−0.4	1.23	1.32	+0.09	54	60	+6
15	21.5	21.9	+0.4	1.67	1.99	+0.32	65	76	+11
16	25.2	26.8	+1.6	2.43	2.62	+0.19	76	74	−2
17	26.8	30.5	+3.7	1.82	2.49	+0.67	52	57	+5
18	24.6	28.1	+3.5	2.03	2.63	+0.60	66	69	+3
25	23.1	24.6	+1.5	2.06	2.30	+0.24	73	74	+1
Sept. 1	20.2	22.6	+2.4	2.04	2.42	+0.38	86	88	+2
2	23.4	25.3	+1.9	2.25	2.80	+0.60	78	87	+9
3	22.7	25.7	+3.0	1.56	2.16	+0.60	57	65	+8
4	24.1	26.9	+2.8	2.12	2.62	+0.50	71	74	+3
5	18.4	18.5	+0.1	1.25	1.36	+0.11	59	64	+5

[a] After Brown and Rosenberg (1972).

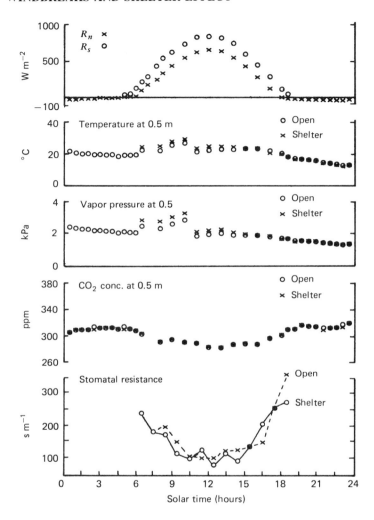

Fig. 9.7 Microclimate and stomatal resistance in exposed and wind-sheltered portions of a sugar beet field on a clear day in mid-August at Scottsbluff, Nebraska (after Brown and Rosenberg, 1970).

spired and evaporated water vapor is not as readily transported away from the source, the evaporating surface, as in an unsheltered field. Vapor pressure remains higher in shelter throughout the night as well, since the surface usually remains the source of vapor, except during periods of dew deposition. Such intensified temperature and vapor pressure gradients in shelter have been observed under a wide range of climatic conditions with many types of vegetative and constructed barriers used to shelter many different types of crops. Studies by Bates (1911), Woodruff et al. (1959), Aslyng (1958), Guyot (1963), Skidmore et al. (1972), and Brown and Rosenberg (1972) sup-

MICROCLIMATE IN SHELTER 345

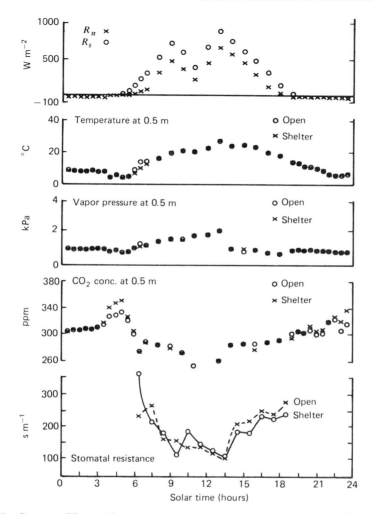

Fig. 9.8 Same as Fig. 9.7 for a cloudy day in mid-August at Scottsbluff, Nebraska.

port this assertion. Much additional information is available in van Eimern et al. (1964), Marshall (1967), and Grace (1977).

Despite the increased temperature the relative humidity is generally greater by day in shelter. The difference in relative humidity between open and shelter is greater still at night because of the lower air temperatures in shelter.

The magnitude of measured differences in daytime mean air temperature, vapor pressure, and relative humidity over a long period in an experiment where corn windbreaks sheltered sugar beets is shown in Table 9.1. Typical examples of diurnal microclimatic patterns in shelter and open parts of the field in that same study are shown in Figs. 9.7 and 9.8. Differences in tem-

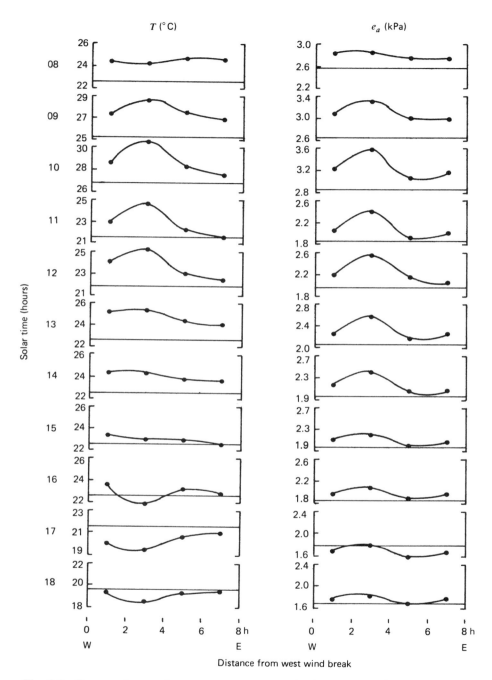

Fig. 9.9 Temperature and water vapor pressure in the open (straight line) and at four locations between parallel windbreaks on a clear day in mid-August at Scottsbluff, Nebraska. The wind direction was consistently SE (from Brown and Rosenberg, 1972).

perature and humidity due to shelter are seen at various times on these 2 days, which differed in radiation and windiness conditions.

It is also important to recognize that the microclimatic differences that develop in shelter vary with distance from the windbreak, with weather conditions and with time of day. This is illustrated in Fig. 9.9 for temperature and vapor pressure measured at four locations equidistantly spaced in a sugar beet field between 2-row corn windbreaks. The shelter effect is generally greatest at midday although differences within shelter may, at any time, be greater than mean difference between sheltered and open fields.

In much of the literature the influence of windbreaks on temperature and humidity has been underestimated. Rosenberg (1966c) points out that placing instruments in standard meteorological screens leads to systematic errors which indicate a smaller than real shelter effect. On the other hand, the nocturnal temperature depression in shelter, especially, is exaggerated where screens are used.

9.4.3 Carbon Dioxide Concentration

There has been very little study of the influence of windbreaks on carbon dioxide concentration. Lemon (1970) suggested that, at low wind speeds, the photosynthetic rate in the field may be decreased because of a shortage of carbon dioxide in the air surrounding the leaves. The photosynthetic rate of individual leaves (Gaastra, 1959; Bierhuizen and Slatyer, 1964), as well as of plant communities (Thomas and Hill, 1949; Rosenberg, 1981), depends directly on carbon dioxide concentration, which generally ranges between 280 and 500 ppm. If a windbreak reduces the carbon dioxide supply because of reduced air movement, the rate of photosynthesis in a windbreak-protected crop could be adversely affected.

Ruesch (1955) measured the carbon dioxide concentration in open and sheltered alfalfa, using chemical absorption techniques. He found a small decrease in the sheltered crop during the day. This may have been due to greater CO_2 uptake by the more luxuriant plants growing there.

Data from the studies of Brown and Rosenberg (1972), described above, are still probably the most complete with regard to the influence of shelter on carbon dioxide concentration. Daytime and nighttime differences in carbon dioxide concentration were measured during a 2-month period of the growing season. Air samples were collected in the open and in the shelter at 3 h from the west windbreak. The daytime (0600–1800 hours) concentration differences (shelter minus open) ranged from +9 to −12 ppm. The average daytime concentration between July 15 and September 11 was 1 ppm less in the sheltered plot. Figures 9.7 and 9.8 show that daytime differences in CO_2 concentration are very slight.

The differences in the mean nocturnal (2000–0400 hours) carbon dioxide concentration between sheltered and open sites ranged from +16 to −2

348 WINDBREAKS AND SHELTER EFFECT

Fig. 9.10 A view of the windbreak in place. The lysimeter is near the large white mast (after Miller et al., 1973).

ppm. The carbon dioxide concentration was, on the average, 3.5 ppm greater in shelter during the night. The nocturnal carbon dioxide concentration is especially sensitive to wind speed. Nocturnal temperature inversions lead to strong atmospheric stability. Under these conditions, the carbon dioxide liberated by respiring plants and soil tends to accumulate over and within the crop canopy. Since shelter reduces turbulent mixing the effect is especially great when winds are very light. Thus, major increases in carbon dioxide concentration occur on calm nights (Figs. 9.8 and 9.9).

9.4.4 Gradients and Profiles

The changes in microclimatic conditions occur at more than one level above and within the sheltered crop. Since turbulent mixing is affected it is reasonable to expect that gradients of temperature, humidity, carbon dioxide, and so on, will also be changed. For example, Skidmore et al. (1972) found that largest daytime and smallest nighttime vertical temperature gradients occurred where wind speeds were least in the lee of solid, 40% and 60% porous barriers.

Miller et al. (1973) observed profiles in a field in which a 50% porous slat fence windbreak was moved frequently between two precision weighing lysimeters. This was done to avoid development of vegetative differences between locations (Fig. 9.10). Turbulent mixing was reduced in shelter, causing intensified vapor pressure and air temperature gradients (Table 9.2).

Table 9.2 Mean morning (0800–1145 hours) and afternoon (1200–1545 hours) energy balance, temperature, and vapor pressure gradients (2.0–1.0 m) over soybeans in the open and in shelter at Mead, Nebraska[a]

Date	Time	R_n (W m^{-2})	S (W m^{-2})	LE (W m^{-2}) Open	LE (W m^{-2}) Shelter	H (W m^{-2}) Open	H (W m^{-2}) Shelter	ΔT (°C m^{-1}) Open	ΔT (°C m^{-1}) Shelter	Δe (kPa m^{-1}) Open	Δe (kPa m^{-1}) Shelter
July 14	AM	544	−7	−726	−565	188	28	0.2	−0.3	−0.15	−0.17
	PM	502	0	−802	−628	300	265	0.3	−0.3	−0.11	−0.15
July 15	AM	481	0	−649	−328	167	−154	0.2	−0.2	−0.10	−0.12
	PM	489	−7	−726	−544	244	63	0.2	0.0	−0.14	−0.19
July 17	AM	230	0	−188	−174	−41	−56	0.0	−0.1	−0.05	−0.17
	PM	258	0	−286	−279	28	21	0.1	0.0	−0.08	−0.16
July 18	AM	321	0	−335	−314	14	−7	−0.4	−0.5	−0.13	−0.18
	PM	425	−7	−516	−523	98	105	−0.1	−0.4	−0.22	−0.26
July 22	AM	537	−7	−495	−440	−35	−90	—	−0.6	−0.09	−0.14
	PM	509	−7	−663	−530	160	28	0.1	−0.2	−0.26	−0.19
July 24	AM	481	−7	−628	−502	153	28	−0.4	−0.7	−0.15	−0.29
	PM	537	−7	−670	−551	140	21	0.0	−0.2	−0.29	−0.36
Mean[b]				−558[b]	−447	119[b]	21				

[a] After Miller et al. (1973).
[b] Mean difference significant at the 5% level of probability.

350 WINDBREAKS AND SHELTER EFFECT

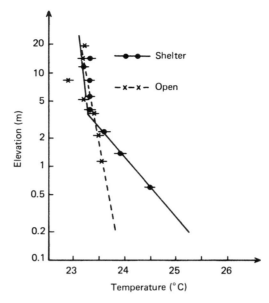

Fig. 9.11 Simultaneous daytime temperature profiles in a bocage (shelter) and open zone in Brittany (after Guyot and Seguin, 1978).

Guyot and Seguin (1978) also show distinctly steeper air temperature gradients in a hedge-protected area in Brittany as compared with a zone from which the hedges had been cleared (Fig. 9.11).

9.5 PLANT PHYSIOLOGICAL RESPONSES TO SHELTER

The evidence is clear that microclimate differences develop in shelter and that the shelter microclimate influences crop growth and production. How do the shelter microclimate and plant response relate? This question is very difficult to answer, since only at the very beginning of any shelter experiment can it be said that soil moisture and microclimatic differences are causal. Soon after the beginning of an experiment the shelter effect may lead to radical differences in plant growth and morphology. Differences in plant form also interact with and affect the microclimate. Often plants in shelter show more rapid vegetative growth and increase in size. This is manifested by greater total LAI and plant height. Shah (1962) and Rosenberg et al. (1967), for example, have demonstrated large differences in the growth of corn and bean plants caused by shelter.

Plant growth and performance in shelter may also be favored because the use of windbreaks reduces the incidence of mechanical injuries such as are caused by "sandblasting" (Skidmore, 1966; Armbrust et al., 1974). Mechanical injuries may cause loss of production by defoliation, that is, loss of viable tissue, or by imposing short-term, high-intensity moisture stress on the injured plant. Both factors may combine to reduce productivity.

Recent studies suggest, however, that the major influence of shelter on plant behavior is due to a greater turgidity and a lower stomatal resistance in the sheltered plants, especially during periods of water stress or strong evaporative demand. Wind tunnel studies (e.g., Shah, 1962) show that transpiration increases with increasing wind speed. Kalma and Kuiper (1966), on the other hand, found, in a greenhouse study, that water use by beans and overall plant growth decreased when winds increased above 1 m s^{-1}. Such conflicting results are not unexpected.

It is rarely possible to simulate in wind tunnels and greenhouses the radiation, temperature, humidity, and turbulence regimes that lead to such levels of moisture stress as occur in the field, especially in semiarid and subhumid climates. Therefore, field studies are probably more instructive.

Rosenberg (1966a) showed, for example, that the minimum relative turgidity of dry bean leaves at Scottsbluff, Nebraska, was 90% in shelter and 85% in the open field. During most of the daylight hours, the relative turgidity was at least 4–5% lower in unprotected plants. Leaf impressions made simultaneously indicated that the average stomatal aperture was also greater in leaves of sheltered than exposed plants.

Rosenberg et al. (1967) provide additional evidence of shelter effects on water relations in two cultivars of snap beans: Bush Blue Lake and Tendercrop. Sheltered plants of both cultivars maintained wider stomatal aperture on each of 4 days when observations were made.

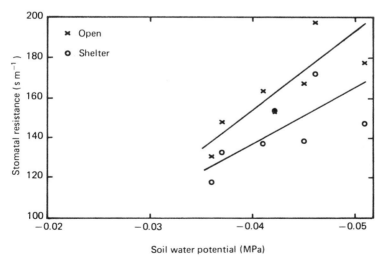

Fig. 9.12 Mean daily (0830–1530 hours) stomatal resistance in sheltered and exposed sugar beets as a function of soil water potential at Scottsbluff, Nebraska (after Brown and Rosenberg, 1970).

More detailed observations of stomatal behavior were made in the study where sugar beets were sheltered by corn windbreaks (Brown and Rosenberg, 1970). The data on stomatal resistance in Figs. 9.7 and 9.8 are typical of the results obtained throughout the sugar beet growing season at Scottsbluff, Nebraska. Generally, the stomatal resistance was lower in shelter, although there was considerable variability in the hourly means. Stomatal resistance oscillated considerably during the day, both in shelter and in the open. This may indicate that the beet plants underwent temporary stress periods when roots were unable to supply water as rapidly as it was evaporated into the air from the leaves. Such an "out-of-phase" effect could cause stomates to close partially until internal plant water deficits were relieved.

Figure 9.12 illustrates the dependency of stomatal resistance in the leaves of sheltered and exposed sugar beets on decreasing soil water potential (increasing dryness). The mean daily stomatal resistance was usually greater in the unsheltered plants. The difference in stomatal resistance increased with decreasing soil moisture content. These data permit us to speculate that the influence and benefits of shelter will probably be greatest in the arid regions. Marshall's review (1967) indicates that, indeed, yield increases in the presence of shelter are greater in continental than in oceanic climates. He attributes this to the effects of drier summers in the former, since soil moisture is often the limiting factor in crop production.

9.6 POTENTIAL AND ACTUAL WATER USE

Experimental reports give a wide range of answers to the question of whether soil water use is actually affected in shelter. We have shown that evaporation from evaporimeters (pans, atmometers) will be reduced in shelter in proportion to the reduction in the wind speed. This effect can be easily explained as being due to an increased aerial resistance to the transport of vapor in shelter. The complications that may occur as differential growth rates lead to differences in plant size between shelter and exposed sites are discussed in Section 9.5.

Brown and Rosenberg (1971) found, in western Nebraska, that daily water use rates of sugar beets did not differ greatly between sheltered and exposed sites. Figure 9.13 shows water use and radiation balance on a typical growing season day. Because of lower stomatal resistance, water use was slightly greater in shelter than in the exposed area. With the onset of sensible heat advection by midafternoon, however, water use by the sugar beet crop in the open was increased sharply. Water use in shelter was also increased during the warm midafternoon hours, but remained considerably lower than in the open.

Miller et al. (1973) measured the water use by irrigated soybeans with precision weighing lysimeters in an experiment cited above (see Fig. 9.10).

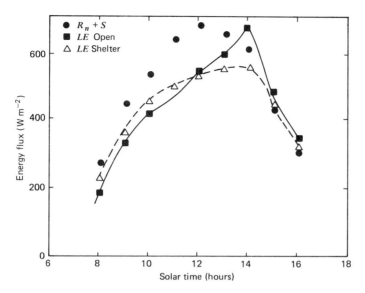

Fig. 9.13 Diurnal patterns of latent heat flux from sugar beets in sheltered and open plots and the synchronous energy balance on a clear day in mid-August at Scottsbluff, Nebraska (after Brown and Rosenberg, 1971).

The energy balance of sheltered and exposed soybeans on days of strong and slight advection are shown in Figs. 9.14 and 9.15. The water-saving effect was greatest on days when sensible heat advection was strong. On days when sensible heat advection was slight, the water savings were small. Shelter reduced soybean water use by about 20% over a 10-day period in July.

The net radiation was 15.40 MJ m^{-2} during the daylight hours of July 15 (Fig. 9.14). Evapotranspiration consumed 24.91 MJ m^{-2} day^{-1} (10.2 mm H$_2$O) in the exposed area and 18.96 MJ m^{-2} day^{-1} (7.8 mm H$_2$O) in the sheltered area. The difference in energy consumed by evapotranspiration LE and that supplied by the net radiation and soil heat flux ($R_n + S$) is the advective contribution of sensible heat H. That contribution was markedly reduced by the shelter throughout the entire day as a result of the diminished passage of wind.

July 22 (Fig. 9.15) was a moderately windy day. The evapotranspiration for exposed and sheltered soybeans was 19.38 MJ m^{-2} day^{-1} (8.0 mm H$_2$O) and 17.66 MJ m^{-2} day^{-1} (7.3 mm H$_2$O) on this day. ($R_n + S$) was 16.95 MJ m^{-2} for the daylight hours. Sensible heat was generated at the surface of the crop in both sheltered and open sites during the morning, and the evapotranspiration rates were similar. In the early afternoon, evapotranspiration rates were greater in the exposed area due to the greater consumption of sensible heat.

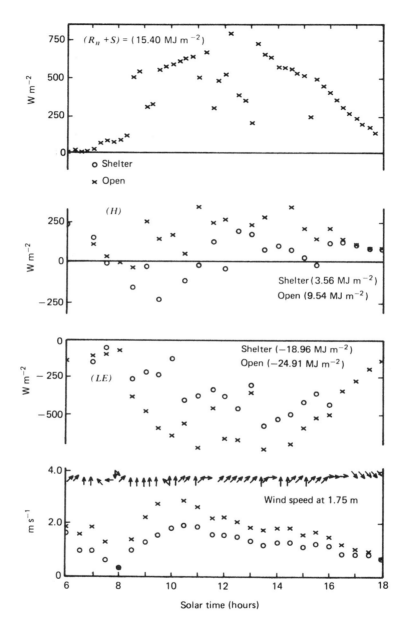

Fig. 9.14 Components of the energy balance and daytime (12 h) integrations in sheltered and open soybeans on a cloudy day with strong sensible heat advection. Mid-July at Mead, Nebraska (after Miller et al., 1973).

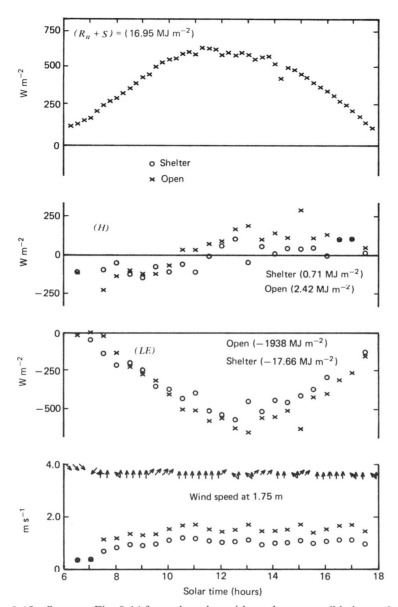

Fig. 9.15 Same as Fig. 9.14 for a clear day with moderate sensible heat advection.

356 WINDBREAKS AND SHELTER EFFECT

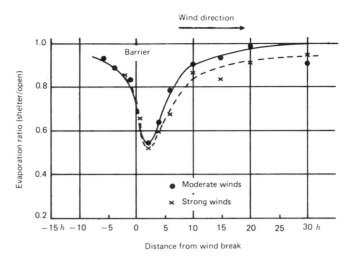

Fig. 9.16 Reduction in evaporation by a 1.75 m high barrier of 50% porosity at two ranges of wind speed. Evaporation was measured with Piche evaporimeters (after Bouchet et al., from Augmentation de l'efficience de l'eau et amelioration des rendements par reduction de l'evapotranspiration potentielle au moyen de brise-vent, © UNESCO 1966. Reproduced by permission of UNESCO).

Distance from the windbreak is also a factor determining actual evaporation and evapotranspiration. Skidmore et al. (1969) measured the influence of windiness on evaporation downwind of a constructed windbreak and described the dependency of the potential evapotranspiration on wind speed as calculated with several formulas. Marshall (1967) reported that evaporation suppression occurs to a distance of about 16 times the height of the windbreak (16 h) downwind. Bouchet et al. (1966) made field measurements of evaporation at distances ranging from 15 h upwind to 30 h downwind of a 50% porous barrier (Fig. 9.16). They show the reduction in potential evapotranspiration (as measured with evaporimeters) is altered as a function of windiness and distance from a windbreak. Other studies indicating reductions of actual water use by crops as a function of distance from the windbreak are to be found in van Eimern et al. (1964).

We have seen that sensible heat advection may play a role in determining the shelter effect on evapotranspiration. Brown and Rosenberg (1971) and Miller et al. (1973) show that, under nonadvective conditions, the differences between sheltered and exposed sites may be small. In fact, water use in shelter may actually exceed that in the open at times. With advection, however, experiments show reduced water use in shelter. Advective conditions may occur at any time but are usually strongest in the mid- to late afternoon.

The causes for the difference in evapotranspiration rates between sheltered and exposed crops are difficult to evaluate. On the one hand, the lower stomatal resistance, higher temperature and vapor pressure gradient, and (sometimes) larger plants should cause greater water use in shelter. On the other hand, the reduced wind speed and turbulent mixing and the consequently increased aerial resistance to diffusion, as well as the higher humidity, should reduce evapotranspiration in shelter. A rational theory to explain these observed differences requires a mathematical model, which can consider these parameters and their interactions one at a time.

A number of attempts have been made to predict, by modeling, the effects of a windbreak on the potential and real evapotranspiration. One approach to the problem is that of Brown and Rosenberg (1973) by means of their so-called resistance model:

$$LE = \left\{ \frac{f[(R_n + S + LE)r_a/\rho_a C_p + T_a] - e_a}{r_a + r_c} \right\} \frac{\rho_a C_p}{\gamma} \qquad (7.50)$$

A number of meteorological and physiological factors control evapotranspiration. To understand how shelter affects each of these factors a reasonable range of conditions for open and sheltered situations can be entered (simultaneously) into the equation, which must be solved by iteration after a first, reasonable value of LE is chosen.

On the basis of data already presented we may assume the following:

1. r_a, the aerial resistance, is greater in shelter.
2. r_c, the crop resistance, is lower in shelter.
3. T_a, the air temperature, is greater in shelter (during the daytime).
4. e_a, the vapor pressure, is greater in shelter.
5. R_n, the net radiation, is identical in sheltered and exposed sites.

Table 9.3 shows the results of resistance model calculations of the influence of a hypothetical windbreak. Part A of the table shows the change from LE in the open that follows from a one-factor change in shelter. For example, the decrease in canopy resistance r_c in shelter leads to a predicted increase in LE from 621 W m^{-2} in the open to 676 W m^{-2} in shelter. Increased aerial resistance r_a decreases LE. Higher air temperature alone increases LE, whereas higher vapor pressure alone decreases LE. When all the changes in shelter have been considered, the model predicts essentially no change in LE from that in the open.

Part B of Table 9.3 deals with changing weather influences superimposed on an already existing shelter effect. For example, in cloudy conditions R_n is uniformly decreased in shelter and open sites. The resultant decrease in LE is greater in shelter. Greater stomatal resistance caused, for example, by moisture shortage in the soil reduces LE, especially in shelter. Similarly, greater windiness and increased air temperature as well as decreased humidity cause an increase in LE, but in each case the increase is greater in the open than in shelter. The input data are those developed in the corn-sheltering sugar beet study in western Nebraska. The model predictions of

Table 9.3 Calculations of the influence of a windbreak on LE, using hypothetical data in the resistance model of Brown and Rosenberg (1973)

		Net radiation, R_n (W m^{-2})	Canopy resistance, r_c (s m^{-1})	Aerial resistance, r_a (s m^{-1})	Air temperature, T_a (°C)	Air vapor pressure, e_a (kPa)	Latent heat flux, LE (W m^{-2})
A.	Open	700	200	100	23	1.8	621
	r_c changes	700	150	100	23	1.8	676
	r_a changes	700	200	200	23	1.8	586
	T_a changes	700	200	100	25	1.8	718
	e_a changes	700	200	100	23	2.2	516
	Shelter	700	150	200	25	2.2	613
B.	Cloudy						
	Open	350	200	100	23	1.8	453
	Shelter	350	150	200	25	2.2	384
	Increased r_c						
	Open	700	300	100	23	1.8	537
	Shelter	700	250	200	25	2.2	516
	Increased wind						
	Open	700	200	50	23	1.8	670
	Shelter	700	150	150	25	2.2	635
	Increased T_a						
	Open	700	200	100	33	1.8	1130
	Shelter	700	150	200	35	2.2	941
	Dry air						
	Open	700	200	100	23	0.8	906
	Shelter	700	150	200	25	1.2	774

the influence of windbreaks have been found consistent with measured effects in a number of cases (e.g., Miller et al., 1973; Brown and Rosenberg, 1973).

9.7 THE EFFECT OF SHELTER ON PHOTOSYNTHESIS

Plant growth and yield are usually greater in shelter. Therefore, the net assimilation of carbon dioxide must somehow be increased. A greater net assimilation may result from a number of possible causes. These include (1) a higher carbon dioxide concentration in the ambient air, (2) greater carbon dioxide flux rates from above and below in the active part of the canopy, (3) a longer daily duration of photosynthesis, and (4) lower nocturnal respiration and/or photorespiration. Indirect effects on soil nutrient status or on nutrient uptake may also occur (e.g., Shah and Kalra, 1970) but will not be discussed in detail here.

The **carbon dioxide concentration**, if different from that in the open, is slightly lower during the daytime in shelter. This would decrease rather than increase the photosynthetic rate. Carbon dioxide concentrations at night are normally higher in shelter. The additional carbon dioxide would be very rapidly consumed and/or dispersed at daybreak and probably does not contribute significantly to the total daily net assimilation in shelter.

Photosynthetic flux rates increase with increasing steepness of the carbon dioxide concentration gradients directed to the active part of the crop canopy. Brown and Rosenberg (1972) report that all gradients are generally steeper over sheltered sugar beets. Differences in the magnitude of the gradients decrease with increasing elevation above the crop. Miller et al. (1973) also report that carbon dioxide concentration gradients are steeper over sheltered than over exposed soybeans. No other comparable measurements of carbon dioxide gradients in sheltered and open sites are available.

The volume of evidence with respect to exchange coefficients in shelter is extensive (see Section 9.3). The work of a number of scientists, Russians especially, has been reviewed by van Eimern et al. (1964) and by Brown and Rosenberg (1971). These show consistently that turbulent mixing is reduced in shelter. Brown and Rosenberg (1972) and Miller et al. (1973) have also shown decreased exchange coefficients in shelter.

The carbon dioxide flux is the product of the carbon dioxide concentration gradient and exchange coefficient. Brown and Rosenberg (1972) and Miller et al. (1973) concluded that shelter had no significant effect on carbon dioxide flux rates to sugar beets or soybeans, respectively. Ogbuehi and Brandle (1981) show, indirectly, that soybeans grown in shelter have a capacity for more rapid photosynthesis. Mean CO_2 exchange rates measured in closed chambers coupled to the upper canopy were greater during much of the day in sheltered plants in six soybean cultivars tested. Differences were greatest at midday. Since turbulence was cancelled, essentially, by the use of chambers, the effect on photosynthesis was probably due to improved internal water relations in the sheltered plants. Ogbuehi and Brandle also found greater penetration of PAR into the sheltered canopy which could have influenced photosynthetic rate.

The **duration of photosynthetic activity** may be longer in shelter. This follows from the observation that the stomatal resistance to carbon dioxide diffusion is usually lower in shelter (e.g., Rosenberg et al. 1967; Brown and Rosenberg, 1972). The delay or avoidance of wilting in shelter (Rosenberg, 1966b) also suggests that a greater photosynthetic opportunity accounts for the greater yield production.

Ogbuehi and Brandle (1981) demonstrated that leaf water potential and stomatal conductance were greater throughout the day in sheltered than in exposed soybeans of the Bonus cultivar. Another cultivar, Wayne, maintained a greater water potential throughout the growing season. Skidmore et al. (1975) in Kansas have shown that shelter has little effect on leaf water potential and stomatal resistance in wheat when moisture stress is low. At intermediate levels of stress leaf water potential was higher in the sheltered plants and stomatal resistance was lower. Under high stress leaf water potentials were identical in sheltered and exposed plants but stomatal resistance was greater in the exposed plants.

Daytime temperatures are normally higher in shelter. Photosynthesis is little affected by such temperature differences as occur in shelter unless extremes of cold or heat are approached. Photorespiration is, however, affected by small temperature differences. Thus, higher temperatures would tend to cause a reduction in the net assimilation of carbon dioxide in those crops that have photorespiratory mechanisms. The higher temperatures in shelter may extend to the soil as well. Higher soil temperature will result in more rapid root respiration and organic matter decomposition and may, thus, cause a greater release of carbon dioxide from below. The lower nocturnal temperatures in shelter should, conversely, cause a reduction in plant and root respiration. This latter effect should increase the daily and seasonal net assimilation of carbon dioxide.

Of the factors described above it seems likely that the improved plant internal water status and the increased duration of photosynthetic opportunity may be most important in increasing photosynthesis in shelter.

9.8 THE EFFECT OF SHELTER ON WATER USE EFFICIENCY

The literature on shelter effect identifies a great many experiments and observations in which water conservation is demonstrated. When plant growth is markedly altered because of shelter, more water may actually be consumed, however. Yields are generally superior in shelter. Generally, then, shelter leads to improved water use efficiency (dry matter or harvestable yield/unit of water evaporated and transpired).

It is possible to make direct measurements of changes in soil water content and in plant growth over periods of a week or longer, so that the long-term effects of shelter on water use efficiency are demonstrable. To learn the mechanisms by which shelter affects photosynthesis and evapotranspiration, it is necessary, however, to conduct micrometeorological experiments. Thus, Bowen ratio–energy balance methods and precision weighing lysimeters have been used in recent years to determine evapotranspiration rates over periods as short as 15 min. Carbon dioxide flux measurements have also been made, using exchange coefficients derived from energy balance and lysimetric observations on the assumption that $K_w = K_c$ (see Chapter 4).

Micrometeorological and lysimetric measurements have generally demonstrated that evapotranspiration is reduced in shelter. The data on carbon dioxide flux obtained by micrometeorological methods suggest that photosynthetic flux rates are not affected as dramatically. Thus, data obtained with micrometeorological methods support and refine the conclusions of the far more extensive agronomic experimentation, showing that wind shelter does result, generally, in improved water use efficiency.

In arid and semiarid regions the principal benefit of windbreaks and shelterbelts appears to be the reduction in water use under conditions of sensible heat advection. In these regions and in more humid regions, as well, rapid seed germination, vigorous vegetative growth and mechanical protection of the plants may be factors of equal importance.

9.9 SOME INTEGRATIVE SCHEMES OF THE SPATIAL DIFFERENCES IN SHELTER EFFECTS

As shown above, the literature of shelter effect is reasonably consistent in the following ways:

1 Shelter alters microclimate.
2 Shelter reduces potential evapotranspiration.
3 Shelter reduces actual evapotranspiration.
4 Shelter improves internal water relations: for example, greater internal water potential, lower stomatal resistance.
5 Shelter provides improved opportunity for photosynthesis.
6 Shelter generally increases yield.

The literature also shows that these generalities are subject to variation depending upon availability of soil moisture: the benefits may be most dramatic in dry years or when moisture shortages are critical. Curiously, the literature also suggests that benefits of shelter may be more consequential (in terms of actual yields) under irrigation than on dry land. Perhaps the

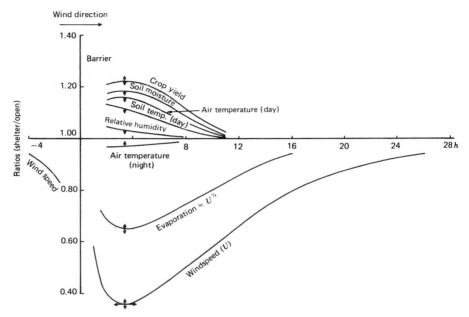

Fig. 9.17 Summary diagram of the effect of barriers on micrometeorological factors. The arrows indicate the directions in which values of different factors have been found to vary relative to the control values measured in unsheltered areas. h = height of barrier (after Marshall, 1967, *Field Crop Abstracts* **20**:1–14. Copyright Commonwealth Agricultural Bureaux, Farnham Royal, England).

greatest uncertainty concerning the use of shelter relates to the "spatial problem," that is, the behavior of plants as related to their distance from the shelter.

A number of schemes for generalizing research results on shelter effect have been proposed. Marshall (1967) shows in Fig. 9.17 how the proportional decrease of windspeed with distance from the shelter is mimicked by the reduction in evaporation. The nocturnal temperature decrease, daytime air and soil temperature increase, and relative humidity increase (the microclimatic effects) are also seen to vary with distance from the barrier, declining to nearly no effect at about $12\ h$. The greatest soil moisture availability and crop yield are found in the zone $2–4\ h$ and these decline with increasing distance from the barrier.

Radke (1976) proposes another scheme based on his observation of soybeans grown in the shelter of a number of kinds of barriers. Figure 9.18 is his schematic description of the turbulent mixing regime (and hence the microclimate) and the resultant plant growth across a zone sheltered by uniformly spaced east–west windbreaks. The crop response is generally of the kind shown in the upper curve. The slope of the mid-region is conditioned by the impact of the south and north barriers on turbulent transport. Negative

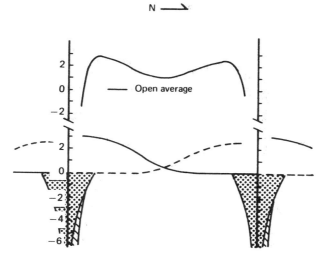

Fig. 9.18 Schematic diagram of sheltered crop response to porous wind barriers with east–west row orientation. The barriers are represented by the two vertical lines and the numbers represent relative increases of plant growth or yields. The solid and dashed curves in the lower part of the figure represent plant response to southerly and northerly winds, respectively. The light shaded areas represent negative plant response to root competition and the hatched areas represent additional decreases due to shading. The upper curve is the sum of the lower curves and represents the expected yield (after Radke, 1976).

plant responses due to root competition and to shading (on the north side) account for the reduction in yield near the barriers.

Schemes such as those of Marshall (1967), Radke (1976) and others (e.g., Guyot, 1963) are useful in providing some qualitative integrations of the information gathered through much experimentation over many years in many lands. But the variety in climates, crops to be protected, and types of windbreaks (and their seasonal as well as long-term changes) make the matter of predicting shelter effects on water use and yield a most complicated one. Adaptive research is needed wherever windbreak and shelterbelt systems are considered for introduction and/or for improvement. But the research should be guided by the principles and experiences in this chapter and in the many useful reviews that are cited within.

REFERENCES

Aase, J. K., and F. H. Siddoway. 1976. Influence of tall wheatgrass wind barriers on soil drying. *Agron. J.* **68**:627–631.

Albrektsson, L., J. Alemo, B. Landgren, and S. A. Svensson. 1978. Windbreaks for greenhouses. Institutionen for Lantbrukets Byggnadsteknik. 22 pp.

Armbrust, D. V., G. M. Paulsen, and R. Ellis, Jr. 1974. Physiological responses to wind- and sandblast-damaged winter wheat plants. *Agron. J.* **66**:421–423.

Aslyng, H. C. 1958. Shelter and its effect on climate and water balance. *Oikos* **9**:282–310.

Bates, C. G. 1911. Windbreaks, their influence and value. *USDA For. Serv. Bull.* **86**. 100 pp.

Bates, E. M., and R. L. Phillips. 1980. Effect of a solid windbreak in a cattle-feeding area. *Oreg. Agric. Exp. Stn. Bull.* **646**:1–9.

Bean, A., R. W. Alperi, and C. A. Federer. 1975. A method for categorizing shelterbelt porosity. *Agric. Meteorol.* **14**:417–429.

Bierhuizen, J. F., and R. O. Slatyer. 1964. Photosynthesis of cotton leaves under a range of environmental conditions in relation to internal and external diffusive resistances. *Australian J. Biol. Sci.* **17**:348–359.

Black, A. L., and F. H. Siddoway. 1976. Dryland cropping sequences within a tall wheatgrass barrier system. *J. Soil Water Conserv.* **31**:101–105.

Bouchet, R. J., G. Guyot, and S. de Parcevaux. 1966. Augmentation de l'efficience de l'eau et amelioration des rendements par reduction de l'evapotranspiration potentielle au moyen de brise-vent. Proc. UNESCO Symp. on Methods in Agrometeorology, July 23–30, pp. 167–173.

Brown, K. W., and N. J. Rosenberg. 1970. Effect of windbreaks and soil water potential on stomatal diffusion resistance and photosynthetic rate of sugar beets (*Beta vulgaris*). *Agron. J.* **62**:4–8.

Brown, K. W., and N. J. Rosenberg. 1971. Turbulent transport and energy balance as affected by a windbreak in an irrigated sugar beet (*Beta vulgaris*) field. *Agron. J.* **63**:351–355.

Brown, K. W., and N. J. Rosenberg. 1972. Shelter-effects on microclimate, growth and water use by irrigated sugar beets in the Great Plains. *Agric. Meteorol.* **9**:241–263.

Brown, K. W., and N. J. Rosenberg. 1973. A resistance model to predict evapotranspiration and its application to a sugar beet field. *Agron. J.* **65**:341–347.

Caborn, J. M. 1957. Shelterbelts and microclimate. *Forestry Commun. Bull.* **29**:1–35, H. M. Stationery Office, Edinburgh, Scotland.

Frank, A. B., D. G. Harris, and W. O. Willis. 1975. Influence of windbreaks on crop performance and snow management in North Dakota. *Shelterbelts on the Great Plains: Proceedings of the Symposium*, R. W. Tinus, ed. Great Plains Agric. Council Publ. No. 78, pp. 41–48.

Frank, A. B., D. G. Harris, and W. O. Willis. 1977a. Growth and yield of spring wheat as influenced by shelter and soil water. *Agron. J.* **69**:903–906.

Frank, A. B., D. G. Harris, and W. O. Willis. 1977b. Plant water relationships of spring wheat as influenced by shelter and soil water. *Agron. J.* **69**:906–910.

Gaastra, P. 1959. Photosynthesis of crop plants as influenced by light, carbon dioxide, temperature and stomatal diffusion resistance. *Meded. Landbouwhogeschool (Wageningen, Netherlands)* **59**:1–68.

General Accounting Office/Controller General of the United States. 1975. Report to the Congress: Action needed to discourage removal of trees that shelter cropland in the Great Plains. RED-75-375, 31 pp.

Grace, J. 1977. *Plant Response to Wind*. Academic Press, New York. 204 pp.

Guyot, G. 1963. Shelter belts—Modification of microclimate and improvement of agricultural production. *Ann. Agron.* **14**:429–488.

Guyot, G., and B. Seguin. 1978. Influence du bocage sur le climat d'une petite region: Resultats des mesures effectuees en Bretagne. *Agric. Meteorol.* **19**:411–430.
Jensen, M. 1954. *Shelter Effect.* Danish Tech. Press, Copenhagen. 266 pp.
Kalma, J. D., and F. Kuiper. 1966. Transpiration and growth of *Phaseolus vulgaris* L. as affected by wind speed. *Meded. Landbouwhogeschool (Wageningen Netherland)* **66**:1–8.
Kaminski, A. 1967. The effect of shelterbelts on the yield of plants in a permanent crop rotation. *Ekol. Polska Ser. A* **15**:426–441.
Kaminski, A. 1968. The effect of a shelterbelt on the distribution and intensity of ground frosts in cultivated fields. *Ekol. Polska Ser. A* **16**:1–11.
Lemon, E. 1970. Gaseous exchange in crop stands. *Physiological Aspects of Crop Yield* (F. A. Haskins, C. Y. Sullivan, and C. H. M. van Bavel, eds.), Am. Soc. Agron., Madison, Wis., pp. 117–137.
Lynch, J. J., and J. K. Marshall. 1969. Shelter: A factor increasing pasture and sheep production. *Australian J. Sci.* **32**:22–23.
McClure, G. 1981. Shelterbelts—Wooded islands of wildlife. *Extension Rev.*, pp. 17–19.
Maki, T., and Allen, L. H., Jr. 1978. Turbulence characteristics of a single line pine tree windbreak. *Proc. Soil Crop Sci. Soc. Florida* **37**:81–92.
Marshall, J. K. 1967. The effect of shelter on the productivity of grasslands and field crops. *Field Crop Abstr.* **20**:1–14.
Miller, D. R., N. J. Rosenberg, and W. T. Bagley. 1973. Soybean water use in the shelter of a slat-fence windbreak. *Agric. Meteorol.* **11**:405–418.
Miller, D. R., N. J. Rosenberg, and W. T. Bagley. 1975. Wind reduction by a highly permeable windbreak. *Agric. Meteorol.* **14**:321–333.
Moysey, E. B., and F. B. McPherson. 1966. Effect of porosity on performance of windbreaks. *Trans. Am. Soc. Agr. Engr.*, **9**:74–76.
Ogbuehi, S. N., and J. R. Brandle. 1981. Influence of windbreak-shelter on light interception, stomatal conductance and CO_2-exchange rate of soybeans. *Trans. Nebr. Acad. Sci.* **9**:49–53.
Pelton, W. L., and A. V. Earl. 1962. The influence of field shelterbelts on wind velocity and evapotranspiration. Can. Dept. Agric. Exp. Farm Annu. Report, Swift Current, Saskatchewan. Mimeo. 4 pp.
Plate, E. J. 1971. The aerodynamics of shelterbelts. *Agric. Meteorol.* **8**:203–222.
Radke, J. K., and W. C. Burrows. 1970. Soybean plant response to temporary field windbreaks. *Agron. J.* **62**:424–429.
Radke, J. K., and R. T. Hagstrom. 1974. Wind turbulence in a soybean field sheltered by four types of wind barriers. *Agron. J.* **66**:273–278.
Radke, J. K. 1976. The use of annual wind barriers for protecting row crops. Proc. Symp. Shelterbelts on the Great Plains, Denver, Colorado. Great Plains Agric. Council Publ. No. 78, pp. 79–87.
Read, R. A. 1958. The Great Plains shelterbelt in 1954. *Nebr. Agric. Exp. Sta. Bull.* **441**:1–125.
Read, R. A. 1964. Tree windbreaks for the Central Great Plains. Agric. Handbook No. 250, U.S. Dept. of Agric. 68 pp.
Rosenberg, N. J. 1966a. Microclimate, air mixing and physiological regulation of transpiration as influenced by wind shelter in an irrigated bean field. *Agric. Meteorol.* **3**:197–224.
Rosenberg, N. J. 1966b. Influence of snow fence and corn windbreaks on microclimate and growth of irrigated sugar beets. *Agron. J.* **58**:469–475.

Rosenberg, N. J. 1966c. On the study of shelter effect with sheltered (screened) meteorological sensors. *Agric. Meteorol.* **3**:167–177.
Rosenberg, N. J. 1967. The influence and implications of windbreaks on agriculture in dry regions. *Ground Level Climatology* (R. H. Shaw, ed.), Am. Assoc. Advancement of Science, pp. 327–329.
Rosenberg, N. J., D. W. Lecher, and R. E. Neild. 1967. Response of irrigated snap beans to wind shelter. *Proc. Am. Soc. Hort. Sci.* **90**:169–179.
Rosenberg, N. J. 1975. Windbreak and shelter effects. *Progress in Plant Biometeorology: The Effect of Weather and Climate on Plants* (L. P. Smith, ed.), Swets and Zeitlinger, Amsterdam, pp. 108–134.
Rosenberg, N. J. 1979. Windbreaks for reducing moisture stress. *Modification of the Aerial Environment of Plants* (B. J. Barfield and J. F. Gerber, eds.), Am. Soc. Agric. Engin. Monogr. 2, ASAE, St. Joseph, Michigan, pp. 394–408.
Rosenberg, N. J. 1981. The increasing CO_2 concentration in the atmosphere and its implication on agricultural productivity. I. Effects on photosynthesis, transpiration and water use efficiency. *Climatic Change* **3**:265–279.
Ruesch, J. D. 1955. Der CO_2-gehalt bodennaher luftschicten unter dem einfluss des wind-scheutzes. *Z. Pflanzenernaehr. Dueng. Bodenk.* **71**:113–132.
Seginer, I. 1975. Flow around a windbreak in oblique wind. *Boundary-Layer Meteorol.* **9**:133–141.
Seguin, B., and N. Gignoux. 1974. An experimental study of wind profile modification by a network of shelterbelts. *Agric. Meteorol.* **13**:15–23.
Shah, S. R. H. 1962. Studies on wind protection. Institut voor Toegepast Biologisch Onderzoek in de Natuur, Netherlands Mededeling No. 60, 113 pp.
Shah, S. R. H., and Y. P. Kalra. 1970. Nitrogen uptake of plants affected by windbreaks. *Plant Soil* **33**:573–580.
Skidmore, E. L. 1966. Wind and sandblast injury to seedling green beans. *Agron. J.* **58**:311–315.
Skidmore, E. L., H. S. Jacobs, and W. L. Powers. 1969. Potential evapotranspiration as influenced by wind. *Agron. J.* **61**:543–546.
Skidmore, E. L., H. S. Jacobs and L. J. Hagen. 1972. Microclimate modification by slat-fence windbreaks. *Agron. J.* **72**:160–162.
Skidmore, E. L., L. J. Hagen, and I. D. Teare. 1975. Wind barriers most beneficial at intermediate stress. *Crop Sci.* **15**:443–445.
Staple, W. J., and J. J. Lehane. 1955. The influence of field shelterbelts on wind velocity, evaporation, soil moisture and crop yield. *Can. J. Agric. Sci.* **35**:440–453.
Stoeckeler, J. H. 1962. Shelterbelt influence on Great Plains field environment and crops. USDA Production Res. Rpt. No. 62, 26 pp.
Sturrock, J. W. 1969. Aerodynamic studies of shelterbelts in New Zealand—I. Low to medium height shelterbelts in mid-Canterbury. *New Zealand J. Sci.* **12**:754–776.
Sturrock, J. W. 1975. The control of wind in crop production. *Progress in Biometeorology: The Effect of Weather and Climate on Plants* (L. P Smith, ed.), Swets and Zeitlinger, Amsterdam, pp. 349–368.
Thomas, M. D., and G. R. Hill. 1949. Photosynthesis under field conditions. *Photosynthesis in Plants* (Plant Physiology Monograph) (J. Frank and W. E. Loomis, eds.), Iowa State College Press, pp. 19–52.

van der Linde, P. J. 1962. Trees outside the forest. Forest Influences. FAO Forestry Forest Production Studies No. 15, Rome, pp. 141–208.
van Eimern, J., R. Karschon, L. A. Razumova, and G. W. Robertson. 1964. Windbreaks and shelterbelts. WMO Tech. Note No. 59, 188 pp.
Woodruff, N. P., R. A. Read, and W. S. Chepil. 1959. Influence of a field windbreak on summer wind movement and air temperature. *Kansas Tech. Bull.* **100**:1–24.

CHAPTER 10 – FROST AND FROST CONTROL

10.1 INTRODUCTION

Protecting plants from the effects of lethally low temperature is a matter of considerable importance in agriculture generally, but especially in the horticultural production of high-value fruits and vegetables. Costs of frost protection and benefits in terms of crop condition and marketing advantage have always guided the grower in deciding whether or not frost protection is justified. The great increase in energy prices that has occurred since 1973 has seriously affected the economics of frost protection and has led to an intensification of efforts to forecast frosts correctly, to assess the monetary value of frost forecasts, and to monitor the onset of frosts and pinpoint areas most likely to be affected. Remote sensing provides some unique opportunities in this regard. The "energy crisis" has also led to intensified efforts to improve the energy efficiency of the many types of devices used for frost protection as described in Section 10.4. Advances in computer modeling of the energy balance of fields under frost conditions have helped in improving design and operation of frost control systems.

Katz and Murphy (1979) point out that the greatest difficulty in calculating the value of a frost forecast to Washington apple growers lies not in the prediction of minimum temperature, but rather in the lack of adequate data relating fruit yield loss and low temperature. Indeed, prediction of the specific degree and type of damage likely to occur in a particular frost depends on a great many factors of which the type of crop is only one. In the case of apples, for example, Rollins et al. (1962) cite a wide range of factors that determine degree of frost damage. These include variety, tissue maturity (which is influenced by fertilization, cultivation, and defoliation), size of crop and pruning history, rate of temperature drop, minimum temperature, and length of time that the tissue remains at the minimum temperature.

Whiteman (1957) gives a detailed listing of the freezing points of most fruits, vegetables, and flowers, based on laboratory determinations. A selection of data from Whiteman given in Table 10.1 shows that plant species differ greatly in their susceptibility to cold injury. Ventskevich (1958) also lists species according to their ability to tolerate low temperatures at various stages of growth. Selected data from Ventskevich are given in Table 10.2.

Table 10.1 Freezing points of various fruits and vegetables[a]

Kind, variety	Freezing point (°C)	
	Lowest	Highest
Apple, Jonathon	−2.5	−1.9
Apricot, Perfection	−1.9	−1.5
Avocado, Pollock	−0.9	−0.8
Banana, Guatemala	−1.1	−1.1
Cherry, Bing	−3.3	−3.1
Fig, Mission	−3.2	−2.4
Grape, Tokay	−3.2	−2.9
Grapefruit, Foster King	−2.4	−1.7
Lemon, Eureka	−1.5	−1.4
Mango, Keitt	−2.0	−1.3
Olive, fresh green	−2.4	−1.4
Orange, Jaffa	−3.2	−1.5
Pear, Anjou	−2.1	−1.6
Pineapple, Pernambuca	−1.3	−1.0
Strawberry, Redstar ripe	−1.1	−1.0
Asparagus	−1.1	−0.6
Bean	−1.2	−0.8
Broccoli	−0.6	−0.6
Cabbage	−1.2	−0.9
Celery	−0.8	−0.5
Cucumber	−0.8	−0.7
Eggplant	−0.9	−0.8
Garlic	−3.4	−2.9
Lettuce	−0.4	−0.2
Mushroom	−1.3	−0.9
Melon, Crenshaw	−1.4	−1.1
Onion, Bermuda	−1.2	−1.0
Pea	−0.9	−0.6
Raddish, white root	−1.1	−0.7
Tomato	−0.9	−0.5
Watermelon	−0.9	−0.8

[a] Selected data from Whiteman (1957).

The actual onset of frost has been difficult to forecast for a number of reasons. Shaw (1954) showed that minimum air temperatures measured in standard weather shelters provide little information on the actual temperature of the leaves during a frost situation. Under conditions favorable for strong upward radiative loss at night leaf temperatures as much as 4–4.5°C lower than air temperatures were found. The problem of relating screen temperature to tissue temperature can be even more complicated. Rahn and

Table 10.2 Resistance of crops to frost in different development phases[a]

	Harmful temperature (degrees below 0°C)		
	Germination	Flowering	Fruiting
Highest resistance			
Spring wheat	9–10	1–2	2–4
Oats	8–9	1–2	2–4
Barley	7–8	1–2	2–4
Peas	7–8	2–3	3–4
Resistant			
Vetch	6–7	3–4	2–4
Beans	5–6	2–3	3–4
Sunflower	5–6	2–3	2–3
Safflower	6–4	2–3	3–4
Flax	5–7	2–3	2–4
Sugar beet	6–7	2–3	—
Carrot	6–7	—	—
Medium resistance			
Cabbage	5–7	2–3	6–9
Soybean	3–4	2–3	2–3
Italian millet	3–4	1–2	2–3
Low resistance			
Corn	2–3	1–2	2–3
Millet	2–3	1–2	2–3
Sudan grass	2–3	1–2	2–3
Sorghum	2–3	1–2	2–3
Potato	2–3	—	1–2
No resistance			
Buckwheat	1–2	1–2	0.5–2
Castor bean	1–1.5	0.5–1	2
Cotton	1–2	1–2	2–3
Melons	0.5–1	0.5–1	1
Rice	0.5–1	0.5–1	0.5–1
Sesame	0.5–1	0.5–1	—
Peanut	0.5–1	—	—
Cucumber	0–1	—	—
Tomato	0–1	0–1	0–1
Tobacco	0–1	0–1	0–1

[a] Selected data from Ventskevich (1958).

Brown (1971) show that different crops experience different temperatures in a frost. Temperatures within a corn canopy, for example, were found higher than those measured at the level of grass during the same frost.

Generally, the tropical and subtropical crops have little or no resistance to freezing and may be damaged when exposed, even for a short time, to temperatures at or near freezing. For example, tea in the Nilgiris region of south India suffers frequent defoliation because of low temperatures during the winter season (Von Lengerke, 1978). Rubber trees in south China can be seriously injured if the trunk is exposed to temperatures below freezing for 4 h or longer.*

Many of the common vegetable crops and succulent vine fruits of the temperate zone are also easily damaged by light frost. The extent of the damage that occurs in a frost may be determined by the duration of freezing temperatures as may the rapidity with which freezing takes place. Large ice crystals, which form under rapid freezing, are more destructive to plant tissue than are small crystals.

Schnell (1976) reports that bacteria may play a role in fostering frost sensitivity in plants by providing nuclei for the formation of ice at temperatures below 0°C. By spraying an antibotic, streptomycin, on the leaves of corn, lettuce and beans, damage caused by exposure to a temperature of $-2°C$ was reduced by 70% compared with unsprayed plants. Yankofsky et al. (1981) found that bacteria that affect ice nucleation in plants can also be active in cloud nucleation. Only small numbers of the cells in the total population are active nucleation agents, however.

Where frosts are frequent they can affect the form and distribution of native vegetation. A good example is provided by Schlegel and Butch (1980) who describe an area in Pennsylvania known as the "Barrens." Because of its topography and the lack of an outlet for cold air, frosts lead to severe and prolonged low temperatures in this area and trees are stunted in appearance. Soils in this area are sandy and frequently dry so that the thermal conductivity is poor, a factor that increases the severity of the frosts.

10.2 TYPES OF FROST

Frost is defined in the *Glossary of Meteorology* as the condition that exists when the temperature of the earth's surface and earthbound objects falls below freezing, 0°C.

Radiation frosts occur on calm clear nights when terrestrial radiation to space is relatively unimpeded because of the absence of clouds and heavy concentrations of water vapor. The severity of radiation frosts varies considerably with general atmospheric conditions as well as with local differences in topography and vegetation. **Hoarfrost** or **white frost** is caused by the sublimation of ice crystals on objects such as tree branches and wires when these objects are at a temperature below freezing. **Black frost** occurs when vegetation is frozen because of a reduction in the temperature of air that does not contain sufficient moisture for the formation on the surface of hoarfrost.

---

* Personal communication from Professor Hueng Ping Wei. Institute of Geography, Academia Sinica, Peking, PRC.

372 FROST AND FROST CONTROL

Biel (1961) differentiated radiation frosts, which are essentially local in extent, from **advection frosts**, which result from large-scale air mass transportation. The advection frost is often termed a "hard freeze." The terms radiation frost and advection frost are somewhat arbitrary. Cool dry air advected into a region sets the stage for unobstructed radiation of heat from soil and plants. Similarly, radiative processes contribute to the heat exchange during an advection frost.

There is one major difference between radiation and advection frosts. Under radiation frost conditions, winds are normally light and temperature inversions develop as air in contact with cold radiating surfaces becomes chilled and heavy. Advection frosts often occur with strong winds, and distinct temperature inversions do not develop under these conditions.

Profiles of air temperature shown in Fig. 10.1 are typical of a radiation frost, one that did considerable damage in the central and eastern Great

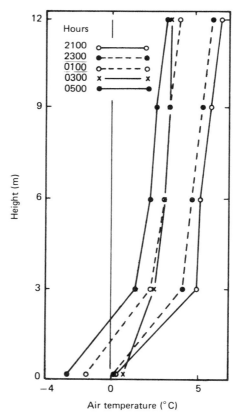

Fig. 10.1 Air temperature above the ground at various times during a radiation frost night, mid-September at Columbus, Nebraska (after Rosenberg, 1963).

THE CLIMATOLOGY OF FROST INCIDENCE 373

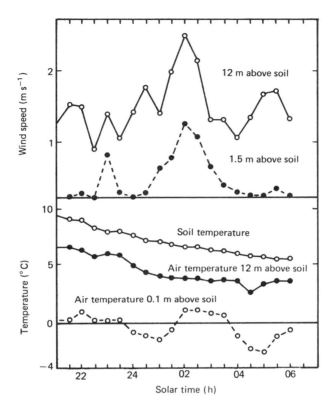

Fig. 10.2 Course of air and soil temperature and wind speed during the same radiation frost as in Fig. 10.1 (after Rosenberg, 1963).

Plains. Temperature profiles were measured at 2-h intervals during the night at a site near Columbus, Nebraska. A general cooling of the air at all levels occurred from 2100 to 0100 hours. By 0100 hours the temperature near the ground was below freezing but at 0300 hours it rose above freezing. By 0500 hours temperatures fell again, this time to well below freezing near the ground. The rise in temperature at 0300 hours was caused by increasing windiness which began about 0200 hours and led to a mixing of warmer air from above with the mass of chilled air near the ground (Fig. 10.2). The existence of strong temperature inversions such as those in Fig. 10.1 are characteristic of conditions during radiation frosts.

10.3 THE CLIMATOLOGY OF FROST INCIDENCE

The length of the growing season for any specified crop is fixed, essentially, by the occurrence in spring and fall of frosts with sufficiently low temperature to kill the crop. The mean dates of the last 0°C minimum temperature in spring and the first such event in fall are often used as a reasonable measure of frost occurrence and to define the length of the frost-free season. Thom and Shaw (1958) demonstrated that frost-dates are randomly and normally distributed so that the mean and the standard deviation are valid statistics. With these parameters, the probability that the last spring frost and the first fall frost will occur before or after a given date may be calculated. Thom and Shaw were also able to show that the dates of the last spring and the first fall frosts are independent of one another. Thus, the probability of a specified length to the growing season can also be calculated.

Rosenberg and Myers (1962) considered that knowledge of the type of frost occurrence might be more useful for agricultural planning purposes than would be the mean date of a specified minimum temperature. Therefore, they developed climatic statistics based on the dates of the last spring and first fall occurrence of radiation and advection frosts.

It is generally thought that the last frosts in the spring and the first frosts in the fall are of the radiation type. Rosenberg and Myers (1962) found, however, in a study of five locations within the Platte River Valley of Nebraska that 7–30% of last spring frosts and 17–42% of first fall frosts are actually advective in nature. They also found that between two and five radiation frosts occur in the spring after the last advective frost in the region studied and between one and three radiation frosts occur in fall before the first advection frost. Furthermore, their statistics show that, if a grower in the Platte River Valley chooses to protect against the effects of radiation frost, the growing season can be lengthened by an average of 15–32 days depending on location.

In mountainous regions it is very difficult to maintain adequate networks of observing stations. Thus, long frost date records are usually unavailable. The prediction of actual frost severity (minimum temperatures and their duration) is a very complicated matter. Topography complicates the interpretation of the limited data available since large differences in nocturnal temperatures develop because of sharp inversions. Caprio (1961) suggested that in mountainous regions daytime temperature maxima provide a reliable basis for predicting frost dates. For example, the average date of last spring frost for locations in Montana can be related by regression equations with the mean date when average daily maximum temperature reaches 21°C.

Bootsma (1976) attempted to predict frost incidence and severity on farm land of hilly and varied topography on Prince Edward Island, Canada, using standard climatological stations. He found that cloud cover and wind speed could explain up to about 80% of the variation in temperature between a base station and field sites 20–40 km away. Standard errors of estimate ranged from 0.8°C for hilltop sites to 2.3°C for valley bottoms. However, differences in minimum temperature between base station and field site were a function of topography, increasing from about 1°C on hilltops to 2–3°C on midslopes and 5–6°C in hollows.

10.4 METHODS OF FROST PROTECTION

If the economics of the situation so warrant, any crop can be protected against any frost. On the average, advection frosts occur earlier in the spring and later in the fall than do radiation frosts, and ambient temperatures are often much lower. Advection frosts may be of long duration. Since they usually involve strong windiness, the number of practical techniques for use against advection frost is limited. Most of the methods of frost protection discussed below are practical and effective against radiation frosts only, although some are applicable to both types of frost.

Methods of frost protection are based on the following principles: (1) site selection, (2) radiation interception, (3) thermal insulation, (4) air mixing, (5) direct convective air heating, (6) radiant heating, (7) release of the heat fusion, and (8) soil manipulation.*

10.4.1 Site Selection

As air cools at night and becomes dense, it drains into the lower topographic levels. Hocevar and Martsolf (1971) made temperature transects across a broad valley in Pennsylvania on radiation frost nights. They found a consistent increase in the air temperature of about 6.2°C for each 100 m of elevation above the bottom of the basin confining the cold air. Figure 10.3 illustrates the relation they found between air temperature and topography on three traverses during the frost season.

Bergen (1969) considers that the rate of down-slope flow of cold air can be predicted. In a forested mountain site of known topography, he found a direct relation of the flow rate with the nocturnal net radiation and its effect in lowering the potential temperature of the air.

The importance of proper site selection for planting high-value frost-sensitive fruits and vegetable crops is stressed by Schaal et al. (1961). Since cold air drains into low areas and pockets, planting sites should provide for natural air drainage into lower areas. They consider further that fence rows, windbreaks, earth fills, or heavy vegetation may block air drainage and should thus be avoided in establishing plantings in frost hazard locations.

The proximity to bodies of water is another important factor in site selection. Fruit orchards are located on the eastern shore of Lake Michigan, for example, because prevailing west winds are warmer after passing over the lake and thus frost frequency and severity are decreased in this zone (Bartholic and Sutherland, 1975).

---

* For a very thorough and authoritative review of these principles and their application the reader is referred to Barfield and Gerber (1979), Section 4: Environmental Modification for Frost Protection.

376 FROST AND FROST CONTROL

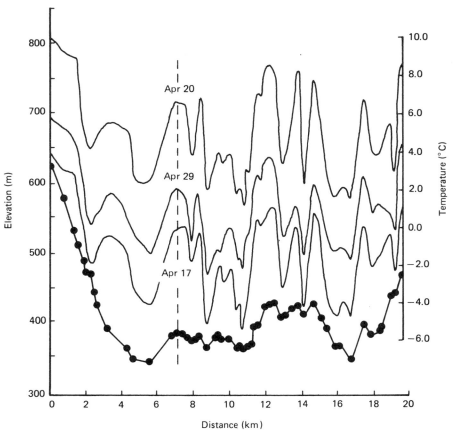

Fig. 10.3 Variation of temperature for three mornings in relation to relief in a mountainous area of Pennsylvania. The dots on the relief indicate the observation points (after Hocevar and Martsolf, 1971).

Site selection can be assisted by remote sensing. Nixon and Hales (1975) made traces of surface temperature in Texas using an airborne thermal radiometer on a night when freezing conditions were anticipated. Bartholic and Sutherland (1976) used NOAA thermal imagery from a satellite-borne scanner to prepare a general map of surface temperatures during an extensive freeze in Florida. Bartholic and Sutherland (1976) also reduced the scale of thermal mapping and were able to relate temperature to distance from a group of lakes in Florida. The temperature ameliorating effects of the lakes were detectable for as far as 8 km.

10.4.2 Radiation Interception

Radiation frosts occur on clear nights because of the absence of clouds that would otherwise absorb some longwave radiation emitted by the surface and reradiate it back to the surface. Clouds also reflect a portion of the long wave radiation back to the earth's surface. The generation of artificial clouds by injecting water mists into the air above a field offers one possible method for closing the "atmospheric window" to infrared terrestrial radiation.

De Boer (1965) developed a theoretical solution to the problem of how many droplets and what size distribution is necessary to effectively eliminate longwave radiation to space from the earth's surface. A vertical column of air 1 m^2 cross-sectional area would require the presence of 8.2×10^9 drops with radii of 7.5–8.5 μm to reduce by 99% the penetration of 8–12 μm radiation. Droplet size is important since, for reflection, the particle or droplet must be about the same as the wavelength of radiation impinging on it.

Artificial clouds of smoke or smoke screens are sometimes used to intercept thermal radiation from the surface. Smudge pots burning oil and fires of burning rubber tires or other waste materials have frequently been used for this purpose. The Russians are apparently adept at generating smoke screens. Ventskevich (1958) describes a number of methods requiring simple equipment. The use of smudging and smoke screens appears to be declining, in part because of environmental and energy conservation considerations, but primarily because these methods are inefficient and require large quantities of materials in order to be effective in trapping radiation (Bouchet, 1965). According to Mee and Bartholic (1979) smoke particles are usually less than 1 μm in size and reflect visible radiation well but are permeable to longwave radiation and so are ineffective in preventing rapid cooling of the surface and the air near the ground. Furthermore, winds transport the smoke screens easily out of the zones they are intended to protect.

A number of methods of artificial fog generation have been developed. In one such cited by Mee and Bartholic (1979) droplets are coated with a monomolecular film of a compound that inhibits evaporation in order to stabilize the fog. In another cited by Mee and Bartholic (1979) mechanical atomization produces fog droplets in the desired size range (10–20 μm).

10.4.3 Thermal Insulation

Radiation frost nights are generally preceded by clear days during which the soil receives some heat from the sun. Where frost-threatened plants are small enough, it is possible to protect them by covering the plants with certain kinds of material. These covers, usually termed "hot caps," are placed over small plants in the late afternoon and removed on the following morning. When the frost is not protracted or of great severity protection by this method is usually effective.

Blanc et al. (1963) thoroughly reviewed the principles and practices of frost protection and the use of covering material of many kinds. Hot caps should be made of materials with low emissivity in the thermal infrared wave band and with low thermal conductivity as well. Figure 10.4a,b show bean

Fig. 10.4 (a) Black plastic supported by straw protected bean plants during a severe radiation frost in late May at Lincoln, Nebraska. (b) Kraft paper used for the same purpose.

METHODS OF FROST PROTECTION   379

plants that survived a sharp radiation frost after being covered with black plastic and kraft paper during a frost night in late May at Lincoln, Nebraska.

Another approach to insulating plants against the effects of frost is reported by Canadian scientists (Desjardins and Siminovitch, 1968; Siminovitch et al., 1972). They covered plants with a nontoxic protein-based foam material that was found very effective under radiation frost conditions but less effective in advective frosts when winds can blow it away.

The foam tends to damp the amplitude of the daily temperature wave. It reduces radiative loss from the vegetation to the sky and conserves heat released by the soil. Areas treated with foam appeared to benefit from a residual fertility effect due to the protein content of the foam material. The technology of foam production and application and the results of studies in a number of Southern states are described in detail by Bartholic and Braud (1979).

Another form of thermal insulation is accomplished through the use of wrapping materials. These are used to prevent freezing in the trunk and lowest branches of fruit and other commercially important trees. Although

Fig. 10.5 Rubber trees with double wrapping protection against freezing (courtesy of Prof. Hueng Ping Wei, Institute of Geography, Academia Sinica, PRC).

a severe freeze may kill the upper part of the tree, a protected trunk may enable rapid regeneration of the branches. Fucik (1979) gives a detailed description of techniques and materials used for tree wrapping. Among materials most commonly used are standard fiberglass insulation batting, polyurethane foam, and foil-covered corrugated cardboard. A double-wrapping technique involving a polyethylene plastic outer cover has been tested in the Peoples Republic of China as a defense against trunk freezing in the very sensitive rubber tree (Fig. 10.5).

The effectiveness of trunk wrappings lies in their ability to slow the rate of temperature change in the trunk during freezing weather. Hence they also reduce the time that the trunks are exposed to below-freezing temperature. If the temperature lag induced by the wrap is great enough, the minimum temperature reached before warming begins will be higher than that of exposed trunks. That wraps may lead to disease and pest problems must also be considered. Certain wrapping materials have been found to harbor rodents and insects. If the materials remain damp for long periods of time they may foster the activity of plant pathogens as well. The reader is referred to Fucik (1979) for a thorough discussion of these problems.

10.4.4 Air Mixing

During radiation nights in "frost season," the net radiation normally ranges from about -60 to -100 W m^{-2}. The air will cool as it comes in contact with the surfaces that radiate skyward. Convective mixing is normally limited under the circumstances of radiation frost and so cooling occurs from the ground up. At some point above the crop surface, air temperatures will be above the freezing point. If so, frost can be disrupted by agitation and mixing of the warm air above with the colder air below. Sometimes wind movement will occur spontaneously and disrupt the frost, as is shown in Fig. 10.2. To rely on such spontaneous occurrences where valuable crops are at stake would, however, be unwise.

Large tower-mounted engine-driven propellers and fans are used in citrus orchards in California and Arizona to drive warm air from within the inversion layer downward. Bates (1972) describes the use of a propeller system in a cherry orchard in Oregon where temperature gradients between 1.2 and 18 m varied from 1 to 9°C. He found some effect on temperature over an area of about 5 ha. The system was able to maintain an area about 2 ha in size at a safe temperature on all frost nights. Of course, protection diminishes with distance from the tower so that a grid of towers is needed within the orchard.

Helicopters have been used occasionally to mix the air in fields where permanent installations of towers are not feasible or where the value of the crop does not justify the use of such installations. Gerber (1979) reports that helicopters can, in fact, be more efficient than tower-mounted propellers since they produce mixing by vertically directing warmer air downward into the orchard.

It has also been demonstrated that the stronger the inversion, the more efficient is the protection accomplished with wind machines in terms of energy input per unit of land area protected. Wind machines must rotate continually so as to reach each area at least once in 5 min, since without agitation the inversions are quickly reestablished.

The theory of wind machine usage is given by Crawford (1965). For an interesting review of the history of wind machines see Gerber (1979).

10.4.5 Direct Air and Plant Heating

The existence of the temperature inversion during radiation frosts provides the physical basis for yet another means of frost protection. Air heated near the ground will become buoyant and rise to a level where ambient air is at the same temperature. If air temperatures are raised to a few degrees above freezing by controlled heating, the heated air will remain under a roof provided by the temperature inversion. Then the volume of air below that level will be warmed. The air temperature can be maintained above freezing only by a continued input of heat, however, since radiation to space of energy from the surface and the air continues. Wind movement will also result in the horizontal transfer of heat out of the protected zone, and some vertical diffusion and turbulent transport of heat are also inevitable.

Crawford (1965) computed heating requirements for frost protection, using a steady-state energy balance approach applied to a defined volume of air including the field or orchard. This "box model" approach considers radiation balance, advection, and induced flow components of the energy balance. The induced flow is due to rising heated air and the subsequent intake of colder air from the surroundings. A wide range of meteorological conditions was considered. The height of the convective plume emanating from the heaters is a function of the intensity of the inversion and the intensity of the fire, which can be measured by fuel consumption (Crawford, 1965).

The use of box models for designing heating strategies has advanced. Martsolf (1979) reviews the theory of the box model and describes advances that consider the generation of turbulent as well as laminar components of flow in the system (Martsolf and Panofsky, 1975). Computer programs for rapid calculation of heating requirements are now available. These are based on models such as are described above and consider the characteristics of the heating systems, as well.

Many types of heaters have been developed for semipermanent placement in orchards and groves. These heaters usually burn oil or kerosene. Natural gas fueled heating systems can be installed and supplied from a central point in regions where that fuel is easily accessible. Valli (1970) recommended such a system because of the cleaner burning characteristics of natural gas and because the heating rate can be altered easily by controlling the pressure with which the gas is supplied to the burners. One hundred burners per hectare were used in the system that Valli described.

It is important that many small heaters be used in the orchard rather than a few large ones. This is so because large fires may heat small volumes of air excessively. Extreme buoyancy would cause these heated eddies to escape through the "roof" of the inversion.

Aside from constructed orchard heaters of the type mentioned above, other methods of air heating have been proposed. Briquettes of compressed petroleum materials (often waste products) can be ignited and dropped, as the need arises, from vehicles moving through an orchard. The briquettes need no refueling and will burn until totally consumed or until extinguished after a frost.

The objective of frost control is to maintain vegetative tissues above the lethal temperature. Often this can be accomplished by simply warming the air in which the crop grows. Radiant heating permits the crop tissues to be warmed directly. Oil and gas heaters heat the air by convection, primarily, but it is also important to recognize that fueled heaters radiate significantly to the vegetation and a considerable portion of their effect is due to direct radiation as well as to convective warming under the inversion lid. Depending on the type of heater, some 10–30% of the heat conveyed into the orchard may be radiated by the hot surfaces of the burners. For a detailed analysis of this component of the heating see Welles et al. (1977).

In principle, ir heating systems can be specially designed for orchard frost protection. Dish-shaped surfaces can be heated and focused at a group of trees. With a sufficient number of such radiators placed strategically in an orchard, complete coverage should be possible. Experimental equipment for this purpose was developed in Michigan by Hassler et al. (1948), but such systems are not widely used.

10.4.6 Application of Water

Water can be used effectively to prevent frost injury. Irrigation water drawn from the ground is often at a uniform temperature significantly above freezing. In Nebraska, for example, groundwater is generally at about 12°C. Thus, each kg of water applied provides about .050 MJ to the air with which it comes in contact. This thermal effect is small, however, when compared to the liberation of heat which occurs when water freezes. The heat of fusion of water at 0°C is about 0.334 MJ kg^{-1}. Thus, irrigation water may contribute about 0.384 MJ kg^{-1} in the process of cooling and freezing.

Water is applied in frost protection by sprinkling systems and by flood irrigation. Each method has its advantages and disadvantages. The physics and energy relations underlying the sprinkling method are given in detail by

Businger (1965). Plants may be sprinkled with the onset of freezing temperatures. As water freezes onto the plant parts the heat of fusion is liberated. As long as the freezing continues, the temperature of the ice will remain at about 0°C. Gerber and Martsolf (1979), however, point out that in order for adequate cold protection to be accomplished by the application of water by sprinkling enough heat must be released to compensate for heat lost by radiation, convection, and evaporation. According to Gerber the evaporative heat loss is far greater than the radiative and convective losses. Hence, some 7.5 times as much water must be frozen as is evaporated or the plants will actually be chilled below ambient temperature.

Sprinkling must be continued until well after dawn or until temperatures have risen sufficiently to melt all the ice. If sprinkling is discontinued prematurely, heat will be drawn from the plant parts to melt the ice and serious frost damage, avoided until that instant, may occur. Care must be exercised in sprinkling tall growing plants and trees. Heavy ice loads may develop and cause serious breakage.

Davis (1955) reported that the sprinkling method is especially applicable to low-growing vegetable crops such as tomatoes, cucumbers, peppers, beans, and squash and to cranberries and strawberries. He reported, too, that in Michigan these crops have been successfully protected against temperatures as low as $-5°C$. Gerber and Martsolf (1979) list many other examples of the use of sprinklers in frost protection. Sprinklers must be carefully chosen to provide uniform and nearly constant coverage. If ice is not accreting in a constant fashion (or nearly so) the plant may freeze as heat is withdrawn from the tissue.

Fields may also be flood irrigated as a means of protecting crops from frost. Such a measure is extreme and is likely to be ineffective in advective frost since winds would rapidly remove the liberated heat from the flooded fields and freeze the unprotected vegetation. However, during periods of positive net radiation preventive irrigation by flooding can be used to increase heat storage in the soil by increasing soil thermal conductivity (Georg, 1979). Georg also lists a number of crops that are protected by flooding. These include cranberries in the Northeast United States, citrus in California and Florida, and truck and sugar crops in parts of Florida. Air in the vicinity of pimento peppers in Florida have been warmed by as much as 3°C by means of flooding according to Georg (1979).

Only those layers of soil that undergo temperature change during a diurnal cycle can contribute heat to the air during the period when the net radiation is negative. Thus there is no point in irrigating more deeply than 0.3–1.0 m in general. In Chapter 2 information is given on the actual depth of soil involved in diurnal heat exchange.

Sprinkling and furrow irrigation to protect late-fall tomatoes were compared in an experiment conducted in Lincoln, Nebraska, by Rosenberg. The layout of plots is seen in Fig. 10.6a. Replicate plots were furrow irrigated in anticipation of a forecasted frost. This furrow irrigation continued through-

Fig. 10.6 (*a*) Field plots of tomatoes before a severe radiation frost in fall at Lincoln, Nebraska. One-third of the plots were furrow irrigated beginning a few hours before the frost and throughout the night. One-third of the plots were sprinkler irrigated

Fig. 10.6 (*Continued*)

out the night. Sprinkling began only when air temperatures had dropped to near freezing. Figure 10.5*b,c*, taken on the following morning, shows the complete destruction of the unprotected plants and the survival of the sprinkled crop. Flooding was less successful than sprinkling, but crop damage was minor in this case too.

An unusual use of sprinkling to protect crops from frost is that of Cary (1974). He constructed a screen to totally cover an orchard and sprinkled the screen during freezing weather. Sprinklers were also used within the screen intermittently when temperatures dropped too low. Ice that formed on the screen protected the inner area from cold winds.

10.4.7 Chilling to Prolong Dormancy

In Section 10.4.6 the uses of sprinkling to protect against frost were described. In another innovative use of sprinklers in the fight against frost, fruit trees are sprinkled in late winter and early spring to *cool* rather than

---

from the onset of the freezing temperatures. One-third of the plots were unprotected in any way. (*b*) Unprotected tomato plants on the morning after a severe radiation frost. (*c*) Tomato plants that had been sprinkler irrigated from the onset of freezing temperature until after the ice had melted off the plants.

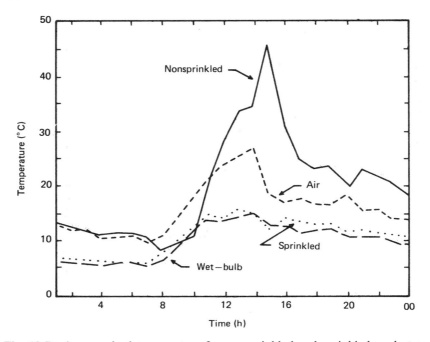

Fig. 10.7 Average bud temperature for nonsprinkled and sprinkled apple trees, as compared with air and wet-bulb temperatures in May at Logan, Utah (after Griffin and Richardson, 1979; reprinted with permission from the American Society of Agricultural Engineers).

warm them. The purpose of chilling is to prevent an early break in the dormancy or "rest period" of the trees so that bud development is delayed and buds do not become exposed to freezing temperatures. Lombard and Richardson (1979) and Griffin and Richardson (1979) explain the physical principles involved in controlling the phenological development of fruit trees. Essentially the trees must be exposed to a certain number of hours of temperatures below some critical threshold after growth cessation occurs in fall. The end of the rest period occurs after a fixed number of chill units has been achieved, usually sometime between mid winter and early spring. Budding can occur anytime thereafter. If early spring temperatures are low, blossoming is delayed. When spring temperatures are high development accelerates and trees blossom early. Frosts at this time can severely reduce yield potential by destroying the buds.

Research from a number of regions was summarized by Griffin and Richardson (1979) who found that bud formation has been delayed by means of sprinkling by from 8 to 17 days for various varieties of apples and peaches. Sprinklers of various kinds have been used in this application. Water should be applied uniformly and continuously when temperatures are high so that bud temperature is lowered by evaporative cooling. The efficiency of the cooling varies with air temperature and humidity, wind speed, net radiation, and with the rates and timing of application. In Fig. 10.7 the results of a test with apple trees in Logan, Utah, are shown. Contact thermometry indicated very high temperatures in midafternoon. These were reduced by nearly 30°C by means of sprinkling. Hamer (1980) has designed a sprinkler system that controls application rate according to simulations of bud temperatures in a model that considers net radiation, wind speed, and air temperature.

10.4.8 Soil Manipulation

Soil temperatures reach their peak in the fall. Thus, especially under conditions of radiation frost in fall, it may be possible to use the heat stored in the soil to mitigate the effects of frost. Methods of management that increase the thermal conductivity and diffusivity of the soil will be helpful in this regard, and methods that decrease them will be detrimental. One of the appropriate techniques, flood irrigation, was discussed above.

Compaction has the effect of increasing conductivity by reducing the volume of air-filled voids in soil. Bridley et al. (1965) attempted to reduce frost damage by rolling soil in an Australian vineyard to compact it. On clear nights with minimum temperatures ranging from -5.5 to $17.7°C$, rolling clean cultivated soil led to an average 0.4°C temperature increase, rolling plus moderate irrigation caused a 0.6°C temperature increase, and heavy irrigation alone caused an 0.4°C increase. As the air temperature declined below the average, the effectiveness of the latter two treatments increased so that air temperature increases as great as 1.1–1.4°C were experienced.

The existence of a grass or shrub cover in an orchard may reduce the effectiveness of the soil as a source of heat on frost nights. Ashcroft and Richardson (1962) suggest that the vegetation cover entraps heat emitted from the soil and returns much of it to the soil. The soil remains warmer than it would if bare, thus maintaining higher temperatures for the roots. A grass or shrub cover, however, prevents heat from reaching the aboveground portions of orchard trees, vines, and so on.

Mulches and covers may also be used to manage the soil as a source of heat for frost protection. Fritton and Martsolf (1981) covered bare soil and grass covered soil with clear plastic. The plastic was in place during the day. In late afternoon an aluminum polyester laminate was placed over the plastic to conserve stored heat. The covers were removed in the early morning hours to release heat at the time that frost severity normally becomes serious. Heat output from the plastic covered soils was twice that from bare soil that had not been covered. Surface temperature of the soil before removal of the plastic was nearly 10°C greater than on bare soil. Hence, as expected, the flux of heat out of the soil after uncovering was quite rapid.

REFERENCES

Ashcroft, G. L., and E. A. Richardson. 1962. Climatic problems in orchard heating. Proceedings of the Utah State Horticultural Society, pp. 29–36.

Barfield, B. J., and J. F. Gerber, eds. 1979. *Modification of the Aerial Environment of Plants*. Am. Soc. Agric. Engin. Monograph 2, St. Joseph, Missouri. 538 pp.

Bartholic, J. F., and R. A. Sutherland. 1975. Modification of nocturnal crop temperature by a lake. 12th Conf. on Agricultural and Forest Meteorology, Am. Meteorol. Soc., Tucson, Arizona, April 14–16.

Bartholic, J. F., and R. A. Sutherland. 1976. Cold climate mapping using satellite high resolution thermal imagery. 7th Conf. on Aerospace and Aeronautical Meteorology and Symposium on Remote Sensing from Satellites. Am. Meteorol. Soc., Boston, Massachusetts, Nov. 16–19.

Bartholic, J. F., and H. J. Braud. 1979. Foam insulation for freeze protection. *Modification of the Aerial Environment of Plants* (B. J. Barfield and J. F. Gerber, eds.), *Am. Soc. Agric. Engin. Monogr.* **2**:353–363.

Bates, E. M. 1972. Temperature inversion and freeze protection by wind machine. *Agric. Meteorol.* **9**:335–346.

Bergen, J. D. 1969. Cold air drainage on a forested mountain slope. *J. Appl. Meteorol.* **8**:884–895.

Biel, E. R. 1961. Microclimate, bioclimatology and notes on comparative dynamic climatology. *Am. Scientist* **49**:327–357.

Blanc, M., H. Geslin, I. A. Holzberg, and B. Mason. 1963. Protection against frost damage. Technical Note No. 51, WMO, 133, TP60. 62 pp.

Bootsma, A. 1976. Estimating minimum temperature and climatological freeze risk in hilly terrain. *Agric. Meteorol.* **16**:425–443.

Bouchet, R. J. 1965. Problèmes des gelées de printemps. *Agric. Meteorol.* **2**:167–195.

Bridley, S. F., R. J. Taylor, and R. T. J. Webber. 1965. The effects of irrigation and rolling on nocturnal air temperatures in vineyards. *Agric. Meteorol.* **2**:373–383.

Businger, J. A. 1965. Protection from the cold. Frost protection with irrigation. *Agricultural Meteorology* (P. E. Waggoner, ed.), Chap. 4, Sec. 1. *Meteorol. Monogr.* **6**:74–80. Am. Meteorol. Soc., Boston.

Caprio, J. M. 1961. A rational approach to the mapping of freeze dates. *Bull. Am. Meteorol. Soc.* **42**:703–714.

Cary, J. W. 1974. An energy-conserving system for orchard cold protection. *Agric. Meteorol.* **13**:339–348.

Crawford, T. V. 1965. Protection from the cold. Frost protection with wind machines and heaters. *Agricultural Meteorology* (P. E. Waggoner, ed.), Chap. 4, Sec. 2. *Meteorol. Monogr.* **6**:81–87. Am. Meteorol. Soc., Boston.

Davis, J. R. 1955. Frost protection with sprinkler irrigation. *Michigan State College Extension Bull.* **327**:1–12.

de Boer, H. 1965. Comparison of water vapor, water and water drops of various sizes, as a means of preventing night frost. *Agric. Meteorol.* **2**:247–258.

Desjardins, R. L., and D. Siminovitch. 1968. Microclimatic study of the effectiveness of foam as protection against frost. *Agric. Meteorol.* **5**:291–296.

Fucik, J. E. 1979. Protecting citrus from freezing with insulating wraps. *Modification of the Aerial Environment of Plants* (B. J. Barfield and J. F. Gerber, eds.), *Am. Soc. Agric. Engin. Monogr.* **2**:364–367.

Fritton, D. D., and J. D. Martsolf. 1981. Solar energy, soil management and frost protection. *Hort. Sci.* **16:**295–296.

Georg, J. G. 1979. Frost protection by flood irrigation. *Modification of the Aerial Environment of Plants* (B. J. Barfield and J. F. Gerber, eds.), *Am. Soc. Agric. Engin. Monogr.* **2:**368–370.

Gerber, J. F. 1979. Mixing the bottom of the atmosphere to modify temperatures on cold nights. *Modification of the Aerial Environment of Plants* (B. J. Barfield and J. F. Gerber, eds.), *Am. Soc. Agric. Engin. Monogr.* **2:**315–324.

Gerber, J. F., and J. D. Martsolf. 1979. Sprinkling for frost and cold protection. *Modification of the Aerial Environment of Plants* (B. J. Barfield and J. F. Gerber, eds.), *Am. Soc. Agric. Engin. Monogr.* **2:**327–333.

Griffin, R. E., and E. A. Richardson. 1979. Sprinklers for micro-climate cooling of bud development. *Modification of the Aerial Environment of Plants* (B. J. Barfield and J. F. Gerber, eds.), *Am. Soc. Agric. Engin. Monogr.* **2:**441–455.

Hamer, P. J. C. 1980. An automatic sprinkler system giving variable irrigation rates matched to measured frost protection needs. *Agric. Meteorol.* **21:**281–293.

Hassler, F. J., C. M. Hanson, and A. W. Farrell. 1948. Protection of vegetation from frost damage by use of infrared energy. *Mich. Agric. Exp. Stn. Q. Bull.* **30:**339–360.

Hocevar, A., and J. D. Martsolf. 1971. Temperature distribution under radiation frost conditions in a central Pennsylvania valley. *Agric. Meteorol.* **8:**371–383.

Katz, R. W., and A. H. Murphy. 1979. Assessing the value of frost forecasts to orchardists: A decision-analytic approach. 14th Conf. on Agricultural and Forest Meteorol., Am. Meteorol. Soc., Minneapolis, Minnesota.

Lombard, P., and E. A. Richardson. 1979. Physical principles involved in controlling phenological development. *Modification of the Aerial Environment of Plants* (B. J. Barfield and J. F. Gerber, eds.), *Am. Soc. Agric. Engin. Monogr.* **2:**429–440.

Martsolf, J. D., and H. A. Panofsky. 1975. A box model approach to frost protection research. *Hort. Science* **10:**108–111.

Martsolf, J. D. 1979. Heating for frost protection. *Modification of the Aerial Environment of Plants* (B. J. Barfield and J. F. Gerber, eds.), *Am. Soc. Agric. Engin. Monogr.* **2:**291–314.

Mee, T. R., and J. F. Bartholic. 1979. Man-made fog. *Modification of the Aerial Environment of Plants* (B. J. Barfield and J. F. Gerber, eds.), *Am. Soc. Agric. Engin. Monogr.* **2:**334–352.

Nixon, P. R., and T. A. Hales. 1975. Observing cold night temperatures of agricultural landscapes with an airplane-mounted radiation thermometer. *J. Appl. Meteorol.* **14:**498–505.

Rahn, J. J., and D. M. Brown. 1971. Corn canopy temperatures during freezing or near-freezing conditions. *Can. J. Plant Sci.* **51:**173–175.

Rollins, H. A., Jr., F. S. Howlett, and F. H. Emmert. 1962. Factors affecting apple hardiness and methods of measuring resistance of tissue to low temperature injury. *Ohio Agric. Exp. Stn. Res. Bull.* **901:**1–71.

Rosenberg, N. J., and R. E. Myers. 1962. The nature of growing season frosts in and along the Platte Valley of Nebraska. *Mon. Weather Rev.* **90:**471–476.

Rosenberg, N. J. 1963. Climate of the Central Platte Valley. Nebraska Agric. Exp. Station Misc. Publ. No. 9. 76 pp.

Schaal, L. A., J. E. Newman, and F. H. Emerson. 1961. Risks of freezing temperatures—spring and fall in Indiana. *Purdue U. Agric. Exp. Stn. Res. Bull.* **721:**1–20.

Schlegal, J., and G. Butch. 1980. The Barrens: Central Pennsylvania's year-round deep freeze. *Bull. Am. Meteorol. Soc.* **61**:1368–1373.

Schnell, R. C. 1976. Bacteria acting as natural ice nucleants at temperatures approaching −1°C. *Bull. Am. Meteorol. Soc.* **57**:1356–1357.

Shaw, R. H. 1954. Leaf and air temperatures under freezing conditions. *Plant Physiol.* **29**:102–104.

Siminovitch, D., B. Rheaume, L. H. Lyall, and J. Butler. 1972. Foam for frost protection of crops. Agriculture Canada Publ. 1490. 24 pp.

Thom, H. C. S., and R. H. Shaw. 1958. Climatological analysis of freeze data for Iowa. *Mon. Weather Rev.* **86**:251–257.

Valli, V. J. 1970. The use of natural gas heating to prevent spring freeze damage in the Appalachian fruit region of the United States. *Agric. Meteorol.* **7**:481–486.

Ventskevich, G. Z. 1958. *Agrometeorology.* Translated from Russian by the Israel Program for Scientific Translation for National Science Foundation OTS 60-51044, 1961.

Von Lengerke, H. J. 1978. On the short-term predictability of frost and frost protection—A case study on Dunsandle Tea Estate in the Nilgiris (South India). *Agric. Meteorol.* **19**:1–10.

Welles, J. M., J. M. Norman, and J. D. Martsolf. 1977. A model of foliage temperature distribution in heated and non-heated orchards. Proc. 13th Agricultural and Forest Meteorol. Conf. Amer. Meteorol. Soc., pp. 47–48.

Whiteman, T. M. 1957. Freezing points of fruits, vegetables and florist stocks. U.S. Dept. of Agric. Marketing Res. Report. No. 196. 32 pp.

Yankofsky, S., Z. Levin, T. Bertold, and N. Sandlerman. 1981. Some basic characteristics of bacterial freezing nuclei. *J. Appl. Meteorol.* **20**:1013–1019.

CHAPTER 11—WATER USE EFFICIENCY IN CROP PRODUCTION: NEW APPROACHES

11.1 INTRODUCTION

A number of factors are converging to make improved water use efficiency (photosynthesis/evapotranspiration or yield/total water consumption) essential in crop production. These factors include

1. Declining supplies of irrigation water in certain parts of the world.
2. Increasing costs of energy required to deliver irrigation water to where it is needed.
3. Growing demand for food, feed, and fiber.
4. Increasing pressure to expand production into more arid environments.

Windbreaks, described in Chapter 9, provide one well-known and commonly practiced method for modifying the microclimate and improving water use efficiency. A number of other methods have been proposed. Of these antitranspirant materials and reflectants have been extensively tested in recent years. Work with reflectants, especially, has led to some field research on architectural variants that are naturally more reflective for solar radiation. Other architectural changes in important crop plants may also prove effective in improving water use efficiency.

There is much speculation at this time on the question of whether or not climate is changing. This general question is beyond the scope of this book. Climate *is*, however, changing in at least one important way, for the concentration of carbon dioxide in the atmosphere has increased since the start of the industrial revolution (ca. 1865) and is now perhaps 15% greater than it was then. Plants *must* respond to this change in the composition of the atmosphere. We know, for example, that it is possible to enrich or "fertilize" the air in greenhouses to increase production. Fertilization of large fields would likely have a positive effect, although it is not yet practical to do so. A brief discussion of the impacts of the increasing CO_2 concentration in the atmosphere on water use efficiency concludes this chapter.

11.2 ANTITRANSPIRANTS

In Chapters 7 and 8 the "resistance-type" equations for evapotranspiration and photosynthesis were given as

$$LE = \frac{\rho_a L \epsilon}{P} \frac{e_a - e_{ss}}{r_a + r_c} \quad (7.49)$$

and

$$P = \frac{C_a - C_{chl}}{r_a + r_c' + r_m} \quad (8.5)*$$

The magnitudes of the resistances r_a, r_c', and r_m are also discussed in Chapter 8. Generally, when plants are photosynthesizing effectively and stomates are open, the mesophyll resistance is considerably greater than the stomatal resistance. If, by some means, the stomates can be made to close partially, canopy resistance will increase. The total impact of such an increase in r_c will be greater on the evapotranspiration rate than on the rate of photosynthesis. Thus water use efficiency should be increased. This concept was elaborated by Zelitch and Waggoner (1962) and Waggoner et al. (1964) and was reviewed by Gale and Hagan (1966) and Davenport et al. (1969, 1978). Experimental and numerical studies, cited below, have raised questions as to whether the dependency of water use efficiency on canopy resistance is quite so straightforward, however.

Sinclair et al. (1975), for example, state that water use efficiency in maize decreases with increasing stomatal resistance because of an increase in leaf temperature which, in turn, increases the leaf's internal vapor pressure. By means of a numerical model Campbell (1977, p. 125) demonstrates that irradiance is also involved. For a range of r_s values from 100 to 3200 s m^{-1} and for an air temperature of 30°C water use efficiency is shown to *increase* with increasing PAR flux density to about 100 W m^{-2}. With greater irradiance water use efficiency decreases at rates which become steeper the greater the stomatal resistance. It now seems evident from the literature based on single leaf studies that increased leaf temperature due to increased stomatal resistance can lower water use efficiency in at least two ways: by increasing leaf internal vapor pressure, and by increasing rates of dark respiration.

Davenport et al. (1969) described three types of synthetic antitranspirant materials: (1) film-forming substances that block the escape of water vapor from the leaf, (2) chemical materials that induce stomatal closure, and (3) reflectant materials that reduce the energy load on the leaf. There are also natural antitranspirants produced by plants. Ogunkanmi et al. (1974) found that sorghum plants under water stress produce abscisic acid (ABA), an

---

* Canopy resistance r_c' is substituted for stomatal resistance r_s' in this application of (8.5). These terms are used interchangeably as appropriate in the following pages.

endogenous chemical antitranspirant. Natural coatings of waxes, cutin and other materials also act as antitranspirants.

Film-forming materials include long-chain alcohols such as hexadecanol. Woolley (1962) found that hexadecanol reduced transpiration, but only because leaf area was reduced, apparently because growth processes were impeded by the coating. Gale and Hagan (1966) found such materials unsatisfactory for use on plant leaves. Low-viscosity silicone materials were tested by Angus and Bielorai (1965), who found water use by potted sunflowers reduced with no detrimental effects on growth. Lee and Koslowski (1974) found that silicone had a very short-lived antitranspirant effect on certain woody plants and, also, caused some plant damage. Belt et al. (1977) found that application of a silicone antitranspirant on a watershed-sized area led to increased streamflow. Gale et al. (1967) found that monomolecular films of high alcohols, when applied at concentrations sufficient to reduce transpiration, also reduce growth because their resistance to the passage of CO_2 is greater than it is for water vapor. Materials that form monomolecular layers on the surface to which they are applied have also been used to reduce evaporation from open bodies of water.

Thick-film antitranspirants including latex, waxes, and plastics have been applied to plants. To be effective, such materials would have to be more permeable to carbon dioxide than to water vapor. Gale and Poljakoff-Mayber (1967) tested a number of such materials and found these materials also to be considerably more permeable to water vapor.

Maki (1974) used a material called Abion-E, a mixture of 30% paraffin in an emulsion of fatty acid. Wind frictional scratch damage to plant tissue was reduced with Abion-E although effects on water use were uncertain. An emulsified beef tallow has also been applied as a mist to greenhouse and field grown potatoes. Rates of water use were reduced and thus the duration of water availability was extended by the treatment. Stress was less frequent and quality of the potatoes was improved.*

Stomate-closing materials were classified by Waggoner and Zelitch (1965) according to their modes of operation. One type acts as a "pump" by affecting the turgor of stomatal guard cells. Atrazine, a commonly used herbicide, and hydroxysulfonates are in this group. A second type of material is believed to affect the permeability of cell membranes to solutes, acting as a "check valve" and retarding all osmotically induced events. Phenylmercuric acetate (PMA) and alkenylsuccinic acids are believed to work in this way.

Hales and Rosenberg (1970) tested Atrazine applied in combination with a surfactant, Vatsol OT, to alfalfa in greenhouse and field studies. Atrazine reduced alfalfa evapotranspiration but caused some foliar damage. Yadav and Singh (1981) applied Atrazine with and without 6% kaolin solution to

---

* Personal communication Dr. Charles Wendt, Texas Technical University, Lubbock, Texas.

barley plants. Kaolin had no effect. Some reduction in evapotranspiration was noted during the first 2 weeks but afterwards it was increased. As a result, there was no difference in seasonal water use. Nonetheless, yield was improved significantly, perhaps because of a reduction in water stress at a critical time. Fuehring (1975) applied Atrazine as an antitranspirant to sorghum and found that yields increased significantly when nitrogen was in adequate supply and, especially, when plants were stressed for water.

Zelitch and Waggoner (1962) found PMA to be effective on tobacco leaves in the greenhouse. Slatyer and Bierhuizen (1964a,b) with cotton and Shimshi (1963b) with corn found, under greenhouse and growth chamber conditions, that PMA did indeed effect a proportionately greater decrease in transpiration than in photosynthesis. Cole et al. (1971) tested PMA and dodecenyl succinic acid (DSA) on greenhouse and growth chamber-grown alfalfa. Neither material had an effect on water requirement, and DSA caused foliar damage.

The results of field studies with PMA and DSA have not been particularly encouraging. Shimshi (1963a) used PMA on sunflower plants in concrete bins out of doors in Israel and noticed some plant damage. Davenport (1967) in England applied PMA and DSA to small lysimeters in a field of creeping red fescue. Both materials caused about 10% reduction in water use. PMA caused yellowing 3 days after its application, whereas DSA-treated plants appeared normal. Davenport et al. (1971) found transpiration by PMA treated oleander to increase in the dark but decrease in the light with an overall water saving effect.

Waggoner et al. (1964) and Monteith et al. (1965) used methylester of noenyl succinic acid (NSA) on barley. Stomatal resistance was increased in the treated plants only on the first day after treatment. No differences were noted where the radiant flux density was less than $210-280$ W m^{-2}. Neither production nor water use was affected on irrigated or unirrigated potatoes by the appliction of DSA (Fulton, 1967), although the stomatal aperture was reduced by 30%. This result suggests a "feedback" effect, most likely an increase in leaf temperature and leaf internal vapor pressure that could compensate for the stomatal closure. PMA, on the other hand, was reported effective in reducing water stress and transpiration rate in a form of "droopy" potato, which wilts easily because of wide-open stomates (Waggoner and Simmonds, 1966). Pasternak and Wilson (1971) also found PMA effective in reducing stomatal aperture and transpiration rate in greenhouse and field grown tomatoes, maize, peanuts, sorghum, and cotton.

Hales and Rosenberg (1970) applied PMA and a silicone material, Antifoamer A, to large plots of alfalfa. Neither material had any significant effect on water use, although PMA suppressed evapotranspiration for a day or two after its application. A small increase in the leaf temperature also occurred in the treated crop. Moreshet (1975) found that PMA lead to increased cuticular resistance in sunflower plants that had been subjected to water stress.

There are at least a few cases in which an antitranspirant material has been effectively used in a natural situation. Waggoner and Hewlett (1965) applied the glyceryl half-ester of decenylsuccinic acid (GLOSA) to the undersides of broad-leafed trees in small watersheds of a Southern forest. Regression analyses showed that a 12% reduction of transpiration could be detected as a significant increase in streamflow. When the material was sprayed from a heliocopter, little reached the undersides of the leaves and no significant streamflow change occurred. Turner and Waggoner (1968) reduced water use in a red pine stand by 10% after two sprayings with 10^{-3} M PMA. Leaf resistance was increased with the partial stomatal closure and absolute water potential in the needles was increased by 0.1–0.3 MPa at midday. Waggoner and Turner (1971) report a seasonal reduction in evapotranspiration from two red pine plantations in Connecticut of 14–32 mm due to the use of PMA.

Brooks and Thorud (1970) treated ponderosa pine and Douglas fir with 10^{-3} M PMA in a greenhouse. Effects lasted for about 7 days during which no temperature increase was observed. Some damage was done to the plant tissue.

In a wind tunnel setting tobacco plants were treated with PMA, monoglycerol ester of n-decenyl succinic acid, and with two commercial film forming materials (Kreith et al., 1975). The former reduced transpiration by more than 60%; the latter by 35–50% of a control. The film forming materials had a longer lasting effect.

Thus, the literature reports a general lack of field success in the use of antitranspirant materials, while greenhouse results have been more encouraging. The reasons for the lack of effectiveness of this technique in the field remain speculative at this time. Waggoner and Zelitch (1965) suggest one explanation: "since several acres of leaves grow on each acre of land, there may be so many stomata per acre that stomatal resistance may be negligible compared to other resistances, as long as stomata are open the least crack."

Another possible mechanism to explain the lack of effectiveness of antitranspirants in the field lies in their effect on leaf temperatures. Tanner (1963) found a leaf temperature increase of about 1°C when a film-forming material was applied to alfalfa in Wisconsin. Hales and Rosenberg (1970) also report that PMA applied to alfalfa in the field caused leaf temperatures to increase by about 1°C, an effect that lasted for a few days. Leaf temperatures may increase if stomatal closure results in a reduction in transpiration. A leaf temperature increase, however, will increase the vapor pressure in the substomatal cavity, since the air in the cavity is nearly saturated. Thus the vapor pressure gradient between the leaf and air will increase and, if all other factors remain identical, the transpiration rate will increase. Slatyer and Bierhuizen (1964a), after testing a number of transpiration suppressants, concluded that, aside from their antitranspirant effects, all reduced photosynthesis as well. This they attributed to the increased resistance to diffusion of CO_2 and also to metabolic inhibition of photosynthesis.

Although field studies with antitranspirant materials have generally been disappointing, there may nevertheless be situations where the use of antitranspirants can be helpful. In situations where the effects of droughts can be lessened by a reduction in transpiration for even a few days, a crop or other vegetative community may survive until the next rainfall. Antitranspirants may also be used to help in establishing ground cover under conditions of severe moisture stress.

11.3 REFLECTANTS

11.3.1 Physical and Optical Effects

Reflectant materials may be considered to be antitranspirants. However, the mechanism involved is somewhat different than that of the film-forming or stomate-closing materials discussed above. If net radiation is reduced over a plant community, less energy will be available for consumption in latent heat flux, sensible heat flux, and soil heat flux.

Net radiation can be decreased significantly by increasing the albedo of the underlying surface. This principle has been employed in designing floating covers for tanks and ponds (Cooley, 1970). Ramdas and Dravid (cited in Geiger, 1965) successfully reduced the temperature of a dark Indian cotton soil by the application of lime to the surface. Stanhill (1965) noted reduced evaporation as well as reduced soil temperatures following the application of a whitening material (magnesium carbonate) to a light-colored loessial soil in the Negev Desert of Israel. Weathering reduced the effectiveness of the coating but treated soil was significantly cooler for more than 30 days.

Fritschen (1967) speculated that altering the optical properties of plants may also provide a way of reducing water use and increasing water use efficiency. Seginer (1969) predicted, on the basis of a numerical analysis or "model" that doubling the albedo of plant surfaces could reduce evapotranspiration by 15%. Seginer's analysis presupposed nonadvective conditions. Where the advection of sensible heat is an important source of energy for evapotranspiration, the proportion of total moisture saved through an increase in the albedo would, necessarily, be less.

In the case where leaves or canopies are treated, any increase in reflectivity should decrease the net radiation. Evapotranspiration and sensible heat flux generation should be reduced, but photosynthesis should not be directly affected unless the availability of photosynthetically active radiation becomes critical, as it well might with certain types of crops. Barring this effect, the net result of reflectant application should be an improvement in water use efficiency.

Aboukhaled et al. (1970) determined the natural spectral reflection of bean plants and other species. They also analyzed changes in spectral reflectivity of plants treated with various reflectant materials. In growth chamber studies, they demonstrated that shortwave reflectivity could be effectively increased and transpiration reduced. The apparent photosynthesis of individual leaves was reduced at low visible irradiance because of the shading of the leaf by the reflectant material. At high irradiance, treated leaves continued to photosynthesize effectively while stomatal closure occurred in untreated leaves. The application of kaolinite to the leaves of rubber and bean plants increased shortwave reflectivity (400–700 nm) from about 10 to 65%. A smaller but significant increase in reflectivity in the near ir was also noted.

Doraiswamy and Rosenberg (1974) applied kaolinite in a mixture containing plant gum and a surfactant to a large plot of soybeans at Mead, Nebraska. The soybean was chosen for this study since its photosynthetic apparatus is normally light saturated at a solar radiation flux density of about 400 W m^{-2} in the climate of the eastern Great Plains and with the prevailing populations and cultivars (Sakamoto and Shaw, 1967; Jeffers and Shibles, 1969; Egli et al., 1970). The kaolinate was applied at a rate of 196 kg ha^{-1}. During 10 days of detailed observation that followed, the reflectant coating persisted with no apparent sign of being washed away despite occasional light rains.

A comparison of the reflection spectra from the coated and uncoated crop canopies was obtained in a subsequent study with a spectroradiometer inverted over the crops (Fig. 11.1). Results of such a comparison are given in

Fig. 11.1 Spectroradiometer (background), albedometer (center), and net radiometers (foreground) used to measure components of the radiation balance over reflectant-treated and -untreated crops (after Doraiswamy and Rosenberg, 1974).

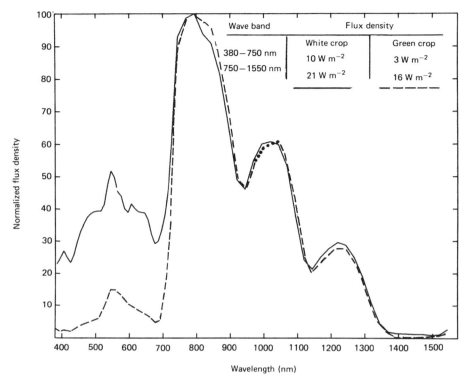

Fig. 11.2 Reflection spectra for kaolinite-coated and untreated soybeans at Mead, Nebraska (after Doraiswamy and Rosenberg, 1974).

Fig. 11.2. Since measurements were made over a period of about 2 h the flux densities of radiation in the treated and untreated plots were not strictly comparable. To minimize differences, the data for each period were normalized with respect to flux density of radiation at 800 nm.

Reflection was increased in the visible wave band (corresponding to approximately 380–750 nm) from 80 to 300%, depending on the crop condition, flux density of shortwave radiation, and time of day. No major difference in the near ir (corresponding approximately to 750–1550 nm) reflection was found to occur. The influence of chlorophyll absorption in the blue-violet region around 450 nm and in the red around 650 nm and the relative prominence of green reflection at about 550 nm are evident, showing no major qualitative difference in the reflection spectra of the treated and untreated canopies.

The effect of a single coating with kaolinite reflectant on the radiation balance is shown in Fig. 11.3. The figure shows an increase in the shortwave reflection of about 56 W m^{-2} under clear skies at midday and a reduction in the net radiation of about the same magnitude.

In subsequent experiments Lemeur and Rosenberg (1976) applied coatings of kaolinite and a diatomaceous earth product (Celite) to soybeans. Celite increased reflection from soybeans even more effectively than kaolinite. A second coating had little effect on the visible reflectance, whereas near infrared reflectance was significantly improved. A secondary effect of the treatment with both kaolinite and Celite reflectants was to increase light penetration deep within the canopy, where the crop is usually light unsaturated. It was also shown in these experiments that soil reflectance was increased as a result of kaolinite application to the surface.

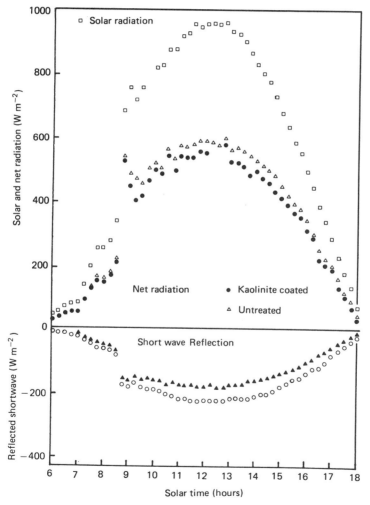

Fig. 11.3 Daytime patterns of solar radiation, shortwave reflection, and net radiation over kaolinite-coated (circles) and uncoated (triangles) soybeans. Mid-August at Mead, Nebraska (after Doraiswamy and Rosenberg, 1974).

400 WATER USE EFFICIENCY IN CROP PRODUCTION: NEW APPROACHES

11.3.2 Plant Physiological Factors

In crops that are not light saturated under their normal growing conditions (e.g., corn, sorghum, and sugar cane) the application of materials that reflect away visible radiation could cause a reduction in photosynthesis as well as in evapotranspiration. Since most available reflectant materials that can be applied directly to plants (kaolinite, diatomaceous earths, aluminum silicates) reflect most effectively in the visible waveband, the practice would seem most appropriate for crops that are light saturated in their regions of adaptation (e.g., soybeans, alfalfa, and sugar beets).

Yet, in a series of experiments conducted in Israel, significant increases in sorghum water use efficiency have been achieved by the application of reflectants. Fuchs et al. (1976) report one experiment where kaolin was applied to the soil, to the crop and to both soil and crop in large field plots (Fig. 11.4). Throughout the day the fully grown untreated sorghum canopy was intermediate in its absorption of solar radiation. Where the soil alone was treated with reflectant, plants growing on it absorbed radiation from below. Where the plant alone was treated, absorption was least (Fig. 11.5). Kaolin applications increased reflectivity of the foliage by 24% at first and by 17% after 5 weeks. Net solar radiation absorption was reduced by 6%. Where the soil was treated the foliage absorbed an additional 9% (half of the incremental reflection from the soil). In another case both crop and soil

Fig. 11.4 Kaolin-coated sorghum plots. (Lower left) untreated; (upper left) plants alone coated; (right) plants and soil coated (courtesy of Dr. Gerald Stanhill, Volcani Institute, Bet Dagan, Israel).

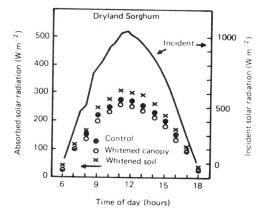

Fig. 11.5 Diurnal variation of solar radiation absorbed by sorghum canopies where the plants alone were coated with kaolin; the soil alone was coated; and the field was untreated (from Fuchs et al., 1976; reproduced from *Agronomy Journal*, Volume 68, pages 865–871, by permission of the American Society of Agronomy).

were treated. The crop absorption of solar radiation did not differ from that of the control, but the field as a whole had a higher albedo.

Stanhill et al. (1976) report, for the same set of experiments, that yield of sorghum was increased as a result of kaolin application to the crop by 446 kg ha^{-1} (or 11%) during three seasons. The best results were obtained when the material was applied during the 10 days beginning 7 weeks after emergence, that is, immediately before panicle emergence. Although total water use was unaffected in these experiments, water use efficiency, because of the greater yield, was improved considerably.

Explanations for these effects are complex. Stanhill et al. (1976) report a slightly increased diffusion resistance due to the physical presence of the reflectant material on the leaf. Moreshet et al. (1977) report that the rate of uptake of labeled CO_2 was reduced on the kaolin-coated leaves. A 23% decrease in apparent photosynthesis accompanied a 26% reduction in solar radiation absorption. They advance the hypothesis that the positive effects on yield were due to a speeding of the senescence of the sorghum leaves. This may have led to an increased rate of translocation from the senescing leaves to the developing grain by an amount greater than the reduction in current rate of CO_2 uptake. An observed increase in the number of panicles per treated plant may also have provided an increased sink strength for photosynthate.

Moreshet et al. (1979) report similar results for dryland cotton coated with kaolinite reflectant. In one study yield was increased by 126%. In a second experiment the reflectant caused a significant increase in number of flowers but subsequent excision rates were high and prevented any significant increase in yield. Labeled CO_2 uptake rate was reduced due to reduction in PAR absorption and diffusion resistance was increased slightly by the coating. No direct measurements of water use are reported but leaf internal water potential decreased more slowly in the treated plants as water stress increased.

11.3.3 Interactions with Plant Temperatures

When exposed to direct solar radiation, a surface with high albedo should be cooler than a surface of low albedo. This effect is simply demonstrated in Fig. 11.6 where a green board and a Celite coated board are compared for temperature. One would expect similarly, that a reflectant-coated plant leaf (or crop canopy) would be cooler than an untreated one. Perhaps surprisingly, however, observations of soybean leaf and canopy temperature in the Nebraska studies cited above revealed that treated leaves and canopies were often warmer by 1–2°C. These observations were made with thermocouples attached to the leaves and with infrared thermometers held a few meters above the crop (Baradas et al., 1976a). Thermal scans made with airborne equipment also demonstrated this effect. Outgoing longwave radiation over treated Amsoy (erectophile) and Beeson (planophile) soybean canopies was increased by about 6–18 W m^{-2} after reflectant application. Outgoing longwave radiation appears to have been slightly greater over both treated and untreated canopies of the Amsoy than over the Beeson cultivar.

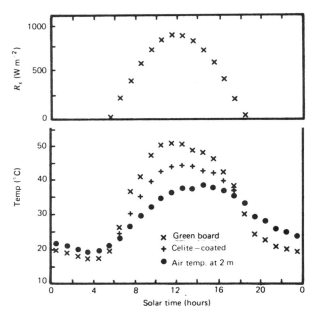

Fig. 11.6 Global radiation, air temperature 2.00 m aboveground, and temperature of green-painted and Celite 209-coated wooden boards (from Baradas et al., 1976a; reproduced from *Agronomy Journal*, Volume 68, pages 843–848, by permission of the American Society of Agronomy).

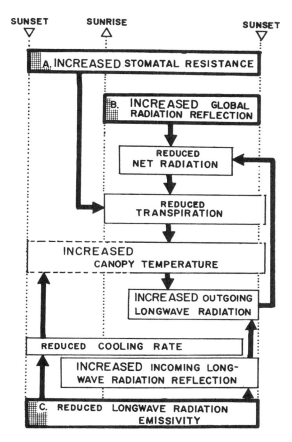

Fig. 11.7 Hypothetical mechanism for reflectant influence on the soybean canopy radiation balance. (from Baradas et al., 1976b; reproduced from *Agronomy Journal*, Volume **68**, pages 848–852, by permission of the American Society of Agronomy).

The finding of increased temperature in the reflectant treated crops held true for canopies of erectophile and planophile cultivars alike. Baradas et al. (1976b) show that longwave emissivity of the treated leaves is lower as is actual latent heat flux. Both factors overweigh the increased shortwave reflectivity and lead to a higher plant temperature.

Higher plant temperature may have some physiological effects on photosynthesis and respiration and, ultimately, on yield (which was not significantly improved in these experiments). In the studies reported by Baradas et al. (1976a) stomatal resistance was also found to be increased by the reflectant.

In an attempt to reconcile the complex of changes induced by application of a reflectant coating to a soybean canopy, Baradas et al. (1976b) proposed a set of hypotheses, shown in Fig. 11.7. The reduction in net radiation over the treated crops was due, about equally, to the increase in reflection of global radiation and to the increase in upward longwave flux. The increase in reflection of incoming longwave radiation, resulting from the reduction in absorptivity of the treated canopy, contributed slightly more to the total upward flux of longwave radiation than did the higher temperature of the treated canopy.

An increase in the temperature of the treated crop occurred coincident with the reduction in R_n. The reduced absorption of R_n was more than compensated by reduced rates of transpiration from the coated plants.

11.3.4 Other Reflectants and Specialized Applications

There are other, more specialized applications of reflectants in agriculture. Lipton and Matoba (1971) report that finely ground aluminum silicate applied to Crenshaw melons in California reduced sunburn of the fruit and reduced the surface temperature significantly (3.8–5.5°C according to Lipton, 1975). The market quality of the melons was much improved. Tomatoes are also damaged by sunburn and have been treated on a field scale with various claylike materials (Elam, 1971) and with whitewash (Anonymous, 1970). Ordinary whitewash has also been used to increase the albedo of a cotton field in Israel (Seginer, 1969). Tomato plants and grapefruit trees were coated with white paint once each season in a Texas experiment (Gerard, 1970). Yields of tomatoes were increased 30% and transpiration reduced about 10%. On citrus the water savings and yield effects were small, and fruit was reduced in size.

There are a number of ways in which reflectants may, in the future, be made more widely applicable. If a cheap means of delivery can be developed, reflectants may be applied even to light-unsaturated crops as a drought avoidance technique, to ensure crop survival over a short period through transpiration reduction. Reflectants may cause a greater penetration of visible radiation into the lower canopy which may be most useful to light-saturated (and perhaps even to light-unsaturated) crops. If so, the effect on photosynthesis will be beneficial. If reflectants that are more effective in the near ir can be developed, greater reduction in the energy load on the crop can result with less direct interference in photosynthesis. Although these advances await research and development, reflectant materials in use thus far already offer one very important advantage over most of the chemical antitranspirants in that they are usually inert materials that pose no danger to the health of man or of domestic and wild animals.

11.4 PLANT ARCHITECTURE

As shown above, there is theoretical support for the idea that increased reflectance should reduce evapotranspiration. There is also experimental evidence that, by artificially increasing reflectance, evapotranspiration may be reduced. Here we will consider ways by which reflectance and other physical properties of plants can be modified naturally.

Albedo varies from species to species and within species according to age of the leaf, turgidity, presence of waxes or other materials on the surface, and concentration of chlorophyll. For example, Ferguson et al. (1972) have reported on barley plants bred isogenically* for increased albedo through a reduction in chlorophyll concentration. In the greenhouse photosynthetic rates were greater in the pale plants on a per unit chlorophyll basis. Light saturation point was also higher in these plants. Ferguson et al. (1973) and Aase (1971) showed that light-colored barleys were cooler in the field. In these tests Aase found that the light-colored barley was slower to develop and yielded less under dry land conditions.

Woolley (1964), Ghorashy et al. (1971b), and Gausman and Cardenas (1973) have found, in the case of soybeans, that leaf pubescence increases reflectivity slightly in the visible waveband, but more significantly in the near infrared. Woolley (1964) and Ghorashy et al. (1971a) also found, for single leaves, that pubescence decreases transpiration, both because of reduced radiation absorption and because of an increased boundary layer resistance. Ehleringer and Bjorkman (1978) and Ehleringer and Mooney (1978) observed a greater visible reflectance, a reduced transpiration, and a slightly reduced photosynthetic rate in *Encelia farinosa* due to increased pubescence. However, Ghorashy et al. (1971b) found no reduction in photosynthetic rate in soybeans and Hartung et al. (1980) found increased photosynthetic rate associated with leaf pubescence.

Isogenically paired Harosoy cv. soybeans which differ only in the degree of pubescence on the leaves and stems were grown in experiments conducted during 1980 at the University of Nebraska's Agricultural Meteorology Laboratory near Mead, Nebraska. Detailed measurements of radiation balance, energy balance, photosynthesis, and evapotranspiration were made in the field. ET was reduced overall by approximately 7% in the densely pubescent isoline. The water use efficiency in terms of CWFR was greater in the pubescent isoline. The pubescence greatly altered the partitioning of R_n with deeper penetration into the canopy of the pubescent isoline (Baldocchi et al., 1983). In agronomic trials over several years Hartung et al. (1980) found the pubescent isolines to outyield the normal isolines.

In practical terms, the research on pubescence, described above, provides an important opportunity for strengthening and stabilizing agriculture where the subhumid and semiarid zones meet. Certain crops such as corn now grown at the margins of semiarid and subhumid regions are very sensitive to dry and hot weather, especially as they enter the reproductive stage. Soybeans are less so, since indeterminate cultivars predominate. The agron-

---

* Isogenes are plants that differ in only one character that is controlled by a single, identifiable gene.

omic trials of Hartung et al. (1980) and the microclimate–physiology studies of Baldocchi et al. (1983) support the idea that soybeans can be introduced to regions thought too arid until now.

Architectural features of crop plants other than reflectance can also influence water use efficiency. Benci et al. (1973) studied the aerodynamic characteristics and energy balance of isogenic barley lines with and without awns. Awns, it was found, increased surface resistance for horizontal wind momentum and increased roughness and turbulence. Net radiation was greater over fully awned barley. The excess energy was dissipated by increased sensible heat transfer to the air. Full-awned barley had plumper grain indicating a possible photosynthetic advantage.

Leaf shape is another architectural feature subject to manipulation in plant breeding. Mandl and Buss (1981) found that soybean isogenes with narrow leaflets gave yields similar to those obtained with normal, oblate leaflets. Narrow-leaf types were shorter and earlier to mature, but did not resist lodging any better than the normal leaf shape types. Baldocchi et al. (unpublished), on the other hand, found that a narrow-leafed isogene of Clark cv. soybeans transpired less water than did its normal oblate-leafed pair. Net radiation over the narrow-leafed isoline was unaffected, but sensible heat flux was increased. Air in the narrow-leafed canopy was hotter and dryer than in the oblate-leafed canopy. These effects can be explained by differences in the

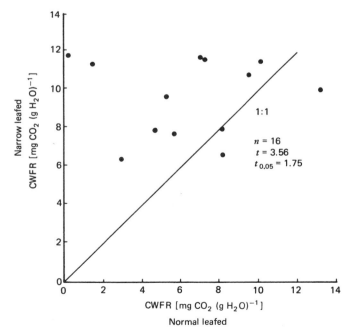

Fig. 11.8 CO_2–water flux ratios (CWFR) for the normal (CN) and narrow-leafed (CLN) canopies of Clark cv. soybeans (from Baldocchi et al., unpublished).

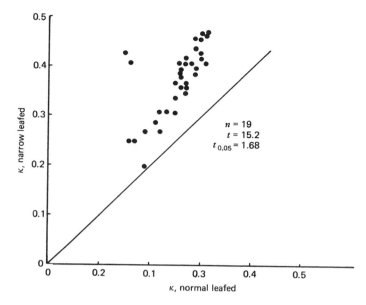

Fig. 11.9 Extinction of net radiation in normal and narrow leafed soybeans (expressed in terms of the attenuation coefficient κ; see Section 1.7 and (1.14) (from Baldocchi et al., unpublished).

partitioning of net radiation within the canopy. Radiation penetrated more deeply in the narrow leafed isogene since its total leaf area was less (Fig. 11.8). Photosynthesis per unit leaf area was increased and, overall, water use efficiency (in terms of the carbon–water flux ratio) was improved in the narrow-leafed isogene (Fig. 11.9).

Lodging is a phenomenon that can greatly reduce yield potential in crops. Heavy rains and strong winds, singly or in combination, can cause plants to lodge (fall over). Microclimatic conditions are generally altered then because of reduced ventilation, changes in radiation penetration, and in aerodynamic roughness. This may cause changes in temperature and humidity that can be conducive to disease. Light penetration may also be severely restricted. Cooper (1971) found that soybean plants suspended to prevent lodging yielded twice as much as did lodged plants.

Plant height is one factor that determines susceptibility to lodging. Height can be manipulated in plant breeding. Figure 11.10 illustrates a range in height that can be achieved with soybeans. The shorter types are determinate in nature, that is, they flower for a certain, limited period and cease adding nodes when flowering ceases. The taller plants are indeterminate and continue adding nodes, leaves and flowers until late in the growing season. The shorter types may yield as well as the taller ones (Hartung et al., 1980) and are much less susceptible to lodging.

There are many other types of architectural changes that should be ben-

408 WATER USE EFFICIENCY IN CROP PRODUCTION: NEW APPROACHES

Fig. 11.10 A range of soybean plant heights achieved by plant breeding (courtesy of Professors J. E. Specht and J. H. Williams, Dept. of Agronomy, Univ. of Nebraska–Lincoln).

eficial to crop production and/or to water use efficiency. Blad and Lemeur (1979) review the role of arrangement, shape, and number of leaves in the plant canopy as these affect the interception, penetration, distribution, and reflection of light. The photosynthetic mechanism as well as the water demand and canopy microclimate are influenced by light penetration.

Openness of the canopy can affect the incidence of disease. Blad et al. (1978) found, in western Nebraska, that the dense canopy of dry edible bean cultivar *Tara* contributed to high incidence of white mold infection under irrigation conditions because of the cool, wet microclimate fostered. A cultivar with a more open canopy, *Aurora*, had the warmest, driest microclimate and the lowest incidence of infection.

It is important to realize that the expressions of architecture bred into plants are altered somewhat by exposure to the environment. Heliotropic plants alter their leaf exposure to solar radiation during the course of the day. Plants that are grown in high population may increase in height or show altered branching and leaf display patterns. Coyne et al. (1974), for example, showed that the same Aurora cultivar noted above, when widely spaced in the row, produced a denser branching habit than where the plants were spaced closely in the row. Availability of water can also influence the expression of architectural features. Fors (1976) showed that canopy shapes differed in each of three varieties of oats when grown with moisture supply greater and less than the natural rainfall.

For a very thorough review of plant architectural changes accomplished by breeders in search of yield improvement of a wide variety of crops see Coyne (1980).

11.5 CARBON DIOXIDE ENRICHMENT

It is now well known that the concentration of carbon dioxide in the atmosphere has increased significantly since preindustrial times (NAS, 1979) and is continuing to increase at a rate of about 1 ppm per year. This fact is cause for concern since CO_2 is an effective absorber of longwave radiation emitted by the atmosphere and the earth's surfaces. Because of this property of CO_2, it is expected that the mean temperature near the earth's surface will increase and this should lead, ultimately, to significant climatic changes (NAS, 1979, 1982). Since the earth's temperature varies from year to year and since natural cooling and warming trends occur over varying periods of years, it is not yet possible to detect, with certainty, any heating effect that may have been caused by the increase in CO_2 concentration to date. If CO_2 concentration continues to increase at current or accelerated rates, the warning signal may emerge from the noise of natural variability by the end of this century. Whether or not climate will change as a result of the increasing CO_2 concentration in the atmosphere, it seems very likely that agriculture will be directly affected through changes in photosynthesis and evapotranspiration and, hence, in water use efficiency.

We have learned in Chapters 7 and 8 that transpiration and photosynthesis can be described as processes analogous to electrical current flow:

$$LE = \frac{\rho_a L \epsilon}{P} \frac{e_a - e_{ss}}{r_a + r_c} \qquad (7.49)$$

and

$$P = \frac{C_a - C_{chl}}{r_a + r'_c + r_m} \qquad (8.5)$$

where all symbols have been previously defined.

11.5.1 Photosynthesis

Figures 8.8 and 8.6 illustrate the fact that augmentation in CO_2 concentration in the ambient air increases the rate of photosynthesis. The increase is greater in sugar beet, a C_3 species, than in corn, a C_4 species. This increase is explained by a number of phenomena. The simplest explanation is that by increasing C_a the gradient or "driving force" for photosynthesis is increased (8.5). This will be true regardless of species. However, the effect is of relatively greater significance in C_3 species because C_{chl} is higher in plants having photorespiration (Table 8.2) and thus the concentration difference is normally smaller than in C_4 species.

Figure 8.8 for sugar beets shows that an increase in ambient CO_2 concentration leads to a direct increase in photosynthetic rate. The increase is especially marked in the range 200–600 ppm CO_2. In the case of C_3 species, the increase in ambient CO_2 concentration may also act to suppress photorespiration since that process proceeds at a rate that depends on competition between oxygen molecules and CO_2 molecules for enzymatic sites (Chollet, 1977; Ehleringer and Bjorkman, 1977).

Other direct effects of ambient CO_2 concentration on photosynthesis will occur through its impact on r'_c. CO_2 leads to stomatal closure in both C_3 and C_4 species. However, r_a is not directly affected by ambient CO_2 and r_m may be slightly responsive. Since r'_c is relatively small compared to the sum of r_a and r_m, its influence on photosynthesis will also be relatively small, unless an almost complete stomatal closure is induced. The influence of CO_2 on stomatal mechanics is not yet well understood and a considerable amount of physiological research on the subject is now under way.

11.5.2 Transpiration

The influence of CO_2 on stomatal closure is more consequential in the process of transpiration than it is in photosynthesis. Reference to (7.49) shows that, except in almost windless conditions, r_s is the primary determinant of the resistance to vapor transport from plant to atmosphere. Any significant increase in r_s should, then, lead to a reduction in transpiration rate.

Experimental evidence supports this hypothesis. Figure 11.11 (from Akita and Moss, 1972) illustrates the relative decrease in transpiration rate for three C_3 and three C_4 species that occurs with increasing CO_2 concentration. The transpiration of these plants was observed in a leaf chamber in the dark and under strong illumination. Clearly, stomates of the C_4 species respond more sharply in the range of CO_2 concentrations currently found in the field air and anticipated in the foreseeable future. In Fig. 11.11 it is seen that, in the light, the response of the C_3 species to realistic ambient CO_2 concentration is very slight. The response is considerably greater in the dark. Since

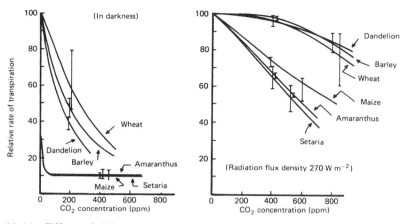

Fig. 11.11 Effect of CO_2 concentration on transpiration of C_3 and C_4 species in light and darkness (after Akita and Moss, 1972; reproduced from *Crop Science*, Volume 12, 1972, pages 789–793, by permission of the Crop Science Society of America).

most transpiration occurs during the daytime, however, the response in darkness is probably of minor importance.

11.5.3 Effects on Water Use Efficiency

Data summarized above indicate that increasing CO_2 concentration in the ambient air will lead to an increase in photosynthetic activity, especially in C_3 species, and to a decrease in transpiration, especially in C_4 species. Thus, the water use efficiency should, if the physiological responses observed under controlled conditions hold true in the field, be increased in both C_3 and C_4 species, although for different reasons in each case. This predicted effect on water use efficiency may be of particular importance in the semiarid and arid regions where limitations in natural rainfall or irrigation limit current agricultural productivity.

11.5.4 Complicating Factors

It is important to realize that the analyses given above, although optimistic concerning the overall impact of a CO_2 concentration increase in the ambient air, are incomplete. It is possible, for example, that elevated CO_2 concentration could affect the timing of phenological events (e.g., time of flowering, maturation) in certain species or affect certain developmental processes (e.g., root ramification, numbers of florets). Such phenological or morphological changes might increase the vulnerability of crops to certain hazards such as late spring frost or prolonged drought. Evidence of such effects at concentrations considered possible within the next century or so is not yet available. At very extreme concentrations, however, there is evidence of deleterious effects. Aoki and Yabuki (1977) found, for example, that dry matter production and photosynthetic rate of cucumber increased with exposure to CO_2 concentrations up to 2400 ppm. Exposure to 5000 ppm of CO_2 decreased dry weight gains below those achieved at lower concentrations.

Others (e.g., Botkin et al., 1973) have argued that a changing CO_2 regime could lead to changes in species composition and succession in forests or other unmanaged ecological associations. Such effects, however, ought not to be consequential in most agricultural ecosystems except, perhaps, where certain weed species may be favored, for example, broad-leafed weeds in corn.

One of the strongest arguments against the hypothesis that an atmospheric CO_2 increase will lead to improved photosynthesis and/or water use efficiency has been proposed by Lemon (1976) and is supported by Goudriaan and Ajtay (1979). They argue that net primary productivity (photosynthesis less respiratory loss) is now limited by shortages in water supply and nutrient availability, not by the CO_2 concentration of the atmosphere. The shortage of water and nutrients, they argue, and such climatic limitations as insufficient length of the growing season explain the fact that the world's vegetation does not even now achieve its potential net primary productivity. On a global scale it does, indeed, seem unlikely that radical increases in rate of photosynthesis and in net primary productivity will occur quickly as the CO_2 concentration continues to increase.

These arguments assume a static ecology, however. One may speculate on what might happen in a regime of CO_2-enriched air if nutrients are not strongly limiting. Assuming no beneficial climatic change induced by the increasing CO_2 concentration but depending on the evidence of physiological experiments alone, it seems likely that CO_2 fertilization would cause an incremental increase in the rate of photosynthesis in, say, an association of C_3 species such as occur in northern forests. This would, in turn, lead to a greater dry matter production, perhaps to larger trees and greater standing biomass. Larger trees should produce denser and, perhaps, deeper root systems. In turn these root systems should senesce to yield greater accumulations of soil organic matter. A heavier leaf litter might also develop.

Soil forming processes, according to Jenny (1941) depend on a number of factors: parent materials, topography, climate, biology (vegetation, soil flora, and fauna), and time. Thus, it is not inconceivable that, even if climate does not change, CO_2 "fertilization" may initiate or stimulate an acceleration of soil formation through enhanced biological activity. A more rapid release of many essential nutrients might follow. Further, a reduction in transpiration rate may have the effect of reducing the severity of moisture shortages, at least occasionally. This, too, might favor biological activity in the soil. Soil forming processes are perpetual and occur in soils used for agriculture as well as in natural ecosystems.

At this point predictions of the long-term direct effects on agriculture of increased CO_2 concentration in the atmosphere are, at best, premature.

11.5.5 Greenhouse and Growth Chamber Experience

The evidence given above strongly suggests that photosynthesis and water use efficiency may be increased in vegetative stands of virtually all kinds. Such an improvement may already be occurring since CO_2 concentration has increased by some 30–40 ppm (about 15%) and perhaps more since the beginning of the industrial revolution.

Fertilization or enrichment of air in commercial greenhouses has been practiced for a considerable time. Wittwer (1967) reviewed the history of this practice and showed that increased yield and improved quality has been achieved with a number of crops: lettuce, tomato, cucumber, and a wide range of flower crops. Allen (1979) has also assembled data from a wide range of greenhouse and growth chamber experiments. It is seen in Table 11.1 from Allen (1979) that, in almost every case, plant growth is significantly

Table 11.1 Responses of some crops to CO_2 enrichment under greenhouse and growth chamber conditions[a]

Crop	Enrichment (ppm CO_2)	Increase in yield (%)
Vegetables crops		
Lettuce	350–600	16–44
	800–2000	23–100
	1000	43
Tomato	800–2000	25–71
	900	30–31
		6–18
	600–1200	2–10
	800–1200	743–755[b]
	600–1400	26–39
	1200	−1–5
	1000–3500	0[c]
	650–3200	−60–60[d]
Cucumber	1000	43
Potatoes	650	43–75
Strawberry	400–3000	18–26
Radish	1200	105–359
Okra	800–1600	56–312
Agronomic crops		
Soybean	1000–2000	500–550
	1350	41–57

Table 11.1 (*Continued*)

Crop	Enrichment (ppm CO_2)	Increase in yield (%)
Sunflower[e]	500–5000	144–550
Rice	1000–2400	19–650
Wheat	1200–2400	100–138
Sorghum	1000–2000	168–232
Flower crops		
Carnation	600–1200	−0.5–3
	400–1700	27–34
	550	11
Rose		8–27
Chrysanthemum	1000–3000	−2–20
Poinsettia	1500–3000	10–50
Seedings		
Crabapple[e]	2000	157 (4 wk)
		128 (20 wk)
Cucumber[e]	2000	370 (15 days)
Tomato[e]	2000	460 (15 days)
Lettuce[e]	2000	200 (15 days)

[a] For sources of the information see Allen, L. H., Jr., In: *Modification of the Aerial Environment of Plants* B. J. Barfield and J. F. Gerber (eds.). Copyright 1979 by the American Society of Agricultural Engineers.
[b] Unenriched control treatment failed.
[c] Some enriched plants died.
[d] Decreases above 1000 ppm.
[e] Growth chamber.

increased by CO_2 enrichment. Kimball (1982) tabulated many experiments where plants have been grown in CO_2-enriched air. Figure 11.12 shows a distribution of the ratio of yields achieved in CO_2-enriched air to that obtained with control (normal) air. Experiments with flower crops constitute a large fraction of the total. With or without flower crops included a mean increase in yield of about 50% is indicated. Very few studies indicate a yield depression, and those are primarily with flower crops. Doubling and tripling of yields has occurred in a significant number of studies.

11.5.6 Field Experiments

There is evidence that yields can be increased in closed environments such as greenhouses and growth chambers where CO_2 can be injected into the air to enrich it. It also appears likely that, as a result of the 15% or so increase

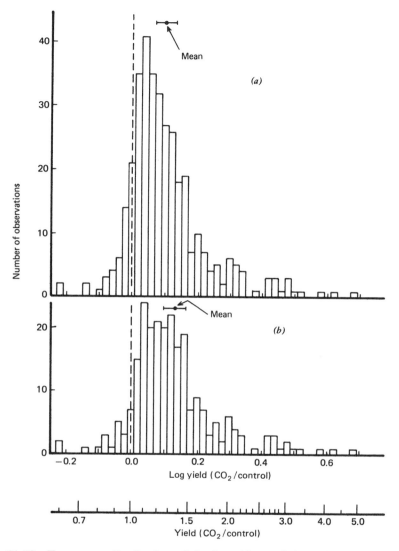

Fig. 11.12 Frequency distribution of the logarithms of the ratios of the yields of CO_2-enriched plants to their respective unenriched controls for (a) all crops and (b) all crops excluding flowers. The points and associated brackets are the means and their 99.9% confidence intervals (from Kimball, 1982, p. 3).

in its concentration in modern times, a global CO_2 fertilization effect is already occurring. What is the likelihood that further benefits can be achieved by artificially enriching ambient air with CO_2 in open crop fields?

This question has been approached by simulation studies in which crop features, environmental conditions, and CO_2 release rates have been con-

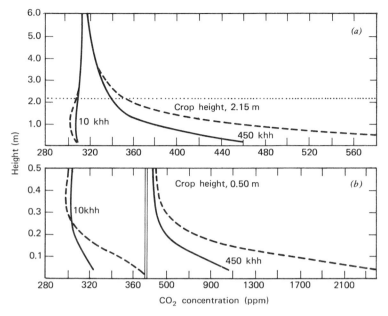

Fig. 11.13 Simulated CO_2 concentration profiles for the cases with a leaf angle of 40° for (a) the tall, open crop (height 2.15 m; LAI = 4) and (b) the short dense crop (height 0.50 m; LAI = 10) for normal soil CO_2 flux density rates of 10 kg ha^{-1} h^{-1} (khh) and enrichment rates of 450 kg ha^{-1} h^{-1}. Dashed lines, wind speed, 1.0 m s^{-1}. Solid lines, wind speed, 6.0 m $^{-1}$. Note the scale changes in (b) (after Allen et al., 1971).

sidered (e.g., Allen et al., 1971; Waggoner, 1969). Allen et al. (1971), for example, mathematically simulated the plant response to carbon dioxide enrichment rates of 225 and 450 kg ha^{-1} h^{-1}. Their simulation considered a wide range of wind speed, crop height, and leaf area conditions. A normal soil carbon dioxide flux rate of 10 kg ha^{-1} h^{-1} was used for comparison. The results of these simulations are shown in Fig. 11.13 for a tall crop (such as corn) and a short crop (such as bermuda grass). On the left-hand sides of the upper and lower figures the natural profiles of carbon dioxide with a slight wind (1 m s^{-1}) and a strong wind (6 m s^{-1}) are shown. On the right-hand sides are the profiles predicted to develop under calm or windy conditions when 450 kg ha^{-1} h^{-1} of carbon dioxide is released. According to the simulation, the carbon dioxide concentration decreases rapidly with height within the crop canopy in spite of the high enrichment rate. The figure shows that wind disperses the added carbon dioxide. The high concentration of carbon dioxide at the bottom of the crop would not be strongly beneficial since photosynthesis is greatest in the top layers where light is plentiful and not at the bottom of the crop where it is usually in short supply. The efficiency of the uptake of artificially supplied carbon dioxide was defined by Allen et al. (1971) as

$$\text{efficiency (CO}_2) = \frac{\text{increased CO}_2 \text{ uptake rate}}{\text{CO}_2 \text{ enrichment rate}}$$

They conclude from the simulations that the uptake efficiency would be very low (<10%) because of the rapid loss of carbon dioxide to the atmosphere and the relatively small buildup of carbon dioxide near the top illuminated leaves.

In later simulations Allen (1974, 1975a) considered a wide range of CO_2 release rates from line sources spaced closely in the field and concluded that the maximum efficiency of capture was 6.5%. In actual field experiments with a line source Allen et al. (1974) found that vertical turbulent diffusion and horizontal mass flow quickly dissipated the released CO_2. CO_2 remained in the canopy only under stable stratification. Since this condition is common only at night the potential benefits are lost except perhaps for plants with the CAM photosynthetic pathway. The CO_2 diffusing out of the fertilized area is wafted downwind where some small portion of it may be captured by the adjacent crops.

In an actual field study of the influence of carbon dioxide enrichment, Harper et al. (1973a) found more encouraging results. They released carbon dioxide into a cotton field in Georgia at the rate of 222.6 kg ha^{-1} h^{-1}. The cotton had a leaf area index of 2.34. Actual recovery rates ranged from 7 to 33% over a range of solar radiation flux density from 200 to 1100 W m^{-2}. A comparison of a normal carbon dioxide profile and one which existed under release conditions is shown in Fig. 11.14. The carbon dioxide concentration remained at least 100 ppm greater in the upper canopy under release conditions. Even at 4 m the concentration of the carbon dioxide remained about 50 ppm greater when carbon dioxide was being released. These high concentrations above the crop remained surprisingly stable when winds were less than 2 m s^{-1} (measured at 1.5 m elevation). Harper et al. (1973a,b) conclude from their field study that their open canopy crop, which intercepted only 65% of the incident radiation, frequently captured 23% of the released carbon dioxide. Dense crop canopies that intercept about 95% of the incident light should capture about 33% of the carbon dioxide released. Harper et al. (1973a) noted considerably higher concentrations of CO_2 at the top and above the crop canopy than did others (Allen et al., 1974; Takami and van Bavel, 1975) who have conducted CO_2 release studies in the field.

Evidence of the efficiency of direct release of CO_2 into the field crop is still inconclusive. Allen (1979) has reviewed the results of a number of studies in which the capture of released CO_2 has been "assisted" by the use of barriers of different kinds. For example, in one such study (Allen, 1975b) a shade cloth cover over a soybean field in New York state reduced eddy

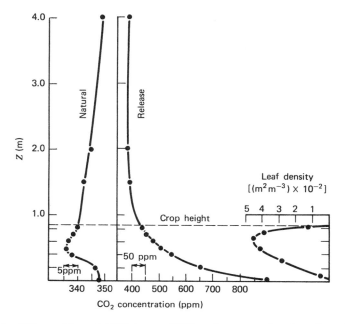

Fig. 11.14 CO_2 profiles at 1030 hours CDT under natural and under release conditions and leaf area density vs. height (after Harper et al., 1973a).

diffusivity by about 25% but CO_2 emitted from lines spaced in every second row was still transported rapidly away by turbulent diffusion. Soybean yields were not reduced by shading, however, which suggests that the CO_2 provided some compensation for the reduction in irradiance.

Allen (1979) reviews studies of the uses of windbreaks of various kinds in conjunction with CO_2 release systems. Clear plastic barriers and corn, wheat, and snowfence barriers have improved the CO_2 capture efficiency for a number of crops such as soybean, cucumber, canteloupe, tomato, and potato. Allen (1979) also reviews the release of CO_2 into polyethylene row covers or tunnels used in vegetable crop production. Since these covers must be lifted periodically during the course of the day, only small yield increases have been achieved.

It appears that, if economically justifiable yield increases are to be obtained, CO_2 release into field crops must be done in conjunction with windbreaks or other techniques that reduce the rapid turbulent diffusion of the released gas away from the crop. Further modeling of the process, considering canopy architectural variations and turbulence suppressors such as windbreaks, is justified. Additional field experiments and measurements with more sensitive and accurate devices than were available in the early 1970s is also justified.

One may speculate at this time, however, that a global fertilization effect due to the increasing CO_2 concentration in the atmosphere will be more significant in increasing crop yields and water use efficiency than will the direct release of CO_2 into the crop.

REFERENCES

Aase, J. K. 1971. Growth, water use and energy balance comparisons between isogenic lines of barley. *Agron. J.* **63**:425–428.

Aboukhaled, A., R. M. Hagan, and D. C. Davenport. 1970. Effects of kaolinite as an effective antitranspirant on leaf temperature, transpiration, photosynthesis and water use efficiency. *Water Resour. Res.* **6**:280–289.

Akita, S., and D. N. Moss. 1972. Differential stomatal response between C_3 and C_4 species to atmospheric CO_2 concentration and light. *Crop Sci.* **12**:789–793.

Allen, L. H., Jr. 1974. Line source carbon dioxide release. II. Two-dimensional numerical diffusion model. *Agron. J.* **66**:616–620.

Allen, L. H., Jr. 1975a. Line source carbon dioxide release. III. Predictions by a two-dimensional numerical diffusion model. *Boundary-Layer Meteorol.* **9**:39–79.

Allen, L. H., Jr. 1975b. Shade-cloth microclimate of soybeans. *Agron. J.* **67**:175–181.

Allen, L. H., Jr. 1979. Potentials for carbon dioxide enrichment. *Modification of the Aerial Environment of Plants* (B. J. Barfield and J. F. Gerber, eds.), Am. Soc. Agric. Engin. Monogr. **2**:500–519.

Allen, L. H., Jr., S. E. Jensen, and E. R. Lemon. 1971. Plant response to carbon dioxide enrichment under field conditions: A simulation. *Science* **173**:256–258.

Allen, L. H., Jr., R. L. Desjardins, and E. R. Lemon. 1974. Line source carbon dioxide release. I. Field experiment. *Agron. J.* **66**:609–615.

Angus, D., and H. Bielorai. 1965. Transpiration reduction by surface films. *Australian J. Agric. Res.* **16**:107–112.

Anonymous. 1970. Whitewashing shields tomatoes from sun, heat injury. *Vegetable Crop Management*, pp. 8–9.

Aoki, M., and K. Yabuki. 1977. Studies on the carbon dioxide enrichment for plant growth, VII. Changes in the dry matter production and photosynthetic rate of cucumber during carbon dioxide enrichment. *Agric. Meteorol.* **18**:475–485.

Baldocchi, D. D., S. B. Verma, N. J. Rosenberg, B. L. Blad, A. Garay, and J. E. Specht. 1983. Influence of leaf pubescence on the mass and energy exchange between soybean canopies and the atmosphere. *Agron. J.* (in press).

Baradas, M. W., B. L. Blad, and N. J. Rosenberg. 1976a. Reflectant induced modification of soybean (Glycine max L.) canopy radiation balance. IV. Leaf and canopy temperature. *Agron. J.* **68**:843–848.

Baradas, M. W., B. L. Blad, and N. J. Rosenberg. 1976b. Reflectant induced modification of soybean (Glycine max L.) canopy radiation balance. V. Longwave radiation balance. *Agron. J.* **68**:848–852.

Belt, G. H., J. G. King, and H. F. Haupt. 1977. Augmenting summer streamflow by use of a silicone antitranspirant. *Water Resources Res.* **13**:267–272.

Benci, J. F., J. K. Aase, and A. H. Ferguson. 1973. Aerodynamic and energy balance comparisons between awned and nonawned barley. *Agron. J.* **65**:373–377.

Blad, B. L., J. R. Steadman, and A. Weiss. 1978. Canopy structure and irrigation influence white mold disease and microclimate of dry edible beans. *Phytopathology* **68**:1431–1437.

Blad, B. L., and R. Lemeur. 1979. Miscellaneous techniques for alleviating heat and moisture stress. *Modification of the Aerial Environment of Plants* (B. J. Barfield and J. F. Gerber, eds.), *Am. Soc. Agric. Engin. Monogr.* **2**:409–425.

Botkin, D. B., J. R. Janek, and J. R. Wallis. 1973. Estimating the effects of carbon fertilization on forest composition by ecosystems simulation. *Carbon and the Biosphere* (G. M. Woodwell and E. V. Pecan, eds.), U.S. Atomic Energy Comm., pp. 328–344.

Brooks, K. N., and D. B. Thorud. 1970. Transpiration of Ponderosa pine and Douglas fir after treatment with phenylmercuric acetate. *Water Resour. Res.* **6**:957–959.

Campbell, G. S. 1977. *An Introduction to Environmental Biophysics.* Springer-Verlag, New York. 159 pp.

Chollet, R. 1977. The biochemistry of photorespiration. *Trends Biochem. Sci.* **2**:155–159.

Cole, D. F., A. K. Dobrenz, and M. A. Massengale. 1971. Effect of growth regulator and antitranspirant chemicals on water requirement and growth components of alfalfa (Medicago sativa L.). *Crop Sci.* **11**:582–584.

Cooley, K. R. 1970. Energy relationships in the design of floating covers for evaporation reduction. *Water Resour. Res.* **6**:717–727.

Cooper, R. L. 1971. Influence of early lodging on yield of soybean (Glycine max L. Merr.). *Agron. J.* **63**:449–450.

Coyne, D. P., J. R. Steadman, and F. N. Anderson. 1974. Effect of modified plant architecture of Great Northern dry bean varieties (*Phaseolus vulgaris*) on white mold severity and components of yield. *Plant Disease Reporter* **58**:379–382.

Coyne, D. P. 1980. Modification of plant architecture and crop yield by breeding. *Hort Science* **15**:244–247.

Davenport, D. C. 1967. Effects of chemical antitranspirants on transpiration and growth of grass. *J. Exp. Bot.* **18**:332–347.

Davenport, D. C., R. M. Hagan, and P. E. Martin. 1969. Antitranspirant research and its possible application in hydrology. *Water Resour. Res.* **5**:735–743.

Davenport, D. C., M. A. Fisher, and R. M. Hagan. 1971. Retarded stomatal closure by phenylmercuric acetate. *Physiol. Plant* **24**:330–336.

Davenport, D. C., R. M. Hagan, L. W. Gay, B. E. Kynard, E. K. Bonde, F. Kreith, and J. E. Anderson. 1978. Factors influencing usefulness of antitranspirants applied on phreatophytes to increase water supplies. California Water Resources Center, ISSN 0575-4941, Contribution 176. 181 pp.

Doraiswamy, P. C., and N. J. Rosenberg. 1974. Reflectant induced modification of soybean canopy radiation balance. I. Preliminary tests with a kaolinite reflectant. *Agron. J.* **66**:224–228.

Egli, D. B., J. W. Pendleton, and D. B. Peters. 1970. Photosynthetic rate of three soybean communities as related to carbon dioxide levels and solar radiation. *Agron. J.* **62**:411–414.

Ehleringer, J. R., and O. Bjorkman. 1977. Quantum yields for CO_2 uptake in C_3 and C_4 plants. *Plant Physiol.* **59**:86–90.

Ehleringer, J. R., and O. Bjorkman. 1978. Pubescence and leaf spectral characteristics in a desert shrub, *Encelia farinosa*. *Oecologia (Berlin)* **36**:151–162.

Ehleringer, J. R., and H. A. Mooney. 1978. Leaf hairs: Effects on physiological activity and adaptive value to a desert shrub. *Oecologia (Berlin)* **37**:183–200.

Elam, F. L. 1971. Snow job for tomatoes. *Farm J.* July issue.
Ferguson, H. C., S. Cooper, J. H. Brown, and R. F. Eslick. 1972. Effect of leaf color, chlorophyll concentration and temperature on photosynthetic rates of isogenic barley lines. *Agron. J.* **64**:671–673.
Ferguson, H. C., R. F. Eslick, and J. K. Aase. 1973. Canopy temperatures of barley as influenced by morphological characteristics. *Agron. J.* **65**:425–428.
Fors, L. 1976. Bioclimate of three varieties of oats. *Agric. Meteorol.* **17**:401–431.
Fritschen, L. J. 1967. Net and solar radiation relations over irrigated field crops. *Agric. Meteorol.* **4**:55–62.
Fuchs, M., G. Stanhill, and S. Moreshet. 1976. Effect of increasing foliage and soil reflectivity on the solar radiation balance of wide-row grain sorghum. *Agron. J.* **68**:865–871.
Fuehring, H. D. 1975. Yield of dryland grain sorghum as affected by antitranspirant, nitrogen and contributing micro-watershed. *Agron. J.* **67**:255–257.
Fulton, J. M. 1967. Stomatal aperture and evapotranspiration from field grown potatoes. *Can. J. Plant Sci.* **47**:109–111.
Gale, J., and R. M. Hagan. 1966. Plant antitranspirants. *Annu. Rev. Plant Physiol.* **17**:269–282.
Gale, J., and A. Poljakoff-Mayber. 1967. Plastic films on plants as antitranspirants. *Science* **156**:650–652.
Gale, J., E. B. Roberts, and R. M. Hagan. 1967. High alcohols as antitranspirants. *Water Resour. Res.* **4**:437–441.
Gausman, H. W., and R. Cardenas. 1973. Light reflectance by leaflets of pubescent, normal and glaborous soybean lines. *Agron. J.* **65**:837–838.
Geiger, R. 1965. *The Climate near the Ground.* Transl. from the fourth German edition. Harvard Univ. Press, Cambridge, Massachusetts. 611 pp.
Gerard, C. J. 1970. Influence of transpiration suppressants, sprinkler irrigation and moisture levels on transpiration and evapotranspiration. Water Resources Inst., Texas A&M Univ., TR-27.
Ghorashy, S. R., J. W. Pendleton, D. B. Peters, J. S. Boyer, and J. E. Beuerlein. 1971a. Internal water stress and apparent photosynthesis with soybeans differing in pubescence. *Agron. J.* **63**:674–676.
Ghorashy, S. R., J. W. Pendleton, R. L. Bernard, and M. E. Bauer. 1971b. Effect of leaf pubescence on transpiration, photosynthetic rate and seed yield of three near-isogenic lines of soybeans. *Crop Sci.* **11**:426–427.
Goudriaan, J., and G. L. Ajtay. 1979. The possible effects of increased CO_2 on photosynthesis. *The Global Carbon Cycle (Scope 13)* (B. Bolin, E. T. Degens, S. Kempe and P. Ketner, eds.), Wiley, New York, pp. 237–249.
Hales, T. A., and N. J. Rosenberg. 1970. Antitranspirant influence on energy balance components in field grown alfalfa. Agronomy Abstracts, 1970 Annual Meeting of the American Society of Agronomy, p. 185.
Harper, L. A., D. N. Baker, J. E. Box, Jr., and J. D. Hesketh. 1973a. Carbon dioxide and the photosynthesis of field crops: A metered carbon dioxide release in cotton under field conditions. *Agron. J.* **65**:7–11.
Harper, L. A., J. Box, Jr., D. N. Baker, and J. D. Hesketh. 1973b. Carbon dioxide and the photosynthesis of field crops: A tracer examination of turbulent transfer theory. *Agron. J.* **65**:574–578.
Hartung, R. C., J. E. Specht, and J. H. Williams. 1980. Agronomic performance of

selected soybean morphological variants in irrigation culture with two row spacings. *Crop Sci.* **20**:604–609.

Jeffers, D. L., and R. M. Shibles. 1969. Some effects of leaf area, solar radiation, air temperature and variety on net photosynthesis in field grown soybeans. *Crop Sci.* **9**:762–764.

Jenny, H. 1941. *Factors of Soil Formation: A System of Quantitative Pedology*. McGraw-Hill, New York. 281 pp.

Kimball, B. A. 1982. Carbon dioxide and agricultural yield: An assemblage and analysis of 430 prior observations. U.S. Water Conserv. Lab. U.S. Dept. of Agriculture, Phoenix, Arizona. Report 11, May, 1982.

Kreith, F., A. Taori, and J. E. Anderson. 1975. Persistence of selected antitranspirants. *Water Resour. Res.* **11**:281–285.

Lee, K. J., and T. T. Kozlowski. 1974. Effects of silicone antitranspirant on woody plants. *Plant Soil* **40**:493–510.

Lemeur, R., and N. J. Rosenberg. 1976. Reflectant induced modification of soybean canopy radiation balance. III. A comparison of the effectiveness of Celite and kaolinite reflectants. *Agron. J.* **68**:30–35.

Lemon, E. R. 1976. The land's response to more carbon dioxide. *The Fate of Fossil Fuel CO_2 in the Oceans* (N. J. Anderson and A. Malahoff, eds.), Plenum, New York, pp. 97–103.

Lipton, W. J., and F. Matoba. 1971. Whitewashing to prevent sunburn of "Crenshaw" melons. *Hort Science* **6**:343–345.

Lipton, W. J. 1975. Whitewashing Crenshaw and cantaloupe melons to reduce solar injury. Marketing Research Report No. 1045, USDA. 16 pp.

Maki, T. 1974. The prevention of wind-frictional scratch on citrus fruit and the suppression of transpiration from eggplant by a surface coating agent. In Japanese, English summary. Source unknown.

Mandl, F. A., and G. R. Buss. 1981. Comparison of narrow and broad leaflet isolines of soybean. *Crop Sci.* **21**:25–27.

Monteith, J. L., G. Szeicz, and P. E. Waggoner. 1965. The measurement and control of stomatal resistance in the field. *J. Appl. Ecol.* **2**:345–355.

Moreshet, S. 1975. Effects of phenylmercuric acetate on stomatal and cuticular resistance to transpiration. *New Phytol.* **75**:47–52.

Moreshet, S., G. Stanhill, and M. Fuchs. 1977. Effect of increasing foliage reflectance on the CO_2 uptake and transpiration resistance of a grain sorghum crop. *Agron. J.* **69**:246–250.

Moreshet, S., Y. Cohen, and M. Fuchs. 1979. Effect of increasing foliage reflectance on yield, growth and physiological behavior of a dryland cotton crop. *Crop Sci.* **19**:863–868.

National Academy of Sciences/National Research Council/Climate Research Board. 1979. Carbon Dioxide and Climate: A Scientific Assessment. National Academy Press, Washington, D.C. 22 pp.

National Academy of Sciences/National Research Council. 1982. Carbon Dioxide and Climate: A Second Assessment. National Academy Press, Washington, D.C. 72 pp.

Ogunkanmi, A. B., A. R. Wellburn, and T. A. Mansfield. 1974. Detection and preliminary identification of endogenous antitranspirants in water-stressed sorghum plants. *Planta (Berlin)* **117**:293–302.

Pasternak, D., and G. L. Wilson. 1971. Some factors responsible for varying effectiveness of stomatal closing antitranspirants. *Australian J. Exp. Agric. Anim. Husb.* **2**:48–52.

Sakamoto, C. M., and R. H. Shaw. 1967. Apparent photosynthesis in field soybean communities. *Agron. J.* **59**:73–75.

Seginer, I. 1969. The effect of albedo on the evapotranspiration rate. *Agric. Meteorol.* **6**:5–31.

Shimshi, D. 1963a. Effect of chemical closure of stomata on transpiration in varied soil and atmospheric environment. *Plant Physiol.* **38**:709–712.

Shimshi, D. 1963b. Effect of soil moisture and phenylmercuric acetate upon stomatal aperature, transpiration and photosynthesis. *Plant Physiol.* **38**:713–721.

Sinclair, T. R., G. E. Bingham, E. R. Lemon, and L. H. Allen, Jr. 1975. Water use efficiency of field grown maize during moisture stress. *Plant Physiol.* **56**:245–249.

Slatyer, R. O., and J. F. Bierhuizen. 1964a. The influence of several transpiration suppressants on transpiration, photosynthesis and water use efficiency of cotton leaves. *Australian J. Biol. Sci.* **17**:131–146.

Slatyer, R. O., and J. F. Bierhuizen. 1964b. The effect of several foliar sprays on transpiration and water use efficiency of cotton plants. *Agric. Meteorol.* **1**:42–53.

Stanhill, G. 1965. Observations on the reduction of soil temperature. *Agric. Meteorol.* **2**:197–204.

Stanhill, G., S. Moreshet, and M. Fuchs. 1976. Effect of increasing foliage and soil reflectivity on the yield and water use efficiency of grain sorghum. *Agron. J.* **68**:329–332.

Takami, S., and C. H. M. van Bavel. 1975. Distribution of carbon dioxide released in a field crop. *J. Agric. Meteorol. (Japan)* **31**:29–42.

Tanner, C. B. 1963. Plant temperature. *Agron. J.* **55**:210–211.

Turner, N. C., and P. E. Waggoner. 1968. Effects of changing stomatal width in a red pine forest on soil water content, leaf water potential, bole diameter and growth. *Plant Physiol.* **43**:973–978.

Waggoner, P. E., J. L. Monteith, and G. Szeicz. 1964. Decreasing transpiration of field plants by chemical closure of stomata. *Nature (London)* **20**:97.

Waggoner, P. E., and I. Zelitch. 1965. Transpiration and the stomata of leaves. *Science* **150**:1413–1420.

Waggoner, P. E., and J. D. Hewlett. 1965. Test of a transpiration inhibitor on a forested watershed. *Water Resour. Res.* **1**:391–396.

Waggoner, P. E., and N. W. Simmonds. 1966. Stomata and transpiration of droopy potatoes. *Plant Physiol.* **41**:1268–1271.

Waggoner, P. E. 1969. Environmental manipulation for higher yields. *Physiological Aspects of Crop Yield* (J. D. Eastin, F. A. Haskins, C. Y. Sullivan, and C. H. M. van Bavel, eds.), Am. Soc. of Agron., Madison, Wis., pp. 343–373.

Waggoner, P. E., and N. C. Turner. 1971. Transpiration and its control by stomata in a pine forest. *Bull. Conn. Agric. Exp. Stn.* **726**:1–87.

Wittwer, S. H. 1967. Carbon dioxide and its role in plant growth. *Proc. 17th Int. Hort. Congr.* **3**:311–322.

Woolley, J. T. 1962. Soil applied hexadecanol as an evapotranspiration suppressant. *J. Soil Water Conserv.* **17**:130.

Woolley, J. T. 1964. Water relations of leaf hairs. *Agron. J.* **56**:569–571.
Yadav, S. K., and D. P. Singh. 1981. Effect of irrigation and antitranspirants on evapotranspiration, water use efficiency and moisture extraction patterns of barley. *Irrig. Sci.* **2**:177–184.
Zelitch, I., and P. E. Waggoner. 1962. Effect of chemical control of stomata on transpiration and photosynthesis. Proc. Natl. Acad. Sci. U.S.A. **48**:1101–1108.

CHAPTER 12 – HUMAN AND ANIMAL BIOMETEOROLOGY

12.1 INTRODUCTION

Biometeorology is "the study of biological effects of weather and climate on living organisms, plants, animals and man and of their direct physico-chemical environment" (Tromp and Bouma, 1974, 1976). Biometeorology is a relatively new science and one in which interdisciplinary research is common. Because of this, biologists must understand physics and mathematics and meterologists must know some biology. In earlier chapters of this book we have discussed interactions between plants and their environment; in this chapter we address relationships between humans and other animals and their environment.

There are similarities and differences in the manner in which humans and animals, on the one hand, and plants, on the other hand, respond to their environment. The physical principles and laws governing the exchanges of radiation and energy are the same; however, because of their mobility humans and animals are able to avoid uncomfortable environmental situations. The study of human and animal biometeorology is fascinating, and a considerable body of literature documenting research results has accumulated over the past few decades. Since space is limited only a sample of these results are cited in this chapter. The interested reader should consult the following publications for addition detail: Kleiber (1961), Lowry (1970), Brody (1974), Tromp and Bouma (1972, 1974, 1976, 1979), Monteith and Mount (1974), Monteith (1975), Campbell (1977), Esmay (1978), Gates (1980), Hobbs (1980), and Tromp (1980).

In order to understand the interactions of living organisms and their environment knowledge of the energy balance existing within the ecosystem is essential. The food chain provides some of the energy for the organism. Other sources of energy are direct insolation and the heat generated at other surfaces receiving solar energy.

Many factors combine to create a specific climate or microclimate. Climate determines, in large measure, the ecological niche inhabited by specific plants and animals. Where the organism lives, its environment (humid and cold, arid and dry, or somewhere in between), determines in great measure the type and magnitude of environmental factors that affect its behavior. Certain environmental conditions exert little or no stress on a particular

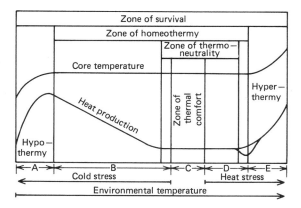

Fig. 12.1 Critical temperature and thermal zones. See text for an explanation of the letters at the bottom of the figure. From Bianca (1976).

organism; other conditions may be so severe that its survival is dependent on its ability to adapt.

Mechanisms of environmental interaction and survival are quite different for the **homeotherms,** warm-blooded animals, and the **poikilotherms,** cold-blooded animals. The homeotherm maintains, by physiological and metabolic mechanisms, a constant body temperature over a wide range of ambient temperature; the poikilotherm cannot modify its body temperature by other than behavioral responses and thus it ranges widely.

An animal's survival depends, largely, on whether or not it can maintain its body temperature within certain limits. Bianca (1976) has defined this zone of survival and illustrated various responses of homeotherms to changes in environmental temperature (Fig. 12.1). Based on the information in Fig. 12.1 Primault (1979) defined five thermal zones:

A. A zone of hypothermia, within which an animal cannot supply sufficient heat to maintain its ideal temperature.

B. A broad range within which the body is at its optimum temperature but the animal must produce heat to maintain that temperature.

C. A zone of thermal indifference where normal metabolism supplies the energy to maintain an optimum temperature.

D. A neutral range within which conditions are no longer ideal and the animal must effect or obtain some cooling.

E. A zone of hyperthermia where the temperature control mechanisms cannot provide sufficient cooling to maintain the body temperature at its normal level.

The impact of the meteorological environment on the animal varies depending on the severity of the climate. Bianca (1976), for example, states that animal response to heat stress may be divided into four arbitrary categories:

1. **Thermal indifference:** a zone within which compensatory responses by an organism are absent (for examples of this zone in different animals see Bianca, 1976; Primault, 1979; Hahn, 1981a).
2. **Mild heat:** thermoregulatory mechanisms of the body completely compensate for the extra heat load and body temperature remains normal.
3. **Moderate heat:** thermoregulatory mechanisms work at a greater intensity and body temperature is stabilized at a higher temperature.
4. **Severe heat:** thermoregulatory mechanisms are overtaxed and body temperature rises leading eventually to exhaustion and death.

Certain anatomical and morphological features such as fur, fins, feathers, and color along with various internal response mechanisms permit a species to survive within a given ecological niche. Wild animals have adapted to their environment through natural selection but domestic animals, because of selective breeding and confinement, are often poorly adapted (Bianca, 1976). Production by domesticated animals (of meat, milk, eggs) is often closely linked to environmental conditions. This has led to considerable study of the responses of animals to environment so that management procedures can be adapted to maximize production and minimize losses.

As we have seen, the maintenance of a proper body temperature is critical for survival. The same is true of the water status of the body. Therefore, those environmental factors that affect body temperature and water status will be emphasized in this chapter. The general principles and laws governing radiation and energy exchange must be dealt with first. Certain body mechanisms for adapting to various kinds of environmental stress will be described. Some of the more common indices that relate environmental conditions to comfort levels will also be discussed.

12.2 RADIATION BALANCE

As with plants a primary source of energy for humans and animals is solar radiation. One need only feel the warning effects of the sun on a cold clear winter day or attempt to escape its effects on a hot summer day to appreciate the importance of solar radiation as a source of energy. In the tropics and deserts the radiation load on an animal may be extremely great, exceeding metabolic heat generated by the organism by a factor of five (Bianca, 1976).

Estimating or measuring the net radiation R_n absorbed by a human or an animal is more complicated than determining the net radiation above a plant canopy. In the former case the radiation streams received, reflected, and emitted by a three-dimensional, irregularly shaped object must be considered. The object itself receives longwave and shortwave radiation from the sky and from nearby surroundings. A general radiation balance for such a body may be written as

$$R_n = (1 - r_b)(R_s + R_r) + \epsilon_b(LW_a + LW_s) - \epsilon_b \sigma T_{bs}^4 \quad (12.1)$$

where r_b is the shortwave reflectivity of the body, R_s is the total direct and diffuse shortwave solar radiation, R_r is the shortwave radiation reflected onto the body from its surroundings, LW_a and LW_s are longwave radiation impinging on the body from the atmosphere and surroundings, respectively, ϵ_b is the longwave emissivity of the body, and T_{bs} is its average surface temperature (Cena, 1974). Each of these fluxes, under certain conditions, may dominate the radiation balance but the relative importance of each is determined by the presence or absence of an intermediate insulating layer, such as fur, feathers, fleece, or clothing, between the object and its surroundings.

Predicting the absorption of radiation is complicated. The position assumed by the object (sitting, standing, lying) determines the area of surface exposed and its orientation to the sun. The changes in solar radiation as influenced by body orientation and solar elevation are illustrated in Fig. 12.2. Procedures for calculating the changes in absorbing area as a man changes position are given by Ward and Underwood (1967). Clapperton et al. (1965) describe the effect of position on solar radiation absorption by sheep. Hutchinson et al. (1976) have computed the ratio of the radiation absorbing area to the total surface area of a sheep and a bull as functions of the animal's orientation to the sun and the solar elevation angle (Fig. 12.3). According to Clapperton et al. (1965) the radiation intercepting surface of a sheep lined up perpendicularly with the sun is reduced by 20% as solar elevation angle

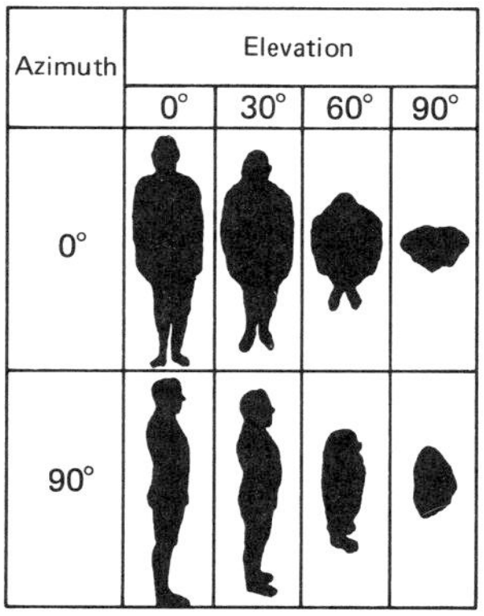

Fig. 12.2 Area of an erect human figure projected in the direction of the direct solar beam at different solar azimuth and solar elevation angles. From Underwood and Ward (1966).

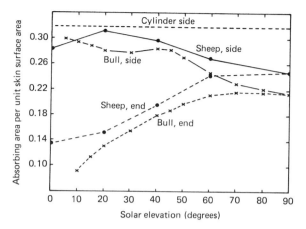

Fig. 12.3 Absorbing area for direct solar beam as a proportion of total surface area for a Sussex bull, shorn British sheep, and open-ended cylinder. For the end orientation the values are an average of the sheep facing toward and away from the sun and for the bull facing the sun only. From Hutchinson et al. (1976).

increases from 0 to 90°. For a sheep facing the sun the intercepting surface area increases from low elevation to a maximum at 60° and decreases slightly with further increase to a solar elevation angle of 90°.

Calculation of the receipt of direct beam solar radiation on a body requires knowledge of the solar elevation angle and the orientation of the body to the sun. The calculation is most easily made by considering the projection of the body onto a plane perpendicular to the solar beam. The projection of the animal's shadow can be traced or, for simple geometric shapes, the area can be calculated. For example, the ratio of the projected area of a sphere A_p to the total area A is

$$\frac{A_p}{A} = \frac{\pi r^2}{4\pi r^2} = \frac{1}{4} \tag{12.2}$$

where r is the radius of the sphere. Calculation of A_p/A can be done for other shapes on the basis of geometric principles. Campbell (1977) gives A_p/A values for several shapes. Monteith (1975) points out, however, that fluxes of radiation are generally measured with respect to a horizontal surface. Thus, the amount of radiation intercepted by an object should be calculated relative to that received on the horizontal surface. To facilitate this analysis, objects such as spheres and cylinders with relatively simple geometry are often used to represent the more irregular shapes of real animals. For example, Stevens et al. (1975) considered that a quadruped might be represented by seven right circular cylinders (Fig. 12.4).

The relationship between A_p, the area projected on a plane normal to the

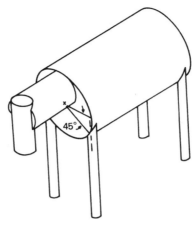

Fig. 12.4 A quadruped represented by seven cylinders. From Stevens et al. (1975).

sun, and A_h, the area projected on a horizontal plane, is illustrated for a sphere in Fig. 12.5. For a given solar elevation, β, A_h is

$$A_h = \frac{A_p}{\sin \beta} \tag{12.3}$$

Thus, combining (12.2) and (12.3), the shape factor A_h/A for the sphere is

$$\frac{A_h}{A} = \frac{A_p}{A \sin \beta} = 0.25 \csc \beta \tag{12.4}$$

Shape factors for other geometric shapes are given by Monteith (1975).

It may be useful at this point to illustrate the calculation of the net radiation balance of an animal in the field. Consider a sheep standing on a grass surface with the axis of its body perpendicular to the sun's rays. The sheep is represented by a horizontal cylinder. The radiation balance of the sheep will be expressed as the average flux density per unit area of the sheep's surface

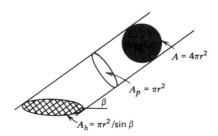

Fig. 12.5 Relationships between the total area A, the area projected in the direction of the solar beam A_p, and the area projected on a horizontal surface A_h with direct beam solar radiation from an elevation angle β.

RADIATION BALANCE

and a bar will be used to distinguish this flux from the flux measured on a horizontal surface. Thus, it is necessary to convert various fluxes of longwave and shortwave radiation to the appropriate average values by multiplying the ratio of the total body surface to the surface area exposed to a particular radiation stream (Monteith, 1975; Campbell, 1977).

For a horizontal cylinder Monteith (1975) gives the shape factor for direct solar radiation as

$$\frac{A_h}{A} = \frac{2rh \csc \beta (1 - \cos^2 \beta \cos^2 \theta)^{1/2} + \pi r^2 \cot \beta \cos \theta}{2\pi rh + 2\pi r^2} \quad (12.5)$$

where r is the radius of the cylinder, h is the length of the cylinder, and θ is the azimuth angle. For the special case of the sheep (cylinder) oriented at right angles to the sun, $\theta = \pi/2$ and (12.5) reduces to

$$\frac{A_h}{A} = \frac{h/r \csc \beta}{\pi(h/r + 1)} \quad (12.6)$$

Assuming the conditions $h/r = 4$ and $\beta = 70°$, $A_h/A = 0.27$.

The area of the sheep that absorbs diffuse shortwave radiation from the sky and the shortwave radiation reflected from the grass surface is the same as that which absorbs the longwave radiation from the sky and the longwave radiation emitted by the grass surface. Assuming all radiation to be isotropic, Monteith (1975) calculated that the upper half of the cylinder receives 82% of the incoming and 18% of the outgoing longwave and shortwave radiation and that these percentages are reversed for the bottom half. For more detail on shape factors for diffuse, reflected and emitted radiation see Monteith (1975) and Campbell (1977).

The average net radiation $\overline{R}_n$ for the sheep may now be calculated using the following equation:

$$\overline{R}_n = \frac{A_h}{A}(1 - r_b)R_D + \frac{1}{2}(1 - r_b)R_d + \frac{1}{2}(1 - r_b)r_g R_s$$

$$+ \frac{1}{2}(\epsilon_b L W_a) + \frac{1}{2}\epsilon_b \epsilon_g \sigma T_g^4 - \epsilon_b \sigma T_{bs}^4 \quad (12.7)$$

where R_D, R_d, and R_s are the flux densities of direct beam, diffuse, and total incoming solar radiation, respectively; r_b and r_g are the reflectivities of the sheep and the grass, respectively; ϵ_b and ϵ_g are the emissivities of the sheep and grass, respectively (from Kirkoffs law $\epsilon_b = \alpha_b$); T_g is the grass temperature; and T_{bs} is the average surface temperature of the sheep.

Given the conditions $r_b = 0.40$, $r_g = 0.23$, $R_D = 800$ W m^{-2}, $R_d = 100$ W m^{-2}, $\epsilon_b = 0.98$, $\epsilon_g = 0.97$, $LW_a = 325$ W m^{-2}, $T_g = 297°$K, $T_{bs} = 307°$K, and $\sigma = 5.67 \times 10^{-8}$ W m^{-2} K^{-4}, $\overline{R}_n$ is calculated to be 98 W m^{-2}. R_n for the grass surface is 580 W m^{-2}. The much lower net radiation for the sheep is a consequence of its high reflectivity and its geometric shape.

Table 12.1 Average reflectivity for humans and selected animals

	Average reflectivities	Reference
Humans		
Caucasian	35	Blaxter (1967)
Negroid	18	Blaxter (1967)
Animals		
Red squirrel	25	Birkebak (1966)
Gray squirrel	31	Birkebak (1966)
Field mouse	14	Birkebak (1966)
Shrew	23	Birkebak (1966)
Mole	19	Birkebak (1966)
Gray fox	9	Birkebak (1966)
Hartebeest	42	Finch (1972)
Eland	22	Finch (1972)
Starling	34	Birkebak (1966)
Glaucous winged gull	52	Birkebak (1966)
Cattle		
White Holstein	61	Stewart (1953)
Black Holstein	8	Stewart (1953)
White Guernsey	60	Stewart (1953)
Tan Guernsey	35	Stewart (1953)
Light brown Jersey	32	Stewart (1953)
Dark brown Jersey	27	Stewart (1953)
Roan Shorthorn	33	Stewart (1953)
Dark red Shorthorn	22	Stewart (1953)
Light tan Hereford	35	Stewart (1953)
Medium red Hereford	24	Stewart (1953)
Dark red Hereford	18	Stewart (1953)
Zulu	51	Blaxter (1967)
Red Sussex	17	Blaxter (1967)
Aberdeen Angus	11	Blaxter (1967)
Sheep		
Weathered fleece	26 (25)	Blaxter (1967); Clapperton et al. (1965)
Newly shorn	42 (50)	Blaxter (1967); Clapperton et al. (1965)

Based on analyses of the sort shown in (12.7) Monteith (1975) evaluated the radiation balance for a grass surface and for a leaf, a sheep, and a human above the grass surface. Calculations were made for a range of solar elevation angles under varying sky conditions. He concluded that (1) the grass absorbs the most radiation during the day while the sheep absorbs least; (2) at low solar elevation angles the human, because of his geometry, has an R_n that is high compared to other objects; (3) R_n is greatest when the sun shines between clouds; and (4) at night the leaf, the sheep, and the human receive longwave radiation from the grass so that their overall radiation loss is less than that of the grass.

Equation 12.7 shows that the emissivity and the reflectivity of an object have an influence on its net radiant energy exchange. For animals and for man the emissivity (or absorptivity) for longwave radiation is 0.95 or greater (Hutchinson et al., 1976; Gates, 1980) and does not vary much between species. The same is not true for reflectivity, however. Reflectivity may range from about 0.10 to 0.50 for animals (Hutchinson et al., 1976; Gates, 1980).

Average reflectivities for man and several other animals are given in Table 12.1. Animal reflectivity is strongly dependent on coat color. The darker coats give reflectivities of about 10% and the lighter coats as much as 60%. Observe also that a white-skinned person reflects about twice as much solar radiation as does a black-skinned person. The disadvantage (at least in a hot environment) of the lower reflectivity of dark skin is compensated by a significant reduction in the depth to which the solar radiation penetrates, a fraction of a millimeter in black skins and several millimeters in white skins (Cena, 1974).

The absorptivity of shortwave radiation $(1 - r)$ varies with wavelength. Absorption is greatest at the shorter wavelengths, followed by a general decrease to a minimum in the near infrared (around 1.3–1.4 μm) followed by a gradual increase in absorption with further increase in the wavelength (Cena, 1974; Hutchinson et al., 1976). It will be remembered that solar elevation angle influences reflectance from plant surfaces (Chapter 1). This is true, too, for animal surfaces with higher values at low sun angles (Riemerschmid and Elder, 1945).

12.3 ENERGY BALANCE

When the sun is shining strongly it usually provides the largest share of the energy input to a homeotherm, but at other times the major energy source is from the production of metabolic heat. For poikilotherms the contribution of metabolic heat is minor; therefore, their body temperature is more closely coupled to ambient environmental conditions.

The survival of homeotherms depends on the maintenance of body temperature within certain acceptable limits. The temperature of an organism under given environmental conditions will adjust until the energy input is equal to the energy output. The equation describing the energy balance of the animal, using the sign convention of the Introduction, may be written as

$$M + R_n + LE + H + C + W + S_t = 0 \qquad (12.8)$$

where M is the metabolic heat, LE is latent heat exchange through evaporation of water from the respiratory tract or from the body surface, H is the sensible heat flux, that is, the energy exchange through convection, C is the heat exchange by conduction, W is the work done by the organism, and S_t

434 HUMAN AND ANIMAL BIOMETEOROLOGY

is the heat stored within the organism. W is the mechanical energy exerted in work by the animal or human. While doing work extra metabolic heat will be generated, body temperature may rise, the extra heat load will need to be dissipated, and the loss of water may increase.

12.3.1 Metabolic Heat Production

An animal's body size, its thermal insulation, the level of its activity and nutrition, and the thermal environment in which it lives are factors that determine the metabolic production of heat. For an animal at rest the rate of metabolic heat production depends on the interaction between nutrition and the thermal environment. The metabolic rate must increase when the ambient temperature is lower than the **critical temperature,** defined as the lower limit of the thermoneutral zone of environmental temperature (Fig. 12.1). In the thermoneutral region the metabolic rate is minimal, constant, and independent of environmental temperature (Mount, 1979).

The **basal metabolic rate** B_m, defined as the minimum metabolic rate required to maintain basic body functions of an organism at rest, is approximated in an empirical relationship given by Kleiber (1932). This relationship shown in Fig. 12.6 is given by

$$B_m = cm^{0.75} \tag{12.9}$$

where c is a numerical factor which for homeotherms is given by Kleiber (1965) as 3.4 and by Hemmingsen (1960) as 1.8 W kg$^{-0.75}$, m is the mass of the animal in kg, and B_m is in W. For poikilotherms the value for c is about 5% of the value for homeotherms (Hemmingsen, 1960). This empirical relationship is supported by theoretical work (McMahon, 1973). Tabulations of metabolic rates for numerous animals and humans of different ages are given by Swan (1974) and Brody (1974). For more detailed discussions of metabolic rates see Brody (1974) and Kleiber (1961).

Metabolic rates are generally reported by physiologists in terms of heat production per unit mass (Campbell, 1977). Since our interest is, generally, in energy exchange per unit of body surface area, ways of estimating surface area must be found. DuBois and DuBois (1915) gave the following formula for humans:

$$A = 0.2m^{0.425}h^{0.725} \tag{12.10}$$

where h is the height in meters. Campbell (1977) gives the following approximate relationship between body surface area and mass:

$$A = 0.2m^{0.66} \tag{12.11}$$

where A and m are in m² and kg, respectively. For computations of the surface area of various geometric and animal shapes see Swan (1974) and

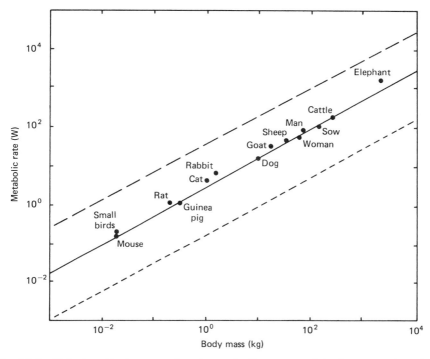

Fig. 12.6 Relationship between basal metabolic rates and body mass of various homeotherms at a body temperature of 39°C (solid line); maximum metabolic rate for sustained work for homeotherms (dashed line) and basal rate for poikilotherms and hibernating mammals (dotted line). Adapted from Kleiber (1932), Benedict (1938), and Hemmingsen (1960).

Brody (1974). Brody (1974) discusses several studies relating surface area to mass and suggests why (12.10) and (12.11) can only be approximations.

The basal metabolic rate per unit area, M_b, is given by

$$M_b = \frac{B_m}{A} \quad (12.12)$$

Monteith (1975) and Campbell (1977) consider that the average basal metabolic rate of man is about 50 W m^{-2}. This rate is comparable to the radiative flux in shade but less than that in the open on clear days. It is similar in magnitude to the net loss of heat by radiation from a body at night when the body surface temperature is near air temperature.

When the level of activity of an animal increases the rate of metabolic heat production also rises. For example, a man walking at a speed of 5.5 km h^{-1} expends about 250 W m^{-2} of energy. According to Monteith (1975) the efficiency with which men and animals use additional metabolic energy is about 30%. Campbell (1977) points out that about two units of heat will be produced for each unit of work done. The maximum metabolic rate that an animal can sustain is, as a rule of thumb, about 10 times the basal rate. Thus, if the maximum metabolic rate can be estimated, the metabolic rate can be determined from

$$M = M_b(1 + 9a/a_m) \tag{12.13}$$

where a/a_m is the ratio of the actual activity to the maximum sustainable activity (Campbell, 1977).

12.3.2 Latent Heat Exchange

The evaporation of water from a human or animal occurs by **insensible perspiration,** that is, the latent heat exchange from the lungs and through the skin in the absence of sweating and by the evaporation of sweat from the body. The latent heat exchange in both cases provides a major mechanism for the dissipation of energy and is very important for homeotherms as a way to prevent overheating in warm environments. In cold environments, however, the loss of heat through "insensible perspiration" can be detrimental to both man and animal (Robertshaw, 1976).

Moisture loss by evaporation from a body is subject to the physical laws described in Chapter 7. Chapter 7 shows that the rate of evaporation is directly affected by atmospheric humidity. Specifically, the rate of evaporation depends on the gradient of vapor pressure between the evaporating surface and the bulk air and on the resistances in the diffusion pathway. The resistance to water vapor diffusion from an animal's surface may or may not depend on the rate of air movement. In the absence of sweating the skin may offer a strong resistance to the diffusion of water vapor and wind will have little effect on water lost by "insensible perspiration." The secretion of water onto the skin surface eliminates the epidermal barrier to diffusion. When this occurs the rate of evaporation increases with increasing wind speed (Robertshaw, 1976).

The rate at which sweat is evaporated from the surface is controlled, ultimately, by the activity of the sweat glands, except at very high sweat rates. When the rate of sweat discharge is less than or equal to the maximum possible evaporation rate the skin dries (McLean, 1974; Robertshaw, 1976). The actual evaporation rate will increase until the rate of secretion exceeds the rate of maximum evaporative capacity from the skin. Then the skin will become visibly wet as is often observed in humid environments. The sweating rate of humans often exceeds the maximum evaporative capacity but, generally, in animals it does not (Robertshaw, 1976).

Sweating and panting dissipate body heat only so long as the heat consumed in vaporization is drawn from the body, rather than from the environment. Thus when sweat falls from the body or saliva drops from the mouth there is no thermal benefit to the animal (Johnson, 1976). When an animal pants, the air passing over wet surfaces in the upper respiratory tract is nearly saturated to a vapor pressure determined by deep body temperature so that the quantity of water evaporated depends only on the vapor concentration of the inspired air and the ventilation rate. Thus the evaporative heat loss from breathing and panting is controlled by one environmental factor, humidity, and one physiological parameter, respiratory ventilation rate (McLean, 1974).

It is reasonable to assume that the respiratory latent heat loss LE_r is a function of the metabolic rate since increased metabolic heat production leads to increased oxygen consumption and increased breathing rate. The ratio of LE_r/M is about 0.10 for dry air and 0.08 for air with a vapor pressure of 1.2 kPa (Monteith, 1975). Certain animals have very small nasal passages and exhale air at temperatures much lower than that of the body. For example, the kangaroo rat exhales air at a temperature approaching the wet-bulb temperature of the ambient air. Its LE_r/M ratio is only about 0.02 for air exhaled at a temperature of 18°C. This low respiratory water loss is very important for survival of kangaroo rats in their arid habitat (Campbell, 1977).

According to Monteith (1975) the "insensible perspiration" loss from dry human skin, LE_s, is about twice the respiratory loss. The total evaporative loss from dry skin may be as great as 20 or 30% of M. LE_s can be determined from the following expression:

$$LE_s = \frac{\rho_a C_p}{\gamma} \frac{e_{ss} - e_a}{r_s + r_c + r_a} \tag{12.14a}$$

where e_{ss} is the saturation vapor pressure at the skin surface temperature, e_a is the vapor pressure in the air, r_s and r_c are the resistances to the diffusion of vapor through the skin and coat, r_a is the aerial resistance, ρ_a is the density of air, C_p is the specific heat of air at constant pressure, and γ is the psychrometric constant. Since certain of the equations and figures used later in this chapter are best given in terms of vapor densities (12.14a) may also be written as

$$LE_s = \frac{L(\rho_{vs} - \rho_{va})}{r_s + r_c + r_a} \tag{12.14b}$$

where ρ_{vs} is the saturation water vapor density at the skin surface and ρ_{va} is the air water vapor density in air. When the skin is dry r_c and r_a are generally negligible as compared to r_s. When the skin is wet or in the case of animals with moist skins such as amphibians, r_a becomes the controlling resistance. Table 12.2 lists values of r_s for several species of animals. There is no single acceptable approach for calculating latent heat losses from heat-stressed animals since physiological responses to heat stress vary greatly with species (Campbell, 1977).

Evaporative heat losses in sweating animals are much greater than the losses that occur through "insensible perspiration." A profusely sweating

Table 12.2 Skin resistance to vapor diffusion for non-heat-stressed animals[a]

Animal	r_s (s m^{-1})	Reference
Human	7,700	Kerslake (1972)
White rat	3,900	Schmidt-Nielsen (1969)
Camel	13,000	Schmidt-Nielsen (1969)
White footed mouse	14,000	Schmidt-Nielsen (1969)
Spring mouse	15,000	Schmidt-Nielsen (1969)
Caiman	5,500	Schmidt-Nielsen (1969)
Water snake	8,800	Schmidt-Nielsen (1969)
Pond turtle	15,000	Schmidt-Nielsen (1969)
Box turtle	33,000	Schmidt-Nielsen (1969)
Iguana	36,000	Schmidt-Nielsen (1969)
Gopher snake	40,000	Schmidt-Nielsen (1969)
Chuckawalla	120,000	Schmidt-Nielsen (1969)
Desert tortoise	120,000	Schmidt-Nielsen (1969)
Sparrow	7,600	Robinson et al. (1976)
Budgerigar	8,500	Bernstein (1971)
Zebra finch	10,200	Bernstein (1971)
Village weaver	12,400	Bernstein (1971)
Painted quail	20,000	Bernstein (1971)
Poor-will	13,200	Lasiewski et al. (1971)
Roadrunner	17,200	Lasiewski et al. (1971)
Ostrich	56,000	Schmidt-Nielsen et al. (1969)

[a] Compiled by Campbell (1977).

man or animal can lose far more heat by evaporation than is produced metabolically (Monteith, 1975). According to Monteith, sheep and cattle lack glands of the type that allow man to sweat so freely. Robertshaw (1976) points out that the sweating rate of man often reaches the maximum evaporative capacity but the sweat rate of animals is much lower and a wet skin is observed only when the air is near saturation. The lack of observable wet skin has led to the erroneous conclusion that many animals do not sweat.

Panting is a compensatory mechanism for increasing evaporative losses in animals that do not sweat at all or sweat only little. Gates (1980) suggests that animals with a high capacity for sweating have a low capacity for panting, and vice versa. Birds, dogs, and cats exhibit high panting rates whereas humans, horses, and cattle sweat freely.

The rate at which sweat evaporates is controlled by environmental factors. Evaporation from wet skin is described by (12.14) with the condition that $r_s = 0$. According to Monteith (1975) the maximum sweating rate for a normal man is about 1 kg h^{-1} although Newburgh (1949) reported rates as high as 4 kg h^{-1} over short periods of time. Macfarlane (1974) gives a maximum sweating rate for cattle of about 0.4 kg h^{-1}. For the camel the rate is 0.35 kg h^{-1}. Assuming a surface area of about 1.8 m^2 for a man, the

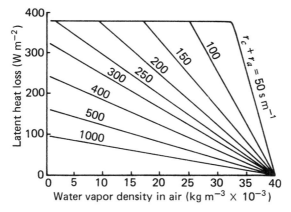

Fig. 12.7 Latent heat loss from the skin of heat stressed humans as a function of water vapor density in air and the combined resistance caused by clothing and air. From Campbell (1977).

latent heat loss would be about 380 W m^{-2} or the value predicted by (12.14), whichever is smaller. Values of latent heat loss from the skin under varying conditions of humidity and as a function of the combined resistances through clothing and air (r_c and r_a) are given in Fig. 12.7.

Animals vary according to species in the ways in which they control their evaporative heat losses. Some may seek the wind to increase evaporative cooling; other may avoid wind to conserve water. Some animals, especially

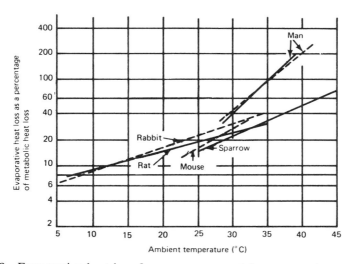

Fig. 12.8 Evaporative heat loss from a sparrow and four mammals expressed as a percentage of the total metabolic heat production under a range of ambient temperatures. From Precht et al. (1973).

insects and other very small animals that cannot afford to lose much water, find ways to reduce water loss during periods of heat stress, even though evaporative cooling is an effective way to reduce body temperature.

Animals that live in deserts are very economical in their water use. Hadley (1972) describes certain water conserving tactics used by desert-living species. Small animals, because of their large surface area to body mass ratio, can lose too much water in desert macroenvironments. Hence, they seek out milder microclimates. Evaporative cooling prevents body temperatures of such desert animals as the camel, the kangaroo rat, and other vertebrates from reaching lethal levels. The urine and feces of such animals contain very little water. Some desert animals are so efficient in conserving water that metabolic water, that is, the water produced when an animal oxidizes food to produce heat, supplies a major part of their water requirements. Thus, most or all of their water needs are met with water contained in their food (Campbell, 1977). Arthopods such as scorpions are very efficient at reducing water losses, and evaporative cooling does not play an important role in dissipating their heat loads.

The moist skin of amphibians can lead to a substantial evaporative cooling effect, which enables them to maintain body temperatures as much as 13°C below the ambient (Gates, 1980). Amphibians seek moist microhabitats, under leaves or in the soil, to prevent too great a water loss. Reptiles generally lack evaporative mechanisms for regulating body temperature (Gates, 1980).

Evaporative cooling through "insensible perspiration," panting, and/or sweating provides a major energy-dissipating mechanism for homeotherms. The relative importance of evaporating cooling depends on the species and the environmental temperature (Fig. 12.8). Evaporative cooling is much more important for a man than, for example, a rat or a rabbit. Some mammals actually lick their fur and add water to the surface, thereby increasing evaporative cooling (Gates, 1980).

12.3.3 Conductive Heat Exchange

The conduction of heat between the animal and its environment involves two distinct processes. First is the conduction of heat to or from the air through the skin and coat or clothing. This heat generally is brought to the outer surface or removed from it by convection. Second is the direct gain or loss of heat by conduction to or from a surface with which the body is in direct physical contact. The exchange of heat in this case is generally small if the animal is standing but it may constitute a significant portion of the energy balance if the animal is sitting or lying on a surface or when the body is partially or completely submerged in a medium other than air, for example, water. Thus a man standing in a swimming pool would exchange a significant amount of heat with the water by conduction.

The heat transfer between the body and its environment is analogous to the flow of heat across an insulating surface layer. Following Kreith (1965) we may treat the animal as a hollow cylinder with radii r_1 and r_2. The heat flux Q to the outer surface is given by

$$Q = 2\pi l K \frac{T_b - T_{bs}}{\ln(r_2/r_1)} \quad (12.15)$$

where K is the thermal conductivity, in units of W m^{-1} °C^{-1}, l is the length of the cylinder, T_b is the internal body temperature, and T_{bs} is the temperature at the outer surface. K for skin is 0.502 W m^{-1} °C^{-1} and for fat it is 0.205 W m^{-1} °C^{-1} (Gates, 1980). Expressing Q on the basis of the outer surface (12.15) becomes

$$Q = K \frac{T_b - T_{bs}}{r_2 \ln(r_2/r_1)} \quad (12.16)$$

which is the form given by Monteith (1975). Slightly different approaches for calculating Q may be found in Campbell (1977) and Gates (1980).

In calculating the heat loss from animals we are generally concerned with the thermal conductance of the whole animal, that is, the conductance of the fat, skin, fur, fleece, or feathers. The **thermal conductance** K_c for a flat surface is defined as the conductivity K divided by the thickness d or

$$K_c = K/d \quad (12.17)$$

and for the cylindrical surface Kreith (1965) gives

$$K_c = \frac{K}{r_2 \ln(r_2/r_1)} \quad (12.18)$$

The thermal conductance is the heat flow per unit area in response to a unit temperature gradient and has units of W m^{-2} °C^{-1}. For an animal it is average conductivity of the various parts of the body divided by the average thickness of the skin and any covering (Gates, 1980). The concept of **insulation** is frequently used for calculating heat losses. The insulation I is the reciprocal of K_c. Insulation is sometimes also called **specific resistance** (Monteith, 1975; Gates, 1980). When heat flows through a series of materials with different thermal properties, the insulation values of the materials are accumulated and termed the **equivalent insulation.**

Animal and human physiologists generally use the term insulation. Micrometeorologists, ecologists, plant physiologists, and agronomists generally use the resistance concept for the same purpose. Monteith (1975) suggest that I be multiplied by $\rho_a C_p$ to convert I to units of resistance (s m^{-1}). Table 12.3 gives average values of thermal resistances r_c for man and several animals.

The effects of wind on coat insulation are complex and difficult to quantify. Campbell (1977) suggests that the average coat resistance decreases

Table 12.3 Thermal resistances of animal skins and coats

Animal	Resistance (s m^{-1}) Vasoconstricted	Dilated	Reference
Skin			
Steer	1700	500	Blaxter (1967)
Calf	1100	500	Blaxter (1967)
Pig	1000	600	Blaxter (1967)
Down sheep	900	300	Blaxter (1967)
Human	1200	300	Blaxter (1967)
Coats	Resistance (s m^{-1})		Reference
Red fox	3300		
Lynx	3100		Hammel (1955)
Skunk	3000		Hammel (1955)
Husky dog	2900		Hammel (1955)
White fox	1400		Scholander et al. (1950)
Moose	1300		Scholander et al. (1950)
Grizzly bear	1250		Scholander et al. (1950)
Reindeer	1250		Scholander et al. (1950)
Wolf	1250		Scholander et al. (1950)
Caribou	1100		Scholander et al. (1950)
Beaver	1100		Scholander et al. (1950)
Polar bear	1050		Scholander et al. (1950)
Eskimo dog	900		Scholander et al. (1950)
Rabbit	900		Scholander et al. (1950)
Red fox	900		Scholander et al. (1950)
Marten	850		Scholander et al. (1950)
Lemming	500		Scholander et al. (1950)
Squirrel	450		Scholander et al. (1950)
Weasel	300		Scholander et al. (1950)
Shrew	200		Scholander et al. (1950)
Dall sheep	1450		Blaxter (1967)
Merino sheep	2800		Blaxter (1967)
Down sheep	1900		Blaxter (1967)
Ayshire cattle	1000		Blaxter (1967)
Galloway cattle	900		Blaxter (1967)

with the square root of wind speed. Wind penetrates the coat and disrupts part of the insulating effect by decreasing the thickness of the still air layer trapped in the hairs. However, the relation between insulation and wind speed is complicated by the orientation of the animal with respect to the wind. Heat transfer by conduction through sheep fleece is independent of wind speed so long as the wind is parallel to the surface of the fleece (Cena and Monteith, 1975). If the wind blows in the direction of the natural hair set the thickness of the coat is decreased but wind in the opposite direction may fluff out the hair (Monteith, 1975). If the wind is perpendicular to the body heat transfer will be increased (Ames and Insley, 1975), although much of the loss in this case is the result of convective processes (Section 12.3.4).

When an animal lies down and its body makes direct physical contact with the surface the insulating properties of the coat will likely be diminished since the thickness of the coat will decrease and much of the insulating air layer will be removed. Gatenby (1977) calculated, for example, that head conduction from a recumbent sheep to the cold ground may dissipate as much as 30% of the animal's heat production.

12.3.4 Convective Heat Exchange

Convection is involved in the transfer of heat between animals and their environment. Yet surprisingly little quantitative information is available on convective contributions to the energy balance of animals, except in the case of humans. There is some heat transfer by convection as air passes through the animal's respiratory tract but most convective heat transfer occurs at the outer surface of the body. At very low wind speeds free convective processes predominate. Under certain environmental conditions both free and forced convective transfer take place. At wind speeds greater than about 0.2–0.5 m s^{-1} forced convection predominates. An equation describing convective heat transfer H was given in Chapter 3:

$$H = \rho_a C_p \frac{T_a - T_s}{r_a} \tag{3.4}$$

where T_a is the air temperature, T_s is the surface temperature, and r_a is the aerial resistance to the transfer of heat. In the literature of biometeorology and heat transfer a convective heat transfer coefficient h_c is used instead of the resistance r_a. r_a and h_c are related by

$$h_c = \frac{r_a}{\rho_a C_p} \tag{12.19}$$

Using h_c the convective heat transfer from an animal with body temperature T_b is

$$H = h_c(T_a - T_b) \tag{12.20}$$

The variables on which h_c depends are numerous, but they have been combined into various dimensionless groups including the Nusselt number Nu; the Grashof number Gr; the Prandtl number Pr; and the Reynolds number Re. A brief discussion of these dimensionless numbers, except for Gr, is given in Chapter 3.

The **Grashof number** is

$$\text{Gr} = \frac{D^3 \rho_a^2 g \beta (T_a - T_b)}{\mu^2} \tag{12.21}$$

where D is a characteristic dimension of the object, g is the acceleration due to gravity, μ is the dynamic viscosity of air, and β is the coefficient of volumetric expansion. The Grashof number is, therefore, the ratio of a buoyant force times an inertial force to the square of a viscous force (Campbell, 1977).

The convective heat transfer coefficient is given by

$$h_c = \frac{\text{Nu} K}{D} \tag{12.22}$$

where K is the thermal conductivity of the air (2.57×10^{-2} W m^{-1} °C^{-1}).

Respiratory Convection. Because air must enter and exit through well-defined body orifices the measurement of respiratory convective heat loss H_r is quite simple. For man H_r can be calculated from

$$H_r = C_1 V(T_i - T_e) \tag{12.23}$$

where H_r is in units of W m^{-2} and V is the ventilation rate in l min^{-1}, T_i and T_e are the temperatures of the inhaled and expired air, respectively, and C_1 is a proportionality factor that incorporates the density and specific heat of air (Mitchell, 1974). H_r is generally a very minor component of the animal's energy balance. Even at very low temperatures H_r for a man would be only about 25 W m^{-2} or 10% of his metabolic rate while walking (Mitchell, 1974).

Free Convection. Free convection occurs indoors or outdoors under very calm conditions. The Nusselt number for free convection can be expressed as

$$\text{Nu} = B\text{Gr}^n \tag{12.24}$$

where B and n are numerical constants that depend on the shape of the object. Values for B and n can be obtained from Birkebak (1966) and Monteith (1975). Campbell (1977) shows that Nu can be obtained for upward-facing heated surfaces by

$$\text{Nu} = 0.54(\text{GrPr})^{1/4} \tag{12.25}$$

and for downward-facing heated surfaces by

$$\text{Nu} \simeq 0.27(\text{GrPr})^{1/4} \tag{12.26}$$

The shape of humans and most domesticated animals may be approximated by large vertical cylinders for the calculation of free convective heat transfer (Mitchell, 1974). Table 12.4 gives some values for h_c for objects of various shapes and for various ranges of $D^3(T_a - T_b)$.

There have been few direct measurements of free convective heat loss from humans or animals. Indirect measurements suggest that a reasonable approximation of free convective losses from humans can be made using

Table 12.4 Coefficients of free convective heat h_c for vertical plates and cylinders[a]

Object	Characteristic dimension D (m)	Range of $D^3 (T_a - T_b)$	h_c (W m^{-2} °C^{-1})
Vertical plate or cylinder	Height	1.5×10^{-4} to 15	$1.4[(T_a - T_b)/D]^{1/4}$
	Height	15 to 1.5×10^4	$1.5(T_a - T_b)^{1/3}$
Horizontal cylinder	Diameter	1.5×10^{-4} to 15	$1.2[(T_a - T_b)/D]^{1/4}$
	Diameter	15 to 1.5×10^4	$1.2(T_a - T_b)^{1/3}$

[a] Modified by Mitchell (1974) from Birkebak (1966).

cylinders as an approximation, particularly at small values of $(T_a - T_b)$ (Mitchell, 1974).

For $(T_a - T_b)$ of 10°C free convection may account for 30–40 W m^{-2}, a great enough rate to dissipate a significant proportion of the metabolic heat production. Thus, when winds are light, free convective heat transfer from man or animal should not be disregarded as a component of the energy balance (Mitchell, 1974).

Forced Convection It is easier to determine forced convection than free convection and many estimates of forced convective heat losses from humans and animals have been reported. Even for forced convection, however, relatively few direct measurements have been made of h_c from animals (Hutchinson et al., 1976). Exceptions are Joyce et al. (1966) and Nishi and Gagge (1970), who worked with real animals. Most calculations have been made assuming that animals behave as smooth cylindrically shaped objects. Mitchell (1974) cites work of Kerslake (1972) and Birkebak (1966) which supports the use of the "cylindrical assumption." Mitchell (1976) concluded, however, that the best estimates of forced convective heat transfer from man and animals are obtained using convective relationships developed for the sphere. Procedures for the calculation of forced convective heat losses from objects of many different shapes in both laminar and turbulent flow are given by Monteith (1975) and Gates (1980).

For forced convection, the Reynolds number Re is used instead of the Grashof number for calculating Nu. For smooth cylinders

$$\text{Nu} \propto \text{Pr}^m \text{Re}^n \tag{12.27}$$

or using appropriate values for Pr in normal air environments:

$$\text{Nu} = N\text{Re}^n \tag{12.28}$$

Where N and n are numerical constants. Monteith (1975) gives Nu = $0.65\text{Re}^{0.5}$ for sheep and for man Nu = $0.78\text{Re}^{0.5}$. Mitchell (1974) gives a procedure for obtaining N and n based on varying values of Re.

Criteria for Free and Forced Convection. It is not always obvious whether heat transfer from a body is due to forced or free convection. The Reynolds and Grashof numbers provide a criterion for determining which predominates. When Gr is much larger than Re^2 buoyancy forces dominate and heat transfer is governed by free convection. When $0.1Re^2 \geqslant Gr$ forced convection is the major mechanism for heat transfer and h_c is calculated using the Re number. From experimental evidence h_c is calculated using the Gr number when $Gr \geqslant 16Re^2$. For the range $0.1Re^2 \geqslant Gr \geqslant 16Re^2$ Nu is calculated for both forced and free convection and the largest value is used to determine h_c (Monteith, 1975; Gates, 1980).

12.3.5 Heat Storage

The energy balance equation contains a heat storage term S_t, which accounts for the change in heat content of an animal. The internal body temperature of most homeotherms is sufficiently constant so that S_t can be neglected. There are, however, certain desert animals whose body temperature rises by several degrees during the day. This heat is dissipated at night. The temperature of a camel, for example, may rise to 40°C in the afternoon and drop to 34°C by morning. This temperature change is equivalent to an energy flux of about 40 W m^{-2} or about 10–20% of the maximum R_n. This "deferred heat transfer" may be an adaptive mechanism for conserving body water. Monteith (1975), however, suggests that about one-half the benefit of this heat storage is lost because metabolism increases with the increased body temperature.

When a substantial change occurs in the ambient temperature the body of a homeotherm may gain or lose stored heat even though T_b remains constant. Monteith calculates that a human may lose 20 W m^{-2} as the air temperature decreases at the rate of 5°C h^{-1} from 30 to 0°C.

In both examples, camel and man, heat released as the body cools represents a substantial part of the organisms nocturnal energy balance. The heat released from storage at night would otherwise have to be provided by metabolic energy. By absorbing energy during warm periods and releasing it in cool periods animals may reduce their requirement for food.

12.4 THE CLIMATE SPACE

Acceptable environments for specific animal species can be described with **climate space diagrams.** The climate space diagram provides a convenient way to visualize the equivalent impact on the animal of varying conditions of temperature and radiation. Acceptable environmental limits on the climate space diagram defines the **climate space.** A procedure for deriving a climate space diagram follows.

For the conditions where C, W, and $S_t = 0$ (12.8) can be rewritten as

$$R_n + H + M + LE = 0 \tag{12.29}$$

The average surface temperature term T_{bs} is a determinant of both H and R_n. Rewriting (12.29) using (12.7) and (3.4) yields

$$R_{abs} - \epsilon_b \sigma T_{bs}^4 + \frac{\rho_a C_p (T_a - T_{bs})}{r_a} + M + LE = 0 \tag{12.30}$$

R_{abs} is the net amount of shortwave and longwave radiation absorbed [all except the final term in (12.7)]. Next, a solution for T_{bs} is achieved using an expression for the conduction of heat from the animal core to an exchange surface. Assuming that all latent heat loss is from inside the animal, then

$$M + LE = \frac{\rho_a C_p (T_b - T_{bs})}{r_c} \tag{12.31}$$

where r_c is the resistance to heat transfer through the tissue and the coat. For nonsweating animals T_{bs} is given by

$$T_{bs} = T_b - \frac{r_c}{\rho_a C_p}(M + LE) = T_b - I(M + LE) \tag{12.32}$$

Solutions for T_{bs} with (12.32) and substitution of the values into (12.30) permits development of a climate space diagram by selecting values for R_{abs} and solving (12.30) for T_a.

An example of a climate space diagram for a sheep is given in Fig. 12.9. The sheep has the characteristics given in Section 12.2 where $\bar{R}_n$ was calculated using (12.7). Other necessary information: at the maximum body temperature for survival (thermal maximum), $M = 90$ W m^{-2}, $LE = -95$ W m^{-2}, $T_b = 41.7°C$, and $I = 1.2$ m^2 °C W^{-1}; at the thermal minimum $M = 70$ W m^{-2}, $LE = -25$ W m^{-2}, $T_b = 39.5°C$, and $I = 1.0$ m^2 °C W^{-1}; the animal has a diameter D of 0.25 m. r_a will be calculated at wind speeds of 0.1 m s^{-1} (approximating free convective conditions) and 2.0 m s^{-1} (forced convective conditions). Assuming a cylindrical shape, r_a is calculated using a relationship due to Gates (1980) so that

$$r_a = \frac{\rho_a C_p}{k_1(U^{0.5}/D^{0.5})} \tag{12.33}$$

where k_1 for a cylinder is 3.89 W m^{-2} s$^{1/2}$ °C^{-1} (Gates, 1980). By (12.33) r_a at low wind speed is 493 s m^{-1}; at the higher wind speed r_a is 110 s m^{-1}. By (12.32) T_{bs} at the thermal maximum is 47.7°C and at the thermal minimum it is $-5.5°C$.

Curves for clear night, blackbody, and full sun on the diagram help define

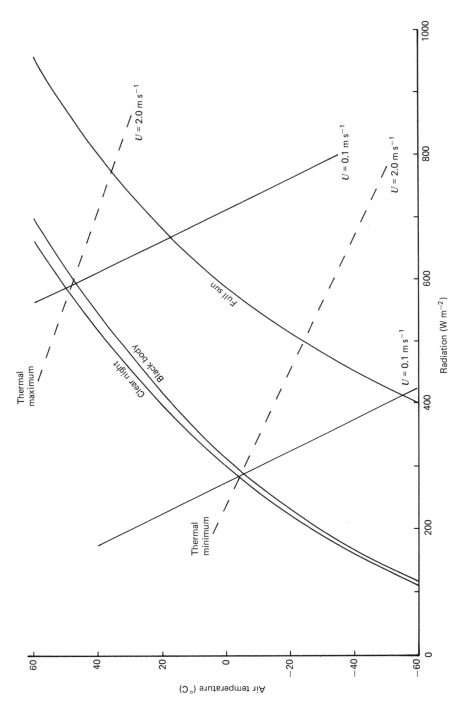

Fig. 12.9 Climate space diagram for a sheep. See text for details.

the limits of the physically tolerable environments for the sheep. The blackbody line is given by

$$R_{abs} = \sigma T_a^4 \tag{12.34}$$

where T_a is the air temperature.

The clear night curve is obtained from

$$R_{abs}(\text{night}) = \bar{\epsilon}\epsilon_b \sigma T_a^4 \tag{12.35}$$

where $\bar{\epsilon}$ is the average emissivity of the surroundings (clear sky and ground).

The full sun line was calculated from

$$R_{abs}(\text{sun}) = \bar{\epsilon}\epsilon_b \sigma T_a^4 + \frac{A_h}{A}(1 - r_b)(R_D)$$
$$+ \frac{1}{2}(1 - r_b)(R_d) + \frac{1}{2}(1 - r_b)r_g R_s \tag{12.36}$$

where $R_D = 1000$ W m^{-2} and $R_d = 200$ W m^{-2}. For such conditions $R_{abs}(\text{sun}) = \bar{\epsilon}\epsilon_b \sigma T_a^4 + 287$ W m^{-2}. That portion of the diagram bounded by the lines for clear night and full sun and by the lines for wind speed defines the conditions of radiation and temperature that are energetically acceptable and environmentally tolerable for the animal (Campbell, 1977). The sheep would not survive outside this climate space, unless, of course, physiological responses such as sweating, panting, restriction of blood flow to the surface, and increased metabolic output change its energy balance. Examples of climate space diagrams for other animals (shrew, desert iguana, cardinal, alligator, and chipmunk) are given by Campbell (1977) and Gates (1980).

12.5 EFFECTS OF CLIMATE ON HUMANS

The weather affects our physical well-being and influences our mental outlook as well. Bright, crisp, cool days make us feel good; extended periods of cloudy or humid weather make us grumpy and depressed. Studies have shown that accidents at work increase during periods of changing weather (Dodds, 1979).

Despite protection from the elements, humans still react involuntarily to the weather. For example, with approaching storms some people experience increased arthritic pain. Asthmatics may begin to wheeze; others may become depressed or experience headaches and other ailments. Research has shown that temperature, humidity, and atmospheric pressure are implicated in these human responses. Molecules of one or more of the gases composing the atmosphere occasionally take on a positive or negative electrical charge and these ions appear to affect our health. When positive ions predominate, some people develop headaches, nasal obstructions, hoarseness, sore throat, dizziness, and other forms of discomfort (Sulman et al., 1974). Such conditions are associated with Chinook, Santa Ana, sirocco, and Sharav-type winds and may help explain the increased incidence of depression, suicide, crimes of passion, and so on, that occurs when these winds blow. When negative ions become dominant these symptoms disappear (Richardson et al., 1968).

The human is unique among the animals in the possession of an intellect, which enables him to develop a cultural world and a modified environment detached from his purely biological condition. By trial and error humans have found ways to diminish direct impacts of the total environment on themselves. In industrialized countries people are much further detached from their natural environment than are such groups as the Australian aborigines (Weihe, 1976). Therefore, treating human adjustment to environmental conditions simply in terms of physical–chemical body processes can be misleading.

Humans have explored the ocean depths, walked in space, and landed on the moon. These actions have been possible only through the creation of artificial environments for these explorers. We create artificial environments, as well, in our homes, offices, and factories.

There is a fairly narrow range of environmental conditions in which man is confortable. To express how one "feels" in a given environment, ranging from very cold to very hot, certain meteorological indices have been developed, generally based on one or more of the following meteorological factors: air temperature, air humidity, wind speed, and solar radiation. Outdoors all these factors are usually important; indoors temperature and humidity are generally the only important considerations.

12.5.1 Warm Season Indices

Effective Temperature Indices. When a human is exposed to a warm environment the first physiological response is dilation of the blood vessels which increases the flow of blood near the skin. The next response occurs through sweating, panting, and evaporative cooling. Since individuals differ in their physiological responses to environmental stimuli, it is difficult to develop heat stress indices based solely on meteorological variables. Nonetheless, certain indices have been found useful.

One such index, used for many years by heating and ventilating engineers, is the **effective temperature** T_{ef}. T_{ef} is defined as "the temperature at which motionless, saturated air would induce in a sedentary worker wearing indoor clothing the same sensation of comfort as that induced by the actual conditions of temperature, humidity and air movement" (Huschke, 1959). The effective temperature is plotted as a function of wind speed and wet- and dry-bulb temperature as a family of curves on a comfort chart (Fig. 12.10).

Givoni (1974) found that T_{ef} is not particularly reliable as an index for

EFFECTS OF CLIMATE ON HUMANS

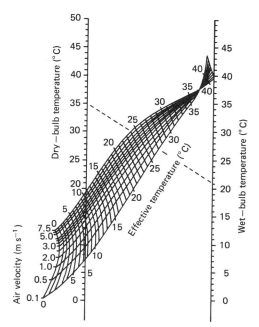

Fig. 12.10 Chart of the effective temperature as a function of air velocity, wet-bulb and dry-bulb temperatures. Reprinted with permission from *ASHRAE Handbook* (1967).

predicting physiological and sensory responses under both comfortable and heat stress conditions. He suggests, instead, that the **resultant temperature index** developed by Missenard (1948), although similar in form to T_{ef}, gives values which agree better with human physiological responses.

Temperature and Humidity Indices. People complaining of discomfort will often suggest that "its not the heat, it's the humidity." Winterling (1979) explains the feeling of discomfort that accompanies high humidity. One can escape from intense solar radiation by seeking shade and one can create a breeze by fanning, but there is no escape from the effects of high humidity coupled with high temperature. A summary of human physiological responses to temperature and humidity is shown in Fig. 12.11.

Winterling (1979) used the term **humiture** coined originally by Hevener (1959) to describe the combined effects of temperature and humidity on comfort levels for Jacksonville, Florida, a hot, humid place in the summer. The sensation of increased heat with high humidities is related to decreased evaporative cooling. At 35°C the vapor pressure of skin moisture should be about 5.6 kPa and the evaporative cooling power can be represented by (5.6 − e)kPa. Winterling established that, if the vapor pressure differential is less than about 3.5 kPa about 0.5°C should be added to the air temperature for each 0.1 kPa of vapor pressure differential less than 3.5 kPa and 0.5°C

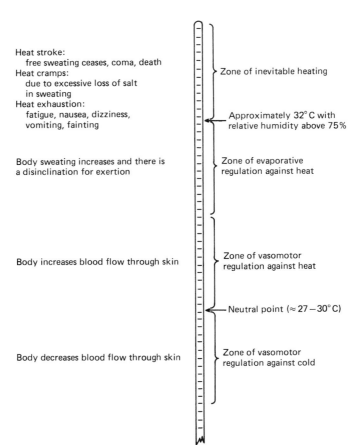

Fig. 12.11 The physiological responses of humans to varying conditions of heat and humidity. From Masterton and Richardson (1979).

should be subtracted for each 0.1 kPa differential greater than 3.5 kPa. Winterling proposes a simplified way to calculate the humiture T_h:

$$T_h = T_a + (T_d - 18) \quad \text{for } T_a \geq 30°C \quad (12.37)$$

where T_d is the dew-point temperature. Figure 12.12 is a plot of T_h as a function of T_a and T_d.

Another commonly used index is the **temperature humidity index** THI (originally defined by Thom, 1959), which can be derived from the T_{ef} diagram (Fig. 12.10). THI is given by

$$\text{THI} = 0.4(T_a - T_w) + 4.78 \quad (12.38)$$

where T_a and T_w are the dry-bulb and wet-bulb air temperatures in °C, respectively. Segal and Pielke (1981) suggest that few people will feel uncom-

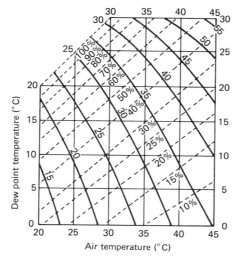

Fig. 12.12 Value of the humiture index (solid lines) and relative humidity (dashed lines) as functions of air and dew point temperature. From Winterling (1979) by permission of the American Meteorological Society.

fortable when THI = 21°C or less, about half the population will be uncomfortable when THI = 24°C and almost everyone will be uncomfortable at THI ⩾ 26.5°C. When THI values are greater than 33°C the body temperature may exceed 40°C and heat stroke leading to death may occur (Hobbs, 1980).

Another index relating temperature and humidity to human comfort was derived by Anderson (1965) for Canadian conditions. This index is called **humidex** H_u and is written as

$$H_u = T_a + \frac{5}{9}(e_a - 10) \tag{12.39}$$

where e_a is the vapor pressure of water in air in kPa × 10. The relationship between various values for H_u and comfort levels is given in Table 12.5.

Steadman (1979a) suggests the use of an apparent temperature based on

Table 12.5 Ranges of humidex values and their relationship to human comfort

Humidity range (°C)	Degree of comfort
20–29	Comfortable
30–39	Varying degrees of discomfort
40–45	Almost everyone uncomfortable
46 and over	Many types of labor must be restricted

SOURCE: Masterton and Richardson (1979).

Table 12.6 Temperature humidity scale: Apparent temperature (°C) corresponding to each combination of dry-bulb temperature and relative humidity[a]

Dry-bulb temperature (°C)	Relative humidity (%)										
	0	10	20	30	40	50	60	70	80	90	100
20	16	17	17	18	19	19	20	20	21	21	21
21	18	18	19	19	20	20	21	21	22	22	23
22	19	19	20	20	21	21	22	22	23	23	24
23	20	20	21	22	22	23	23	24	24	24	25
24	21	22	22	23	23	24	24	25	25	26	26
25	22	23	24	24	24	25	25	26	27	27	28
26	24	24	25	25	26	26	27	27	28	29	30
27	25	25	26	26	27	27	28	29	30	31	33
28	26	26	27	27	28	29	29	31	32	34	(36)
29	26	27	27	28	29	30	31	33	35	37	(40)
30	27	28	28	29	30	31	33	35	37	(40)	(45)
31	28	29	29	30	31	33	35	37	40	(45)	
32	29	29	30	31	33	35	37	40	44	(51)	
33	29	30	31	33	34	36	39	43	(49)		
34	30	31	32	34	35	38	42	(47)			
35	31	32	33	35	37	40	(45)	(51)			
36	32	33	35	37	39	43	(49)				
37	32	34	36	38	41	46					
38	33	35	37	40	44	(49)					
39	34	36	38	41	46						
40	35	37	40	43	49						
41	35	38	41	45							
42	36	39	42	47							
43	37	40	44	49							
44	38	41	45	52							
45	38	42	47								
46	39	43	49								
47	40	44	51								
48	41	45	53								
49	42	47									
50	42	48									

Source: Steadman (1979a) by permission of the American Meteorological Society.
[a] Values in parentheses correspond to humidities above 90% at the skin surface and are approximate.

temperature and humidity to assess **sultriness,** the extent to which humidity aggravates the physiological effects of high temperature. Sultriness is a measure of how a typical human will feel when exposed to a given combination of temperature and humidity. The index considers the amount of clothing needed to achieve thermal comfort and the reduction in the skin's resistance needed to reach thermal equilibrium. Conditions of equal sultriness are referred in Table 12.6 to a vapor pressure of 1.6 kPa in order to calculate the apparent temperature corresponding to various combinations of temperature and humidity.

Table 12.6 is useful for comparing indoor conditions but for outdoor conditions the factors of wind and radiation must also be taken into account. Steadman (1979b) provides an approach that incorporates these factors. Lee and Vaughn (1964) also describe a procedure that accounts for the effects of solar radiation.

Indices Based on Evaporation and Sweat Rate. Belding and Hatch (1955) were the first to construct a comprehensive biophysical model involving calculations of external heat stress acting on a man in a given thermal environment. Metabolic heat production and the evaporative capacity of the air are considered in this model (Givoni, 1974). The resultant **heat stress index** HSI is defined by

$$\text{HSI} = \frac{E}{E_{cm}} \qquad (12.40)$$

where E is the evaporation rate required to balance the energy of radiation, convection, and metabolic processes; and E_{cm} is the maximum evaporative capacity of the air, calculated from the wind speed and the difference in vapor pressure between the ambient air and the saturation vapor pressure at the skin temperature.

An index similar to the HSI was developed by Lee and Henschel (1966) which they call the **relative strain index.** This index incorporates a term for the resistance of clothing to heat and vapor transfer.

Givoni (1969) developed a biophysical model called the **thermal stress index** TSI describing heat exchange between the body and the environment. The total thermal stress and the sweat rate required to maintain thermal equilibrium can be predicted from this index.

$$\text{TSI} = S = (M - W \pm H \pm R_n)/f \qquad (12.41)$$

where S is the required sweat rate in W m^{-2}, f is the cooling efficiency of sweating (dimensionless), and other terms have previously been defined. This index has been shown to provide reliable estimates of sweat rates under a wide range of climatic conditions, work rates, and types of clothing. Schiller and Karschon (1973) used this index in an interesting study to evaluate the effects of forests and open areas for recreational purposes.

Indices Based on Dry and Wet Equivalent Temperatures. Campbell (1977) suggests that the energy balance approach can be used to define comfortable conditions for humans by incorporating appropriate physiological parameters to derive an **equivalent blackbody temperature** T_{eq}:

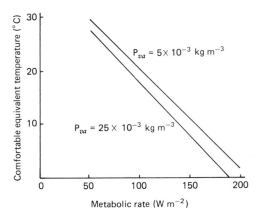

Fig. 12.13 Comfortable equivalent temperature as a function of metabolic rate for two water vapor densities in air. From Campbell (1977).

$$T_{eq} = \frac{\rho_a C_p}{r_c + r_e} T_b - \frac{r_c + r_e}{\rho_a C_p} (M + LE) \qquad (12.42)$$

where $1/r_e = 1/r_a + 1/r_r$ and $r_r = \rho_a C_p 4\epsilon\sigma T^3$. Using (12.42) with LE evaluated from $LE = (2.3 \times 10^{-3}M + 0.31)(37 - \rho_{va})$ Campbell calculated T_{eq} values representing comfort for ρ_{va} values of 5×10^{-3} and 2.5×10^{-2} kg m^{-3} kg m^{-3} (Fig. 12.13).

If the room wall temperature equals T_a then $T_{eq} = T_a$. For normal metabolic activity ($M = 90$ W m^{-2}) at the low vapor density a comfortable room temperature would be 22°C while at the higher humidity 19°C would be comfortable. This example supports recommendations of the U.S. National Weather Service that buildings should be humidified as a way to save

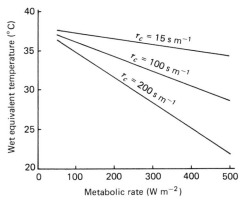

Fig. 12.14 Maximum tolerable wet equivalent temperature as a function of metabolic rate at three values of r_c—the tissue plus coat resistance. From Campbell (1977).

energy and money. One should keep in mind, however, that it takes energy to evaporate the water into the air so that savings may be less than expected. There are other advantages, however, to humidifying the air under cold conditions.

Campbell (1977) suggests a procedure for combining all the environmental factors into a single parameter indicative of heat stress. His heat stress index is termed the **equivalent wet blockbody temperature** T_{eqw} defined by

$$T_{eqw} = T_a + \frac{\gamma^*}{s + \gamma^*}\left(r_e \frac{R_{abs} - \epsilon\sigma T_a^4}{\rho_a C_p} - \frac{\rho'_{va} - \rho_{va}}{\gamma^*}\right) \quad (12.43)$$

where $\gamma^* = \gamma(r_s + r_c + r_a)/(r_c + r_e)$, s is the slope of the saturation vapor pressure as a function of temperature, ρ'_{va} is the saturation water vapor density of air, and all other terms have been previously defined. With some air movement, a wet skin surface, and equal resistances for vapor and heat transfer, $\gamma^* = \gamma$. At body temperature, $\gamma/(s + \gamma) = 0.17$; the last term in (12.43) becomes small and T_{eqw} differs from T_a by only a few degrees.

Campbell (1977) calculated the maximum tolerable T_{eqw} for a range of r_c and a range of metabolic rates with results as shown in Fig. 12.14. As expected T_{eqw} increases with increases in M and r_c. As yet there is no direct way to measure T_{eqw}. Campbell suggests that T_{eqw} may be correlated with the surface temperature of a water-filled copper sphere covered with black, moistened cloth.

12.5.2 Cold Season Indices

Of the environmental factors that affect human feelings of comfort in cold weather, wind and temperature are far more important than humidity and solar radiation (Steadman, 1971). Two indices of the combined effects of wind and low temperature are based on the work of Siple and Passel (1945). These are the **windchill index** and the **windchill equivalent temperature.**

Wind Chill Index. The wind chill index WCI is a measure of the cooling effect created by any combination of temperature and wind. WCI is expressed in terms of the loss of body heat per unit area of skin surface (kcal m^{-2}). WCI is calculated for a nude body in the shade and is an approximation only since individuals vary in shape, size, and metabolic rate (Huschke, 1959).

Most wind chill information is based on the empirical wind chill formula of Siple and Passel (1945). The formula is based on observed freezing rates of water in a plastic cylinder. Their equation is

$$\text{WCI} = (10.45 + 10U^{1/2} - U)(33 - T_a) \quad (12.44)$$

where U is wind speed in m s^{-1} and T_a is air temperature in °C. Court (1948)

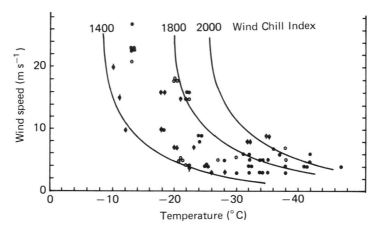

Fig. 12.15 Occurrence of frostbite in relation to temperature and wind speed. The conditions are: (●) stationary meteorological conditions; (○) changing meteorological conditions (means of wind and temperature extremes used); and (◆) estimated wind speed during sledding. The curved lines are for wind chill indices of 1400, 1800, and 2000 kcal m^{-2} hr^{-1}. From Wilson (1967).

suggested that the data of Siple and Passel could be better fit with the equation

$$\text{WCI} = (9.0 + 10.9 U^{1/2} - U)(33 - T_a) \qquad (12.45)$$

Wilson (1967) suggested that exposed skin will experience frostbite at WCI values of about 1400 k cal m^{-2} h^{-1} (Fig. 12.15). Equation 12.45 predicts that heat loss will reach a maximum at 25 m s^{-1}. According to Steadman (1971), in strong winds the sensation of cold depends more on the wind speed than is suggested by (12.44) and (12.45).

Wind Chill Equivalent Temperature. The most generally accepted way to express the sensation caused by the combined effects of wind and low temperature is the wind chill equivalent temperature T_{wc} (sometimes called the **wind chill factor**). T_{wc} is calculated by reference to the Siple and Passel formula. The heat loss predicted for a given combination of T_a and U is equated to the same heat loss at low wind speed and some equivalent air temperature T_{wc}.

A practical problem occurs in selecting the low wind speed conditions since neither the atmosphere nor the person experiencing intense cold is completely at rest (Griffiths and Driscoll, 1982). Values of 5–8 km h^{-1} (1.39–2.22 m s^{-1}) have been used in calculating T_{wc}. The lower the wind speed threshold the lower will be T_{wc}. Thus available charts and graphs of T_{wc} may differ from those of the U.S. National Weather Service. Table 12.7 gives values of T_{wc} calculated by the U.S. National Weather Service (National Oceanic and Atmospheric Administration, 1980) and is based on a threshold wind speed of about 1.8 m s^{-1}.

Table 12.7 Wind chill chart[a]

Wind-speed (m s^{-1}) | | | | | | Equivalent temperature (°C) | | | | | | | | | |
---|---|---|---|---|---|---|---|---|---|---|---|---|---|---|---|---
Calm | 1.7 | -1.1 | -3.9 | -6.7 | -9.4 | -12.2 | -15.0 | -17.8 | -20.6 | -23.3 | -26.1 | -28.9 | -31.7 | -34.4 | -37.2 | -40.0 | -43.9
2.2 | 0 | -2.8 | -5.6 | -8.9 | -11.7 | -14.4 | -17.8 | -20.6 | -23.3 | -26.1 | -29.4 | -32.2 | -35.0 | -37.8 | -41.1 | -43.8 | -46.7
4.5 | -5.6 | -8.9 | -12.2 | -16.1 | -19.4 | -22.8 | -26.1 | -30.0 | -32.8 | -36.7 | -40.0 | -43.3 | -46.7 | -50.0 | -53.3 | -57.2 | -60.6
6.7 | -8.9 | -12.8 | -16.7 | -20.6 | -23.9 | -27.8 | -31.8 | -35.0 | -38.9 | -42.3 | -50.0 | -53.9 | -53.8 | -57.8 | -61.1 | -65.0 | -68.9
8.9 | -11.1 | -15.6 | -19.4 | -23.3 | -27.2 | -31.1 | -35.0 | -39.4 | -43.3 | -47.2 | -51.1 | -55.0 | -58.9 | -63.8 | -66.7 | -70.6 | -75.0
11.2 | -13.3 | -17.2 | -21.7 | -26.1 | -30.0 | -33.9 | -37.8 | -42.2 | -46.1 | -50.6 | -54.4 | -58.9 | -63.8 | -68.9 | -71.1 | -75.0 | -78.9
13.4 | -14.4 | -18.9 | -23.3 | -27.8 | -31.7 | -36.1 | -40.6 | -45.0 | -48.9 | -53.3 | -57.2 | -61.7 | -65.5 | -69.4 | -73.9 | -78.3 | -82.2
15.6 | -15.6 | -20.0 | -24.4 | -28.9 | -32.8 | -37.2 | -41.7 | -46.7 | -50.0 | -55.0 | -58.9 | -63.3 | -67.2 | -71.7 | -76.1 | -80.6 | -84.4
17.9 | -16.1 | -20.6 | -25.0 | -29.4 | -33.9 | -38.3 | -42.8 | -47.2 | -51.1 | -56.1 | -60.0 | -64.4 | -68.9 | -73.3 | -77.2 | -81.7 | -86.1
20.1 | -16.7 | -21.1 | -25.6 | -30.0 | -34.4 | -38.9 | -43.4 | -47.8 | -52.2 | -56.7 | -61.1 | -65.0 | -69.5 | -74.4 | -78.3 | -82.8 | -87.2

[a] For equivalent temperature read right and down from calm-air line. Values on the calm air line are ambient temperatures. Adapted from National Oceanic and Atmospheric Administration (1980).

In interpretation of these indices certain facts should be remembered: actual heat loss experienced by a person exposed to the cold depends on the amount and type of clothing worn; human comfort depends on sensations in the exposed extremities rather than on the heat balance of the entire body; even when the body is protected by clothing, heat is transferred more rapidly from the finger than from the limb (Munn, 1970). Steadman (1971) describes procedures for calculating wind chill indices for clothed persons.

Since no sensible person would expose much uncovered flesh under extremely cold, windy conditions, Griffiths and Driscoll (1982) suggest that the wind chill indices for clothed people are more appropriate than those based on bare flesh. It should be noted that the formula for bare flesh always gives lower wind chill equivalent temperatures than does the more realistic formula for clothed people. At 0°C with a wind speed of 7.2 m s^{-1} (26 km h^{-1}) T_{wc} for a clothed person is -5°C compared to -10°C for an unclothed person.

12.6 EFFECTS OF CLIMATE ON ANIMALS

Study of the effects of weather on animals should consider the ecological niche involved. A forest-dwelling rodent may respond little to a storm that mauls the forest canopy; yet a small variation in temperature or light on the forest floor may create conditions in which it cannot survive.

Certain environmental conditions are found optimum for growth and production by farm animals. Bianca (1976) states that the greatest and most efficient production of meat, milk, and eggs occurs when the meteorological elements are within a certain range. Outside this range the animal has to combat meteorological stress. This requires energy that would otherwise be available for productive processes. A discussion of the meteorological conditions affecting animal growth and production of animal products will not be attempted here. The interested reader is referred to Lowry (1970), Bianca (1970, 1976), Tromp and Bouma (1972, 1976, 1979), and Monteith and Mount (1974). Considerable effort has been made in recent years to develop shelters and procedures that reduce environmental stress on animals in order to increase production (e.g., Hahn and Osburn, 1969, 1970; Hahn and McQuigg 1970; Monteith and Mount, 1974; and Hahn, 1976, 1981b).

Most livestock animals do not adjust readily to heat stress. They may be especially vulnerable when closely confined in a building or pen or while being transported. A study by the U.S. National Weather Service (1976) indicated that, even out of doors, when temperatures are above about 26.5°C death loss can be great if humidity is high. Death losses are predictable on the basis of a **humidity temperature index** HTI developed by U.S. National Weather Service. The index is given by

$$\text{HTI} = 0.55(1.8T_a + 32) + 0.20(1.8T_d + 32) + 17.5 \qquad (12.46)$$

HTI in the range 75–78 should alert livestock producers that precautions

Table 12.8 Livestock Weather Safety Index[a]

Dry-bulb temperature (°C)	\multicolumn{20}{c}{Relative humidity (%)}
	5	10	15	20	25	30	35	40	45	50	55	60	65	70	75	80	85	90	95	100
23.9									70	70	71	71	72	72	73	73	74	74	75	75
24.5								70	70	71	72	72	72	73	74	74	74	75	76	76
25.0						70	70	71	71	72	72	73	73	74	74	75	75	76	76	77
25.5					70	70	71	71	72	72	73	74	74	75	75	76	76	77	78	78
26.1				70	70	71	72	72	73	73	74	74	75	76	76	77	77	78	78	79
26.7			70	70	71	72	72	73	73	74	74	75	76	77	77	78	78	79	79	80
27.2		70	70	71	71	72	73	73	74	75	75	76	77	77	78	78	79	80	80	81
27.8		70	71	71	72	72	73	74	75	75	76	77	78	78	79	79	80	81	81	82
28.3	70	71	71	72	73	73	74	75	75	76	77	78	78	79	80	80	81	82	82	83
28.9	70	71	72	72	73	74	75	75	76	77	78	78	79	80	80	81	82	83	83	84
29.4	71	72	72	73	74	74	75	76	77	78	78	79	80	81	81	82	83	84	84	85
30.0	71	72	73	74	74	75	76	77	78	78	79	80	81	81	82	83	84	84	85	86
30.6	72	73	73	74	75	76	77	77	78	79	80	81	81	82	83	84	85	85	86	87
31.1	72	73	74	75	76	76	77	78	79	80	81	81	82	83	84	85	85	86	88	88
31.7	73	74	74	75	76	77	78	79	80	80	81	82	83	84	85	86	86	87	88	89
32.2	73	74	75	76	77	78	79	79	80	81	82	83	84	85	86	86	87	88	89	90
32.8	74	75	76	76	77	78	79	80	81	82	83	84	85	85	86	87	88	89	90	92
33.3	74	75	76	77	78	79	80	81	82	83	83	84	85	86	87	88	89	90		
33.9	75	76	77	78	79	80	80	81	82	83	84	85	86	87	88	88	89	90		
34.4	75	76	77	78	79	80	81	82	83	84	84	85	86	87	88	89	90			
35.0	76	77	78	79	80	81	82	83	84	84	86	86	87	88	89	90				
35.6	76	77	78	79	80	81	82	84	84	85	87	87	88	89	90					
36.1	77	78	79	80	81	82	83	84	85	86	87	88	89	90	91					
36.7	77	78	79	80	82	82	84	85	86	87	88	88	90	91						
37.2	78	79	80	81	82	83	84	86	87	88	89	89	90							
37.8	78	79	80	82	82	84	85	86	87	88	90	91								
40.6	80	82	83	84	86	87	89	90	91											

[a] The values in the table correspond to each combination of temperature and relative humidity. Values between the lines indicate alert, danger, and emergency categories. See text for explanation. Adapted from U.S. National Weather Service (1976).

are needed to avoid losses. HTI in the range 79–83 means that conditions are dangerous for confined livestock and avoidance of disastrous losses requires that proper safety steps be taken. HTI ≥ 84 means that emergency conditions exist. Then handling of the animals should be minimal, animals should be placed where air circulates freely, shade should be provided, water should be readily available, and buildings (or animals) should be sprinkled to lower temperatures. HTI values for various combinations of T_a and relative humidity are given in Table 12.8.

12.7 ADAPTATION AND ACCLIMATIZATION

Humans and animals often exhibit a remarkable ability to adapt to harsh or rapidly changing environmental conditions. One obvious means of adaptation is to move to where the environmental stresses are less severe. Thus, we observe mass seasonal migrations of birds and certain animals; we find animals that burrow, build, or seek other forms of shelter; and we see other animals seeking shade or sunshine, depending on conditions. The body also has many physiological responses for adaptation. Some responses are short term such as sweating, panting, or the constriction or expansion of blood vessels near the skin surface. Other responses are long term as in the case of hibernation or acclimatization. There is an ample literature on adaptive mechanisms and processes. Many of the references cited at the beginning of this chapter contain discussions of adaptive responses. A few general statements on acclimatization are all that will be made in this chapter. **Acclimatization** is defined as the physiological changes that result in improved performance with successive exposures to or residence in a hot or cold environment (Ladell, 1957). When a person becomes acclimatized the hypothalamus and other body control organs and systems settle into cooperative equilibrium, with certain chemical or hormonal levels that are appropriate for that particular season (Ponte, 1982).

12.7.1 Acclimatization to Heat

An unacclimatized person is unfit but after successive daily exposures to heat becomes more able to work and feels the heat less. Heat regulation is improved, at least in part, by the induction of sweating at a lower internal body temperature (Ladell, 1957). Acclimatization to heat stress sets in very rapidly during the first few days of exposure and is almost complete by the end of 2 weeks (WHO, 1972). Further improvement is detectable up to 23 days after first exposure (Ladell, 1957). Sweating appears to be the main thermoregulatory mechanism operating in hot environments. Thus, Macfarlane (1974) reported that Australian aborigines, who are particularly well adapted to high summer heat, had body sweating rates that were nearly double those of less well-adapted people of European origin.

Acclimatization to heat can be achieved by relatively brief periods of daily exposure. The critical factor seems to be that the heat exposure must provide sufficient stimulus for sweating (Hobbs, 1980).

12.7.2 Acclimatization to Cold

Cold climates also stimulate thermoregulatory mechanisms. One of the first responses of the body is to constrict the blood vessels so that less blood flows to the surface. The warm blood concentrates in the internal organs and protects them from harm, but the reduced conduction of heat to the surface may result in frostbite. Shivering is a mechanism for raising the metabolic heat production but only temporarily. One manifestation of acclimatization to cold, over the long term, is increased metabolic heat production. Another mechanism may be the development of improved insulating properties in the skin.

Studies of people indigenous to frigid areas of the globe have revealed some interesting facts. Some Indians and Eskimos have a significantly higher metabolic rate than do Caucasians and actually show an increase in metabolic rate while they sleep. It has been found that Australian aborigines have extra skin insulation that conducts 30% less heat from the core than does the outer layer of Caucasian people. Although individuals not native to these frigid areas may not acclimatize to the level of the native, they do become much better adapted to living in cold conditions after a period of 1 week to 10 days (Kavaler, 1970).

REFERENCES

Ames, D. R., and L. W. Insley. 1975. Wind chill effect for cattle and sheep. *J. Anim. Sci.* **40**:161–165.

Anderson, S. R. 1965. *Humidex Calculation*. Atmospheric Environment Service of Canada. C.D.S. No. 24–65.

ASHRAE. 1967. *Handbook of Fundamentals*. Am. Soc. Heating, Refrigeration, Air-Conditioning Engineers, New York. 544 pp.

Belding, H. S., and T. F. Hatch. 1955. Index for evaluating heat stress in terms of resulting physiological strains. *Heat Piping Air Conditioning* **27**:129–136.

Benedict, F. G. 1938. Vita energetics: A study in comparative basal metabolism. *Carnegie Inst. Publ.* **503**:1–215.

Bernstein, M. H. 1971. Cutaneous water loss in small birds. *Condor* **73**:468–469.

Bianca, W. 1970. Animal response to meteorological stress as a function of age. *Biometeorology* **4**:119–131.

Bianca, W. 1976. The significance of meteorology in animal production. *Int. J. Biometerol.* **20**:139–156.

Birkebak, R. C. 1966. Heat transfer in biological systems. *Int. Rev. Gen. Exp. Zool.* **2**:269–344.

Blaxter, K. L. 1967. *The Energy Metabolism of Ruminants*, 2nd ed. Hutchinson, London. 332 pp.

Brody, S. 1974. *Bioenergetics and Growth*. Hafner, New York. 1023 pp.
Campbell, G. S. 1977. *An Introduction to Environmental Biophysics*. Springer-Verlag, New York. 159 pp.
Cena, K. 1974. Radiative heat loss from animals and man. *Heat Loss from Animals and Man* (L. J. Monteith and L. E. Mount, eds.), Butterworths, London. pp. 33–58.
Cena, K., and J. L. Monteith. 1975. Transfer processes in animal coats. II. Conduction and convection. *Proc. R. Soc. London Ser. B* **188**:395–412.
Clapperton, J. L., J. P. Joyce, and K. L. Baxter. 1965. Estimates of the contribution of solar radiation to thermal exchanges of sheep at a latitude of 55°N. *J. Agric. Sci. Camb.* **64**:37–49.
Court, A. 1948. Windchill. *Bull. Am. Meteorol. Soc.* **29**:487–493.
Dodds, T. 1979. The weather can psych you. *Family Safety* (Fall 1979):20–21.
DuBois, D., and E. F. DuBois. 1915. The measurement of the surface area of man. *Arch. Intern. Med.* **15**:868–881.
Esmay, M. L. 1978. *Principles of Animal Environment*. AVI, Westport, Connecticut. 348 pp.
Finch, V. A. 1972. Energy exchange with the environment of two East African antelopes, the eland and the hartebeest. Symp. Zool. Soc. London **31**:315–326.
Gatenby, R. M. 1977. Conduction of heat from sheep to ground. *Agric. Meteorol.* **18**:387–400.
Gates, D. M. 1980. *Biophysical Ecology*. Springer-Verlag. New York. 611 pp.
Givoni, B. 1969. *Man, Climate and Architecture*. Elsevier, Amsterdam. 364 pp.
Givoni, B. 1974. Biometeorological indices. *Progress in Biometeorology* (S. W. Tromp and J. J. Bouma, eds.), Swets and Zeitlinger, Amsterdam. pp. 138–145.
Griffiths, J. F., and D. M. Driscoll. 1982. *Survey of Climatology*. Merrill, Columbus, Ohio. 358 pp.
Hadley, N. F. 1972. Desert species and adaptation. *Am. Sci.* **60**:338–347.
Hahn, G. L., and D. D. Osburn. 1969. Feasibility of summer environmental control for dairy cattle based on expected production losses. *Trans. ASAE* **12**:448–451.
Hahn, G. L., and D. D. Osburn. 1970. Feasibility of evaporative cooling for dairy cattle based on expected production losses. *Trans. ASAE* **13**:289–294.
Hahn, G. L., and J. D. McQuigg. 1970. Evaluation of climatological records for rational planning of livestock shelters. *Agric. Meteorol.* **7**:131–141.
Hahn, G. L. 1976. Rational environmental planning for efficient livestock production. *Biometeorology* **6**:106–114.
Hahn, G. L. 1981a. Use of weather data in the rational selection of livestock management practices. Proc. of Workshop on Computer Techniques and Meterological Data Applied to Problems of Agriculture and Forestry. Sponsored by Am. Meteorol. Soc. and Nat. Sci. Found. (A. Weiss, ed.), pp. 362–386.
Hahn, G. L. 1981b. Housing and management to reduce climatic impacts on livestock. *J. Anim. Sci.* **52**:175–186.
Hammel, H. T. 1955. Thermal properties of fur. *Am. J. Physiol.* **182**:369–376.
Hemmingsen, A. M. 1960. Energy metabolism as related to body size and respiratory surfaces. Rep. Steno. Meml. Hosp., Copenhagen, 9, parts 2 and 7.
Hevener, O. F. 1959. All about humiture. *Weatherwise* **12**:56, 83–85.
Hobbs, J. E. 1980. *Applied Climatology*. Dawson Westview, Boulder, Colorado. 218 pp.

Huschke, R. E. 1959. *Glossary of Meteorology*. Am. Meteorol. Soc., Boston, Massachusetts. 638 pp.
Hutchinson, J. C. D., G. D. Brown, and T. E. Allen. 1976. Effects of solar radiation on the sensible heat exchange of mammals. *Progress in Animal Biometeorology, 1963–1973* (S. W. Tromp and J. J. Bouma, eds.), Swets and Zeitlinger, Amsterdam. Vol. 1, Part 2, pp. 61–79.
Johnson, K. G. 1976. Evaporative temperature regulation in sheep. *Progress in Animal Biometeorology* (S. W. Tromp and J. J. Bouma, eds.), Swets and Zeitlinger, Amsterdam. pp. 140–147.
Joyce, J. P., K. L. Blaxter, and C. Park. 1966. The effect of natural outdoor environments on the energy requirements of sheep. *Res. Vet. Sci.* **7**:342–359.
Kavaler, L. 1970. *Freezing Point: Cold as Matter of Life and Death*. John Day, New York. 416 pp.
Kerslake, D. M. 1972. *The Stress of Hot Environments*. Cambridge Univ. Press, London. 316 pp.
Kleiber, M. 1932. Body size and metabolism. *Hilgardia* **6**:315–353.
Kleiber, M. 1961. *The Fire of Life: An Introduction to Animal Energetics*. Wiley, New York. 454 pp.
Kleiber, M. 1965. Metabolic body size. *Energy Metabolism* (K. L. Blaxter, ed.), Academic Press, London. pp. 427–435.
Kreith, F. 1965. *Principles of Heat Transfer*. International Textbook, Scranton, Pennsylvania. 620 pp.
Ladell, W. S. S. 1957. The influence of environment in arid regions on the biology of man. Human and Animal Ecology/Ecologic Humaine et Animals, UNESCO. pp. 43–99.
Lasiewski, R. C., M. H. Bernstein, and R. D. Ohmart. 1971. Cutaneous water loss in the roadrunner and poor-will. *Condor* **73**:470–472.
Lee, D. H. K., and J. A. Vaughn. 1964. Temperature equivalent of solar radiation on man. *Int. J. Biometeorol.* **8**:61–69.
Lee, D. H. K. and A. Henschel. 1966. Effects of physiological and clinical factors on response to heat. *Ann. N.Y. Acad. Sci.* **134**:743–749.
Lowry, W. P. 1970. *Weather and Life*. Academic Press, New York. 305 pp.
Macfarlane, W. V. 1974. Acclimatization and adaptation to thermal stress. *Progress in Human Biometeorology* (S. W. Tromp and J. J. Bouma, eds.), Swets and Zeitlinger, Amsterdam. Vol. 1, Part 1B. pp. 468–473.
Masterton, J. M., and F. A. Richardson. 1979. Humidex—A method of quantifying human discomfort due to excessive heat and humidity. Atmospheric Environment Service of Canada. CLI 1–79, pp. 1–45.
McLean, J. A. 1974. Loss of heat by evaporation. *Heat Loss from Animals and Man* (J. L. Monteith and L. E. Mount, eds.), Butterworths, London. pp. 19–31.
McMahon, T. 1973. Size and shape in biology. *Science* **179**:1201–1204.
Missenard, H. 1948. Equivalences thermiques des ambiences; equivalences de passage; equivalences de sejour. Chaleur et Industrie.
Mitchell, D. 1974. Convective heat loss. *Heat Loss from Animals and Man* (J. L. Monteith and L. E. Mount, eds.), Butterworths, London. pp. 60–76.
Mitchell, J. W. 1976. Heat transfer from spheres and other animal forms. *Biophys. J.* **16**:561–569.
Monteith, J. L., and L. E. Mount. 1974. *Heat Loss from Animals and Man*. Butterworths, London. 457 pp.

Monteith, J. L. 1975. *Principles of Environmental Physics*. Edward Arnold, London. 241 pp.
Mount, L. E. 1979. Recent observations on the influence of weather and climate on energy metabolism and growth in pig, sheep and cattle. *Biometeorological Survey* (S. W. Tromp and J. J. Bouma, eds.), Heyden and Sons, London. Vol. 1, Part B. pp. 23–33.
Munn, R. E. 1970. *Biometeorological Methods*. Academic Press, New York. 336 pp.
National Oceanic and Atmospheric Administration. 1980. Heed the wind chill factor. *NOAA News* **5**:4, 5.
Newburgh, L. H., ed. 1949. *Physiology of Heat Regulation and the Science of Clothing*. Hafner, New York. 457 pp.
Nishi, Y., and A. P. Gagge. 1970. Direct evaluation of convective heat transfer coefficient by naphthalene sublimation. *J. Appl. Physiol.* **29**:830–838.
Ponte, L. 1982. How a change in the weather changes you. *Reader's Digest,* March 1982:55–62.
Precht, H., J. Christopherson, H. Hensel, and W. Larcher, eds. 1973. *Temperature and Life*. Springer-Verlag, New York. 779 pp.
Primault, B. 1979. Optimum climate for animals. *Agrometeorology* (J. Seemann, Y. I. Chirkov, J. Lomas, and B. Primault, eds.), Springer-Verlag, New York, pp. 182–189.
Richardson, E. A., L. M. Cox, and G. L. Ashcroft. 1968. Man-natures recalcitrant anomaly. *Utah Sci.* **29**:1337–1341.
Riemerschmid, G., and J. S. Elder. 1945. The absorptivity for solar radiation of different colored hairy coats of cattle. *Onderstepoort J. Vet. Res.* **20**:223–234.
Robertshaw, D. 1976. Evaporative temperature regulation in ruminants. *Progress in Animal Biometeorology* (S. W. Tromp and J. J. Bouma, eds.), Swets and Zeitlinger, Amsterdam. Vol. 1, Part 1, pp. 132–139.
Robinson, D. E., G. S. Campbell, and J. R. King. 1976. An evaluation of heat exchange in small birds. *J. Comp. Physiol.* **105**:153–166.
Schiller, G., and K. Karschon. 1973. Microclimate and thermal stress of man in an allepo pine plantation and an oak scrub. *Israel J. Agric. Res.* **23**:79–90.
Schmidt-Nielsen, K. 1969. The neglected interface: The biology of water as a liquid-gas system. *Q. J. Biophys.* **2**:283–304.
Schmidt-Nielsen, K., J. Kanwisker, R. C. Lasiewski, J. E. Cohn, and W. L. Bretz. 1969. Temperature regulation and respiration in the ostrich. *Condor* **71**:341–352.
Scholander, P. F., V. Walters, R. Hock, and L. Irving. 1950. Body insulation of some arctic and tropical mammals and birds. *Biol. Bull.* **99**:225–236.
Segal, M., and R. A. Pielke. 1981. Numerical model simulation of human biometeorological heat load conditions—Summer day case study for Chesapeake Bay area. *J. Appl. Meteorol.* **20**:735–749.
Siple, P. A., and C. F. Passel. 1945. Measurements of dry atmospheric cooling in subfreezing temperatures. *Proc. Am. Phil. Soc.* **89**:177–199.
Steadman, R. G. 1971. Indices of windchill of clothed persons. *J. Appl. Meteorol.* **10**:674–683.
Steadman, R. G. 1979a. The assessment of sultriness. Part. I: A temperature-humidity index based on human physiology and clothing science. *J. Appl. Meteorol.* **18**:861–873.
Steadman, R. G. 1979b. The assessment of sultriness. Part II. Effects of wind, extra radiation and barometric pressure on apparent temperature. *J. Appl. Meteorol.* **18**:874–885.

Stevens, D. G., J. R. Bohart, and L. Hahn. 1975. Computer method-obtaining radiant interchange factors for a quadruped. USDA-ARS-NC-29. 10 pp.

Stewart, R. E. 1953. Absorption of solar radiation by the hair of cattle. *Agric. Engr.* **34**:235–238.

Sulman, F. G., D. Levy, A. Levy, Y. Pfeifer, E. Superstine, and E. Tal. 1974. Air-ionometry of hot, dry desert winds (Sharav) and treatment with air ions of weather-sensitive subjects. Int. *J. Biometeorol.* **18**:313–318.

Swan, H. 1974. *Thermoregulation and Bioenergetics.* American Elsevier, New York. 430 pp.

Thom, E. C. 1959. The discomfort index. *Weatherwise* **12**:57–60.

Tromp, S. W., and J. J. Bouma. 1972. *Progress in Human Biometeorology.* Swets and Zeitlinger, Amsterdam. Vol. 1, Part III. 157 pp.

Tromp, S. W., and J. J. Bouma. 1974. *Progress in Human Biometeorology,* Swets and Zeitlinger, Amsterdam. Vol. 1, Parts 1A and 1B. 726 pp.

Tromp, S. W., and J. J. Bouma. 1976. *Progress in Animal Biometeorology, 1963–1973.* Swets and Zeitlinger, Amsterdam. Vol. 1, Part I, 603 pp., and Part II, 301 pp.

Tromp, S. W., and J. J. Bouma, 1979. *Biometeorology Survey, 1973–1978.* Heyden, London. Vol. 1, Part A. Human Biometeorology, 257 pp. and Part B. Animal Biometeorology, 180 pp.

Tromp, S. W. 1980. *Biometeorology: The Impact of the Weather and Climate on Humans and their Environment.* Heyden, London. 346 pp.

Underwood, C. R., and E. J. Ward. 1966. The solar radiation area of man. *Ergonomics* **9**:155–168.

U.S. National Weather Service. 1976. Livestock hot weather stress. Regional Operations Manual Letter C-31-76. 5 pp.

Ward, E. J., and C. R. Underwood. 1967. The effect of posture on the solar radiation area of man. *Ergonomics* **10**:399–409.

Weihe, W. H. 1976. The application of meteorology in medical science. *Int. J. Biometeorol.* **20**:157–165.

Wilson, O. 1967. Objective evaluation of wind chill index by records of frostbite in the Antarctica. *Int. J. Biometeorol.* **11**:29–32.

Winterling, G. A. 1979. Humiture—Revised and adapted for the summer season in Jacksonville, Florida. *Bull. Am. Meteorol. Soc.* **60**:329–330.

World Health Organization. 1972. *Health Hazards of the Human Environment,* WHO, Geneva.

AUTHOR INDEX

Aase, J. K., 111, 334, 405, 406
Abbott, C. G., 12
Abeles, F. B., 37
Aboukhaled, A., 396
Ackerson, R. C., 223, 238
Acosta, M. J. C., 20
Adams, J. E., 195, 199, 201, 217, 266
Akita, S., 410, 411
Albrektsson, L., 331
Alemo, J., 331
Alessi, J., 190
Alexander, D. Y., 26
Allen, C. D., 252
Allen, J. R., 20
Allen, L. D., 269
Allen, L. H., Jr., 66-68, 146, 230, 252, 271, 341, 392, 413-418
Allen, T. E., 428, 429, 433, 445
Al-Nakshabandi, G., 95, 96
Alperi, R. W., 338
Alvo, P., 320
Ambrus, L., 262
Ames, D. R., 443
Anderson, F. N., 409
Anderson, J. E., 392, 395
Anderson, M. C., 68, 77
Anderson, S. R., 453
Andrew, R. H., 199
Angell, J. K., 19, 21, 24
Angus, D. E., 260, 393
Antal, E., 263
Aoki, M., 412
Apple, S. B., Jr., 191
Arkin, G. F., 201, 230, 265
Armbrust, D. V., 350
Armijo, J. D., 261
Ashbel, D., 26
Ashcroft, G. L., 387, 450
ASHRAE, 451
Aslyng, H. C., 52, 176, 245, 246, 344
Aston, A. R., 77
Aston, M. J., 254
Atjay, 289, 290, 412

Atwater, M. A., 42
Aubertin, G. M., 53
Auer, A. H., Jr., 30, 33
Ayers, R. C., 43

Baars, H. D., 241
Badham, R., 44, 45, 47
Bagley, W. T., 339, 348, 349, 352, 354, 356, 359
Baier, W., 176, 177, 212, 263
Bailey, W. G., 251
Baille, A., 204
Baker, D. G., 48, 52, 63, 100
Baker, D. N., 322, 417, 418
Baker-Blocker, A., 38
Baldocchi, D. D., 297-299, 313-319, 405-407
Baldwin, B., 30
Balek, J., 265
Balif, J. L., 199
Baradas, M. W., 229, 242, 402, 403
Barfield, B. J., 4, 126, 375
Barkley, D. G., 199
Barney, C. W., 43
Bartholic, J. F., 375-377, 379
Barton, I. J., 252
Basuk, J., 36, 37
Bates, C. G., 331, 344
Bates, E. M., 380
Bauer, M. E., 405
Baumgardner, M. F., 43, 44
Baumgartner, A., 242, 319
Baxter, K. L., 428, 432
Bazzaz, F. A., 299, 300, 302
Bean, A., 338
Bean, B. R., 186
Beardsell, M. F., 175
Beerwinkle, K. R., 201, 202
Beirsdorf, M. I. C., 20
Belding, H. S., 455
Belt, G. H., 393
Benci, J. F., 406
Benedict, F. G., 435

Bennett, J. M., 322
Benoit, G. R., 200
Benz, L. C., 260
Bergen, J. D., 159, 163, 375
Bernard, R. L., 405
Berri, G. J., 30
Berry, J. A., 309-312, 322
Beuerlein, J. E., 405
Bianca, W., 426, 427
Biel, E. R., 372
Bielorai, H., 393
Bierhuizen, J. F., 214, 347, 394, 395
Bill, R. J., Jr., 270
Billings, W. D., 191
Bingham, F. T., 191
Bingham, G. E., 392
Birkebak, R. C., 432, 444, 445
Birth, G. S., 48
Biscoe, P. V., 113, 303, 307, 310, 311, 319
Bjorkman, O., 296, 297, 405, 410
Black, A. L., 334
Black, J. N., 20
Black, J. R., 43
Black, T. A., 219, 224, 266-268
Blackmer, A. M., 37
Blad, B. L., 48, 52, 63, 64, 167, 175, 210, 213, 223, 228, 232, 254, 248, 254-256, 402, 403, 405-408
Blanc, M., 377
Blanchet, R. N., 215, 216
Blaney, H. F., 244, 270
Blaser, R. E., 199
Blaxter, K. L., 432, 442, 445
Bloeman, G. W., 263
Bloodworth, J. E., 264
Boersma, L., 190, 199, 204-206, 253
Boggess, W. R., 299, 300, 302
Bohart, J. R., 429, 430
Bolsenga, S. J., 44
Bonde, E. K., 392
Bootsma, A., 374
Borucki, W. J., 36
Bosc, M., 215, 216
Botkin, D. B., 412
Bouchet, R. J., 211, 212, 377
Bouma, J. J., 425, 460
Bowen, I. S., 255-256
Box, J. E., Jr., 181, 417, 418
Boyer, J. S., 241, 294, 295, 405
Braaten, M. O., 68
Brach, E. J., 320
Bradley, E. F., 141, 143, 233
Brakke, T. W., 233, 236, 237
Brandle, J. R., 359, 360

Branton, C. I., 269
Braud, H. J., 379
Brazel, A. J., 29, 30
Bremner, J. M., 37
Bridley, S. F., 387
Brill, G. D., 218
Brock, P., 242
Brody, S., 425, 434, 435
Brooks, C. E. P., 154
Brooks, F. A., 121
Brooks, K. N., 395
Brown, D. M., 370
Brown, G. D., 428, 429, 433, 445
Brown, J. H., 405
Brown, K. W., 43, 45, 66-68, 75, 179, 180, 223, 258, 260-261, 293, 303, 306, 307, 319, 324, 335, 338, 339, 343, 344, 346, 347, 351-353, 356-360
Brun, L. J., 224, 225
Brunt, D., 50
Brust, K. J., 242
Brutsaert, W., 254
Bryson, R. A., 29, 45, 46
Buchele, W. F., 192, 193
Buck, A. L., 186
Burman, R. D., 261
Burnett, E., 225, 226, 261, 266
Burrage, S. W., 176
Burrows, W. C., 194, 195, 335
Burt, J. E., 250
Busch, N. E., 141, 157
Businger, J. A., 35, 141, 143, 255, 383
Buss, G. R., 406
Butch, G., 371
Butler, J., 379
Buttimor, P. H., 44, 45, 52
Byrne, G. F., 248

Caborn, J. M., 333, 336
Calder, I. R., 219, 251, 267, 268
Callander, B. A., 224
Campbell, G. S., 3, 124, 125, 146, 164, 181, 392, 425, 429, 431, 434-441, 444, 449, 455-457
Campbell, R. B., 52, 53, 262, 263
Cannell, G. H., 191
Caprio, J., 114, 374
Cardenas, R., 405
Carder, A. C., 263
Carlson, R. E., 48
Carson, H. W., 320
Carson, J. E., 101, 102
Cartee, R. L., 195
Cary, J. W., 107-109, 217, 385

Cass, A., 181
CAST, 37
Catsky, J., 322
Cena, K., 428, 433, 442
Chamberlain, A. C., 126
Chang, J. H., 309
Changnon, S. A., 19
Charlson, R. J., 28
Charney, J. G., 43
Charney, J. P., 43
Chartier, P., 62, 68
Chepil, W. S., 344
Chilton, R., 175
Chin Choy, E. W., 229
Chollet, R., 297, 410
Choudhury, P. N., 214
Christopherson, J., 439
Cicerone, R. J., 37
Cipra, J. E., 43, 44
Clapperton, J. L., 428, 432
Clayton, R. N., 26
Clegg, N. D., 322
Clough, B. F., 264
Cochrane, J., 199, 200
Cohen, Y., 401
Conaway, J., 254
Connor, G. I., 40
Cook, A. F., 271
Cooley, K. R., 396
Cooper, R. L., 407
Cooper, S., 405
Corruthers, N., 154
Coulson, K. L., 46
Court, A., 174, 457, 458
Cowan, I. R., 240, 241
Cowley, W. R., 264
Cox, L. M., 450
Coyne, D. P., 409
Craker, L. E., 37
Crawford, T. V., 233, 381
Criddle, W. D., 244
Crutzen, P. J., 36
Cull, P. O., 216, 250
Cutchis, P., 36

Da Costa, J. M. N., 112, 300, 301, 304, 305, 325
Dale, R. F., 214
Dalton, F. N., 190
da Mota, F. S., 20
Danielson, R. E., 260, 262
Daudet, F. A., 181
Davenport, D. C., 392, 394, 396
Davidson, J. M., 242

Davies, J. A., 7, 44, 50, 52, 210, 251-252
Davis, D. R., 187
Davis, J. R., 383
de Boer, H., 377
de Bruin, H. A. R., 270, 271
Decker, D. L., 183
Decker, W. L., 52, 53, 212
de Jager, J. M., 181
De Luisi, J. J., 29
Denison, E. L., 198, 199
Denmead, O. T., 37, 52, 53, 227, 256, 319
De Rao, H. C., 196-198
Desjardins, R. L., 62, 320, 379, 417
De Walle, D. R., 45, 202, 269
de Wit, C. T., 213, 214
Dilley, A. C., 257, 263
Dirmhirn, I., 41, 43
Doering, T. J., 260
Domonkos, S. K., 28
Doorenbos, J., 249, 250, 251
Doraiswamy, P. C., 43, 45, 75, 175, 219, 268, 397-399
Doty, C. W., 52, 53, 218
Dreibelbis, F. R., 260
Driscoll, D. M., 458, 460
Drummond, A. J., 12
DuBois, D., 434
DuBois, E. F., 434
Duncan, C. H., 12
Duncan, P. R., 124
Durand, R., 21
Dusek, D. A., 214, 215
Dutil, P., 199
Dyer, A. J., 28, 143, 144, 233, 255
Dykeman, W. R., 326
Dzerdzeevski, B. L., 232

Eagleman, J. R., 212
Earl, A. V., 335
Eaton, F. D., 43
Eddy, J. A., 11
Egli, D. B., 397
Ehleringer, J. R., 63, 297, 309-312, 322, 405, 410
Ehrler, W. L., 223
Eisenlohr, W. S., Jr., 269
Ekern, P. C., 199
Elam, F. L., 404
Elder, J. S., 433
Ellis, H. T., 27
Ellis, R., Jr., 350
Ellsaesser, H. W., 27, 37
El Nadi, A. H., 225, 227
El Sharkawy, M., 293

Emerson, F. H., 375
Emmert, F. H., 368
Endrodi, G., 137
Eschner, A. R., 44
Eslick, R. F., 405
Esmay, M. L., 425
Evans, D. D., 98, 199
Evans, G. N., 251
Ewing, P. A., 269

Fairbourn, M. L., 200
Fang, S. C., 191
Farrell, A. W., 382
Federer, C. A., 44, 52, 210, 267, 338
Feldhake, C. M., 261
Ferguson, H. C., 405, 406
Finch, V. A., 432
Findlay, W. I., 260
Fischbach, P. E., 216
Fish, B. R., 26
Fisher, J. B., 61
Fisher, M. A., 394
Fjeldbo, G., 33
Fleagle, R. G., 35
Fleming, P. M., 248
Flocker, W. J., 113
Flowers, E. C., 27, 30, 31, 32
Follett, R. F., 260
Forrence, L. E., 37
Fors, L., 409
Forseth, I., 63
Forsgate, S. A., 260
Foskett, L. W., 75
Foster, N. B., 75
Fowler, D. P., 38
Francey, R. J., 143
Frank, A. B., 336
Franklin, W. T., 262
Franzoy, C. E., 246
Fraser, A. B., 82
Freney, J. R., 37
Fritschen, L. J., 3, 52, 53, 77, 175, 219, 221, 228, 230-232, 268, 396
Fritton, D. D., 387
Fritz, S., 21
Fuchs, M., 46, 48, 112, 217, 257, 262, 400, 401
Fucik, J. E., 380
Fuehring, H. D., 394
Fukai, S., 63
Fullerton, T. M., 215
Fulton, J. M., 260, 262, 394
Furnival, G. M., 62, 68

Gaastra, P., 298, 347
Gagge, A. P., 445
Galbally, I. E., 39
Gale, J., 213, 392, 393
GAO, 333
Garay, A., 405-407
Garber, M. J., 191
Gardner, B. R., 214
Gardner, H. R., 107, 110, 111, 200, 230
Gardner, W. R., 190, 217, 241, 267
Garratt, J. R., 126, 144, 320
Garton, J. E., 229
Gash, J. H. C., 268
Gatenby, R. M., 443
Gates, D. M., 3, 9, 15, 17, 34, 66, 124, 425, 433, 438, 440, 441, 445-447, 449
Gausman, H. W., 405
Gay, L. W., 3, 55, 68, 260, 268, 393
Geiger, R., 44, 396
Geist, J., 12
Gelfi, N., 215, 216
Georg, J. G., 383
Gerard, C. J., 404
Gerber, J. F., 4, 375, 380, 381, 383
Geslin, H., 377
Ghorashy, S. R., 405
Gignoux, N., 332, 341
Gillespie, T. J., 187
Gillette, D. A., 26
Gindel, I., 226
Ginn, L. H., 261
Givoni, B., 450, 455
Glaeser, J. L., 43
Glover, J., 52, 260
Goddard, W. B., 260
Goltz, S. M., 184, 233
Goodling, J. S., 270
Goodman, B. M., 29
Gosse, G., 30
Goudriaan, J., 289, 290, 412
Grace, J., 331, 334, 345
Graedel, T. E., 150, 151
Grauze, G., 270
Graves, R. A., 264
Greb, B. W., 200
Green, G. C., 218, 219
Greenstone, R., 20
Gregory, E. J., 205, 206
Griffin, R. E., 386
Griffiths, J. F., 78, 458, 460
Grimes, D. W., 218
Gringorten, I. I., 174
Guyot, G., 333, 342, 344, 350, 363

Habibian, M. T., 243
Häckel, H., 187
Hadas, A., 98
Hadley, N. F., 167, 440
Hagan, L. J., 237, 344, 360
Hagan, R. M., 205, 206, 262, 392-395
Hage, K. D., 255
Hagstrom, R. T., 340
Hahn, G. L., 427, 460
Hahn, L., 429, 430
Haise, H. R., 246, 247
Hales, T. A., 376, 393-395
Hall, A. E., 241
Halverson, H. G., 61, 264
Hamer, P. J. C., 387
Hammel, H. T., 442
Hanchett, A., 37
Hands, R. J., 262
Hanks, R. J., 98, 107, 110, 111, 195, 215, 230, 260
Hansen, G. K., 219, 220, 221, 241
Hansen, J. E., 38
Hanson, C. L., 242
Hanson, C. M., 382
Hanson, K. J., 27, 43
Hare, F. K., 210
Hargreaves, G. H., 244
Harper, L. A., 417, 418
Harris, D. G., 336
Harrison, H., 26
Harrison, L. D., 168, 178
Harrold, L. L., 259, 260
Hart, H. E., 209, 260
Hart, W. E., 261
Hartung, R. C., 405
Hashemi, F., 243
Hassler, F. J., 382
Hatch, T. F., 455
Haupt, H. F., 393
Haurwitz, B., 19
Havens, A. V., 244
Hay, J. E., 41
Hayes, J. T., 251
Hays, J. D., 17
Heatherly, L. G., 260
Heidel, K., 29
Heilman, J. L., 146, 248, 249, 255
Helmond, I., 263
Hemmingsen, A. M., 434, 435
Henderson, D. W., 218
Henderson, R. C., 218
Henry, J. A., 24
Henschel, A., 455
Hensel, H., 439

Herman, B. M., 30
Herman, C., 213
Hesketh, J. D., 239, 293, 417, 418
Hewlett, J. D., 395
Hicks, B. B., 28, 126, 143, 144, 146, 225, 249, 255, 271, 320
Hiler, E. A., 214
Hilgeman, R. H., 221
Hill, G. R., 296, 298, 322, 347
Hillel, D. I., 217
Hinze, G., 63
Hobbs, E. H., 213
Hobbs, J. E., 425, 453, 463
Hobbs, P. V., 26
Hocevar, A., 375, 376
Hock, R., 442
Hoffer, R. M., 48
Hoffman, G. J., 167
Hofstede, G. J., 55
Hofstra, G., 239
Holmes, R. M., 320
Holzberg, I. A., 377
Holzman, B., 143, 255, 319
Honda, H., 61
Horner, D. A., 50
Horowitz, J. L., 68, 73
Houghton, R. A., 319
Hounam, C. E., 210, 250
Howard, F. D., 112
Howard, P. H., 37
Howard, R. G., 28
Howell, T. A., 214
Howlett, F. S., 368
Hoyt, D. V., 23
Hsia, J., 219, 268
Hudson, J. P., 227
Hughes, E., 35
Hughes, J. E., 187
Huschke, R. E., 25, 117, 124, 450, 457
Hutchinson, G. L., 37
Hutchinson, J. C. D., 428, 429, 433, 445
Hutchison, B. A., 68-70

Idso, S. B., 29, 30, 43, 44, 76, 111, 217, 224, 247, 248, 249, 269
Imai, K., 39
Imbrie, J., 17
Impens, I., 52, 53, 55, 76, 77
Ingersoll, R. B., 39
Inman, R. E., 39
Inoue, E., 319
Insley, L. W., 443
Irving, L., 442
Isaksen, I. S. A., 36

Isobe, S., 319
Izumi, Y., 141, 143

Jackman, A. P., 255
Jackson, E. B., 262
Jackson, M. L., 26
Jackson, R. D., 43, 44, 98, 102, 107, 110, 111, 217, 248, 249
Jacobs, H. S., 98, 344, 348, 356
James, D. W., 215
Janek, J. R., 412
Jarman, G. D., 199, 205, 206
Jarvinen, J., 271
Jarvis, P. G., 191, 221, 322
Jeffers, D. L., 397
Jenny, H., 413
Jensen, E., Sr., 52
Jensen, M. E., 216, 232, 246, 247, 333, 335
Jensen, S. E., 250, 416, 417
Johnson, K. G., 436
Johnson, T. B., 40
Johnston, H., 36, 37
Jones, E. P., 320, 321
Jones, F. E., 184
Jones, J. B., 190
Jones, J. N., Jr., 200
Jordan, C. F., 264
Jordan, W. R., 218
Joseph, J. H., 26
Joyce, J. P., 428, 445
Jury, W. A., 253, 254, 266

Kafkafi, V., 48
Kagan, Y., 48
Kalma, J. D., 44, 45, 46, 52, 55, 248, 351
Kalra, Y. P., 359
Kaminski, A., 335, 342
Kanemasu, E. T., 146, 223, 230, 248, 249, 253, 255, 325
Kappen, L., 239
Karsai, H. A., 263
Karschon, K., 455
Karschon, R., 333-337, 345, 356, 359
Kaspar, T. C., 190
Katz, R. W., 368
Kaufmann, M. R., 241
Kavaler, L., 463
Keijman, J. Q., 270, 271
Kelley, J., 36
Kerr, J. P., 175
Kerslake, D. M., 445
Kessler, E., 26
Kidd, G. E., 187

Kimball, B. A., 43, 44, 98, 102, 107, 110, 217, 414, 415
King, J. G., 393
King, K. M., 260
Kirkham, D., 200
Kleiber, M., 425, 434, 435
Kline, J. R., 265
Kliore, A., 33
Knoerr, L. R., 68
Koehler, F. E., 190
Kohnke, H., 95, 96
Komp, M. J., 30, 33
Korshover, J., 19, 21, 24
Korven, H. C., 263
Koslowski, T. T., 393
Kovanda, J. J., 264
Kowsar, A., 199
Kramer, P. J., 191
Kreith, F., 124, 392, 395, 441
Krieg, D. R., 223, 238
Kristensen, K. J., 99, 225, 226, 223, 250, 262
Kristensen, L., 157
Krogman, K. K., 213
Kuhn, P. M., 75
Kuiper, F., 351
Kukla, G. J., 42
Kukla, H. J., 42
Kumar, V., 214
Kung, E. C., 45, 46
Kuo, T., 190
Kurfis, K. R., 30-32
Kynard, B. E., 392

Lacis, A. A., 38
Ladell, W. S. S., 462
Lakshman, G., 270
Lambert, J. R., 240
Lamontagne, R. A., 39
Landgren, B., 331
Landsberg, H. E., 11
Lang, A. R. G., 63, 217, 256
Lange, O. L., 238, 322
Langer, G., 26
Larcher, W., 439
Latimer, J. R., 3
Laurenroth, W. K., 261
Leather, G. R., 37
Lecher, D. W., 221, 335, 350, 351, 360
Ledeboer, F. B., 201
Lee, D. H. K., 455
Lee, K. J., 393
Lee, R., 3, 8, 40, 224
Lehane, J. J., 335
Lembke, G., 322

AUTHOR INDEX 475

Lemeur, R., 52, 53, 55, 62-66, 70, 71, 76, 77, 229, 399, 408
Lemon, E. R., 62, 145, 217, 229, 245, 297, 309, 310, 312, 313, 319, 320, 324, 347, 392, 412, 416, 417
Lenschow, D. H., 45, 46
Leonard, R. E., 44
Leontaris, S. N., 21
Lettau, H. H., 242
Leuning, R., 146, 321, 322
Leverington, K. C., 210, 250
Levy, E. A., 39
Leyton, L., 240, 241
Lillard, J. H., 200
Linacre, E. T., 55, 124, 237, 245, 247, 249, 269
Lindstrom, M. J., 190
Lingle, J. C., 191
Linnenbom, V. J., 39
Liou, K-N., 20
Lipton, W. J., 404
List, R. J., 16, 33
Livingston, B. E., 263
Logan, J. A., 99
Lomas, J., 177, 184
Lombard, P., 386
London, 36
Loomis, R. S., 63, 309
Lorenz, O. A., 105, 106
Losch, R., 238
Lourence, F. J., 143, 144, 211, 232, 256, 260-262
Lowry, W. P., 425, 460
Lugg, D. G., 63
Lukens, D. L., 187
Lull, H. W., 10, 44
Lumley, J. L., 141, 143, 255
Luxmoore, R. J., 77
Lyall, L. H., 379
Lydolph, P. E., 232
Lynch, J. J., 331

McArthur, L. B., 41
McBee, G. G., 201, 202
McCabe, W. J., 271
McCaffery, K., 217, 250
McClure, G., 331
McConnell, J. C., 38
McCormick, R. A., 30-32
McCure, W. E., 201, 202
MacDonald, R. B., 43, 44
MacDonald, T. H., 21
Macfarlane, W. V., 438
McGavin, R. E., 186

McGuiness, L. J., 229
McGuire, S. G., 45
Mach, W., 82
Machta, L., 35
McIlroy, I. C., 210, 229, 233, 251, 256, 257, 319
Mack, H. H., 191, 204-206
McKay, D. C., 269
McKeon, G. M., 248
McKiel, C. G., 201
McLean, J. A., 436, 437
McMahon, T., 434
McMichael, B. L., 261
McMillan, W. P., 260
McNaughton, K. G., 219, 268, 269
McNeil, D. A., 268
McPherson, F. B., 339
McQuigg, J. D., 460
McVey, G. R., 48
Maguire, S. G., 60
Maki, T., 341, 393
Makkink, G. F., 138, 246
Malhotra, G. P., 242
Manabe, S., 255, 271
Mandl, F. A., 406
Manes, A., 26
Mansfield, T. A., 239, 392
Marlatt, W. E., 218
Marotz, G. A., 24
Marsh, P., 252
Marsh, V., 12
Marshall, J. K., 331, 333, 338, 342, 345, 352, 356, 362, 363
Martin, J. R., 264
Martin, P. E., 392
Martsolf, J. D., 53, 375, 376, 381-383, 387
Mason, B., 377
Massee, T. W., 217
Masterson, J. M., 452, 453
Mather, C. W., 244
Matoba, F., 404
Matt, D. R., 68-70
Maurer, R. E., 215
Maxwell, J. C., 213
Mayeda, T. K., 26
Mayer, H., 44
Maynard, D. N., 105, 106
Mederski, H. J., 190
Mee, T. R., 377
Mees, R. T., 26
Meidner, H., 239
Meiman, J. R., 269
Mendonca, B. G., 29
Mercer, J. H., 40

Mermier, M., 204
Meyer, W. S., 218, 219
Meyers, L. E., 260
Miller, D. A., 61
Miller, D. E., 199
Miller, D. R., 339, 348, 349, 352, 354, 356, 359
Miller, E. E., 62
Miller, P. M., 196-198
Millington, R. J., 77
Milthorpe, F. L., 240, 241
Mintz, Y., 228
Missenard, H., 451
Mitchell, D., 444, 445
Mo, T., 38
Moench, A. F., 98
Moermans, R., 76, 77
Mogenson, V. O., 112, 113, 191, 324-326
Molz, F. J., 241
Monin, A. S., 143
Monsi, M., 61, 62, 64
Monteith, J. L., 3, 112, 123-126, 136-138, 175, 176, 210, 237, 249, 251, 257, 303, 307, 308, 310, 311, 319, 324, 392, 394, 425, 429-432, 435, 437, 438, 441, 442, 444-446, 460
Monteny, B., 30
Moody, J. E., 200
Moomen, S. E., 43
Mooney, H. A., 309-312, 322, 405
Moreshet, S., 238, 394, 401
Morgan, D. L., 143, 144
Morton, F. I., 271
Moses, H., 101, 102
Mosier, A. R., 37
Moss, D. N., 294-297, 322, 324, 410, 411
Mount, L. E., 425, 434, 460
Moysey, E. B., 339
Munn, R. E., 255, 460
Munro, D. S., 140, 269
Murphy, A. H., 368
Musgrave, R. B., 297, 322, 324
Musick, J. T., 214, 215
Myers, R. E., 128, 129, 374

Nakayama, F. S., 43, 44, 98, 107, 110, 217
NAS, 216, 409
NCAR, 57, 58
Neild, R. E., 128, 129, 221, 335, 350, 351, 360
Neumann, H. H., 240
New, L. L., 214, 215
Newburgh, L. H., 438
Newman, E. J., 241

Newman, J. E., 221, 269, 375
Nielsen, D. R., 121, 243
Nielson, R. E., 221
Niki, H., 39
Nishi, Y., 445
Nixon, P. R., 376
NOAA, 29, 106, 458, 459
Noffsinger, T. L., 176
Norman, J. M., 36, 38, 62, 79, 82, 156, 223, 224, 225, 382
Nulsen, R. A., 239
Nunn, J. R., 261
NWS, 460, 461

Obukhov, A. M., 143
Ogbuehi, S. N., 359, 360
Ogren, J. A., 28
Ogunkanmi, A. B., 392
Oguntoyinbo, J. S., 43, 45, 46
Ohtaki, E., 320
Oke, T. R., 4, 123, 139, 140, 143, 255
Oliver, H. R., 249, 250
O'Rourke, P. A., 251
Osburn, D. D., 460
O'Toole, J. C., 259
Otterman, J., 43
Ouellet, C. E., 103
Owen, P. R., 126
Qwston, P. W., 265

Page, J. B., 264
Palland, C. L., 262
Palm, E., 269
Palmer, W. C., 244
Paltridge, G. W., 49, 50
Panofsky, H. A., 141, 143, 156, 255, 381
Papendick, R. I., 190
Park, C., 445
Parkhurst, D. F., 124
Parmele, L. H., 229
Parry, H. D., 37, 38
Parton, W. J., 99, 261
Pasquill, F., 255
Passel, C. F., 457, 458
Pasternak, D., 394
Paul, H. A., 260
Paulsen, G. M., 350
Pavlik, O., 265
Pearce, A. J., 267, 268
Pearman, G. I., 124, 146, 321, 322
Peek, J. W., 61
Pelton, W. L., 244, 258, 259, 263, 335
Pendleton, J. W., 61, 397, 405
Penman, H. L., 219, 248, 249, 251, 252

Perry, S. G., 82, 156
Peters, D. B., 53, 61, 263, 264, 397, 405
Peterson, J. T., 30, 33
Peterson, K. M., 191
Peterson, W. A., 41
Peyton, T. O., 38
Phene, C. J., 262
Philip, J. R., 233, 258
Phillips, R. L., 331
Phipps, R. H., 199, 200
Pielke, R. A., 453
Pierson, F. W., 255
Pinker, R. T., 255
Plate, E. J., 141, 336
Platt, R. B., 78
Pochop, L. O., 50
Poljakoff-Mayber, A., 213, 393
Pollack, J. A., 102, 103
Pollack, J. B., 30
Ponte, L., 462
Power, J. F., 190
Powers, W. L., 56, 211, 224, 225, 253, 262, 325, 348, 356
Precht, H., 439
Priestley, C. H. B., 252
Primault, B., 426, 427
Prospero, J. M., 26
Pruitt, W. O., 121, 143, 144, 211, 232, 233, 249, 250, 251, 254, 255, 256, 260, 262
Pueschel, R. F., 26, 27

Qashu, H. K., 199
Quinn, F. H., 270
Quirk, W. J., 43

Radke, J. K., 335, 340, 341, 362, 363
Radke, L. F., 28
Rahn, J. J., 369, 370
Ramanathan, V., 40
Rarick, J. F., 26
Raschke, K., 124
Rasool, S. I., 33
Rawlins, S. L., 181
Razumora, L. A., 333-337, 345, 356, 359
Read, R. A., 331, 332, 344
Redford, T. G., Jr., 155, 156, 180, 181, 186
Reeves, W. E., 221, 232
Regehr, D. L., 299, 300, 302
Reginato, R. J., 43, 44, 98, 107, 110, 217, 249
Reichman, G. A., 269
Reicosky, D. C., 52, 53, 218, 240, 264

Reifsnyder, W. E., 10, 44, 62, 68
Retta, A., 215
Reynolds, C. L., 30
Reynolds, D. W., 46
Rheaume, B., 379
Rhodes, E. D., 227, 267
Richards, L. A., 183
Richardson, C. W., 227, 267
Richardson, E. A., 386, 387, 450
Richardson, F. A., 452, 453
Rider, N. E., 233
Riemerschmid, G., 433
Rijks, D. A., 232, 269
Riley, J. P., 262
Ripple, C. D., 254
Ritchie, J. T., 201, 217, 218, 225-227, 229, 261, 266, 267
Robb, D. C. N., 246
Robelin, M., 211, 212
Roberts, E. B., 393
Robertshaw, D., 436, 438
Robertson, G. W., 333-337, 345, 356, 359
Robinson, E., 26
Robinson, P. J., 7, 20, 50
Rollins, H. A., Jr., 368
Rose, C. W., 248
Rosenberg, N. J., 21, 22, 43, 45, 56, 66, 70, 71, 75, 128, 129, 149, 151-153, 155, 156, 178-181, 186, 209, 211, 219, 222, 223, 224, 228, 229, 233-237, 248, 250, 251, 252, 254-258, 259, 260-262, 297-299, 303, 306, 307, 313-319, 324, 333-335, 338, 339, 343, 344, 346-354, 356-360, 372-374, 393-395, 397-399, 402, 403, 405-407
Rouse, W. R., 252, 253, 270
Rowe, L. K., 267, 268
Rubin, J., 255
Rudisill, J. H., 30
Ruesch, J. D., 347
Russell, G., 224
Rykbost, K. A., 204-206

Saeki, T., 61, 62, 64
Saffell, R. A., 113
Sagan, C., 30
Sainty, G. R., 270
Sakamoto, C. M., 61, 397
Salas, L. J., 39
Salisbury, F. B., 40
Saltzman, B., 102, 103
Samela, H. A., 174
Sammis, T. W., 260
Savage, M. J., 181
Schaal, L. A., 375

Scheupp, P. H., 320
Schiff, H. I., 38
Schiller, G., 455
Schlegel, J., 371
Schlough, D. A., 199
Schmidt, R. E., 199
Schmisseur, W. E., 204-206
Schnell, R. C., 27, 371
Schofield, R. K., 219
Scholander, P. F., 442
Schreiber, M. M., 48
Schulze, E. D., 238, 322
Scott, R. K., 303, 307, 310, 311, 319
Scribner, E., 38
Sebenik, P. G., 264
Segal, M., 453
Seginer, I., 237, 338, 396, 404
Seguin, B., 212, 332, 341, 342, 350
Seidel, B. L., 33
Sellers, W. D., 77, 129, 230
Seo, T., 320
Sestak, Z., 322
Shackleton, N. J., 17
Shah, S. R. H., 350, 351, 359
Shanklin, M. D., 50
Sharp, J., 174
Shaviv, O. G., 66
Shaw, G. E., 26
Shaw, J. H., 39
Shaw, R. H., 48, 52, 53, 61, 192, 193, 198, 199, 214, 369, 374, 397
Shawcroft, R. W., 260
Shell, G. S. G., 63
Shibles, R. M., 292, 397
Shigeishi, H., 38
Shih, S. F., 269
Shimshi, D., 394
Shipe, E. R., 215
Shouse, P., 253, 254
Showalter, A. K., 237
Shukla, J., 228
Shuttleworth, W. J., 267, 268
Siddoway, F. H., 334
Sij, J. W., 325
Sill, B. L., 271
Siminovitch, D., 379
Simmonds, N. W., 394
Simpson, J. R., 37
Sinclair, T. R., 48, 62, 392
Singh, B., 224
Singh, D. P., 393
Singh, H. B., 38
Singh, N. T., 216
Sinha, B. K., 216

Siple, P. A., 457, 458
Skidmore, E. L., 237, 244, 348, 350, 356, 360
Skogley, C. R., 201
Slabbers, P. J., 227, 251
Slade, D. H., 148
Slatyer, R. O., 210, 214, 229, 233, 243, 251, 252, 347, 394, 395
Sly, W. K., 242, 263, 265
Smith, F. M., 261
Smith, J. L., 61, 264
Smith, P. D., 113
Smith, R. C. G., 217, 250
Smith, R. E., 242
Smith, S. D., 320, 321
Solomon, I., 174
Sorey, M. L., 243
Specht, J. E., 405-407
Spittlehouse, D. L., 268
Sridhar, K., 26
Stahl, G. R., 264
Staley, D. O., 30
Stanhill, G., 41, 42, 46, 48, 55, 138, 227, 245, 263, 396, 401
Staple, W. J., 335
Steadman, J. R., 167, 175, 408, 409, 453-455, 457, 458, 460
Stern, W. R., 233
Stevens, D. G., 429, 430
Stevenson, K. R., 240, 241
Stewart, D. W., 229
Stewart, J. B., 219, 268
Stewart, J. I., 262
Stewart, R. B., 252, 270
Stewart, R. E., 432
Stigter, C. J., 180, 182, 249
Stirk, G. B., 242
Stoeckeler, J. H., 332
Stoffel, T. L., 33
Stolarski, R. S., 37
Stolzy, L. H., 253, 254
Stone, J. F., 229
Stone, L. R., 243, 253
Stone, P. H., 43
Stoner, E. R., 43, 44
Stout, D. G., 253
Stromberg, L. K., 218
Sturrock, J. W., 333, 336
Stutler, R. K., 216
Suckling, P. W., 260
Sullivan, C. Y., 322
Summers, A., 30
Suomi, V. E., 57, 75, 260
Sutherland, R. A., 375, 376
Sutton, O. G., 25, 44, 95, 97

AUTHOR INDEX 479

Svensson, S. A., 331
Swan, H., 435
Swinbank, W. C., 49, 50, 143, 258
Swinnerton, J. W., 39
Sze, N. D., 39
Szeicz, G., 112, 137, 224, 225, 324, 392, 394
Szwarcbaum, I., 66

Tajchman, S., 137
Takami, S., 224, 417
Tan, C. S., 262
Tani, N., 319
Tanner, C. B., 62, 79, 113, 124, 154, 155, 184, 214, 229, 245-246, 253, 256-258, 260, 266, 314, 395
Taori, A., 395
Tarmy, B. L., 43
Tarpley, J. D., 24
Taylor, H. M., 190
Taylor, R. J., 252, 387
Taylor, S. A., 181
Teare, I. D., 223, 360
Tenpas, G. H., 199
Terjung, W. H., 251
Thom, A. S., 126, 135-137, 249, 250, 374, 452
Thomas, J. L., 264
Thomas, M. D., 296, 298, 322, 347
Thompson, E. S., 21, 25
Thompson, O. E., 252, 255
Thompson, T. L., 216
Thomson, W. R., 126
Thornthwaite, C. W., 210, 243, 244, 255, 259, 319
Thorpe, M. R., 52
Thorud, D. B., 395
Thurtell, G. W., 79, 154, 155, 184, 217, 240, 241, 267, 269
Tillman, J., 179
Tinus, R. W., 228
Todhunter, P. E., 251
Tomar, V. S., 259
Toon, O. B., 30
Tromp, S. W., 425, 460
Tuller, S. E., 175
Turco, R. P., 36
Turner, N. C., 191, 395
Tveitereid, M., 269
Twitchell, G. A., 261
Twomey, S. B., 20, 30

Uhlenhopp, P. B., 186
Underwood, C. R., 428

Unger, P. W., 200
Unsworth, M. H., 164
USDA, 106
USGS, 270

Valancogne, C., 181
Valli, V. J., 381, 382
van Bavel, C. H. M., 217, 218, 223, 224, 230-232, 237, 242, 251, 254, 261, 417
Van Camp, W., 30
Vance, B. F., 198, 199
van der Linde, P. J., 333
van Eimern, J., 333-337, 345, 356, 359
van Hylckama, T. E. A., 227, 259, 260
van Ryswyk, A. L., 253
Van U'u, N., 107
van Wijk, W. R., 98
Vaughn, J. A., 455
Ventskevich, G. Z., 368, 370, 377
Verma, S. B., 126, 155, 156, 180, 181, 186, 224, 232-237, 248, 256, 258, 297-299, 313-319, 405-407
Viebrock, H. J., 27, 43
Vonder Haar, T., 57
Von Lengerke, H. J., 371

Waco, D., 174
Waggoner, P. E., 196-198, 392-395, 416
Wahua, T. A. T., 61
Walker, J. M., 190
Wallender, W. W., 218
Wallin, J. R., 175
Wallis, J. R., 412
Walters, S., 37
Walters, V., 442
Wang, V., 39
Wang, W. C., 38
Ward, E. J., 428
Watts, D. G., 213, 215
Weaver, H. L., 124
Webb, E. K., 141, 143, 144, 146, 255, 321, 322
Webber, R. T. J., 387
Weeks, E. P., 243
Weertman, J., 17
Weihe, W. H., 450
Weinstock, B., 39
Weisman, R. N., 270
Weiss, A., 167, 175, 187, 408
Welgraven, A. D., 180, 182
Wellburn, A. R., 392
Welles, J. M., 382
Wells, R. F., 215
Went, F. W., 26, 27
Wesely, M. L., 146, 154, 155, 249, 255, 320

Wetherald, R. T., 255, 271
Whiteman, T. M., 368, 369
Whitten, R. C., 36
Wierenga, P. J., 205, 206
Wight, J. R., 242
Wilcox, J. C., 242, 263
Williams, G. D. V., 211
Williams, J. H., 405-407
Williams, R. J., 252, 253
Williams, W. A., 309
Willis, J., 44
Willis, W. O., 336
Willits, N. A., 218
Wilson, R. C., 12
Wilm, H. G., 259
Wilson, G. L., 394
Wilson, O., 458
Winterling, G. A., 451, 453
Wittwer, S. H., 413
Woo, M. K., 253
Woodhead, T., 224

Woodruff, N. P., 344
Woodwell, G. M., 319, 326
Wooley, D. G., 190
Woolhiser, D. A., 242
Woolley, J. T., 393, 405
Wright, J. L., 216, 232
Wyngaard, J. C., 141, 143, 157

Yabuki, K., 112, 324, 412
Yadav, S. K., 393
Yakuwa, R., 104
Yamaguchi, M., 112
Yao, A. Y. M., 52, 174, 212
Yarger, D. N., 48
Youngman, V. E., 63
Yu, S. L., 254, 255
Yung, Y. L., 38

Zelitch, I., 392-395
Zollinger, W. D., 181

SUBJECT INDEX

Abbott pyranometer, *see* Pyranometers
Absorption, 18, 33, 38, 39, 398, 400-401, 405, 433
  selective, 33-40
  of solar radiation, by clouds, 19-20
Absorptivity, 6, 7
Acclimatization:
  to cold, 463
  to heat, 462-463
  see also Animals, adaptation and acclimatization to environment
Adiabatic:
  lapse rate, 118-120
  process, 117-118
  wet lapse rate, 119-120
Advection of sensible heat:
  global, large-scale, regional, 229-230, 232-233, 236
  local, 229-230, 233-236
  within-row, 230
  see also Sensible heat, advection
Aerial resistance, 124, 125, 257-258, 293, 342, 352, 357, 358, 437, 443
Aerodynamic methods, theory, 142, 255, 272, 318
Aerosols, 24, 26, 29, 30
Agricultural burning, 26
Air:
  constituents, 33
  flow, 134, 135, 140
  heating, for frost protection, 375, 381-382
  mixing, for frost protection, 375, 380-381
  moist, density, 168-169
  saturated, 170-171, 182
  specific heat, 141, 142, 210
  temperature:
    measurement, instrumentation, 130-132
    patterns, 126-129
      influence of elevation, 129
    in plant canopies, 126
    profiles:

      inversion (stable), 118, 121, 123, 126, 140, 142-143, 342, 372, 374, 380, 381, 382
      lapse (unstable), 118, 121, 123, 126, 140, 142-143
      neutral, 118-119
  thermal diffusivity, 124
  thermal stability, 118-119, 121, 123, 338, 339
Albedo:
  changing, related to drought, 43
  shortwave, 42, 43, 44-46, 55, 57
  visible, 42, 43
  see also Shortwave reflection
Alfalfa, 43, 45, 112, 215, 218, 223, 227, 232, 234, 235, 250, 252, 253, 261, 264, 289, 296, 298, 313, 335, 347, 394, 395
  net radiation over, 52
Ammonia, effect on ozone layer, 38
Animals:
  adaptation and acclimatization to environment, 462-463
    constriction/expansion of blood vessels, 462
    hibernation, 462
    insulation, 463
    panting, 462
    sweating, 462-463
  body orientation, 428
  body surface area, 434-435
    relationship to mass, 435
  body temperature, 426, 427, 433, 434, 437, 440, 441, 443, 446, 462
    optimum, 426
    surface, 435, 437, 441, 443
    thermal maximum, 447
    thermal minimum, 447
  body water, 446
  conductive heat exchange, transfer:
    calculation, 441-443

SUBJECT INDEX

Animals (*Continued*)
    effect of wind, 441-442
    convective heat exchange, transfer:
      calculation, 443-445
        calculation of respiratory convective heat loss, 444
    critical temperature, 434
    domesticated, adaptation to environment, 427
    effects of climate, *see* Climate, effects on animals, livestock
    environmental conditions, 443. *See also* Environment, animal and human
    evaporation of water, 436-440
      sweating, 436, 437-438, 440, 449, 450
      panting, 436, 438, 439, 449, 450
    evaporative heat loss, 437-440
      from sweating animals, 438
      wind as means of controlling, 439-440
    forced convection, 443, 445
      criteria, 446
      heat transfer, calculation, 445
    free convection, 443, 444-445, 446, 447
      criteria, 446
      heat transfer, 445, 446
    heat storage, 433, 446
    heat stress, 437, 440
      categories, 427
    homeotherms, 426, 433, 434, 440, 446
    hyperthermia, 426
    hypothermia, 426
    insensible perspiration, 436, 437, 440
    insulation (specific resistance), 441-443
      equivalent, 441
      thermal, 434
    latent heat exchange, 433, 436-440
    metabolic energy, 435-436, 446
    metabolic heat, calculation and production, 434
    metabolic rate, basal, 434-436, 437, 444, 463
    metabolism, 426
    poikilotherms, 426, 433, 434
    projected area:
      calculation of, 429-433
      shape factors, 430-431
    radiation balance, 427-433
    resistance:
      thermal, 441
      to vapor diffusion, 436, 437, 439
    respiratory latent heat loss, 437, 439
    respiratory water loss, 437
    sensible heat flux, 433
    thermal conductance, 441
    thermal zones, categories of, 434
    water use:
      in amphibians, 440
      in arthropods, 440
      in reptiles, 440
    wild, adaptation to environment, 427
    work, 433-434
Antitranspirants, 391, 392-396
    natural, 392-393
    purpose, theory, 392
    reflectant, 392
    stomate-closing materials, 393-396
    synthetic, 392-393
Aphelion, 13
Apple, 48, 368, 387
    net radiation over, 53
Aspect, 191-195
    altered, by ridging and shaping, 193
    effect on soil temperature via capture of radiation, 192
    importance of, affected by season, 193
    importance of, influenced by latitude, 192-193
Astronomical factors, to determine quantity of solar radiation, 13-18
Atmospheric extinction coefficient, 18
Atmospheric window, 35, 38, 377
Australian aborigines, affected by environment, 450, 462, 463
Axis-tilt cycles, 18

Barley, 225, 289, 303, 307-308, 310, 319, 335, 393-394, 405
    isogenic, awns, 406
      dissipation of energy, 406
      effect on surface resistance, turbulence, 406
Beans, 225, 371, 378, 383, 396
    dry, 223, 335, 351, 408, 409
    extinction of net radiation within canopy, 53
    net radiation over, 52, 76-77
    snap, 221, 223, 335
Beer-Bouguer law, 18, 60, 61
    application to turbidity, 26
    light extinction coefficient, 68
Bermuda grass, 201, 293, 416
Biometeorology, animal and human, 425-463
Blackbody:
    definition, 6
    temperature:
      equivalent, 455-456
      equivalent wet, 456-457
    see also Radiation

Blaney-Criddle method for estimating ET, 244
Blue haze, 26
Boltzmann constant, see Stefan-Boltzmann law
Boundary layer:
  fully adjusted, 140, 146
  internal, 139-140
  planetary, 134
  resistance, 405
Bowen ratio energy balance method:
  for estimating carbon dioxide flux, 303, 307-308, 319
  for estimating ET, 256-257, 269, 271, 339, 361
Box model calculations of energy balance, 381
Broccoli, 205
Brown and Rosenberg model, 258
Buoyancy, 117, 134, 140, 141, 382, 446

Carbon dioxide:
  as absorber of longwave radiation, 49
  as absorber of thermal (terrestrial) radiation, 38
  absorption spectrum, 34-35
  balance, 302-308
  concentration:
    compensation point, 290
    as function of soil temperature, 112
    in shelter, 347-348
  effects of windbreaks and barriers, 417-419
  field experiments and simulation studies, 414-418
  flux measurements, 321-322, 361
  greenhouse effect, contribution to, 40
  increased concentration, 38-39, 409-419
    effect on photorespiration in C_3 plants, 410
    effect on plant phenology, 411-412
    effect on stomatal closure and transpiration in C_3 and C_4 plants, 409-411
    effects on climate possible, 409
    extreme, deleterious effects, 412
    fertilization, 412-413, 415
      possible benefits, 412
    possibility of changes in species composition, 412
  increase in photosynthesis and water use efficiency, 413
  measured with infrared gas analyzer, 324
  solar radiation receipts, effect on, 33-34, 38-39
  water use efficiency in C_3 and C_4 plants, effect on, 411

Carbon monoxide, 39
  acidic soils as sink, 39
  increase:
    due to combustion, 39
    due to traffic, 39
  ocean as major source, 39
  produced by natural processes, 39
Carbon tetrachloride, effects on ozone layer, 38
Caucasians, affected by environment, 463
Characteristic dimensions, 124
Characteristic velocity, 136
Chlorofluorocarbons, as potential reactant with ozone, 36, 37-38. See also Freons
Chlorofluoromethanes, contribution to greenhouse effect, 40
Citrus crops, 223, 380, 381-382, 404
Civil noon, 14
Climate:
  effects on animals, livestock, 460-461
  effects on humans, 449-460
    effects of winds, 450, 455, 457
    reaction of people to weather, 449-450
  role in determining ecological niche, 425
Climate space, 446-449
  diagrams, 446-449
  example for sheep, 447
  procedures for calculations, 446-449
Climatic change, implications of increased reflection on, 42
Clothesline effect, 229, 232
Cloud cover, effects on longwave radiation flux, 50
Cloudiness:
  of anthropogenic material, 20
  caused by climatic change, 20
  effects on solar radiation, 21
  estimation of solar radiation from, 75
Clouds:
  altostratus, 20
  cirrus, 19, 20
  contrails, 19
  cumulonimbus, 20
  cumulus, 19, 20, 24
  effect on radiation receipts, 19
  low, 20
    as effective absorbers of solar radiation, 20
    as effective emitters in longwave, 20
  middle, 19

Clouds (*Continued*)
  nimbostratus, 20
  stratus, 19
  thickness, 20
  thin stratus, 20
  type, 20
Combination formulae for estimating ET and evaporation, 248-253
Compaction of soil, 98, 387
Condensation, 119-120
Conduction, 123, 124, 129, 433, 441-443, 447. *See also* Animals, conductive heat exchange, transfer
Consumptive use:
  coefficient, 244
  factor, 244
Convection, 117, 123, 129, 173, 269, 382, 383, 443, 445
  forced, 117, 124-125, 140, 443
  free, 117, 140, 443, 444-445
  see also Animals, convective heat exchange, transfer
Coriolis force, 134
Corn:
  azimuthal and angular distribution of leaves, 63
  emergence, 195
  net radiation over, 52, 76-77
    altered by population and row spacing, 52
    depletion of extinction within crop canopy, 52, 53
    reflectors directing light into corn canopies, 61
  sweet, 52, 199
    net radiation over, 52
  as windbreak for carbon dioxide capture, 418
Cosine law, *see* Lambert's cosine law
Cotton, 64, 218, 225, 232, 394, 401, 417
  use of whitewash as reflectant on, 404
Cranberries, 383
Crop canopy resistance, 211, 225, 258
  calculation, 225-226
  effect of environmental factors, 221
  measurement, 211-212
Cropping, effect on soil temperature, 195
Cucumbers, 383, 413, 418
  effects of extreme carbon dioxide concentration, 412

Dalton's law of partial pressures, 168
Day length, 15

Dew:
  biological significance, 168
  contribution to ET, 219
  definition, 175
  duration, 187
    importance in pathology of fungus disease, 175
  effect on growth and development of phytopathogens, 167
  electrical grid, 187
  formation:
    energetics, 175-176
    meteorology, 176-177
  hydrological significance, 168
  influence on physiology and water status of plants, 175
  measurement, instrumentation, 187
    animal membranes, 187
    duration plate recorders, 187
    Duvduvani dew plate, 187
    lysimeters, 187
  quantity, 187
  range of nocturnal dewfall, 175
  salt-impregnated electrodes, 187
  as source of water in arid regions, 175
Dew point, *see* Humidity
Diffuse radiation, *see* Sky radiation
Distillation, *see* Water, distillation
Dust:
  effect on solar radiation receipts, 34
  storms, 26

Earth:
  eccentricity, 17, 18
    of elliptical orbit about sun, 17
    of orbit about sun, 17, 18
  tilt of axis, 17
Eccentricity, 13, 17. *See also* Earth, eccentricity
Eddy correlation:
  for estimating ET, 258, 269, 271
  for estimating photosynthesis, 319-321
  techniques and equation, 144-146, 258, 320
Effective temperature, 450
  indices, 450-457
    heat stress, 450, 455, 457
    resultant temperature, 451
Effective terrestrial radiation, *see* Radiation
Einstein (Ei), 7
Electromagnetic spectrum, 6
Emissivity, 6-7, 8, 82, 112, 433
  of frost protection materials, 377
  of quartz sand, 6

of vegetation, 6
of water, 6-7
  effect of oil spills on, 6-7
of wet soil, 6
Encelia farinosa, photosynthetic rate affected by pubescence, 405
Energy balance, 425, 433-446, 449, 455
  equation, 228, 235, 248, 249, 256, 269, 361, 446
Energy flux, 144-146
Environment, animal and human, 425-426, 427, 440
  meteorological indices:
    cold season, 457-459
    warm season, 450-457
  warm, dilation of blood vessels as response to, 450
Equinox, 14, 15
Eskimos, affected by environment, 463
Evaporation:
  definition, 210
  drying, stages of, 217
  effect of soil water availability, 217-218
  equilibrium, 252
  estimation, 241-259
    climatological methods, 243-253
      air-temperature, 243-245
      combination, 248-249
      solar radiation, 245-248
    hydrologic method, 242-243, 270
    micrometeorological methods, 253-259
  free water, 211
  importance of, in plants and soils, 213-216
  from soils, 217
  of water, from animals, *see* Animals, evaporation of water
  from water bodies, 269-271, 393
Evaporative cooling, 213, 386-387, 450, 451
  by sweating, 168, 436-440
  see also Animals, evaporation of water
Evaporative demand, 218
Evapotranspiration (ET):
  actual, 211-212, 334-335, 361
  climatic influences, 228-239
  definition, 210
  effect of increased carbon dioxide, 409
  effect of meteorological conditions, 228-239
  effect of stomatal and canopy resistance, 211
  effect of temperature, 239
  equilibrium, 210
  estimation, 241-259

climatological methods, 243-253
  air temperature, 243-245
  combination, 248-249
  solar radiation, 245-248
hydrologic method, 242-243, 270
micrometeorological methods, 253-259
  aerodynamic, 255-256
  Bowen ratio energy balance, 255-257
  eddy correlation, 258
  mass transport, 254-255
  resistance, 257-258
intercepted water, contribution of, 219
in irrigation management, 216-217, 218
maximum, 211
measurement:
  atmometers, 263, 334
  chamber techniques, 263-264
  evaporimeters, 261-263
    evaporation brush, 262-263
    evaporation pans, 261-263, 334
  lysimeters, 258-261, 263, 268, 361. *See also* Lysimeters
  potential, 334, 341, 354, 361
  sap flow techniques, 264-265
plant influences, 219-228
potential, 210-213
soil influences, 217-219
Exchange coefficient, 146, 339-340, 359
  for carbon dioxide transport, 361
  for heat transport, 142, 256
  for momentum transport, 142
  for water vapor transport, 142, 256
Extinction coefficient, 18-19
  for light in plant canopies, 62

Fescue, creeping red, antitranspirant effect on, 394
Fetch requirements, 139-140, 230, 233, 259, 339
Flower crops, 413-414
Flux-gradient (profile) methods and relationships, 318-319, 142-144
Fog interception, 176
Forests:
  coniferous, 68, 219
  control of light regime, 61
  deciduous, 68
  evapotranspiration, 268-269
  Liriodendron, 68
  Loblolly pine, 68
  measurement of surface carbon dioxide exchange over, 320
  net radiation over, 52
  radiation penetration, 68-71

Forests (Continued)
  seasonal change in, 68
  red pine, 395
  reflectivity of shortwave radiation, 44, 45
  shade tolerance, 62-63
Freons (chlorofluorocarbon compounds):
  possible effects on ozone layer, 37-38
  reaction with ozone, 37
Friction velocity, 125, 135, 141, 298
Frost, 368-387
  climatology, 373-374
    in hilly terrain, 374
    in mountainous terrain, 374
    statistical description of growing season, bounded by spring and fall frosts, 374
  control, importance in horticultural production, 368
  duration, 371
  as ecological determinant, 371
  effect on tropical crops, 370
  factors affecting degree of damage, 371
    bacteria as nuclei for ice formation, 371
    duration of freezing temperatures, 371
    rapidity of freezing, 371
  freezing points of fruits, vegetables, and flowers, 368
  frost damage susceptibility factors, 368
  onset, 369
  probability, 374
  protection methods, 375-387
    computer modeling to improve design, uses of, 368
    covers as sources of heat, 387
    energy prices, affected by, 368
    flood irrigation, 382, 383-384
    foam, 378
    furrow irrigation, 383-385
    heaters, 381-382
    hot caps, 377-378
    sprinkling, 382-385
    thermal insulation, 375, 377-380
  resistance, 368
  types:
    advection, 371, 372, 374, 375, 379, 385
    black, 371
    hoarfrost (white frost), 371
    radiation, 371, 372-373, 374, 375, 378, 379, 380, 381

Global radiation, 11, 21, 57, 71-73, 77
Global warming, 38-39
Grapefruit trees, use of whitewash on, as reflectant, 404
Grashof number, 124, 443-444, 445, 446
Grass (turf, pasture), 223, 225, 228, 261-262, 268, 335, 370
  net radiation over, 52, 56
Greenhouse effect, 40
Greenhouse production, 413-414
  experience of fertilization in, 413, 414
Ground fog, as water vapor, attenuation of UV by, 38
Growth chamber studies, 413-414
  of reflectants, 396-397

Hargreaves method for estimating ET, 244-245
Heat dissipation, by plant leaves, 124
Heat of fusion, 167
  released for frost protection, 375, 382, 383
Heat stress, 450, 455, 457, 460, 462. *See also* Effective temperature, indices
Heliotropism, 61, 63, 408-409
  influence on leaf orientation, 63
Hibernation, *see* Animals, adaptation and acclimatization to environment
Hour angle, 15
Humidity (atmospheric):
  absolute, 170, 182
  biological importance, 167
  dew point, 171-172, 174, 176, 179, 183, 239, 245
  effect on ET, 168, 237-239
  human and animal adaptation, 167, 209
  indices, 451-455
    humidex, calculation of, 453
    temperature humidity index, calculation of, 452-453
  measurement, instrumentation:
    hygrometry, 146, 178, 182-183, 319. *See also* Hygrometers
    psychrometry, 178-181, 319. *See also* Psychrometers
  mixing ratio, 170
  photosynthesis, role in, 168
  relative, 171-175, 177, 182, 183-184, 215, 237-240, 245, 255, 271, 292, 345, 362, 462
  saturation vapor pressure, 170-171, 179, 455, 457
  in shelter, 342-347
  specific, 169, 255
  structure of air, 172-174
  temperature index, 460, 462
  vapor pressure, 451, 453, 455
  vapor pressure deficit, 171, 179, 214, 252
  vapor pressure profiles, 174-175

Humiture, calculation of, 451-452
Hydrologic cycle, 212-213, 242, 271
Hydrologic (water balance) method for estimating ET, 242-243, 270
Hygrometers:
  adsorption of water vapor, 183-184
  Dewcell and Dewprobe, 183
  dewpoint, 183
  hair, 181-182
  humidity probe, 184
  Lyman-alpha, 186
Hyperthermia, in animals, see Animals, hyperthermia
Hypothermia, in animals, see Animals, hypothermia

Ice:
  cloudy, effective in absorbing PAR, 60
  evaporation from, 268-269
  light penetration, 60
  low albedo, 60
  opaque, transmission of visible light, 60
  shortwave reflectivity, 43, 44
Indians, affected by environment, 463
Industrial emissions, effect on solar radiation receipts, 33
Infrared gas analyzer, 184-185, 320, 324
Infrared radiation, 36, 40, 184, 342, 377, 405
  absorption by water, 60
  depletions in water vapor and carbon dioxide absorption, 34
  see also Radiation
Insolation, 13, 17, 18, 19, 25, 310, 425
Intermittency, 62-63
Irrigation, 216, 218, 219, 242, 243, 331, 336, 361, 391
  effect on yield, 218
  limited water for, 218
  scheduling, 216, 218
  water, for frost protection, 382-385

Jensen and Haise method for estimating ET, 246
Jerusalem artichoke, azimuthal distribution of leaves, 63

Kirchoff's law, 7-8

Lambert's cosine law, 12-13, 193
Laminar flow, 124, 125, 445
Laminar sublayer, 123, 134
Land drainage, 37
Latent heat exchange in animals, see Animals, latent heat exchange
Latent heat flux, 229, 230, 233, 236, 248-249, 252, 256-257, 258, 269, 292, 313, 319, 339, 353, 357, 396, 403
Latent heat of vaporization, 167
Latitude, 15, 36, 45, 52, 192-193, 245
Leading edge, 139-140, 233
Leaf:
  angles, 63-64, 82
    influence on light penetration, 63
  architecture, measurement, 82
  orientation:
    in desert species, 63
    influence of heliotropism, 63
    influence on light penetration, 63
  shape, 406
  temperature, 392, 394, 395
Leaf area index (LAI), 82, 225-227, 266-267, 350
  influence on light penetration, 71
Lettuce, 371, 413
Light:
  compensation point, 289, 290-291, 293
  fluctuations in plant canopy, effect of wind on, 62
  penetration:
    models, 64-66
      geometrical approach, 65-66
      statistical approach, 65-66
    into plant canopies, 59-66, 399
      effect of leaf orientation, 63
    as factor limiting productivity, 60-61
    into water bodies, 59-60
  quality:
    measurement, 78-79
    in plant canopies, 66, 68
  saturation point, 289
  transmission, 178
  visible, depletion of, 66, 68
Linacre method for estimating ET, 245
Linke's turbidity factor, 25
Longwave exchange, radiation, see Radiation
Lysimeters, 258-261, 263, 265-266, 268, 348, 352-353, 361, 394
  floating, 259-260
  for measuring photosynthesis, 319
  potential evapotranspirometers, 259
  weighing, 261, 352

Makkink method for estimating ET, 246
Mass flux, 144-146
Mass transport (Daltonian) method for estimating ET, 254-255, 270-271
Maunder minimum, 11, 12

Melons, 404, 418
Mesophyll resistance, 292-293, 392
Methane:
   absorption of radiation, 34, 39, 49
   effect on ozone layer, 38
   as natural constituent of atmosphere, 39
Methochloroform, effects on ozone layer, 38
Microclimate:
   of animals, 425, 440
   in shelter, 341-350
Micrometeorology, 117, 130-131, 146, 186
   methods for estimating ET, 361
   methods for estimating photosynthesis, 315-322
Milankovitch ice ages theory, 17
Molecular diffusion, 123, 134, 176
Momentum, 135, 137
   exchange, 135-139
   exchange coefficient, see Exchange coefficient, for momentum transport
   flux, 135, 140, 142, 145-146
   sink for, 125
   transfer, 125, 142
Monin-Obukhov parameter, 140-141, 143, 270
Monsi-Saeki model, 61-62
Monte Carlo approach, 66
Mulching, 195-201, 217, 266, 387
   to damp diurnal soil temperatures, 199
   principles, 196-199
   to reduce evaporation from soil, 199
   types, 195-201
      aluminum foil, 197, 199
      cinders, 196
      coal, 196
      dust, 196
      gravel, 200
      hay, 197, 199
      paper, 197-199
      petroleum by-products, 196
      plastic, 197-199
      sawdust, 199
      straw, 199
      stubble, 196
      weeds, 196
      wood fiber cellulose, 199
   to warm soil, 199

Net radiation, see Radiation
Net radiometers, see Radiation, measurement
Nitrous oxide:
   effect on ozone layer, 36-38
   radiation, absorption of, 49

   sinks of, 37
   sources of, in atmosphere, 36-37
Nusselt number, 124, 443, 444

Oasis effect, 229, 232-233
Oats, 409
   net radiation over, 52
   extinction of, within oat canopy, 53
Ohm's law, 240, 292
Oleander, 394
Orographic phenomena, 120
Oxygen, 33, 34, 35
   absorption spectrum, 34-35
Ozone, 36-38
   absorption spectrum, 34-35
   anthropogenically induced alterations, 20, 33
   effect on solar radiation receipts, 33-34
   natural variation in concentration, 36
   potential reactants with, 36-38
   ultraviolet, absorption of, 36
Ozonosphere, possible destruction of, 36

Panting, see Animals, adaptation and acclimatization to environment; Animals, evaporation of water
Parallelism, 13
Pasture, see Grass
Path length, see Beer-Bouguer law
Peanuts, 394
Penman method for estimating ET, 249-251, 252, 263, 267, 268, 270, 334
   modified by Monteith, 251
Peppers, 383
Percent possible sunshine, 20-22, 55
   seasonal effects, 21
   secular trends, 21
Perihelion, 13, 17
Photorespiration, 290-291, 359, 410
   in C_3 plants, 290-291, 410
Photosensors, see Radiation, measurement
Photosynthesis, 7, 10, 63, 229, 288-324, 342, 347, 361, 392, 394, 395, 396, 400, 401, 403, 404, 405, 407, 412, 413, 416
   apparent, 292
   chlorophyll, role of, 288
   classification of green plants:
      C_3, 288-290, 293-294, 302
      C_4, 289-290, 293-294
      crassulacean acid metabolism (CAM), 289
   definition of, 288
   effect of environmental factors, 293-299
      carbon dioxide concentration, 297-299

increase in soil and root respiration, 296-297
light, 293-294
temperature, 296-297
water, 293-295
wind and turbulence, 298-299
gross, 291-292, 303, 310
and apparent, 292
irradiance, response to, 289
measurement, 315-324
micrometeorological methods, 315-322
nonmeteorological methods, 322-324
chambers, 322
cuvettes, 322
radiant energy conversion, 308-312
actual photosynthetic efficiencies, 309-312
in *Camissonia claviformis,* 310-311, 322
calculation of photosynthetic efficiency, 309
PAR as source of energy for photosynthesis, 308-309
shelter, effects of, 359-360
carbon dioxide concentration, 359
daytime temperatures, 360
duration of photosynthesis activity, 360
photosynthetic flux, 359, 361
Photosynthetically active radiation (PAR), 62, 71, 222, 360, 392, 396, 401
Pineapple:
soil temperature increased by mulch, 199
water use reduced by mulch, 199
Planck's particle concept, 7, 8-9
Planetary reflection, affected by aerosol loading, 42
Plant architecture and morphology, 391, 404-409
adaptations, adjustments, 405-406, 408-409
breeding, for greater reflectivity, 405
of canopy, 408
and disease, 408
leaf shape, 406-407
lodging, 407
plant height, 407
Plant internal resistance, 219-224, 241, 270
Pollens, 33
effect on solar radiation receipts, 34
Potatoes, 262, 314, 335, 394, 418
Prandtl number, 124, 443
Precession, 17
Precipitation, interception of, 267
Priestley-Taylor model for estimating ET, 252-253, 268. 270

Psychrometers:
fine-wire thermocouples, 179-180
for humidity measurement, 179-181, 319
miniature thermocouples, 180-181
for soil moisture content measurement, 181
for water potential measurement, in plant tissue, 181
Pubescence, 405-406
as adaptation to aridity, 405-406
effect on boundary layer resistance, 405
effect on photosynthetic rate, 405
effect on radiation absorption, 405
effect on soybean reflectivity, 405
effect on transpiration, 405
soybean isolines, 405
Pyranometers:
Abbott, 12
albedometers, 74-75
Bellani, 73
characteristics, 72-73
cosine response, 72
response time, 72
sensitivity, 72
temperature dependence, 72
Eppley, 71, 73, 74-75
filtered, 78
nonelectrical, 72-73
occulting discs, rings (shadow band), 73-74
Pyrheliometers, *see* Radiation, measurement

Quantum:
concept, 7
efficiency in photosynthesis, 308-309
requirement in photosynthesis, 309

Radiant flux density, 5-6, 8, 12-13, 18, 25, 27, 46, 48, 49, 52, 63, 66, 72, 78, 82, 99, 186, 192, 289, 298, 309, 394, 397-398, 417
Radiant heating, for frost protection, 375, 382
Radiation, 5-82
absorption, 428-429
balance, 341-342, 352, 398, 405, 427-433
blackbody, 4, 7, 8-10, 76, 82, 447, 449
blue:
depletion, 66
enrichment, 39-40
penetration in water, 60
diffuse, distribution of, 68
direct beam, 72, 73, 429
distribution, 68
proportion of diffuse to, 68, 193

Radiation (*Continued*)
  effective terrestrial, 50
    effects of clouds, 50
    effects of water vapor, 50
    emission of, hemispheric and global, 58
  frequency, 5
  green, 398
    enrichment, 66
    extinguished by water, 60
  intercepting surface, calculation, 428-433
  interception, for frost protection, 374, 376-377
    artificial fog, 377
    use of smoke and smokescreens, 377
  interference, 7
  longwave, 10, 29, 49-52, 55-57, 76, 77, 82, 126, 197, 199, 376-377, 402, 403-404, 428, 431, 432, 447
    absorbers of, 49
    atmospheric, 57
    back radiation, 49
    emissivity, 428
    sources, 49
  measurement, instrumentation:
    net radiometers, 76-77
    photosensors, 79
    pyranometers, 24, 45-46. *See also* Pyranometers
    pyrheliometers, 72
    selenium cell sensors, 79
    spectroradiometers, 79
    of sunshine duration, 20, 75-76
  net:
    balance, satellite mapping of, 57
    components, 51
    crop canopies, within, 53
    crop surfaces, over various, 52
    diurnal course, 51-52
    effect on ET, 229
    extinction, 53-54
    measured with net radiometer, 76-77
    related to solar radiation, 54-55
    seasonal patterns, 52
  period, 5
  photosynthetically active, *see* Photosynthetically active radiation
  red enrichment, 18, 39-40, 48, 398
    depletion, 66
  reflection, 7
    diffuse, 43, 79
    of natural surfaces, 45
    specular, 43, 46
  refraction, 7

  scattering, increased by isolated cumulus clouds, 20
  shield, 132
  shortwave, 10, 45, 50-51, 66, 74, 76, 197, 398, 428, 431, 447
    diffuse, 428, 431
    simulation, 66
  solar:
    diurnal course, 51-52
    extinction, 18
    net absorption, hemispheric and global, 58
    spectrum, 9, 72
    total, 57, 71-73, 77
  terrestrial, 57
  thermal, 49-51, 248
  total hemispherical:
    air temperature, relation to, 49
    humidity, relation to, 49
    measurement, 77
  transmission, 7, 309, 310
  wavelength, 5-10, 12, 18, 19, 38, 39, 42, 46, 48, 49, 50, 62, 66, 72, 76, 78, 79, 308, 309, 320
Rayleigh's scattering law, 18
Reflectants, as antitranspirants, 391, 396-404
  aluminum silicates, 400, 404
  celite, 399
  diatomaceous earths, 400
  effectiveness on C_3 plants, 400, 401
  effectiveness on C_4 plants, 400-402
  evaporation from soil and water, to reduce, 396
  field experiments:
    with cotton, 401-402
    measured reflection spectra in, 397-398
    with sorghum, 400-401
    with soybeans, 402-403
  kaolinite, 397-399, 400, 401
    effect on light penetration, 399
    effect on soil reflectance, 399
  mechanism of reflectant action, 396
  penetration into C_3 canopies, 404
  plant physiological condition, interaction with, 400-402
  plants, experimentation applied to, 396-397
  plant temperatures, interactions with, 402-404
  specialized applications, 404
  transpiration on crop surfaces, to reduce, 396
Reflection, diffuse, *see* Shortwave reflection
Reflectivity, 7, 74-75, 433
  changing land use effects, 43

SUBJECT INDEX

of earth, intentional changes, 42-43
measurements, 43
varied by season, time of day, 45
of water, effect of oil spills on, 43
Regression methods for estimating ET and evaporation, 245-246
Relative strain index, 455
Remote sensing, 25, 79, 368
frost forecasting, 376
Resistance approach:
aerial, *see* Aerial resistance
for carbon dioxide transport, 292-293
for estimating ET, 257-258
mesophyll, *see* Mesophyll resistance
for sensible heat flux estimation, 123-126, 257-258
stomatal, *see* Stomatal resistance
Respiration:
dark, 290, 291, 300, 303, 324
definition, 288
environmental influences, 299-300, 302
soil moisture availability, 300-301
temperature, 299-300
measurement, 324-326
chambers, 324-326
in forests, without chambers, 326
Reynolds analogy, 142
Reynolds number, 124, 443, 445, 446
Richardson number, 140-143
Roughness:
elements, 135, 183
parameter, 135-137, 140, 252
Row orientation, effect on soil temperature, 195
Ryegrass, 191, 211, 233, 261

St. Augustine grass, 201
Satellites, observation and mapping of global net radiation balance, 57-59
Scattering, 34, 39
of blue light, 18
Selenium cell sensors, *see* Radiation: measurement, instrumentation
Sensible heat:
advection, 129, 211, 228, 230, 232-233, 246, 248, 250, 252, 253, 255, 256, 259, 269, 314, 352, 361, 381, 396
definition, 229
flux, transfer, 117, 123-126, 141, 142, 144, 145, 269, 292, 396, 406, 433
turbulent exchange coefficient, 123
sources and sinks, 125, 126, 342
Shearing stress, 135
Shelterbelts, 332. *See also* Windbreaks

Shelter effects, *see* Windbreaks
Shortwave radiation, *see* Radiation
Shortwave reflection, 42-49, 57, 71, 75
effect of ground cover, 49
effect of leaf moisture status, 48
effect of soil fertility, 48
of natural surfaces, 44
Site selection, for frost control, 375-376
effects of topography, 375
effects of water bodies, 375
Sitka spruce, 191, 219
effects of cold hardening on photosynthesis, 191
Sky radiation, 40-42, 73-74
biological importance, 40
effect of cloudiness, 40
effect of latitude, 40
effect of season, 40
isotropic nature, 41
measured with pyranometers, 73-74
total radiation, fraction of, 42
Slatyer and McIlroy method for estimating ET, 251-252
Slope, 191-195, 362
affected by season and latitude, importance of, 192-193
altered by ridging and shaping, 193-195
effect on soil temperature, 193
Snow:
evaporation, 268-269
fence, as barrier to assist carbon dioxide capture, 418
light penetration, transmission, 60
powder, effective in absorbing PAR, 60
reflectivity, 43-44
Soil:
density, 94-95, 103
drying:
combined effects of row spacing and straw mulching, 200-201
stages, 217
evaporation, 217, 266
cumulative, 267
separation from transpiration, 265-267
heat, 111-112
capacity, 94, 95, 103
release to air, during frost, 383
heat flux, 94-115, 229, 230, 270, 353, 387, 396
by conduction, 94, 117, 129
as indicator of evaporation, 111
measurement, 98, 112-115
calorimetric procedure, 98
"null alignment" method, 98

Soil (*Continued*)
    soil heat flux plates, 112-113
        controlled by radiation balance, 111
        effect of soil texture, 103, 105-106
        effect of water content, 107-111
    heating, artificial, 201-206
        heating cables, 201
        irrigation with warm (industrial cooling) water, 206
        waste industrial heat, 201-202
    manipulation, for frost protection, 387
    mass specific heat, 94
    moisture:
        content, 195, 218
        manipulation of, 191-206
        regime, modification of, 190-206
    penetration of heat into, 99
    physical properties, determination of, 98
    respiration, 111-112
    shortwave reflectivity, 44-45
    sink for industrial waste heat, 206
    sink for nitrous oxide, 37
    source of carbon dioxide, 111
    source of heat during frost, 387
    specific heats, 94-95, 103
    as storage for heat, 94
    temperature, 94-115
        amplitude, 99-100, 103
            and range of temperature at given depths, 99-100
        changes, effect on crop growth and nutrition, 190
        crop yields, effect on, 191
        daily and seasonal patterns, 99-101
        emergence, effect on, 195
        freezing, depth of, 114-115
        and germination, 106, 190
        land shaping, affected by, 193-195
        measurement instrumentation, 113-115. *See also* Temperature, measurement
        nitrification, effect on, 191
        nutrient uptake, effect on, 190
        observations, 106
        optimal, 190
        profiles, 101-103, 196
            influence on moisture distribution patterns, 107
        regime, modification of, 190-206
        root development, effect on, 191
        soil composition, effect on, 191
        soil moisture, interactions with, 191
        soil respiration, effect on, 191
        time lag with depth, 101
        transpiration, effect on, 191
        water flow in soils, effect on, 107
    thermal conductivity, 102, 112
        dependence on soil moisture content, 95
        dependence on texture, 95
        methods to increase, 383, 387
    thermal diffusivity, 95-96, 98-99, 103
        function of compaction, 98
        methods to increase, 387
        function of moisture content, 96
        function of soil organic matter, 98
    thermal properties, 94-98
    tillage, 98
    vapor flow, controlled by thermal gradients, 107
    volume specific heat, 94, 98
    warming, 205
    water, 267
        balance, 241-243
        content, 98
        extractable, 267
        flux, 243
        potential, 218, 352
    water-repellent clods, 200
Soil-plant-atmosphere continuum, 239-241
Solar:
    altitude, 15, 46
    azimuth, 15, 19, 61, 73, 82, 83
    constant, 11-12, 18, 19
    declination, 14-15
    elevation, 13, 15, 19, 26, 43, 44, 46, 62, 68-69, 73, 428-429, 432, 433
    energy receipts, at earth's surface:
        effect of atmospheric constituents, 33-34
        qualitative effects, 33-40
    noon, 14, 15
    radiation, *see also* Radiation
        estimates, by geosynchronous satellite, 24
        prediction, 20
        and thermal radiation method, for estimating evaporation and ET, 248
        thermal unit method, for estimating evaporation and ET, 247-248
Solstice, 14, 15
Sorghum, 48, 199, 200, 215, 218, 223, 225, 289, 394, 400
    azimuthal distribution of leaves, 63
    emergence of, 195
Soybeans:
    extinction coefficients, 62
    leaf distribution, 63-64
    narrow leaves, 406
    net radiation within canopy, 53
    pubescence, effect of, 405

SUBJECT INDEX 493

radiation penetration in narrow-leafed isogene, 407
reflectants, effects of, 402-403
transpiration, 406
Spectroradiometers, see Radiation: measurement
Squash, 383
Stability (atmospheric), 140-141, 142
 function, dimensionless, 143
 neutral, 135-136, 139, 142
 stable, 140
 unstable, 140
 see also Monin-Obukhov parameter, Richardson number
Stefan-Boltzmann constant, 8
Stefan-Boltzmann law, 8
Stomatal resistance, 126, 211, 218, 221, 224-225, 240-241, 251, 258, 261, 292-293, 351-352, 356-357, 360, 361, 392, 403
 calculation, 224-225
 effect of environmental factors, 221
 procedures for averaging values, 224-225
Strawberries, 383
Sublimation, 167, 210, 269, 371
Sudan grass, 221, 232
Sugar beets, 221, 225, 296, 297, 303, 306-307, 310, 311, 319, 335, 339, 345, 347, 352, 357, 359, 410
Sugar cane, 52, 289
Sultriness, 454-455
Sunflecks, 62-63, 82
Sunflowers, 293, 341, 393, 394
 azimuthal distribution of leaves, 63
 extinction coefficients, 62
 heliotropic behavior, 63
 net radiation, 52, 77
  extinction of, 53
 planophile leaf orientation, 63
Sunspot activity, 11-12
Surface, emissivity, 50
 roughness, 135, 139, 140
 temperature, 79-82, 248, 253, 270, 376, 387
  diurnal range, 102-103
  influence, 50
  measurement, 79-82
Swamps, evaporation and ET, 269
Sweating, see Animals: adaptation and acclimatization to environment

Temperature:
 indices, 451-455
  humidex, 453
  temperature humidity index, 452-453
 measurement, 130-132
  temperature-sensitive diodes, 130
  thermistors, 130, 132, 146
  thermocouples, 130-132, 146, 179-180
  thermometers, 130, 132
 virtual, 141
 see also Air; Soil; Surface, temperature
Terrestrial radiation, see Radiation
Thermal radiation, see Radiation
Thermal stress index, 455
Thornthwaite method for estimating ET, 243-244
Tillage, influence on soil temperature, 195
Tobacco, 394, 395
Tomatoes, 191, 200, 205, 262, 383, 394, 404, 413, 418
Transmissivity, 7
Transpiration:
 definition, 210
 effects on crop yield and dry matter production, 213-216
 importance in plants and soils, 213-217
 stomatal influence on, 221
 yield-transpiration relationships, 215
Truck crops, 383
Turbidity, 25-33
 air mass, effects of, 30
 annual, cyclicity in, 27
 atmospheric, 25-26
  effect of dust, pollen, suspended materials, 25-26
  effect of water vapor, 25-26
 and climatic change, 29-30
 depletion of visible light, 60
 fluctuations, 29
 geographic distribution, 30-33
 industrialization, effects of, 30
 natural aerosols, 26
 seasonal effects, 27
 urbanization, effects of, 30, 33
Turbulence, 123, 125, 135, 140, 340-341, 351, 359-360, 406
Turbulent exchange coefficient, see Exchange coefficient
Turbulent mixing, transfer, 125, 134-164, 176, 339, 341, 342, 348, 357, 359, 362
Turbulent surface layer, 134, 144

Ultraviolet, 34-36
 absorption, by ozone and oxygen, 34-35
 depletion, in water, 60
 germicidal effects, 36

Ultraviolet (*Continued*)
screened, 34

van Bavel combination method for estimating ET, 251
Vegetable crops, 205
  soil temperature conditions for germination and seedling emergence, 105
Vegetation:
  shortwave reflectivity, 44, 46, 48, 55
  on soil residues, as source of heat, 387
Vineyards, 387
Volcanoes:
  activity, 27-28, 29
  eruptions, 27-29
    leading to climatic change, 30
Volz turbidity index, 30
von Karman constant, 125, 135, 251

Water:
  application, for frost protection, 383-385
  balance, 233
  distillation, 176, 187
  emissivity, 50
  functions in plants, 209
  of guttation, 176, 178
  infrared radiation, absorption of, 60
  interception of, 268
  penetration, of radiation, 59-60
  potential, 240-241, 360, 361
    gradients, 240
    resistances, 240-241
  properties, 209
  shortwave reflectivity, 44
  use:
    actual, 352-357, 359
    affected by distance from windbreaks, 356
    direct measurements, 352-353
  use efficiency, 52, 312-314, 335
    actual achieved, 312-313
    carbon dioxide – water vapor flux ratio, 313-314
    in crop production, 391-419
    definition, 314
    effect of increased carbon dioxide, 409, 411
    effect of leaf temperature, 392
    effect of shelter, 360-361
    effect of solar radiation, 392
    theoretically possible, 313
  vapor:
    absorption spectrum, 34-35, 178, 184
    contribution to greenhouse effect, 40
    flux, 142, 144, 145-146, 186, 255, 269, 270, 319, 321-322
    optical properties, 184-186
      infrared gas analyzer, 184-185
      microwave refractometers, 185-186
    retention of thermal (terrestrial) radiation, 38
    solar radiation receipts, effect on, 33-34, 38
Wheat, 190, 215, 218, 228, 289, 335, 418
Wien's displacement law, 8-10
Windbreaks, 331-363, 375, 391, 418
  gaps, 338
  gradients and profiles, 348-350
  height, 336
  interrelations of wind shelter, moisture conservation, plant growth, and yield, 334-336
  length, 338
  multiple, 341
  photosynthesis in, 356-360
  plant physiological responses, 350-352
    effect of shelter on plant growth and morphology, 350
    effect on stomatal resistance, 351-352
    effect on turgidity, 351-352
    reduction of mechanical injuries, 350-351
  porosity, 338-339, 341, 356
  resistance model of ET in shelter, 356-359
  reviews, 333
  turbulence, turbulent mixing, 336-341, 342
  uses, 331
    as animal shelter, 331
    in Great Plains, 331, 332
    heat conservation in greenhouses, 331
    for protection against mistral winds, 331
    for protection of plants, 331-332
  water use efficiency, 360-361
  windspeed, 336-341
  yields, effect on, 333-334
Windchill:
  calculation, 457-458
  equivalent temperature, 457, 458
    calculation, 458
  index, 457-460
Wind rose, 148, 150, 151
Windspeed, 134-164, 237, 240, 254, 255, 259, 263, 269, 270, 334, 336-341, 348, 351, 356, 357, 362, 374, 387, 436, 442, 443, 447, 449, 450, 455, 457-458
  boxplots, 150, 151
  constancy, 152-154

within crop canopies, 146-147
daily patterns, 147-148
direction, seasonal patterns of, 148-154
effect on ET, 237
measurement:
  anemoclinometers, 154-156
  anemometers, 154, 320
    cup, 157
    drag, 146, 155-156, 186
    heated thermistor, 159, 163, 164
    hot-film, 158-159, 340-341
    hot-wire, 146, 158-159
    propeller, 157-158
    sonic, 146, 164
    bivanes, 158
  Pitot tubes, 154
  wind vanes, 158
profiles, 125-126, 135-139, 140
  logarithmic law, 136, 139, 140, 142, 255, 338

Zenith distance, 12-13, 25
Zero plane displacement, 125, 135-136, 137-139, 140. *See also* Windspeed, profiles